IRON-SULFUR PROTEINS

METAL IONS IN BIOLOGY

EDITOR: Thomas G. Spiro,
Department of Chemistry
Princeton University, Princeton, New Jersey 08540

IRON-SULFUR PROTEINS

Edited by

THOMAS G. SPIRO
Princeton University

A WILEY-INTERSCIENCE PUBLICATION

JOHN WILEY & SONS

New York • Chichester • Brisbane • Toronto • Singapore

Library of Congress Cataloging in Publication Data
Main entry under title:

Iron-sulfur proteins.

(Metal ions in biology, ISSN 0271-2911; v. 4)
"A Wiley-Interscience publication."
Bibliography: p.
Includes index.
1. Iron sulfur proteins. I. Spiro, Thomas G., 1935-
II. Series. [DNLM: 1. Metalloproteins.
W1 ME9611AU v. 4/QU 55 I711]
QP552.I7I76 574.19′2454 82-2599
ISBN 0-471-07738-0 AACR2

Printed in the United States of America

10 9 8 7 6 5 4 3 2 1

Series Preface

Metal ions are essential to life as we know it. This fact has long been recognized, and the list of essential "trace elements" has grown steadily over the years, as has the list of biological functions in which metals are known to be involved. Only recently have we begun to understand the structural chemistry operating at the biological sites where metal ions are found. This has come about largely through the application of powerful physical and chemical structure probes, particularly X-ray crystallography, to purified metalloproteins. From such studies we have learned that nature has evolved highly sophisticated ways of controlling the relatively flexible stereochemistry of metal ions. In one case after another, the structure and the reactivity of a metalloprotein active site have turned out to be different from anything previously encountered in simple compounds of the metals. Indeed, many a reasonable inference about active-site structure, based on the known properties of metal complexes in solution, have turned out to be erroneous. These surprises have inspired inorganic chemists to expand their vision of metal ion reactivity. The biological studies have spurred much fruitful synthetic and mechanistic work in inorganic chemistry, aimed at elucidating the means whereby nature achieves its stereochemical ends. The terra incognita of the biochemical functions of metal ions has become familiar territory to an increasing number of inorganic chemists and biochemists, and several of the more imposing mountains have been scaled. Vast stretches remain uncharted, and the field is alive with a sense of both accomplishment and new opportunities.

The purpose of this series is to convey some of this excitement, as well as the emerging intellectual shape of the field, to a wide audience of nonspecialists. Individual volumes will cover topics that are current and exciting—

the recently scaled mountains that are still under active exploration. The chapters are not intended to be exhaustive reviews of the subject matter. Rather, they are intended to be readable accounts of the insights and directions that are emerging in active new areas of research. Volumes will appear on an occassional basis as progress in the field dictates.

THOMAS G. SPIRO

Princeton, New Jersey

Preface

Discovered only in the mid-1950s, proteins containing Fe bound to inorganic sulfide and cysteine thiolate are now known to be very widely distributed in nature. The initial period of discovery, spurred by the unique electron paramagnetic resonance (EPR) signals that these proteins display, was followed by the development of their structural chemistry and by the elegant elaboration of synthetic analogs by R. H. Holm and his co-workers. In Chapter 1 R. H. Holm and J. M. Berg provide an authoritative account of the chemistry and properties of these analogs, and their relationship to the various protein classes. In Chapter 2 W. H. and N. R. Orme-Johnson review the important question of how to determine the type of Fe-S complex in a given protein, and describe some of the unique properties of the Fe-S entities in nitrogenase. Crystallography has played a vital role in our understanding of the Fe-S proteins, and in Chapter 3 C. D. Stout reviews the findings from this technique and also describes his own determination of the first structure of a 3-Fe cluster in *Azotobacter* ferredoxin. The very recent discovery of 3-Fe proteins, complementary to the well-known 1-Fe, 2-Fe, and 4-Fe classes, has generated much excitement, as evidenced by several contributions in this volume. In Chapter 4 E. Münck describes the definitive role that Mössbauer spectroscopy has played in identifying 3-Fe proteins and elucidating their electronic structure; new results on sulfite oxidase, which show a coupling between heme and Fe-S centers, are also described. In Chapter 5 J. LeGall, J. J. G. Moura, H. D. Peck, Jr., and A. V. Xavier describe the wealth of Fe-S proteins produced by sulfate-reducing and methane-forming bacteria. These include a variety of hydrogenases, and also ferredoxin II (Fd II) from *D. gigas,* which was the first example of a protein with solely 3-Fe centers.

The Fe-S proteins are ubiquitous electron-transfer agents in biology. The next two chapters describe their involvement in the two most important electron transport chains. M. C. W. Evans assays our current knowledge of Fe-S centers in photosynthetic electron transport in Chapter 6, while T. Ohnishi and J. C. Salerno do the same for mitochondrial electron transport in Chapter 7. In Chapter 8 R. K. Thauer and P. Schönheit present evidence that, at least in bacteria, Fe-S proteins also play a role in the storage of Fe.

The last three chapters describe new applications of physical techniques to Fe-S proteins in the various classes. In Chapter 9 B. K. Teo and R. G. Shulman describe their exploration of Fe-S X-ray absorption spectra, including the fine structure (EXAFS), which gives information on the Fe-S distances. Chapter 10, by M. K. Johnson, A. E. Robinson, and A. J. Thomson, discusses recent results from magnetic circular dichroism (MCD) spectroscopy. Here again the 3-Fe proteins come to the fore, because of their very characteristic MCD behavior. In Chapter 11 the editor, along with J. Hare, V. Yachandra, A. Gewirth, M. K. Johnson, and E. Remsen, describe newly obtained resonance Raman (RR) spectra of Fe-S proteins and their analogs. Vibrational signatures are becoming available for the basic Fe-S structures and their variants in different proteins.

It seems clear that the Fe-S proteins, having been through successive phases of discovery and structural elaboration, are yielding new insights to increasingly sophisticated techniques. It is very likely that many surprises are still in store.

Thomas G. Spiro

Princeton, New Jersey
March 1982

Contents

CHAPTER 1

Structures and Reactions of Iron-Sulfur Protein Clusters and Their Synthetic Analogs

JEREMY M. BERG
R. H. HOLM

Department of Chemistry
Harvard University
Cambridge, Massachusetts

CONTENTS

1 INTRODUCTION

This volume is yet another testimony to the explosive growth of the field of Fe-S biochemistry and the allied area of the inorganic chemistry of Fe-S coordination complexes. As one of us has observed (1), "With the possible exception of heme proteins no class of metalloproteins has been as thoroughly investigated in the last decade as the non-heme iron-sulfur proteins. . . ." These proteins, having passed from scientific near-obscurity to prosperity in the last 15 years, represent one of the major stories in contemporary metallobiochemistry. Their pervasiveness extends from the general problem of electron transport in biology, as they are the most widely dispersed metalloprotein electron carriers in nature, to their role in the metabolism of H_2 and N_2, among other elements.

In assessing the growth of Fe-S biochemistry as related to the contents of this chapter, it is appropriate to recall those points of origin from which certain classes of proteins could be considered as molecules of at least partially defined structures. Some of the essential structural elements of the [2Fe-2S] site **1** of spinach ferredoxin (Fd) were deduced as early as 1966 from spectroscopic information (2, 3). The mononuclear tetrahedral site **2** of clostridial rubredoxin was established by crystallography in 1970 (4), and the more complex [4Fe-4S] cluster structures **3**, also determined crystallographically in the *Chromatium* high-potential protein (HP) (5) and in *P. aerogenes* Fd (6), followed in 1971 and 1972. Though these structures have since been extensively refined and **1** has recently been confirmed by an X-ray study of an algal Fd (7, 8), the significance of these early results can hardly be overestimated. Indeed, the same encomium applies to the crystallographic (9) and spectroscopic (10, 11) detection in 1980 of [3Fe-3S] clusters, whose existence presages another fascinating adventure in the biological and inorganic chemistry of Fe-S clusters.

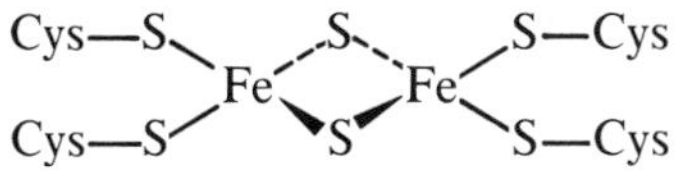

1

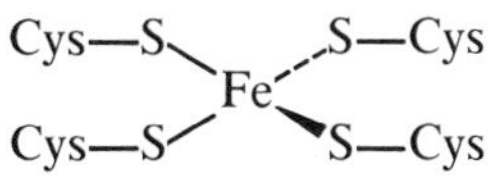

2

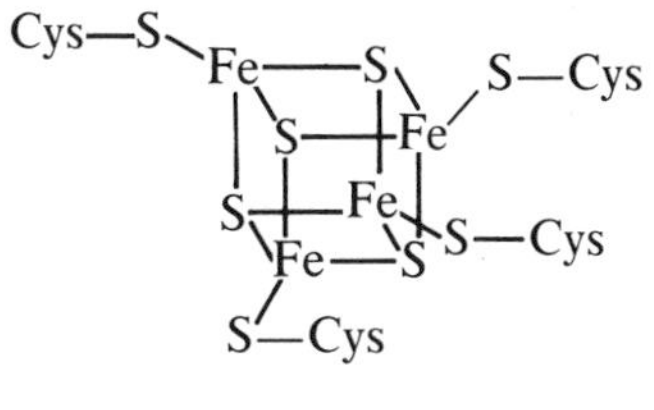

3

Presented in this chapter are an analysis of the structures and an exposition of selected reactions of Fe-S protein sites **1**, **2**, and **3** and their synthetic analogs $[Fe(SR)_4]^{1-,2-}$, $[Fe_2S_2(SR)_4]^{2-,3-}$, and $[Fe_4S_4(SR)_4]^{2-,3-}$. As has been amply demonstrated (1, 12–16), the latter species, which usually contain alkyl- or arylthiolate terminal ligands as simulators of cysteinate binding in proteins, are in general credible structural and electronic representations of isoelectronic protein sites. More specifically, well-designed analogs convey the intrinsic properties of a protein site, that is, properties that are unmodified by whatever perturbations are imposed on a site by protein structure at all levels. The local philosophy of, and the information ideally forthcoming from, the synthetic analog approach to protein sites are set out in more detail elsewhere (1, 14, 17). As the following examples illustrate, analogs have proven effective in the elucidation of certain fundamental properties of protein sites. Because analog chemistry is now quite extensive and well developed (1, 12, 13, 17, 18), it is appropriate to provide a summary of species prepared and their leading properties. Listed in Tables 1–3 are the types of 1-Fe, 2-Fe, and 4-Fe complexes that have been isolated (or in some cases generated in solution), together with original references to properties (14–16, 19–74). Of these properties only ligand-substitution behavior is considered in detail. References to structural features, which are examined in Section 2, are omitted.

2 STRUCTURES OF PROTEIN SITES, ANALOGS, AND RELATED COMPOUNDS

In this section structural data are presented for 1-Fe, 2-Fe, 3-Fe, and 4-Fe protein sites, their synthetic analogs, and certain selected compounds that, although not site analogs, contain related structural features. Shape parameters for the 2-Fe and 4-Fe cases have been devised and are employed as measures of deviations from idealized symmetries. Protein structures in their entireties are not described. The tabulated data provide a convenient single-source reference to all pertinent structural information available through 1980.

2.1 1-Fe Structures

Proteins containing single iron sites are of three types: conventional rubredoxins (Rd) with molecular weights $M_r \sim 6200$ containing a single site **2**, the 2-Fe form of *P. oleovorans* Rd (M_r 19,000) (75) containing two equivalent sites **2** (76), and the dimeric (α_2) 2-Fe protein from *D. gigas* (M_r 7600) (77), described as desulforedoxin. This protein appears to have two equivalent sites **2**, but their spectroscopic properties (78, 79) depart from those of conventional Rd. This behavior may arise from unusual cysteine placements in the primary structure (77) which, alone or in combination with

Table 1 1-Fe Site Analogs $[Fe(SR)_4]^{z}$ [a]

Analog	Isolation	Preparation	AS	V	NMR	EPR	Mb	MS
$[Fe(S_2\text{-}o\text{-xyl})_2]^{1-}$ [b]	+	19	14, 19, 20	14, 19	—	21	14, 19, 21	19
$[Fe(S_2\text{-}o\text{-xyl})_2]^{2-}$	+	14	14	14	—	—	14	14
$[Fe(SPh)_4]^{2-}$	+	22, 23	22, 23	23	23	24	25	22
$[Fe(\text{peptide-1})]^{2-}$ [c]	—	26	26	—	—	—	—	—
$[Fe(\text{peptide-2})]^{1-,2-}$ [d]	—	27	27	—	—	—	—	—

[a] Abbreviations used in this and following tables: AS, absorption spectra; V, voltammetry; NMR, nuclear magnetic resonance; EPR, electron paramagnetic resonance; MB, Mössbauer spectroscopy; MS, magnetic susceptibility; S_2-*o*-xyl, *o*-xylyl-α,α'-dithiolate.

[b] Magnetic circular dichroism (MCD) spectrum (20), magnetization (21).

[c] Peptide-1 = *t*-Boc·Gly-(Cys-Gly-Gly)$_3$-Cys-Gly·NH_2.

[d] Peptide-2 = Ac·Gly-Gly-(Cys-Gly-Gly)$_4$·NH_2.

Table 2 2-Fe Site Analogs $[Fe_2X_2(SR)_4]^z$

R	X	z	Isolation	Preparation	AS	V	NMR	Mb	MS	LS[d]
o-$C_6H_4(CH_2)_2$[a, b]	S	2−	+	28–31	29	28, 29	28, 32	28, 32	28, 32	29, 34
o-$C_6H_4(CH_2)_2$[a, b]	S	3−	−	15	15	15	—	15	—	—
o-$C_6H_4(CH_2)_2$[a, b]	Se	2−	+	31	31	31	31	—	—	—
C_6H_5	S	2−	+	23, 29–31, 33	29	29–31	31	32	32	29
C_6H_5	Se	2−	+	31	31	31	31	—	—	—
p-$C_6H_4CH_3$	S	2−	+	29,31	29	29, 31	31	—	—	29
p-$C_6H_4CH_3$	Se	2−	+	31	31	31	31	—	—	—
p-$C_6H_4CH_3$	S, Se	2−	−	73	—	—	73	—	—	—
p-C_6H_4Cl	S	2−	+	29	29	29	—	—	—	29
m-$C_6H_4CF_3$	S	2−	+	34	34	34	34	—	—	34
p-$C_6H_4CF_3$	S	2−	+	34	34	34	34	—	—	34
p-$C_6H_4N(CH_3)_3$	S	2+	+	29, 35	29	29	—	—	—	29
S–S–S[a, c]	S	2−	+	33	33	—	—	—	—	—
Cl	S	2−	+	36	36	36	—	—	36	36
Br; I	S	2−	+	36	36	36	—	—	—	—

[a] Bischelate complex.
[b] 2R = o-$C_6H_4(CH_2)_2$.
[c] 2R = S–S–S.
[d] Ligand-substitution reactions.

Table 3 4-Fe Site Analogs $[Fe_4X_4(SR)_4]^z$

R	X	z	Isolation	Preparation	AS	V	NMR	EPR	Mb	MS	LS[a]
CH_3	S	2−	+	37	35	35	38	—	—	—	—
CH_2CH_3[b]	S	2−	+	35, 37	35, 41	35, 39, 40	38, 41	—	42	—	43
CH_2CH_2OH	S	2−	+	44, 45	44, 46, 47	44, 47	45	—	47	—	43–45
CH_2CH_2OH	S	3−	—	47	—	—	—	47	—	—	—
CH_2CH_2OH	Se	2−	+	48	—	—	—	—	—	—	—
$CH_2CH_2CH_3$	S	2−	+	38	35	35	38	—	—	—	—
$CH(CH_3)_2$	S	2−	+	38	35	35	38	—	—	—	—
$CH(CH_3)C_6H_5$	S	2−	+	49	49	49	49	—	—	—	—
$CH_2CH_2CO_2^-$	S	6−	+	50	50, 51	51–53	—	—	—	—	—
$CH_2CH_2CO_2^-$	S	7−	—	51	51	—	—	—	—	—	—
$CH_2CH_2CH_2CO_2^-$	S	6−	?	53	—	53	—	—	—	—	—
$C(CH_3)_3$	S	2−	+	37, 38,43, 45	35	35	35, 38, 54	—	42	—	35, 39, 43, 44, 54–57, 60
$C(CH_3)_2CH_2OH$	S	2−	+	49	49	49	49	—	—	—	—
$C(CH_3)_2CH_2NHC_6H_5$	S	2−	+	49	49	49	49	—	—	—	—
1,1-$C_6H_{10}CH_3$	S	2−	+	49	49	49	49	—	—	—	—
$CH_2CH(CH_3)_2$	S	2−	+	53, 58	—	53	—	—	—	—	—
$CH_2Si(OMe)_3$	S	2−	—	59	—	59	—	—	—	—	59
$CH_2C_6H_5$[c, d]	S	2−	+	16, 35, 37, 61	30, 35, 41 61–63	35, 61	38, 64	—	41, 42, 61	16, 61, 65	43, 56
$CH_2C_6H_5$[e]	S	3−	+	16, 30, 66	30, 64, 66	—	64	66, 67	16, 42, 62, 66	16, 64	—

Table 3 *(Continued)*

R	X	z	Isolation	Preparation	AS	V	NMR	EPR	Mb	MS	LS[a]
CH_2-p-$C_6H_4OCH_3$	S	2−	+	67	—	—	—	—	—	—	—
CH_2-p-$C_6H_4OCH_3$	S	3−	+	67	—	—	—	67	67	67	—
$CH_2C_6H_{11}$	S	2−	+	37	35	35	38	—	—	—	—
C_6H_5	S	2−	+	16, 23, 35, 37, 45	30, 35, 39, 62, 68	30, 35, 68	38, 64	—	16, 42	16	27, 39, 43, 56
C_6H_5[e]	S	3−	+	16, 30, 66	30, 62, 66	—	64	65–68	16, 42, 62, 66	16, 64	—
C_6H_5	Se	2−	+	68, 69	68	68, 69	69	—	—	68	—
C_6H_5	Se	3−	—	68	—	—	—	68	—	—	—
$C_6H_5(Se)$[f]	S	2−	+	43, 68	68	68	43	—	—	—	—
$C_6H_5(Se)$	S	3−	—	68	—	—	—	68	—	—	—
$C_6H_5(Se)$	Se	2−	+	68	68	68	—	—	—	—	—
$C_6H_5(Se)$	Se	3−	—	68	—	—	—	68	—	—	—
o-$C_6H_4CH_3$	S	2−	+	64	64	64	64	—	—	—	—
o-$C_6H_4CH_3$[e]	S	3−	+	64	—	—	64	67	67	67	—
m-$C_6H_4CH_3$	S	2−	+	64	64	64	64	—	—	—	—
m-$C_6H_4CH_3$[e]	S	3−	+	64	—	—	64	67	67	67	—
p-$C_6H_4CH_3$	S	2−	+	38	29, 35	35	38, 54, 64	—	—	—	43, 54, 56
p-$C_6H_4CH_3$[e]	S	3−	+	64	64	64	64	67	67	67	—
p-$C_6H_4CH_3$	Se	2−	+	70	70	70	70	—	—	—	—
p-$C_6H_4CH_3$	Se	3−	+	70	70	—	70	—	—	—	—
m-$C_6H_4CF_3$	S	2−	+	34	34	34	34	—	—	—	34

p-$C_6H_4CF_3$	S	2−	+	34	34	34	34	—	—	—	34
p-C_6H_4Cl	S	2−	+	29	29	29	—	—	—	—	—
p-$C_6H_4N(CH_3)_2$	S	2−	—	35, 43	35, 43	35	43	—	—	—	35, 43
p-$C_6H_4N(CH_3)_3$	S	2+	+	29	35	35	43	—	—	—	43
p-$C_6H_4NO_2$	S	2−	+	37	—	35	38	—	—	—	43
p-$C_6H_4CH(CH_3)_2$	S	2−	+	67	—	—	—	—	—	—	—
p-$C_6H_4CH(CH_3)_2$	S	3−	+	67	—	—	—	67	67	67	—
p-$C_6H_4C(CH_3)_3$	S	2−	+	67	—	—	—	—	—	—	—
p-$C_6H_4C(CH_3)_3$	S	3−	+	67	—	—	—	67	67	67	—
p-C_6H_4OMe	S	2−	+	71	71	—	—	—	—	—	—
C_6Cl_5	S	2−	+	37	—	35	—	—	—	—	—
C_6F_5	S	2−	+	37	—	35	—	—	—	—	—
m-$C_6H_4(CH_2)_2$[g]	S	2−	+	38	35	35	38	—	—	—	—
Ac-Cys-$NHCH_3$	S	2−	—	35, 39, 44	35, 39, 44, 54	35, 39, 44	35, 39	—	—	—	35, 39, 43
peptide-3/t-Bu[h]	S	2−	—	39	39	39	39	—	—	—	39
peptide-3/Ac-Cys-$NHCH_3$[h]	S	2−	—	39	39	39	39	—	—	—	39
peptide-1	S	2−	—	39	39	39	39	—	—	—	—
peptide-4[i]	S	2−	—	57	57	—	57	—	—	—	57
peptide-4[i]	S	3−	—	72	—	—	—	72	—	—	—
peptide-5[j]	S	2−	—	46	46	46	—	—	—	—	46
peptide-5[j]	S	3−	—	46	46	—	—	46	—	—	—
peptide-6[k]	S	3−	—	72	—	—	—	72	—	—	—
peptide-2	S	2−	—	27, 57	27	—	57	—	—	—	57
peptide-2	S	3−	—	72	—	—	—	72	—	—	—
Cl	S	2−	+	36	36	36	—	—	—	36	36

Table 3 *(Continued)*

R	X	z	Isolation	Preparation	AS	V	NMR	EPR	Mb	MS	LS[a]
Br; I	S	2−	+	36	36	36	—	—	—	—	—
$[Fe_4S_{4-n}Se_n(S\text{-}p\text{-}C_6H_4CH_3)_4]$ (n = 1-3)		2−, 3−	—	73	—	—	73	—	—	—	—

[a]Ligand substitution reactions.
[b]ESCA spectrum (41).
[c]Resonance Raman spectrum (74).
[d]MCD spectrum (63).
[e]Magnetization (65, 67).
[f]The ligand is $C_6H_5Se^-$.
[g]Chelate ligand (= 2R).
[h]Peptide-3 = t-Boc·Gly-(Cys-Gly-Gly)$_2$-Cys-Gly·NH_2.
[i]Peptide-4 = Ac·Gly-Gly-Cys-Gly-Gly·NH_2.
[j]Peptide-5 = Ac·Gly-Gly-(Cys-Gly-Gly)$_2$·NH_2.
[k]Peptide-6 = Ac·Gly-Gly-(Cys-Gly-Gly)$_3$·NH_2.

other elements of protein structure, could cause distortions in site stereochemistry different from those in clostridial Rd. All such proteins exist in an oxidized [Fe(III)] and reduced [Fe(II)] form, each of which is high-spin.

The structures of two Rd_{ox} proteins from anaerobic bacteria have been determined by X-ray diffraction techniques. *C. pasteurianum* Rd_{ox} was originally solved at 3 Å resolution (4) and has subsequently been refined in several stages, leading to the last published refinement at 1.2 Å resolution (80–82). The structure of *D. vulgaris* Rd_{ox} has been solved by the molecular replacement method starting from the clostridial protein structure and has been partially refined at 2.0 Å resolution (83, 84). The refinement is being extended using data at 1.5 Å resolution (84).

Crystallographic results for the protein site **2** are profitably considered in relation to the structures of synthetic analogs and Fe–S distance data derived from the extended X-ray absorption fine structure (EXAFS) of analogs and proteins (85, 86). Structures of the analogs $[Fe(S_2\text{-}o\text{-}xyl)_2]^{1-,2-}$ (**4, 5**) and $[Fe(SPh)_4]^{2-}$ in Table 1 have been determined by X-ray diffraction (14, 87). Selected structural data are collected in Table 4. All species contain a $[Fe(SR)_4]$ coordination unit distorted to various extents from idealized T_d symmetry. The significance of the structures of most analogs considered in this chapter is that they present essentially unconstrained (i.e., symmetrized) versions of isoelectronic protein sites. Complexes **4** and **5**, which contain seven-membered chelate rings with flexible bite distances and angles, are such cases.

1−, 2−

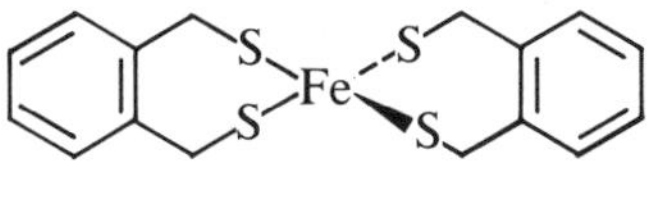

4 (1−), **5** (2−)

2.1.1 Oxidized Forms. Comparison of bond angles of $[Fe(S_2\text{-}o\text{-}xyl)_2]^{1-}$ (**6**) and clostridial Rd_{ox} (**7**) reveal a definite stereochemical similarity. Further, the anomalously short Fe–S distance in early protein structural refinements (88–90) has disappeared in the 1.2 Å refinement, which reveals that the four Fe–S distances are equivalent within experimental uncertainty. The mean value of 2.29 (4) Å is in good agreement with the 2.267 (9) Å value of the analog. The same comment applies to Fe–S distances in crystalline and lyophilized clostridial Rd_{ox} and in lyophilized and dissolved *P. aerogenes* Rd_{ox} from EXAFS. Spreads in bond distances obtained from the latter method as applied to the proteins under these conditions are somewhat smaller than those from X-ray diffraction (Table 4). These collective results

Table 4 Selected Structural Data for Rubredoxins and Their Analogs

Protein/Analog	Fe–S (Å)		S···S (Å)		S–Fe–S (deg)		$V(S_4)$	
	Range	Mean	Range	Mean	Range	Mean	($Å^3$)	Refs.
C. pasteurianum Rd_{ox}[a]	2.24–2.33	2.29 (4)[b]	3.60–3.87	3.74 (11)	103.7–114.3	109.5 (4.8)	6.10	81, 82
Lyophilized (EXAFS)	[c]	2.267 (3)	—	—	—	—	—	85
D. vulgaris Rd_{ox}[a, g]	2.24–2.30	2.27 (3)	3.54–3.79	3.70 (9)	102.5–113.4	109.5 (4.3)	5.92	84
P. aerogenes Rd_{ox} (EXAFS)								
Lyophilized	[d]	2.265 (13)	—	—	—	—	—	86
Solution	[e]	2.256 (16)	—	—	—	—	—	86
$[Fe(S_2\text{-}o\text{-}xyl)_2]^{1-}$[a, h]	2.252–2.279	2.267 (9)	3.617–3.763	3.70 (5)	105.8–112.6	109.5 (2.3)	5.97	14, 19
P. aerogenes Rd_{red}								
Solution (EXAFS)	[f]	2.32 (2)	—	—	—	—	—	86
$[Fe(S_2\text{-}o\text{-}xyl)_2]^{2-}$[a]	2.324–2.378	2.356 (26)	3.669–3.961	3.84 (11)	103.5–114.9	109.5 (4.3)	6.66	14
$[Fe(SPh)_4]^{2-}$[a]	2.338–2.372	2.356 (14)	3.554–4.051	3.84 (20)	97.6–119.5	109.5 (8.4)	6.53	87

[a]Crystal structure.

[b]In this and succeeding tables the numbers in parentheses are standard deviations from the mean calculated from the expression $\sigma = [\Sigma_{i=1}^{n} (x_i - \bar{x})^2/(n-1)]^{1/2}$ if the individual x_i values are available; in other cases they are determined as stated in the references.

[c–f]RMS deviation about mean values: [(c)]$0.032^{+0.013}_{-0.032}$; [(d)]$0.04^{+0.06}_{-0.04}$; [(e)]0.03 ± 0.04; [(f)]≈ 0.08 Å.

[g]These parameters are based on preliminary atomic coordinates from partial refinement of the protein structure at 1.5 Å resolution. Further refinement is in progress (L. H. Jensen, private communication).

[h]The average values are given from two independent anions.

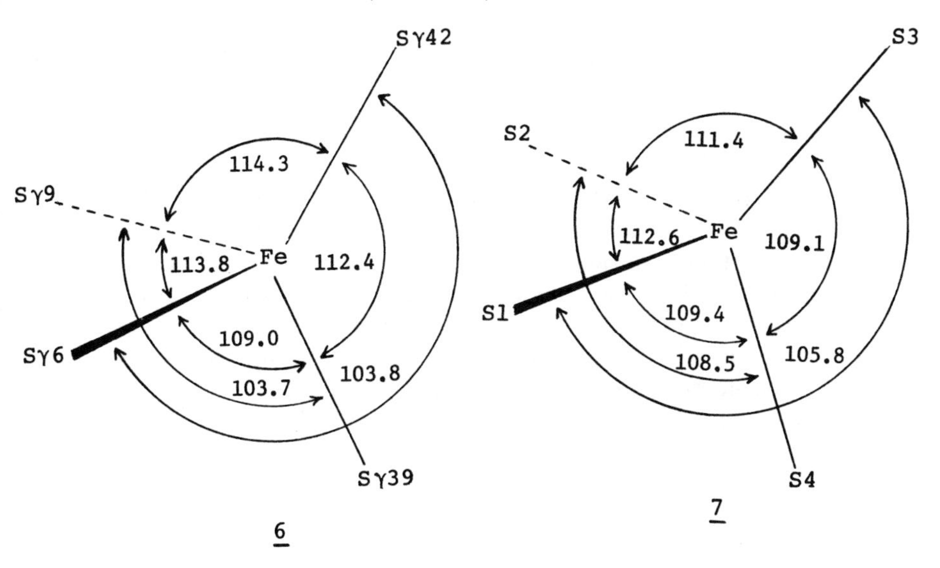

indicate that the Fe–S distances in the protein are indistinguishable from the unconstrained analog values. All species exhibit definite rhombic distortions from T_d symmetry, as may be seen from the appreciable ranges of S · · · S distances and S–Fe–S angles. As these distortions are nearly identical for the two independent anions in crystalline (Et_4N) $[Fe(S_2\text{-}o\text{-}xyl)_2]$ and are similar although slightly larger in clostridial Rd_{ox}, the analog structure appears to be representative of the [Fe(III) $(SR)_4$] unit in the absence of specific stereochemical constraints.

2.1.2 Reduced Forms. Crystallographic examination of Rd_{red} at 4.0 Å resolution "showed no change of the structure *within* the protein molecule [compared to Rd_{ox}], in particular the configuration of the S atoms about Fe had not changed" (90). This preliminary observation indicates retention of a tetrahedral Fe(II) site **2**, consistent with all spectroscopic properties of Rd_{red}, but at such low resolution that no metrical information was obtained. Fortunately, the structures of two Rd_{red} analogs, $[Fe(S_2\text{-}o\text{-}xyl)_2]^{2-}$ (**5**) and $[Fe(SPh)_4]^{2-}$, are available. For both the mean Fe–S distance is 2.356 Å, an increase of 0.089 Å over that of the oxidized analog. This represents an 11.7% volume enlargement of the S_4 tetrahedron on passing from **4** to **5**. The mean Fe–S distance in *P. aerogenes* Rd_{red}, 2.32(2) Å, from EXAFS analysis, is indistinguishable from those of the analogs. The latter two species do, however, differ from each other (and from $[Fe(S_2\text{-}o\text{-}xyl)_2]^{1-}$) in distortions from T_d symmetry as measured by S · · · S distances and S–Fe–S angles. Both have a less regular Fe-S_4 stereochemistry than the oxidized analog, and $[Fe(SPh)_4]^{2-}$ is substantially more distorted than $[Fe(S_2\text{-}o\text{-}xyl)_2]^{2-}$. Neither complex exhibits a structure meaningfully described in terms of a derivative point group of T_d. Because of the ^5E ground state of tetrahedral Fe(II), these complexes are susceptible to Jahn-Teller distortions. Additionally, theoretical studies of $[Fe(SH)_4]^{2-}$, an idealization of the Rd_{red} site and its analogs, indicate that the ^5E state can be appreciably split by metal-ligand orbital interactions dependent on R–S bond orientation (91). These studies also provide an electronic structural model that satisfactorily accounts for spectroscopic properties. Lastly, crystal packing forces may contribute to the distortions, at least for $[Fe(SPh)_4]^{2-}$, inasmuch as isomorphous benzenethiolate complexes of the d^{10} ions Zn(II) and Cd(II) show similar (albeit somewhat smaller) deviations from idealized tetrahedral stereochemistry (92).

2.2 2-Fe Structures

An excellent example of deduction of a correct protein site structure from spectroscopic and magnetic properties is afforded by the case of [2Fe-2S] proteins. Following the original investigations in 1966, analysis of a large body of physicochemical data led to the proposal of site **1** (93) containing tetrahedrally coordinated Fe atoms, with 2 Fe(III) in the oxidized form and a trapped valence Fe(III) + Fe(II) configuration in the reduced form. A "superreduced" oxidation level having 2 Fe(II) has not been identified in vitro. The correctness of site **1** has been persuasively supported as well by the synthetic analog approach (*vide infra*). Recent X-ray crystallographic studies of the chloroplast-type Fd_{ox} from the blue-green alga *Spirulina platensis* have provided final confirmation of the overall structure **1**, but metrical details have not yet been reported. The site was located in electron density maps at 5.0 and 3.5 Å using isomorphous replacement and anomalous scattering methods (94). Subsequently the complete structure has been determined at 2.8 Å resolution, including identification of coordinated cysteinate residues (7), and has been extended to 2.5 Å resolution (8). A similar 2-Fe Fd_{ox} from *Aphanothece sacrum* has been examined at 5.0 Å resolution and the [2Fe-2S] cluster has been located (95). Unfortunately, the crystalline form obtained contains four molecules in the asymmetric unit. Of the 98 residues of *S. platensis* Fd 6 are cysteines. Of these Cys-41 and Cys-46 are bound to 1 Fe atom and Cys-49 and Cys-79 to the other, an involvement previously based on sequence studies of other chloroplast-type Fds (96). Inasmuch as these residues and their positions are largely conserved, it is probable that their coordination in site **1** will prove to be a general occurrence in algal and higher plant proteins, the [2Fe-2S] sites of which exhibit virtually identical physicochemical properties.

Of the synthetic analogs of site **1** listed in Table 2 the structures of $[Fe_2S_2(S_2\text{-}o\text{-xyl})_2]^{2-}$ (**8**), $[Fe_2X_2(S\text{-}p\text{-}C_6H_4CH_3)_4]^{2-}$ (**9, 10**), $[Fe_2S_2(S_5)_2]^{2-}$ (**11**), and $[Fe_2S_2Cl_4]^{2-}$ (**12**) have been accurately determined by X-ray crystallography (29, 33, 97, 98). In addition, the structure of Na_3FeS_3 has been shown to contain the discrete anion **13** (99). In each case the anion contains two tetrahedrally coordinated Fe(III) atoms and therefore corresponds to the $[2Fe\text{-}2S]^{2+}$ oxidation level of a Fd_{ox} protein. Of these complexes **8** is the most meaningful analog owing to terminal coordination by alkylthiolates which effectively simulate cysteinate binding in site **1**. In this sense $[Fe_2S_6]^{6-}$ is not an

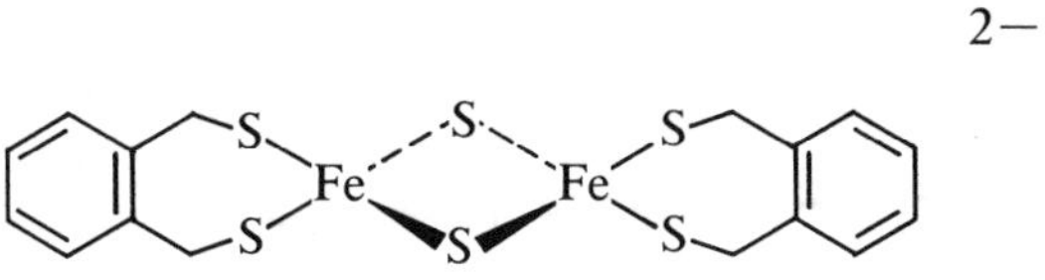

8

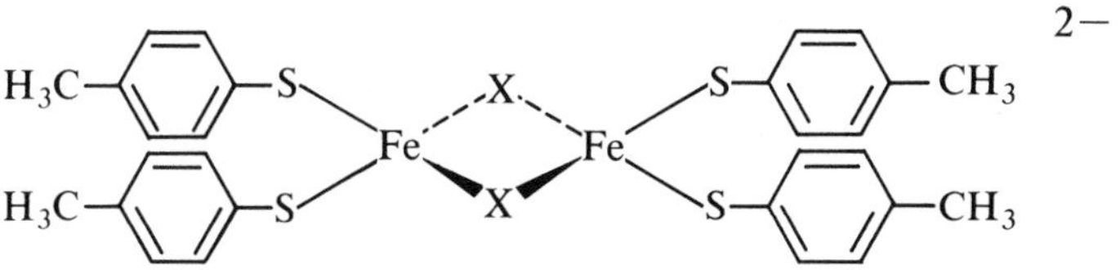

9 (X = S)
10 (X = Se)

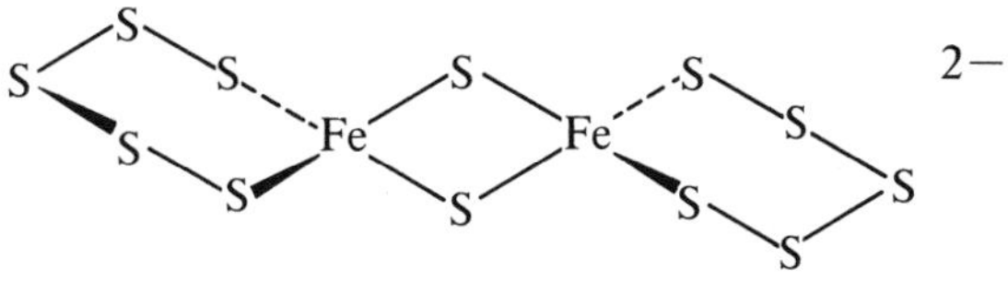

11

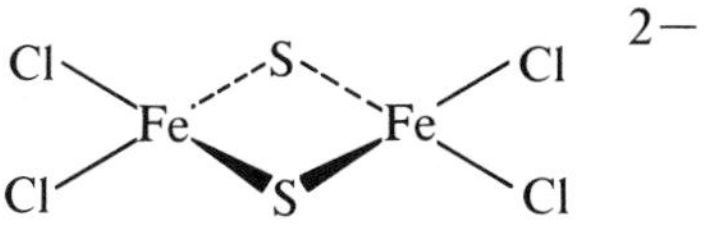

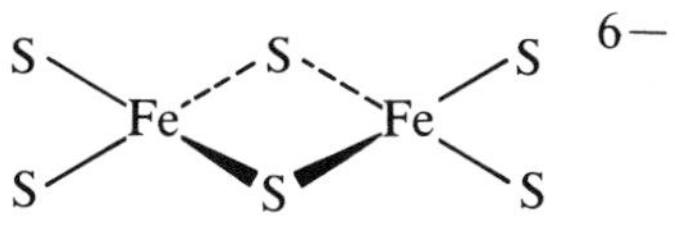

12

13

analog but is included here because of the presence of a Fe_2S_2 core and the altogether remarkable occurrence of *terminal* sulfido ligands. All anions have a crystallographically imposed symmetry center, requiring that their Fe_2X_2 cores be exactly planar. Selected structural parameters are collected in Table 5 for the six complexes hereafter generally formulated as $Fe_2X_2Y_4$ (**14**).

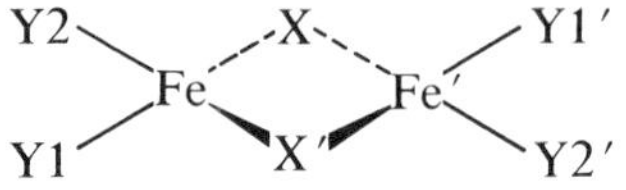

Y2 X Y1′
Fe Fe′
Y1 X′ Y2′

14

The highest idealized symmetry of species **14** is D_{2h}. From the data in Table 5 three types of small but distinct distortions from this symmetry are recognized. The first is a difference between the two independent Fe-X distances observed in **8** and **13**. The remaining two modes of distortion are described by angular shape parameters: θ is the supplement of the angle between the vectors from Fe to the midpoint of the X-X′ and Y1-Y2 segments; ϕ is the complement of the angle between the X-X′ and Y1-Y2 vectors. In D_{2h} symmetry $\theta = \phi = 0$. Calculated values of the parameters are entered in Table 5. For **8, 11,** and **13** θ is significantly different from zero. In these cases θ describes a motion of ligand set Y1, Y2 out of the plane perpendicular to the Fe_2X_2 plane and containing Fe_2, toward bridging atom X. An idealized structure of this type with $\phi = 0$ has C_{2h} symmetry with the twofold axis passing through the center of, and perpendicular to, the Fe_2X_2 plane. Complexes **8, 11,** and **13** have essentially this symmetry. For **9** and **10** $\theta \lesssim 1°$ but $\phi \cong 7°$. Here ϕ describes a twist of the Y1-Fe-Y2 plane relative to the Fe_2X_2 core. Such species with $\theta = 0$ also have C_{2h} symmetry, but the twofold axis contains the two Fe atoms. Distortions represented by θ and ϕ are manifested by differences in X-Fe-Y angles of each complex. These distortions, which presumably arise from solid state effects, appear to be exclusive of each other in that a species with one parameter decidedly different from zero has a near-zero value for the other. Of all complexes in Table 5 $[Fe_2S_2Cl_4]^{2-}$ is the only one with virtually perfect D_{2h} symmetry.

Except for $[Fe_2S_6]^{6-}$, dimensions of the Fe_2S_2 core units are nearly constant. The larger Fe-Se distance of **10** is accountable in terms of the $\approx$0.1 Å difference in selenide vs. sulfide radii (100). The Fe···Fe distances of 2.70-2.88 Å indicate the presence of significant metal-metal bonding, a feature that has emerged in recent theoretical calculations of the electronic structure of $[Fe_2S_2(SH)_4]^{2-,3-}$ (101). In *S. platensis* Fd_{ox} this distance is calculated to be 2.85 Å based on reported atomic coordinates (95), a value within experimental uncertainty of the analog distances given the resolution (2.8 Å) of the protein structure.

No analogs of site **1** in Fd_{red} proteins have yet been structurally characterized. $[Fe_2S_2(S_2\text{-}o\text{-xyl})_2]^{3-}$ has been generated in solution by reduction of **8**, and Mössbauer spectroscopy has shown it to have the Fe(II) + Fe(III) configuration of all (2Fe-2S] Fd_{red} sites (15). This is a significant finding, for it demonstrates that the trapped valence nature of the protein sites is an in-

Table 5 Selected Structural Data for $Fe_2X_2Y_2$ Complexes[a]

Complex	Distances (Å)			Angles (deg)						$V(X_2Y_2)$ ($Å^3$)	Ref.
	Fe-Fe	Fe-X	Fe-Y	Fe-X-Fe	X-Fe-X	Y-Fe-Y	X-Fe-Y	θ[b]	ϕ[b]		
$Fe_2S_2(S_2\text{-}o\text{-xyl})_2]^{2-}$	2.698 (1)	2.185 (2) 2.232 (1)	2.305 (2)[d]	75.3	104.7	106.4	112.2[e] 110.7	2.21	0.05	5.87	29
$[Fe_2S_2(S\text{-}p\text{-}C_6H_4CH_3)_4]^{2-}$	2.691 (1)	2.200 (1) 2.202 (1)	2.312 (1)	75.4	104.6	111.2	115.1[f] 105.4	0.93	7.01	5.83	29
$[Fe_2Se_2(S\text{-}p\text{-}C_6H_4CH_3)_4]^{2-}$	2.789 (2)	2.324 (1) 2.318 (1)	2.304 (1)	73.9	106.1	112.6	114.5[f] 104.6	1.16	7.04	6.29	97
$[Fe_2S_2(S_5)_2]^{2-}$	2.701 (3)	2.197 (3) 2.187 (3)	2.321 (4)	76.1	104.0	108.0	114.4[e] 107.9	6.46	0.03	5.85	33
$[Fe_2S_2S_4]^{6-}$ [c]	2.877 (2)	2.260 (3) 2.298 (2)	2.251 (2)	78.3	101.7	111.1	112.4[e] 109.3	3.91	0.52	5.93	99
$Fe_2S_2Cl_4]^{2-}$	2.716 (1)	2.199 (1) 2.202 (1)	2.252 (9)	76.2	103.8	105.4	112.0[g]	0.43	0.53	5.62	98

[a] X = bridging, Y = terminal ligand.
[b] Defined in text.
[c] Na_3FeS_3.
[d] Mean of two independent values; e.s.d. $\sigma = (x_1 + x_2)/\sqrt{2}$.
[e] θ describes dominant distortion; values are averaged according to X′-Fe-Y1 ≡ X′-Fe-Y2, X-Fe-Y1 ≡ X-Fe-Y2.
[f] ϕ describes dominant distortion; values are averaged according to X′-Fe-Y1 ≡ X-Fe-Y2, X′-Fe-Y2 ≡ X-FeY1.
[g] Mean of four independent values.

trinsic feature rather than a consequence of protein structure. EXAFS analyses of **8** and rhubarb $Fd_{ox,red}$ have been performed (102). Values of Fe ··· Fe and mean Fe–S distances for the oxidized analog are in good agreement with the diffraction results in Table 5, and distances of the former in Fd_{ox} in the solid and in solution are 2.70 and 2.73 Å, respectively. The Fe ··· Fe distance increases slightly to 2.76 Å in dissolved Fd_{red}, consistent with the larger radius of Fe(II) noted earlier.

2.3 3-Fe Structures

The newest structural discovery in Fe-S biochemistry is that of a [3Fe-3S] site, first detected by Mössbauer spectroscopy (10) and crystallographically established (9) in *A. vinelandii* Fd I [previously designated as major Fd (103), nonheme Fe protein III (104), or Fe-S protein III (104, 105)]. This protein (M_r 14,500), now recognized to possess 7 Fe and 7 inorganic sulfur atoms and 9 Cys residues, was found by X-ray diffraction studies at 4.0 Å to contain two different types of Fe-S unit (106). Refinement at 2.5 Å resolution (9, 107) demonstrated the presence of one cubane-type [4Fe-4S] site (*vide infra*) and the [3Fe-3S] cluster **15**, which had not been encountered previously in any natural or synthetic compound. The latter consists of a core six-membered Fe_3S_3 ring in a roughly planar conformation, with two additional ligands on each Fe atom resulting in four coordination, a property common to sites **1-3** also. The terminal ligation pattern is as follows (107): Fe(1), Cys-8, Cys-20; Fe(2), Cys-16, Glu-18 (or a small exogenous ligand such as H_2O or OH^-); Fe(3), Cys-11, Cys-49. Fe ··· Fe distances occur in the range 3.2–3.7 Å, and thus are substantially longer than those in sites **1** and **3** and their analogs. In addition to the Fe_3S_3 ring the cluster structure contains several other unprecedented features. Binding of the noncysteinate ligand at Fe(2) is the first documented instance of nonsulfur coordination in a Fe-S protein site. At the present stage of structural refinement the coordination unit of Fe(3) appears to be largely distorted from tetrahedral, the preferred stereochemistry of unconstrained four-coordinate Fe(II, III), toward a planar geometry.

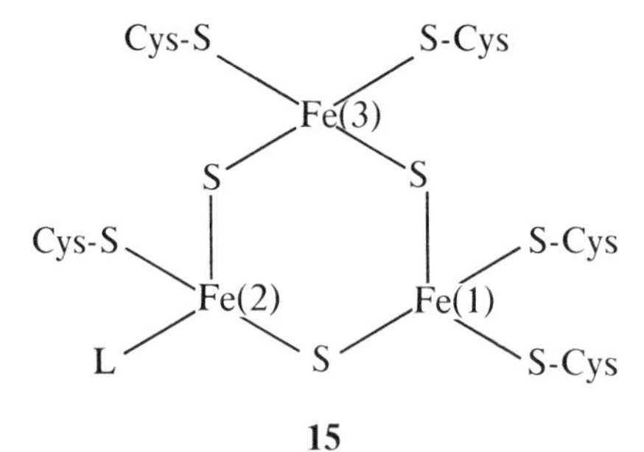

15

Shortly after establishment of the [3Fe-3S] site in *A. vinelandii* Fd I, Mössbauer spectroscopy demonstrated the occurrence of such a site in tetrameric Fd_{ox} II of *D. gigas* (11), earlier described as containing 4 Fe + 4 S per monomeric (M_r ~ 6000) unit (108). The presence of six Cys residues allows the possibility of site **15** with terminal coordination by only cysteinate. There is gratifying agreement between the Mössbauer spectra of this protein and those of the [3Fe-3S] site obtained by deconvolution of the more complex *A. vinelandii* Fd I spectra. Similar spectra are mentioned for beef heart aconitase and *A. vinelandii* glutamate synthase (10), indicating the presence of [3Fe-3S] sites in these enzymes. Oxidized sites exhibit a roughly isotropic EPR spectrum centered at g ~ 2.01; one-electron reduction renders the site EPR-silent. Mössbauer parameters are consistent with the $[3Fe\text{-}3S]^{3+}$ (3Fe(III)) and $[3Fe\text{-}3S]^{2+}$ (2Fe(III) + Fe(II)) core oxidation levels in the oxidized and reduced forms, respectively. Based on combined Mössbauer and EPR properties, demonstration of the occurrence of [3Fe-3S] sites in numerous other proteins and enzymes will doubtless follow in the near future. *R. rubrum* Fd IV (M_r ~ 14,000) (109) and *M. flavum* Fd I (M_r ~ 12,000, 7 Fe, 7 S, 8 Cys) and II (M_r ~ 13,500, 3 Fe, 3 S, 4 Cys) (110) are particularly likely candidates in this regard. Certainly the recognition of [3Fe-3S] sites is one of the major achievements of biologically applied Mössbauer spectroscopy.

A limited number of trinuclear Fe complexes containing sulfide have been structurally characterized; these include $Fe_3S_2(CO)_9$ (111), $Fe_3S(SO)(CO)_9$ (112), $Fe_3S(SN\text{-}t\text{-}Bu)(CO)_9$ (113), and $Fe_3S_2[S_2C_2(CF_3)_2]_4 \cdot \frac{1}{2}S_8$ (**16**) (114). The carbonyl complexes are obviously unrelated to the protein site. Complex **16** is a closer approach, but lacks the correct ring structure and contains the decidedly abiological dithiolene ligands, which permit ligand-based redox reactions that tend to obscure the oxidation state(s) of the Fe atoms. The properties of this species were investigated and interpreted (115) prior to knowledge of the structure, which thus far has been reported in abstract form only. Certain of its properties, most notably an apparently paramagnetic ground state, are difficult to reconcile with the even-electron structure **16**. However, the exact composition of the species examined spectroscopically is not clear and H may be present (115). The only documented example of a Fe_3S_3 ring system is the $Fe_3(SPh)_3$ fragment of $[Fe_4(SPh)_{10}]^{2-}$, which has the adamantanelike structure **17** (116). However, the occurrence of M_3 $(S(R))_3$ rings in discrete compounds of elements M of Groups IIIA–VA is not uncommon. Compounds of this type, together with several trinuclear transition metal complexes, which have been structurally characterized by X-ray diffraction (117–123), are listed in Table 6. All except $Pd_3(SCH_2CH_2\text{-}SCH_2CH_2S)_3$ have essentially tetrahedral coordination at the M sites. From this small set of examples it is found that the twist-boat and chair conformations are adopted by molecules containing S atoms and substituted S ligands,

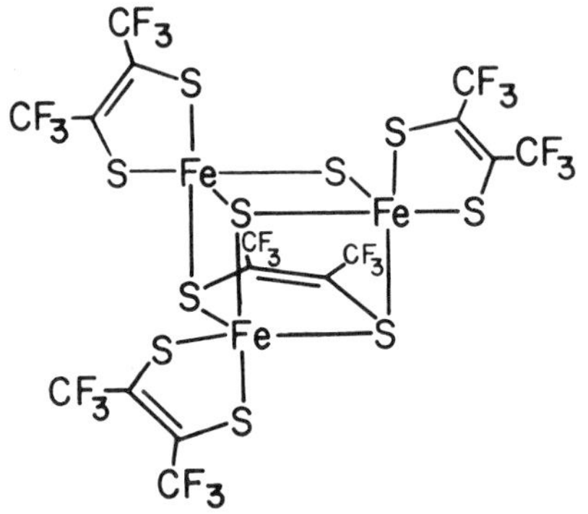

16

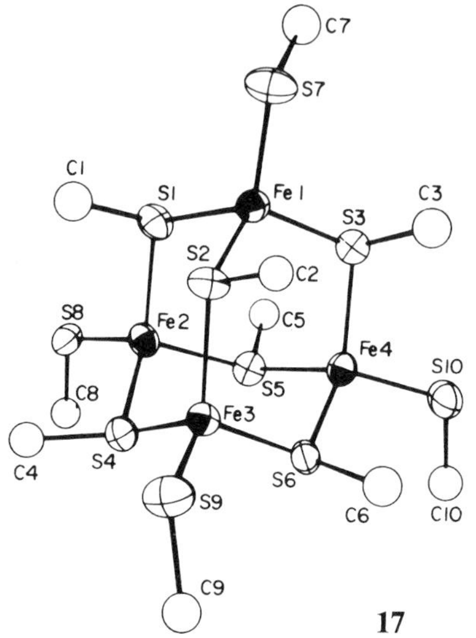

17

respectively, in their rings. With the radius of tetrahedral Fe(III) and Sn(IV) differing by only ≈ 0.06 Å (100), it will be interesting to learn whether further refinement of site **15** reveals a bias toward the twist-boat conformation. Lastly, by way of emphasizing the unanticipated occurrence of the Fe_3S_3 core structure in proteins, it is noted that the triangular $[Fe_3O]^{7+}$ assembly is the only triiron unit observed with any frequency in synthetic coordination complexes of Fe(III). It has been established in species such as $[Fe_3O(RCO_2)_6(H_2O)_3]^+$ (124, 125), $[Fe_3O(H_3NCH(R)CO_2)_6(H_2O)_3]^{7+}$ (126, 127), and basic ferric sulfate (128), all of which have the general structure **18**. While attention has been called to the $[Fe_3O]^{7+}$ unit as a possible structural component of fer-

Table 6 Conformations of Molecules Containing 3 S Atoms in Six-Membered Rings

Molecule	Ring Conformation	Refs.
$(CH_3SiC_6H_5)_3S_3$	twist-boat	117
$[(CH_3)_2Sn]_3S_3$[a]	twist-boat	118, 119
$(Cl_2B)_3(SCH_3)_3$	chair	120
$(Br_2B)_3(SCH_3)_3$	chair	120
$(ClCu)_3[SP(C_6H_5)_3]_3$	chair	121
$(ClCu)_3(CH_3C(S)NH_2)_3$	chair	122
$Pd_3(SCH_2CH_2SCH_2CH_2S)_3$	chair	123

[a] Two crystalline modifications.

ritin (126), its occurrence remains to be established in this or any other protein.

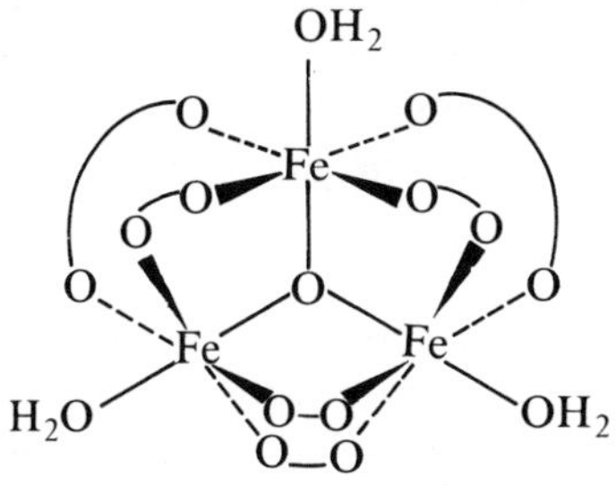

18

2.4 4-Fe Structures

Protein [4Fe-4S] sites possess the cubane-type structure **3** and, compared to 1-Fe, 2-Fe, and 3-Fe sites, present the added complication of three physiologically active core oxidation levels. These are shown in the form of the one-electron transfer series [eqs. (1)]; placed beneath each is that analog with an isoelectronic core. A given protein in vitro operates between two of the three oxidation levels, 2+/1+ or 3+/2+. The former couple is more common and is characterized by $E_m \sim -420$ mV. The latter couple was first detected in proteins that in their as-isolated state could be oxidized but not reduced. These were designated as "high-potential" proteins (HP) in view of their positive potentials ($E_m \sim +350$ mV). This designation is retained here because of its familiarity. However, the common description of all such proteins as ferredoxins is now the recommended nomenclature of the International Union of Biochemistry (129).

$$
\begin{array}{ccccc}
[4Fe\text{-}4S]^{1+} & \rightleftharpoons & [4Fe\text{-}4S]^{2+} & \rightleftharpoons & [4Fe\text{-}4S]^{3+} \\
(Fd_{red}, HP_{s\text{-}red}) & & (Fd_{ox}, HP_{red}) & & (Fd_{s\text{-}ox}, HP_{ox}) \\
[Fe_4S_4(SR)_4]^{3-} & \rightleftharpoons & [Fe_4S_4(SR)_4]^{2-} & \rightleftharpoons & [Fe_4S_4(SR)_4]^{1-}
\end{array}
\tag{1}
$$

High-potential proteins were first obtained from photosynthetic bacteria, that from *Chromatium vinosum* being the initial example (130), but recent work has shown that the occurrence of sites exhibiting the 3+/2+ couple is more widespread. An important case in point is *A. vinelandii* Fd I, which has two Fe-S sites with $E_m = -420$ and $+350$ mV. Both potentials were proposed to correspond to the $[4Fe\text{-}4S]^{3+/2+}$ couple (131), involving in the present notation oxidized and superoxidized (s-ox) levels. This proposal, offered five years before recognition of [3Fe-3S] sites, is partially correct. As recounted previously, structural and spectroscopic results (9, 10) have established that the high-potential site is indeed a 4-Fe cluster with the indicated oxidation levels, but the low potential site is **15** whose redox couple is $[3Fe\text{-}3S]^{3+/2+}$. This finding eliminates the paradox of two isoelectronic redox couples with a 770 mV potential difference. The occurrence and properties of 4-Fe sites is proteins and enzymes have recently been reviewed (132, 133).

2.4.1 Proteins. The structures of three proteins containing site **3** have been determined by X-ray diffraction methods. Early studies of *Chromatium* HP_{red} at 4 Å resolution using anomalous scattering effects revealed that the Fe atoms are present as "a compact, perhaps tetrahedral cluster" (134). Subsequently, HP_{red} and HP_{ox} have been refined at 2.0 Å resolution (5, 135–138). The structure of *P. aerogenes* Fd_{ox}, containing two sites **3**, was originally solved at 2.8 Å resolution (139) and has been further refined using data extending to 2.0 Å resolution (139–142). Finally, the structure of *A. vinelandii* Fd I has been solved at 2.5 Å and partially refined (9, 106, 107, 143). All clusters are terminally coordinated to Cys residues exclusively. In a number of these reports the entire protein structure is described in detail, including cluster environments which, for the $HP_{red,ox}$ and *P. aerogenes* Fd_{ox} proteins, involve intriguing hydrogen-bonding patterns of the polypeptides with the clusters (144). Review treatments are available (133, 138, 140, 145).

Bond distance and angle data for protein sites are summarized in Table 7. Fe–Fe distances have been determined by EXAFS techniques for the same or similar proteins (102) and are in good agreement with the diffraction results. Corresponding data for **3** of *A. vinelandii* Fd I have not yet been reported, nor are any results available for clusters in the $[4Fe\text{-}4S]^{1+}$ (Fd_{red}) oxidation level. At the present stage of resolution the most prominent aspect of the cluster structures in HP_{red} and Fd_{ox} is their distortion from an idealized cubic array toward a compressed tetragonal structure containing four short and eight longer Fe–S bonds. The similarity of the shapes and dimensions of the two clusters with each other and with those of the analog $[Fe_4S_4(SCH_2Ph)_4]^{2-}$ was quickly recognized (146) after the three structures were solved, and form part of a large body of evidence (13) that all three clusters contain an isoelectronic $[4Fe\text{-}4S]^{2+}$ core. As will be seen, the distorted cluster structure is clearly defined in the more accurate analog structures. The structure of the HP_{ox} cluster appears to approach more closely a tetrahedral arrangement and is described as being smaller in certain dimensions than that of the HP_{red} cluster, a feature consistent with the increased Fe(III) character of the core.

2.4.2 Analogs. The structures of five analogs containing the $[4Fe\text{-}4X]^{2+}$ core (X = S, Se) and two containing the $[4Fe\text{-}4S]^{1+}$ core have been

Table 7 Mean Bond Distances and Angles of $Fe_4X_4Y_4$ Clusters

Cluster	Core Oxidation Level	Distances, (Å)				Angles (deg)		Refs.
		Fe–Fe	Fe–X	Fe–Y	X···X	X–Fe–X	Fe–X–Fe	
C. vinosum HP_{ox}	$[4Fe\text{-}4S]^{3+}$	2.73	2.25	2.21	3.56	103.3	74.7	135, 137
$[Fe_4S_4(SCH_2Ph)_4]^{2-}$	$[4Fe\text{-}4S]^{2+}$	2.776 (2)[a] 2.732 (4)	2.239 (4) 2.310 (8)	2.251	3.645 (2) 3.586 (4)	104.1	73.8	37
$[Fe_4S_4(SPh)_4]^{2-}$	$[4Fe\text{-}4S]^{2+}$	2.730 (2) 2.736 (4)	2.267 (4) 2.296 (8)	2.263	3.850 (2) 3.592 (4)	104.3	73.5	43
$[Fe_4S_4(SCH_2CH_2CO_2)_4]^{6-}$	$[4Fe\text{-}4S]^{2+}$	2.778 (2) 2.743 (4)	2.261 (4) 2.300 (8)	2.250	3.613 (2) 3.596 (4)	103.9	74.1	50
$[Fe_4S_4Cl_4]^{2-}$	$[4Fe\text{-}4S]^{2+}$	2.755 (2) 2.771 (4)	2.260 (4) 2.295 (8)	2.216	3.637 (2) 3.562 (4)	103.5	74.6	98
$[Fe_4Se_4(SPh)_4]^{2-}$	$[4Fe\text{-}4Se]^{2+}$	2.773 (2) 2.788 (4)	2.385 (4) 2.417 (8)	2.273	3.901 (2) 3.826 (4)	106.4	70.6	68
C. vinosum HP_{red}	$[4Fe\text{-}4S]^{2+}$	2.81	2.25 (4) 2.36 (8)	2.22	—	103.6	74.5	137
P. aerogenes Fd_{ox}								
#1[d]	$[4Fe\text{-}4S]^{2+}$	2.73	2.19 (4) 2.25 (8)	2.20	3.49	102.7	75.5	141
#2[d]	$[4Fe\text{-}4S]^{2+}$	2.69	2.11 (4) 2.23 (8)	2.23	3.41	102.3	75.9	141
$[Fe_4S_4(SPh)_4]^{3-}$ [b]	$[4Fe\text{-}4S]^{1+}$	2.730 (2) 2.750 (4)	2.351 (4) 2.288 (8)	2.295	3.605 (2) 3.685 (4)	104.8	72.9	16
$[Fe_4S_4(SCH_2Ph)_4]^{3-}$	$[4Fe\text{-}4S]^{1+}$	2.759[c]	2.302 (6) 2.316 (6)	2.297	3.647 (4) 3.702 (2)	104.6	73.1	147

[a] The number of values averaged is given in parentheses where two distances are shown; otherwise the mean of all distances is given; e.s.d. values are not presented in this table.
[b] Average values of two independent anions are given.
[c] Because of the irregular structure the mean of all values is given.
[d] Two clusters per molecule.

characterized. Structural data are collected in Table 7, and structures of five analogs are schematically depicted in Figure 1. These species contain a cubanelike Fe_4X_4 core with Fe and chalcogen atoms at alternate vertices, and one terminal thiolate ligand completing approximately tetrahedral coordination at each Fe atom. The highest idealized symmetry of the Fe_4X_4 unit is T_d. The distortion pattern of the five analogs containing the $[4Fe\text{-}4X]^{2+}$ core is quite uniform. All closely approach idealized D_{2d} core symmetry owing to a *compression* along the $\bar{4}$ symmetry axis, resulting in four short and eight longer Fe–X bonds approximately parallel to and perpendicular to, respectively, this axis. Although isoelectronic protein cores appear to be less regular, they do exhibit a similar type of distortion. Consequently, a reasonable conclusion from structural results thus far is that a tetragonally compressed geometry (or close approaches thereto) is the intrinsically stable structure of the $[4Fe\text{-}4S]^{2+}$ core.

Although results are limited, it is evident that no such core structural uniformity exists in the reduced analogs $[Fe_4S_4(SR)_4]^{3-}$ in the solid state. The two independent anions of $(Et_3MeN)_3[Fe_4S_4(SPh)_4]$ have idealized D_{2d} core symmetry arising from *elongation* along the $\bar{4}$ axis. In this arrangement

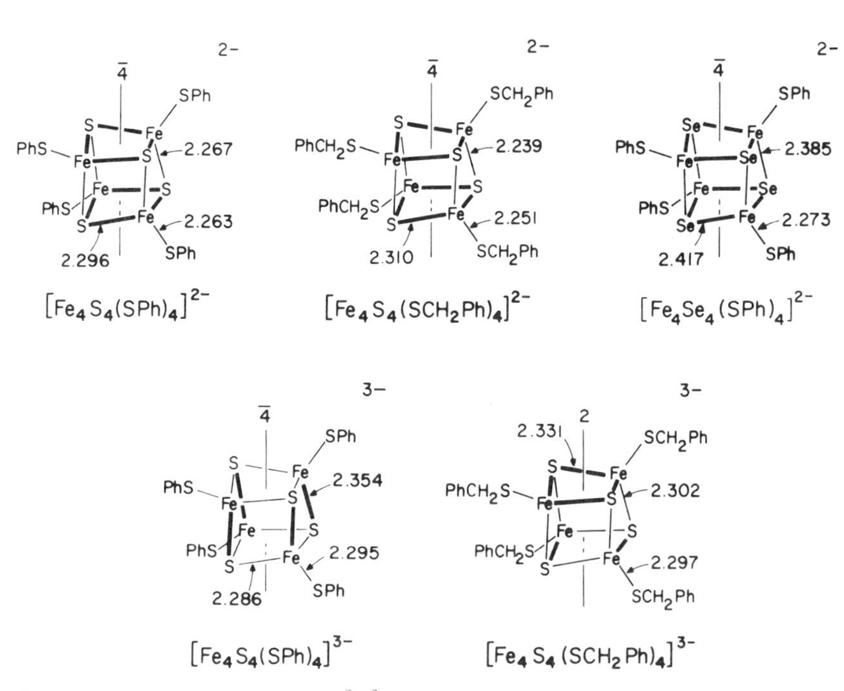

Figure 1. Structures of $[Fe_4X_4(SR)_4]^{2-,3-}$ complexes in the solid state, showing idealized core symmetry axes and mean values of Fe–SR and sets of longer (bold lines) and shorter Fe–X bond distances (X = S, Se).

there are four long bonds approximately parallel to this axis. However, the anion of $(Et_4N)_3[Fe_4S_4(SCH_2Ph)_4]$ displays a different type of distortion that does not preserve tetragonality but instead approaches C_{2v} symmetry (147) (Figure 1). Spectroscopic and magnetic properties of these two trianion compounds are different in the solid state, but in solution because nearly coincident with each other and with those of crystalline $(Et_3MeN)_3[Fe_4S_4(SPh)_4]$ (16). The same behavior has been shown for a number of other trianion compounds (67), leading to the conclusion that a tetragonally elongated geometry (or close approaches thereto) is the intrinsically stable structure of the $[4Fe\text{-}4S]^{1+}$ core, and that $[Fe_4S_4(SR)_4]^{3-}$ species are more susceptible to core distortions in the solid state than are $[Fe_4S_4(SR)_4]^{2-}$ analogs.

A generalized M_4X_4 cluster **19** can be described as two interlocking, concentric tetrahedra (M_4, X_4). If the overall assembly has D_{2d} symmetry, so must each tetrahedron and each must be staggered with respect to the other. With reference to the component tetrahedron **20** the cluster **19** of D_{2d} symmetry is completely defined by four shape parameters: the distances r_M and r_X and the polar angles β_M and β_X ($\beta = 54.74°$ for a perfect tetrahedron). These values are easily calculated using a coordinate origin given by the mean of coordinates of the eight atoms and a $\bar{4}$ axis which passes through the origin and is fit to the centers of M–M and X–X segments on opposite faces using least-squares methods. The four independent values of each parameter are then averaged. The value of any structural property of cluster **19** may be calculated from the shape parameters. Some representative relationships are given by eqs. (2)-(4). The shape parameters afford a concise description of tetragonal clusters, and together with volumes, afford an incisive means of assessing similarities of structures idealized to D_{2d} symmetry.

$$d(M - X, \text{set of } 4) = [r_M^2 + r_X^2 + 2r_M r_X \cos(\beta_M + \beta_X)]^{1/2} \quad (2)$$

$$d(M - X, \text{set of } 8) = [r_M^2 + r_X^2 - 2r_M r_X \cos\beta_M \cos\beta_X]^{1/2} \quad (3)$$

$$V(M_4) = \frac{4}{3} r_M^3 \cos\beta_M \sin^2\beta_M \quad (4)$$

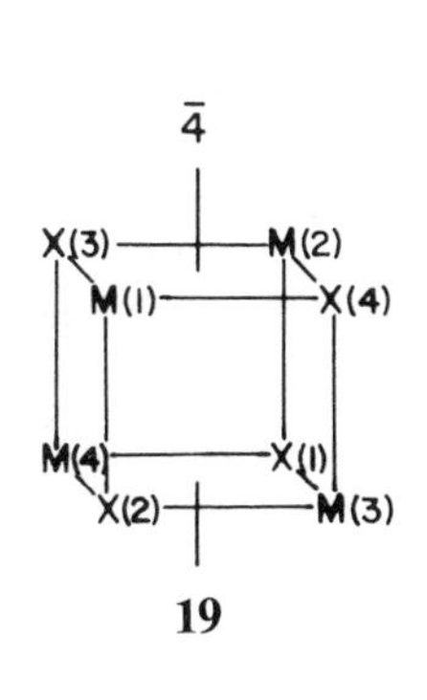

19

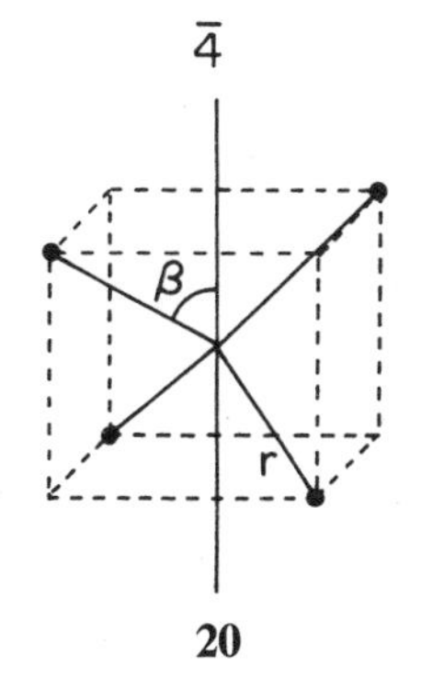

20

Shape parameters and volumes for protein and analog clusters are listed in Table 8. (These values could not be calculated for HP_{red} because of the unavailability of atomic coordinates.) The following features emerge from the data for analogs: (1) Cores of all X = S analogs are essentially congruent when oriented with mutually parallel $\bar{4}$ axes. (2) S_4 units are substantially larger than Fe_4 units, enclosing ≈2.3 times the volume of the latter. (3) Based on the angle β, distortions from T_d symmetry are larger in the S_4 than in the Fe_4 units in three of the five cases, including $[Fe_4S_4(SPh)_4]^{2-}$. (4) On reduction of $[Fe_4S_4(SPh)_4]^{2-}$ the Fe_4 unit retains its near-perfect tetrahedral shape ($\Delta\beta = 0.3°$) and changes very slightly in size ($\Delta V < 1\%$). The substantial changes are localized essentially in the S_4 unit, where β decreases by 2° attendant to adoption of an elongated tetragonal structure. This result is a restatement of our earlier argument that cores of these oxidized and reduced clusters approach maximum congruency when their $\bar{4}$ axes are mutually perpendicular (147). (5) The small overall core structural differences in the pairs $[Fe_4S_4(SR)_4]^{2-,3-}$ is emphasized by the Fe_4S_4 volume increases of 2.6% (R = CH_2Ph) and 1.9% (R = Ph) on reduction. (6) Terminal Fe-SR bond distances in reduced clusters are 0.03–0.05 Å longer than in the oxidized forms (Table 7), a result compatible with the larger Fe(II) core character of the former. Because of the less regular shapes of protein clusters at present refinement levels, their shape parameters are not as meaningful as those of the analogs. Moreover, their calculation requires a choice for the directions of the $\bar{4}$ axes. For the two clusters in Fd_{ox}, these axes are chosen based on the observation that the mean Fe–S distance along one direction of each cluster is shorter than those along the other two (Table 7). For HP_{ox} the axis is chosen to coincide with the approximate $\bar{4}$ axis noted in the HP_{red} cluster (137). It is observed that the clusters in Fd_{ox} appear smaller than the analogs, particularly with regard to S_4 units. It seems likely that this result is artifactual, but any further conclusions concerning relative protein-analog cluster geometries must await the availability of more highly refined protein structures. However, at this point it is clear that all such clusters at the $[4Fe\text{-}4S]^{1+,2+}$ oxidation levels significantly depart from T_d core symmetry. The source of the distortion is not yet established, but theoretical models of electronic structure in this symmetry admit the possibility of orbitally degenerate ground states (148, 149) and, thus, of first-order Jahn-Teller distortions. One model suggests that the $[4Fe\text{-}4S]^{3+}$ (HP_{ox}) level could have a $S = 3/2$ orbital singlet ground state (148) and, therefore, would not be susceptible to operation of the first-order Jahn-Teller effect. All evidence for HP_{ox} species indicates a spin-doublet ground state, perhaps as a consequence of protein-induced distortions. A proper test of the model requires accurate structures of one or more $[Fe_4S_4(SR)_4]^{1-}$ analogs. Although such clusters have been detected electrochemically (35, 49) and apparently by EPR (72), none has as yet been isolated as a fully characterized compound.

Table 8 Shape Parameters of Fe_4X_4 Cores of $Fe_4X_4Y_4$ Clusters

Cluster	r_{Fe}(Å)	r_X (Å)	β_{Fe} (deg)	β_X (deg)	V (Fe_4) (Å^3)	V(X_4) (Å^3)	V (Fe_4S_4) (Å^3)
c. vinosum HP_{ox}	1.66 (6)[a]	2.18 (2)	53.6 (1.8)	53.9 (6)	2.33	5.28	9.18
$[Fe_4S_4(SCH_2Ph)_4]^{2-}$	1.682 (4)	2.208 (5)	55.6 (2)	55.6 (2)	2.44	5.52	9.61
$[Fe_4S_4(SPh)_4]^{2-}$	1.675 (2)	2.211 (7)	54.6 (3)	55.6 (1)	2.41	5.54	9.55
$[Fe_4S_4(SCH_2CH_2CO_2)_4]^{6-}$	1.687 (2)	2.206 (5)	55.4 (2)	55.0 (1)	2.46	5.50	9.66
$[Fe_4S_4Cl_4]^{2-}$	1.694 (5)	2.197 (5)	54.4 (1)	55.9 (2)	2.49	5.43	9.70
$[Fe_4Se_4(SPh)_4]^{2-}$	1.704 (10)	2.358 (8)	54.5 (3)	55.8 (1)	2.54	6.72	10.54
P. aerogenes Fd_{ox}							
#1[d]	1.67 (4)	2.11 (8)	57.0 (8)	53.8 (1.5)	2.39	4.98	9.17
#2[d]	1.65 (6)	2.09 (7)	57.1 (2.2)	55.5 (2.7)	2.28	4.62	8.65
$[Fe_4S_4(SPh)_4]^{3-}$ [b]	1.680 (9)	2.240 (5)	54.3 (4)	53.6 (3)	2.43	5.76	9.73
$[Fe_4S_4(SCH_2Ph)_4]^{3-}$	1.690 (18)	2.244 (6)	[c]	[c]	2.48	5.80	9.86

[a] The standard deviation for each value is estimated by using the equation $s = [\Sigma_{i=1}^{4} (x_i - \bar{x})^2/3]^{1/2}$.

[b] The average values of two independent anions are given.

[c] Distortion angles are not uniquely defined because the cluster lacks D_{2d} symmetry.

[d] Two clusters per molecule.

Lastly, the cubane-type cluster **19** is not an unusual stereochemical array in inorganic chemistry. Over 50 molecules containing such an M_4X_4 unit have been structurally characterized; listings of a majority of such molecules are available (150, 151). Among these are a number of Fe_4S_4 species, including $[(\eta^5 - C_5H_5)_4Fe_4S_4]^{0,1+,2+}$ (152–156), $[Fe_4S_4(NO)_4]$ (157), and $[Fe_4S_4(S_2C_2(CF_3)_2)_4]^{2-}$ (158). These complexes have not been dealt with here for, in addition to their obviously nonphysiological ligands, simple considerations (37, 157) show that their cores are not isoelectronic with those of any species in eqs. (1). Many of the M_4X_4 units approach or possess cubic or tetragonal symmetry, and their structures can be profitably compared using the shape parameters reported here. Other means of comparison have been devised and applied to $[Fe_4S_4(SR)_4]^{2-,3-}$ analogs (37, 147, 159). In addition, more general approaches have been developed for quantitative structural comparisons of molecular fragments or whole molecules of related stereochemistry (160, 161).

3 SPONTANEOUS ASSEMBLY OF CLUSTERS

3.1 Formation of $[Fe_4S_4(SR)_4]^{2-}$

Entry to the field of synthetic analogs of Fe-S protein sites was afforded by the discovery in 1972 (61) that the reaction system $FeCl_3/NaHS/3NaSCH_2Ph$ in methanol yielded the cluster $[Fe_4S_4(SCH_2Ph)_4]^{2-}$, which was readily isolated and purified as its Et_4N^+ salt. Very soon thereafter the procedure was shown to be general for a variety of thiolate salts NaSR (37), producing a portion of the now extensive collection of $[Fe_4S_4(SR)_4]^{2-}$ clusters (Table 3). The proposed reaction scheme (37) consisted of the formation of a polymeric Fe(III) thiolate in reaction (5) followed by cluster formation, reaction (6). The overall process has the limiting stoichiometry of reaction (7). The method remained unchanged until it was shown in 1979 (45) that elemental S in the presence of sufficient thiolate reductant could be substituted for sulfide, as shown in reactions (8) and (9). Both preparative methods are readily extended to $[Fe_4Se_4(SR)_4]^{2-}$ clusters (68, 69). While it has been clear for some time that cluster formation is thermodynamically controlled, the nature of the thermodynamic product could not have been uniquely predicted. This situation is indigenous to most initial cluster syntheses and, in the present case, the term "spontaneous self-assembly" has been applied (17) as a reminder of the thermodynamic origin of cluster formation.

$$4FeCl_3 + 12RS^- \rightarrow 4Fe(SR)_3 + 12Cl^- \qquad (5)$$

$$4Fe(SR)_3 + 4HS^- + 4OMe^- \rightarrow [Fe_4S_4(SR)_4]^{2-} + RSSR + 6RS^- + 4MeOH \quad (6)$$

$$4FeCl_3 + 4HS^- + 6RS^- + 4OMe^- \rightarrow [Fe_4S_4(SR)_4]^{2-} + RSSR + 12Cl^- + 4MeOH \quad (7)$$

$$4FeCl_3 + 14RS^- + 4S \rightarrow [Fe_4S_4(SR)_4]^{2-} + 5RSSR + 12Cl^- \quad (8)$$

$$4FeCl_2 + 10RS^- + 4S \rightarrow [Fe_4S_4(SR)_4]^{2-} + 3RSSR + 8Cl^- \quad (9)$$

Recently a study has been conducted which has identified the principal reaction steps and intermediates in the assembly of $[Fe_4S_4(SR)_4]^{2-}$ clusters from simple reactants (23). Reaction (3), which typically proceeds in $\gtrsim 80\%$ yield, was found to be too rapid to be examined usefully by the conventional spectrophotometric and ^1H NMR spectroscopic methods employed. However, reaction system (8) with R = Ph forms soluble iron-thiolate species and proceeds at a convenient rate. The sequence of reactions was found to be dependent on the molar ratio of reactants and the solvent (protic vs. aprotic). In a reaction system composed of $PhS^-/FeCl_3/S = 3.5:1:1$ in methanol the first detectable product, formed in reaction (10), is $[Fe_4(SPh)_{10}]^{2-}$ (**17**), which was isolated and structurally characterized. In either methanol or acetonitrile this complex reacts quantitatively with sulfur to give $[Fe_4S_4(SPh)_4]^{2-}$. Of particular interest is the finding that this is an "all-or-nothing" reaction; that is, with a deficiency of sulfur ($n < 4$) in reaction (11) $[Fe_4S_4(SPh)_4]^{2-}$ is produced at the expense of unreacted **17**. In this system preparative reaction (8) = (10) + (11) ($n = 4$). Inasmuch as **17** can also be synthesized by reaction (12) in methanol, reaction (9) is diagnosed as the sum of reactions (12) + (11) ($n = 4$). In methanol solution in situ yields of $[Fe_4S_4(SPh)_4]^{2-}$, determined spectrophotometrically and based on $FeCl_3$, are essentially quantitative. Reactions (10) and (12) are also ones of cluster assembly, but as yet no information has been obtained regarding the steps leading to construction of $[Fe_4(SPh)_{10}]^{2-}$ in these rapid processes.

$$4FeCl_3 + 14RS^- \rightarrow [Fe_4(SR)_{10}]^{2-} + 2RSSR + 12Cl^- \quad (10)$$

$$[Fe_4(SR)_{10}]^{2-} + nS \rightarrow \frac{n}{4}[Fe_4S_4(SR)_4]^{2-} + \frac{4-n}{4}[Fe_4(SR)_{10}]^{2-} + \frac{3n}{4} RSSR \quad (11)$$

$$4FeCl_2 + 10RS^- \rightarrow [Fe_4(SR)_{10}]^{2-} + 8Cl^- \quad (12)$$

$$FeCl_3 + 5RS^- \rightarrow [Fe(SR)_4]^{2-} + \frac{1}{2} RSSR + 3Cl^- \quad (13)$$

$$2[Fe(SR)_4]^{2-} + nS \rightarrow \frac{n}{2}[Fe_2S_2(SR)_4]^{2-} + (2-n)[Fe(SR)_4]^{2-} + \frac{n}{2} RSSR + nRS^- \quad (14)$$

$$2[Fe_2S_2(SR)_4]^{2-} \rightarrow [Fe_4S_4(SR)_4]^{2-} + RSSR + 2RS^- \quad (15)$$

In a second reaction system initially containing the molar ratios $PhS^-/FeCl_3/S \geq 5:1:1$ in methanol the sequential reactions (13)–(15) have been established. Here a systematic buildup of the final cluster occurs involving species of smaller nuclearity. The first reaction product, $[Fe(SPh)_4]^{2-}$, had been prepared and structurally characterized earlier (22, 87). It reacts with sulfur in the all-or-nothing process (14) to afford the known binuclear complex $[Fe_2S_2(SPh)_4]^{2-}$ (29), whose structure is doubtless that established for the *p*-tolyl variant **9**. This species spontaneously converts to $[Fe_4S_4(SPh)_4]^{2-}$ by reaction (15) in protic solvents, a process discovered previously (30). In situ yields are quantitative. Because reaction (15) does not occur in acetonitrile solution, the assembly process in this solvent terminates with the formation of the binuclear species. The reaction sum is 4 × (13) + 2 × (14) ($n = 2$) + (15) = (8). Consequently, this reaction system is equivalent to the cluster assembly reaction (8). As such it offers no practical advantage in cluster synthesis compared to that based on the latter reaction. However, resolution of the processes in this system into three distinct stages provides the first demonstration that tetranuclear clusters can be elaborated by a series of spontaneous irreversible reactions commencing with trivial reagents and passing through successive mononuclear and binuclear intermediates. The sequence of events in both cluster assembly systems is schematically represented in Figure 2. Further details, including some evidence for the generality of the assembly scheme as well as a consideration of the possible relevance of these reactions to the biosynthesis of protein site **3**, are included in the original report (23).

3.2 Changes in Nuclearity

Reactions (14) and (15) (R = Ph) are examples of processes whereby one well-defined Fe-S complex is transformed into another with a different number of Fe atoms. Other cases are represented by the spontaneous high-yield reactions (16) (30), (17) and (18) (49), (19) (14), (20) (33), and (21) (30). Reactions (16) and (19) have been observed in methanol or ethanol, and the remainder in an aprotic solvent such as acetonitrile. In the equilibrium (19) the binuclear species is favored in ethanol. Reaction (20), in which the

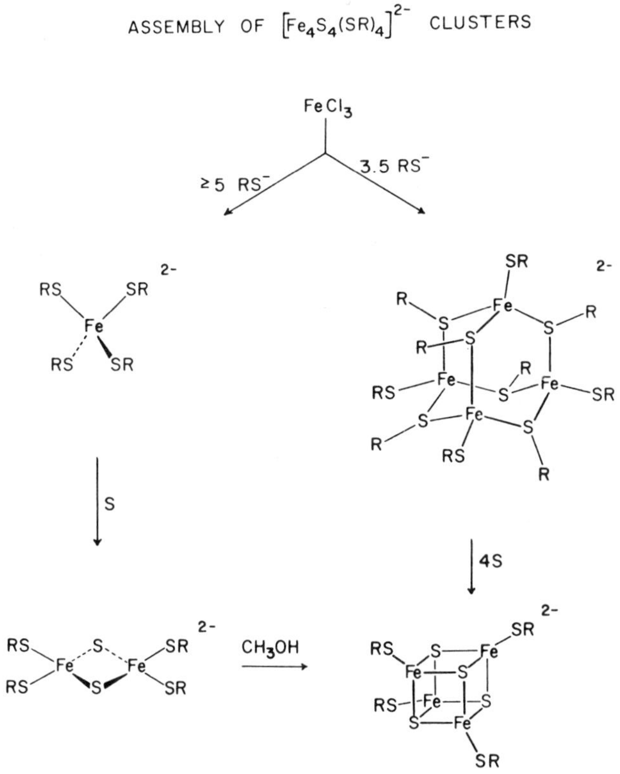

ASSEMBLY OF $[Fe_4S_4(SR)_4]^{2-}$ CLUSTERS

Figure 2. Depiction of the course of reactions resulting in the assembly of $[Fe_4S_4(SR)_4]^{2-}$ clusters via the intermediates $[Fe(SR)_4]^{2-}$, $[Fe_2S_2(SR)_4]^{2-}$, and $[Fe_4(SR)_{10}]^{2-}$ (23).

trisulfide serves as a sulfur carrier, is equivalent to reaction (14) with $n = 2$. Reactions (16)–(18) result in monomer → dimer conversion and offer additional preparative routes to the dianion **8**. Reactions (15) and (21) are the best documented cases of dimer → tetramer conversion. The latter reaction, which has also been found with reduced halide dimers such as $[Fe_2S_2Cl_4]^{3-}$ (36), is the more facile, presumably because it involves dimerization of isoelectronic cores: $2[2Fe\text{-}2S]^{+} \rightarrow [4Fe\text{-}4S]^{2+}$. The only clear instance of the reverse of these reactions observed in this laboratory is the formation of $[Fe_2S_2Cl_4]^{2-}$ from $[Fe_4S_4Cl_4]^{2-}$ by reaction with ferricinium ion (36). Presumably because of its chelate structure, $[Fe_2S_2(S_2\text{-}o\text{-xyl})_2]^{3-}$ is far more stable than reduced dimers with monothiolate ligands. As noted earlier, it has been generated in solution, and certain of its properties have been measured (Table 2).

$$2[Fe(S_2\text{-}o\text{-xyl})_2]^{1-} + 2HS^- + 2OMe^- \rightarrow [Fe_2S_2(S_2\text{-}o\text{-xyl})_2]^{2-} + 2(S_2\text{-}o\text{-xyl})^{2-} + 2\,MeOH \quad (16)$$
(**4** → **8**)

$$2[Fe(S_2\text{-}o\text{-xyl})_2]^{1-} + 2S \rightarrow [Fe_2S_2(S_2\text{-}o\text{-xyl})_2]^{2-} + 2S_2\text{-}o\text{-xyl} \quad (17)$$

$$2[Fe(S_2\text{-}o\text{-xyl})_2]^{2-} + 2S \rightarrow [Fe_2S_2(S_2\text{-}o\text{-xyl})_2]^{2-} + (S_2\text{-}o\text{-xyl})^{2-} + S_2\text{-}o\text{-xyl}) \quad (18)$$
(**5**)

$$2[Fe(S_2\text{-}o\text{-xyl})_2]^{2-} \rightleftharpoons [Fe_2(S_2\text{-}o\text{-xyl})_3]^{2-} + (S_2\text{-}o\text{-xyl})^{2-} \quad (19)$$

$$2[Fe(SPh)_4]^{2-} + 2(PhCH_2S)_2S \rightarrow [Fe_2S_2(SPh)_4]^{2-} + 2(PhCH_2S)_2 + PhSSPh + 2PhS^- \quad (20)$$

$$2[Fe_2S_2(SR)_4]^{3-} \rightarrow [Fe_4S_4(SR)_4]^{2-} + 4RS^- \quad (21)$$

$$FeCl_3 + 4RS^- \rightarrow [Fe(SR)_4]^{1-} + 3Cl^- \quad (22)$$

$$[Fe(SR)_4]^{1-} + RS^- \rightarrow [Fe(SR)_4]^{2-} + \frac{1}{2}RSSR \quad (23)$$

$$4[Fe(SR)_4]^{2-} + 4S^{2-} + RSSR \rightarrow [Fe_4S_4(SR)_4]^{2-} + 14RS^- \quad (24)$$

$$Rd_{ox} + 2Fe^{2+} + Fe^{3+} + 4S^{2-} \rightarrow \text{“}Fd_{ox}\text{”} \quad (25)$$

$$\text{apoadrenal-Fd} + Fe^{3+} \rightarrow \text{Fe(III)}\cdot\text{complex} \xrightarrow{S^{2-}} \text{adrenal } Fd_{ox} \quad (26)$$

A monomer → tetramer conversion has been observed in a system with the reactant molar ratios Cys_4-peptide ($\equiv 4RS^-$)/$FeCl_3$/S^{2-} = 1.5:1.2:2.0 in DMSO (27). Before sulfide is added, reaction (22) occurs, which is followed by decay of the product to a species whose spectrum is more consistent with $[Fe(SR)_4]^{2-}$, as in reaction (23), than with the suggested $[Fe_2(SR)_6]^{2-}$. Introduction of sulfide at this stage results in quantitative formation of the $[Fe_4S_4(SR)_4]^{2-}$ chromophore in a process represented here as the overall reaction (24). Interestingly, the reaction sum is (22) + (23) = (13), raising the possibility that $[Fe(SPh)_4]^{1-}$ is the initial species formed in the assembly system where reaction (13) is the first recognizable step. In fact $[Fe(SPh)_4]^{2-}$ can be reversibly oxidized electrochemically to the monoanion (23), which lacks the stability of the chelated oxidized peptide complex and $[Fe(S_2\text{-}o\text{-xyl})_2]^{1-}$. A similar result has been achieved in 90% DMSO/H_2O starting with *C. pasteurianum* Rd_{ox}, reaction (25) (162). The initial protein is unfolded in this medium, doubtless tending to facilitate the reaction whose product chromophore (λ_{max} 414 nm) is similar to that of native clostridial

Fd_{ox} in 80% DMSO/H_2O (44). Not surprisingly, when the aqueous content of the system was increased, allowing refolding of the polypeptide, some reformation of Rd_{ox} was observed. A related observation is that of reaction sequence (26) in aqueous solution (163). Here apoadrenal Fd was treated with Fe(III), forming an unstable Rd_{ox}-like chromophore that spontaneously converted to the native $[2Fe\text{-}2S]^{2+}$ adrenal Fd_{ox} on treatment with sulfide. The instability of the intermediate species in this reaction and of "Fd_{ox}" in reaction (25) both are examples of the precariousness of an iron site in an "unnatural" peptide when that peptide can assume its normal folded configuration in the presence of its "natural" site constituents. The absence or transitory existence of Fe_2S_2 species in reactions (24) and (25) apparently reflects a similar relative instability under the particular experimental conditions employed. As seen in a subsequent section, the selectivity of several apoproteins for their natural Fe-S clusters is obligatory to one variant of the core extrusion method for identification of protein sites **1** and **3**.

4 REACTIONS OF PROTEINS AND ANALOGS

Reactions of protein Fe-S sites and their analogs may be classified as follows: (1) electron-transfer processes, which alter only the oxidation level; (2) substitution of core atoms with or without change in terminal ligands but with retention of overall cluster structure; (3) substitution of terminal ligands with retention of core structure; (4) disruption of cluster/core structure, as by solvolytic (hydrolytic) processes (52, 58, 164–166), reactions with thiophiles (167, 168) and iron chelators (169, 170), and destructive oxidation (171). This classification does not include reactions of *apo*proteins with potential site components, as in the reconstitution of proteins with an Fe salt and sulfide (169, 172–176) or selenium reagents (176–180), and formation of Co(II) "Rd" (181) and Co and Ru derivatives of apoadrenal Fd (182). These reactions and those in classification (4) are not dealt with in the sections following.

4.1 Electron-Transfer Reactions

The biological function of Fd proteins, in cases where it has been deduced from in vitro reconstitution of an active enzyme system or can be otherwise reasonably inferred, is that of electron transfer. These proteins, which together with the cytochromes and "blue" copper proteins constitute the three main types of metalloprotein electron carriers, are coupled to a wide variety of Fd-dependent enzymes (183) that require electron supply or removal to catalyze substrate transformations. Figure 3 shows an elementary

BIOLOGICAL FUNCTION OF FERREDOXINS

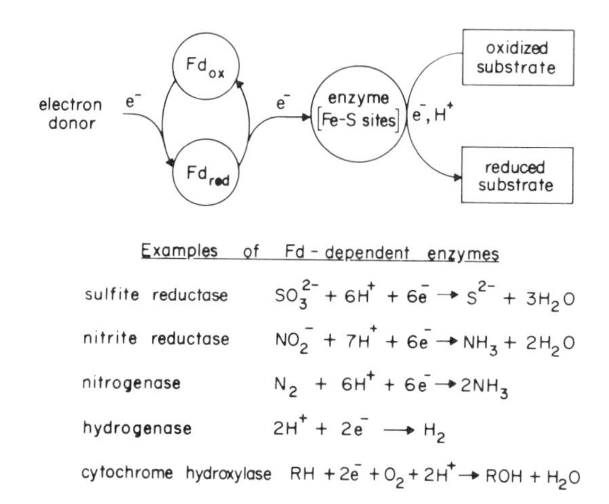

Figure 3. Schematic representation of Fd proteins as electron carriers to Fd-dependent enzymes, and five examples of the process.

representation of an Fd protein acting as a reductant in a generalized electron-transfer scheme. Also indicated are some examples of Fd-dependent enzymes that are of particular current interest. Such enzymes may also contain within their structure Fe-S sites that form part or all of an intramolecular conduit of electron flux to or from the catalytic site; several instances are cited. Intramolecular electron transfer is almost certainly the main function (excluding in this and other cases an obvious structural role) of the single [4Fe-4S] unit in trimethylamine dehydrogenase (71, 184) and the [2Fe-2S] units in xanthine oxidase (34, 185–187). A similar function obtains for some or all of the Fe-S centers in hydrogenases (188) and nitrogenase FeMo proteins (189), which appear to be preponderantly of the [4Fe-4S] type (190–200). In nitrite and sulfite reductases there is now adequate evidence that the siroheme prosthetic group is the catalytic site, thereby allocating to the [4Fe-4S] units in these enzymes (201–203) the role of electron transfer. In executing this function Fe-S centers in all known cases change their total oxidation level by *one* electron, as in adjacent couples of series (1) and the $[2Fe\text{-}2S]^{2+/1+}$ couple operative in protein 2-Fe sites.

4.1.1 Redox Potentials. A minimal understanding of any biological redox center requires a knowledge of those factors, both extrinsic and intrinsic to the center, that exert a significant influence on equilibrium redox potentials and rates of electron transfer. Considering potentials first, extrinsic factors are those set by protein structure at all levels, and include microscopic dielectric (local polarity, distribution of charged groups, etc.) and noncovalent interactions with the center, primarily solvation and hydrogen bonding. Intrinsic factors encompass detailed stereochemistry, nature of ligands, and charge distribution within the center. Both sets of factors, which cannot be precisely the same in two oxidation levels coupled by electron transfer in a given protein, operate in concert to determine the redox potential. In soluble proteins the Fe(III)/Fe(II) potentials of Rds (≈ -0.10 V vs. SHE) and the $[4Fe\text{-}4S]^{3+/2+}$ values of HP centers ($\approx +0.35$ V) do not vary largely, but examples are relatively limited. The $[4Fe\text{-}4S]^{2+/1+}$ and $[2Fe\text{-}2S]^{2+/1+}$ potentials vary over the ranges -0.28 to -0.49 V (44) and -0.23 to -0.46 V (204–206), respectively, but measurement techniques and experimental conditions were not constant. However, the latter couple in algal and higher plant Fds, when determined under constant conditions, does display a 0.15 V range (204). In membrane-bound proteins it is not uncommon for these two couples to exhibit even more negative potentials. These collective results suffice to show a real influence of variant protein on the potentials of isoelectronic couples.

As already indicated, one notable advantage of comparative protein-analog studies is that analog properties are those intrinsic to the related unit in the protein. Deviations from these properties reflect an influence of the protein matrix. Measurement of a clostridial Fd_{ox}/Fd_{red} couple (-0.43 V) and those of two analogs, $[Fe_4S_4(SCH_2CH_2OH)_4]^{2-/3-}$ (-0.51 V) and $[Fe_4S_4(S\text{-}Cys(Ac)NHMe)_4]^{2-/3-}$ (-0.49 V) in aqueous solution under the same conditions (44) indicates that the protein induces a 0.06–0.08 V shift of the unperturbed potential to less negative values. A polarographic method with a dropping Hg electrode was used; certain classical tests for protein adsorption on the electrode were negative, although these do not eliminate any departure from purely diffusion-controlled electrode processes. Other polarographic studies of clostridial Fd have afforded a potential of ≈ -0.33 V (pH 7.0) for the adsorbed Fd_{ox}/Fd_{red} couple (207–209). This result, together with the close agreement of -0.43 V with -0.41 V, the value determined for the protein couple by careful equilibrium methods (210, 211), suggests that the former value is appropriate for comparison with the analog data. Other analog potentials, determined under slightly different conditions, are ≈ -0.52 to -0.58 V for $[Fe_4S_4(S(CH_2)_{2,3}CO_2)_4]^{6-/7-}$ (51–53). Inclusion of these results leads to the conclusion that protein effects displace the unperturbed $[4Fe\text{-}4S]^{2+/1+}$ potential by 0.06–0.15 V to less negative values. In the case of *B. stearothermophilus* Fd, which has the least negative potential reported [-0.28 V (212)], the apparent displacement is much larger, ≈ -0.2 to -0.3 V. These displacements, together with the ranges in protein potentials, make very evident protein modulations of redox potentials. A theoretical model for protein-analog potential differences has been offered (213), but it is not yet possible to dissect potential shifts into contributions from extrinsic and intrinsic effects.

Analogs are of additional utility in determining certain features localized at the redox center that might be subject to protein influence and, therewith, contribute to potential variations between protein and analog and among proteins. A feature of prime importance is the structural difference of the center in its two oxidation levels. In the case of [4Fe-4S] units no Fd_{red} structure is available and EXAFS results (102), though suggestive of a slight increase in mean Fe-S distances in passing from Fd_{ox} to Fd_{red}, are insufficiently sensitive to localize any small structural changes. As discussed earlier, the inherently stable configurations of cores in the $[4Fe\text{-}4S]^{2+}$ and $[4Fe\text{-}4S]^{1+}$ oxidation levels are currently best described as compressed (**21**) and elongated tetragonal (**22**), respectively. Thus the emergent picture for the unconstrained structural change attendant to electron transfer is that schematically depicted in reaction (27). Here the essential feature is expansion and contraction by 0.08 Å along the $\bar{4}$ axis, with a roughly cubic transition state for electron transfer. From the $[Fe_4S_4(SPh)_4]^{2-,3-}$ structures in Figure 1 it can be seen that Fe-S bond distances roughly perpendicular to the axis change much less (≈ 0.01 Å). With the Fe_4 portion as the more constant structural component of the core, the two oxidation levels can be interconverted by, primarily, the indicated S atom displacements. Any protein structural feature that acts to impair this structural change could conceivably shift protein potentials from their unperturbed value. A major unresolved problem is identification of the means of stabilization, in a given protein, of only two core oxidation levels of the three in series (1). This is equivalent to asking why a protein cannot be made to tranverse reversibly all three oxidation levels while maintaining an ordered tertiary structure. Perhaps the most satisfying rationale to date derives from the cluster H-bonding interactions found for *P. aerogenes* Fd_{ox} and *Chromatium* HP (144). The greater number of

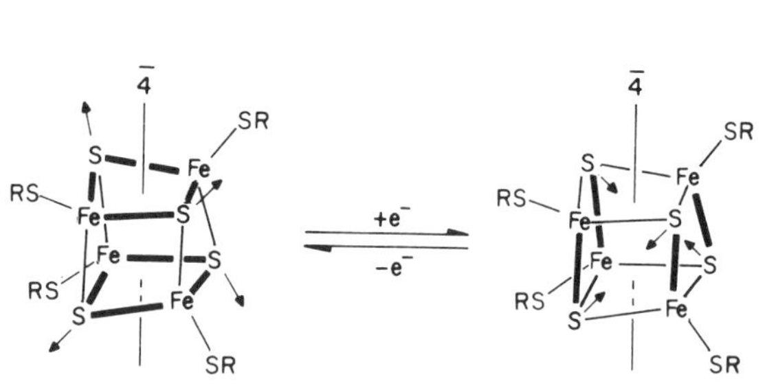

Compressed D_{2d}
21

Elongated D_{2d}
22

N-H · · · S (core) interactions in the former should be effective in stabilizing the more negatively charged cluster. HP_{red} has not been reduced to the $[4Fe\text{-}4S]^{1+}$ level in its normal protein conformation but, in a notable experiment (214), has proven reducible when unfolded in 80% *v/v* $DMSO/H_2O$. To this picture has been added the speculation (16) that, in the native form of HP_{red}, molecular forces may provide a barrier to the anisotropic core dimensional change which is sufficient to displace the $HP_{red}/HP_{s\text{-}red}$ potential to quite negative values. The potential [$\lesssim -0.6$ V (214)] at which the unfolded protein is reduced is in the range of one Fd_{ox}/Fd_{red} and several $[Fe_4S_4(SR)_4]^{2-,3-}$ potentials in the same medium (44).

4.1.2 Electron-Transfer Rates. In seeking a common base for comparison of electron-transfer efficacies of related molecules, recourse to self-exchange rate constants is frequently taken. For the $[4Fe\text{-}4S]^{2+/1+}$ case of analogs and proteins the relevant reaction is (28). ^1H NMR examination of solutions containing equal amounts of *B. polymyxa* Fd $I_{ox,red}$, a protein with 1 [4Fe-4S] unit, revealed spectra that are superpositions of those of the oxidized and reduced proteins (215); that is, a slow exchange situation prevails. In contrast, partially reduced clostridial Fds, which contain 2 [4Fe-4S] units separated by ≈ 12 Å, give exchange-averaged ^1H and ^{13}C spectra (216–219). No rate constants for the protein reactions (28) have been measured.

$$*[Fe_4S_4(SR)_4]^{2-}\ (*Fd_{ox}) + [Fe_4S_4(SR)_4]^{3-}\ (Fd_{red}) \rightleftharpoons *[Fe_4S_4(SR)_4]^{3-}\ (*Fd_{red}) + [Fe_4S_4(SR)_4]^{2-}\ (Fd_{ox}) \qquad (28)$$

To provide a basis for partially interpreting the foregoing qualitative results, the rate constants have been determined for three analog reactions (28) in acetonitrile solutions at ~300 K (70). At the millimolar concentration level these systems exhibit exchange-broadened and shifted ^1H NMR spectra which, together with large chemical-shift differences between the two oxidation levels arising from isotropic hyperfine interactions (38, 64), allow rate measurements by lineshape analysis. The reactions are bimolecular and outer sphere, with rate constants $k \sim 10^6\text{–}10^7\ M^{-1}\ s^{-1}$. For the system $[Fe_4S_4(S\text{-}p\text{-}C_6H_4CH_3)_4]^{2-,3-}$, $k = 2.8 \times 10^6\ M^{-1}\ s^{-1}$, $\Delta G^{\ddagger}_{298K} = 8.7$ kcal/mol, $\Delta H^{\ddagger} = 3.6$ kcal/mol, and $\Delta S^{\ddagger} = -17$ eu. These reactions are among the faster inorganic electron self-exchange processes known, a matter consistent with the small structural change in reaction (27) and the $S = 0$ and $S = \frac{1}{2}$ ground states of the oxidized and reduced clusters, which obviates the problem of electron rearrangement. A very crude estimate of the energy required to adjust reactants to identical nuclear configurations in the transition state (structural reorganization energy) is $\lesssim 2$ kcal/mol. Although other factors also influence electron-exchange rates (220), the small rearrangement barriers and large rate constants are intuitively acceptable features of any assembly biologically selected to store and deliver electrons. These and other properties pertinent to biological redox centers are developed more fully elsewhere (13, 220–222).

The analog rate constants $k \sim 10^6$–10^7 M^{-1} s^{-1} are the best available estimates of the intrinsic value for the $[4Fe\text{-}4S]^{2+,1+}$ protein site electron-exchange reaction, and are $>10^3$ larger than the estimated upper limit rate constant for the *B. polymyxa* Fd I slow-exchange system. It is most unlikely, therefore, that the much decreased rate of protein self-exchange arises from inherently slow reactions of the sites themselves. Rather, the rate decrease is more reasonably interpreted as a rough measure of kinetically retarding steric influences of protein structure in the transition state and whatever effects this structure may have on the details of the electron-transfer mechanism. Slow exchange on the ^1H NMR time scale has also been reported between *Chromatium* HP_{ox} and HP_{red} (215). Because the redox site is ≈ 4 Å from the protein surface, the "slow" rate apparently is also a consequence of the insulating protein structure provided that the intrinsic $[4Fe\text{-}4S]^{3+,2+}$ protein site self-exchange rate is comparable to those of the analog reaction (28). Exchange-averaged spectra of clostridial Fds suggest a component of intramolecular exchange, a reasonable possibility because certain cysteinate sulfur atoms of the *P. aerogenes* Fd_{ox} clusters can approach each other more closely than 12 Å (141,142). If anything, the very fast analog rates tend to enhance this possibility.

Lastly, a developing area of investigation is that of the kinetics of electron-transfer reactions between Fe-S proteins and inorganic complexes, usually of the outer-sphere type. Following the original work on clostridial Rd in 1974 (223), systems containing plant Fds (224–226), HP (227–233), and clostridial Fd (234) have been examined. These studies are not reviewed here, but among the results forthcoming are the recognition that a protein binding site of an inorganic reagent is dependent on charge and ligand type (hydrophobic, hydrophilic), rough estimates of protein self-exchange rate constants from relative Marcus theory (235), and development of a scale of protein site relative kinetic accessibility, which has been most recently expressed as protein site–inorganic reagent site distance parameterized in the protein self-exchange rate constant (233).

4.2 Core Atom Substitution Reactions

The first indication of reactions of this type was found in the base-catalyzed exchange of the Fe and S components of [4Fe-4S] units in clostridial Fd_{ox} with radioactive isotopic Fe(II) salts and sodium sulfide (236). The reactions were found to be ≈ 100 times faster in the presence of urea denaturant than in its absence. Absorption spectra of protein samples before and after exchange were the same, indicating essential identity between the starting material and reisolated Fd_{ox}. These results suggest a substitutional lability of the $[4Fe\text{-}4S]^{2+}$ protein oxidation state that could extend to analogs and, if so, presage interesting core atom substitution chemistry of a more general sort.

Core atom substitution in the form of X = S/Se atom interchange has recently been demonstrated in acetonitrile reaction systems initially containing $[Fe_4X_4(SR)_4]^{3-}$ (I), $[Fe_4X_4(SR)_4]^{2-}$ (II), and $[Fe_2X_2(SR)_4]^{2-}$ (III) clusters (73). In all three systems slow reactions were observed that could be followed by ^1H NMR spectroscopy because m-H and p-CH_3 contact shifts proved to be quite sensitive to the core chalcogenide atom composition. The equilibrium spectrum of one system I is shown in Figure 4. The eight m-H and eight p-CH_3 resonances were assigned on the basis of relative signal intensities and from patterns of signal appearance and decay with time in this system and in others where the mole fractions of initial clusters were varied. The spectrum is fully interpretable on the basis of the presence of five species $(4 - n, n)$, whose structures are depicted. A similar approach resulted in the detection of five species in system II and three in system III, indicating the generality of the substitution process. Reaction rates in system I were the fastest, but with 5 mmolar reactants over 200 hr at $\approx 27°C$ were required to attain equilibrium. Systems I and II involve five species and six equilibria with three independent equilibrium quotients. In system I equilibrium constants evaluated from integrated signal intensities were within experimental uncertainty of statistical values; those of system II deviated from this behavior, with $K_{eq} \lesssim K_{stat}$.

The much faster substitution reactions in system I vs. system II are perhaps caused by the longer and weaker bonds in the reduced clusters (Fig. 1). The occurrence of these reactions with the reduced species is a further indicator of a structural lability or pliability of their cores. This feature is also reflected in the existence of both tetragonal and nontetragonal structures in the crystalline state and the adoption of an elongated tetragonal structure (or at least a shorter range of distortions therefrom) in solutions where environmental constraints are smaller (16, 67, 147). It is not yet clear what biological implications, if any, core substitution reactions present. However, the occurrence of these reactions suggests that Fe atoms may also be substituted, perhaps most easily in reduced analog or protein clusters, by other metals M of the same charge and comparable size and stereochemical preference. By such means rational formation of new hetero- and homometallic clusters may be possible. With analogs such reactions cannot be attempted similarly to those described here inasmuch as the clusters

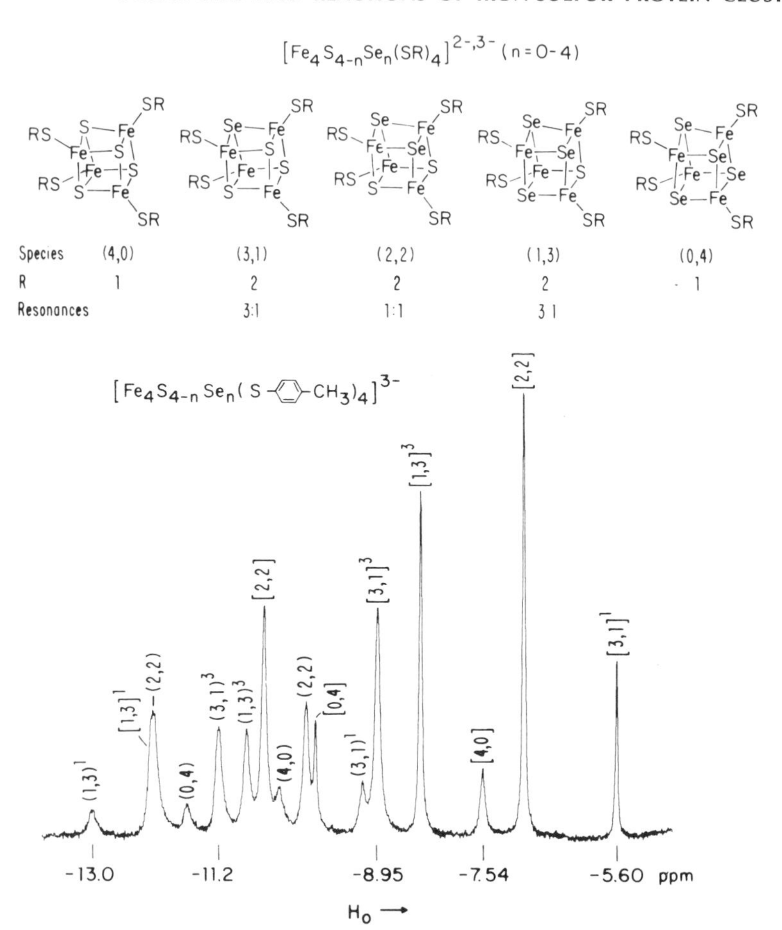

Figure 4. Upper: structures and nomenclature of the species $[Fe_4S_{4-n}Se_n(SR)_4]^{2-,3-}$ (n = 0–4). Lower: ^1H NMR spectrum of the system $(Fe_4X_4(S\text{-}p\text{-}C_6H_4CH_3)_4]^{3-}$ (X = S, Se) at equilibrium (240 hr reaction time). The mole fraction of the initial sulfur cluster was 0.53. Signals in parentheses and brackets are those of m-H and p-CH_3, respectively; superscripts refer to the relative intensities of resonances of (3, 1) and (1, 3) species (73).

$[M_4X_4(SR)_4]^{-z}$ (M $\neq$ Fe) are unknown. The observed chalcogenide atom substitution reactions have no clear precedent in metal cluster chemistry.

4.3 Terminal Ligand-Substitution Reactions

These reactions have been accomplished with both analogs and protein sites, and examples with each are considered in turn. As will be seen, one type of reaction, thiolate substitution, is common to both and has been developed as one means of identifying sites **1** and **3** in proteins.

4.3.1 Analog Reactions. The first evidence for the occurrence of ligand-substitution reactions was obtained in 1974 (54), when it was demonstrated that treatment of $[Fe_4S_4(S\text{-}t\text{-}Bu)_4]^{2-}$ in acetonitrile solution with a small excess of *p*-tolylthiol resulted in quantitative formation of $[Fe_4S_4(S\text{-}p\text{-}C_6H_4\text{-}CH_3)_4]^{2-}$ and liberation of *t*-butylthiol. Experimentation following shortly thereafter (35, 43) revealed that this reaction is a specific example of the general stepwise, reversible equilibria (29) ($n = 1$–4). Equilibrium position is readily manipulated by choice of R, R′, and n. At fixed n the substitution process has the characteristics of an acid-base reaction; the reaction proceeds farther to the right when R′SH is a stronger acid than RSH, the conjugate acid of the coordinated thiolate (43). Therefore arylthiols are more effective than alkylthiols in promoting substitution. The substitution of the first ligand is a biomolecular process, first-order in cluster and thiol; second-order rate constants increase with an increase in thiol (aqueous) acidity. The mechanistic scheme (30) (S* = core atom) is consistent with the kinetic data (55). The rate-determining step is protonation of bound thiolate by R′SH, an expression of the rate-acidity relationship, followed by rapid separation of the weakly bound thiol RSH and capture of the generated $R'S^-$ anion by the Fe atom. The remaining substitution steps were not investigated kinetically but presumably proceed in the same manner.

$$[Fe_4S_4(SR)_4]^{2-} + nR'SH \rightleftharpoons [Fe_4S_4(SR)_{4-n}(SR')_n]^{2-} + nRSH \quad (29)$$

The next stage in the development of substitution reactions is shown in Figure 5. Here cysteinyl peptide cluster complexes are formed from $[Fe_4S_4(S\text{-}t\text{-}Bu)_4]^{2-}$ in initial reactions. In depicting their structures the simplifying assumption has been made that peptides bind to a single core. However, there is no evidence ruling out the formation of oligomeric cluster species by peptide binding to two (or more) clusters. A second substitution reaction using benzenethiol results in removal of peptide structure and recovery in nearly quantitative yield of the starting Fe_4S_4 units in the form of $[Fe_4S_4(SPh)_4]^{2-}$ (39). Very recently the peptides Ac-Gly$_2$-(Cys-Gly$_2$)$_n$-Cys-Gly$_2$-NH$_2$ ($n = 0$–3) have been incorporated around Fe_4S_4 cores by similar reactions (57). An interesting result is that the four successive equilibrium constants for substitution with the monofunctional ($n = 0$) ligand approach statistical values.

$$[Fe_2S_2(SR)_4]^{2-} + 4R'SH \rightleftharpoons [Fe_2S_2(SR')_4]^{2-} + 4RSH \quad (31)$$

The overall reaction (31) has also been demonstrated (29) and has the same characteristics as reaction (29). Both reactions can be monitored by ^1H NMR spectroscopy and absorption spectrophotometry. The most effective

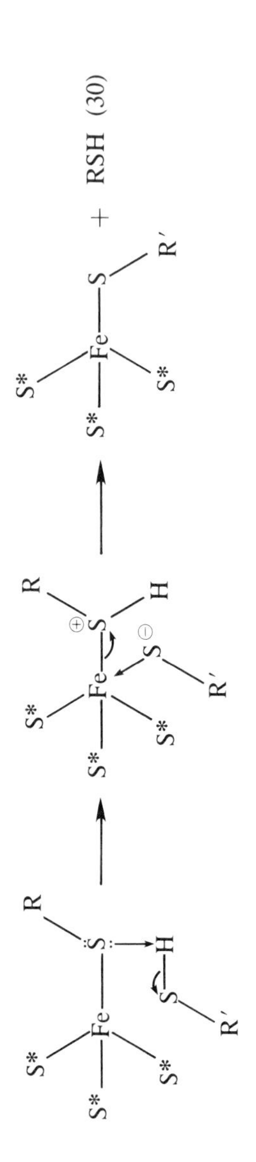
S*
Fe
S
R
H
R′
+ RSH (30)

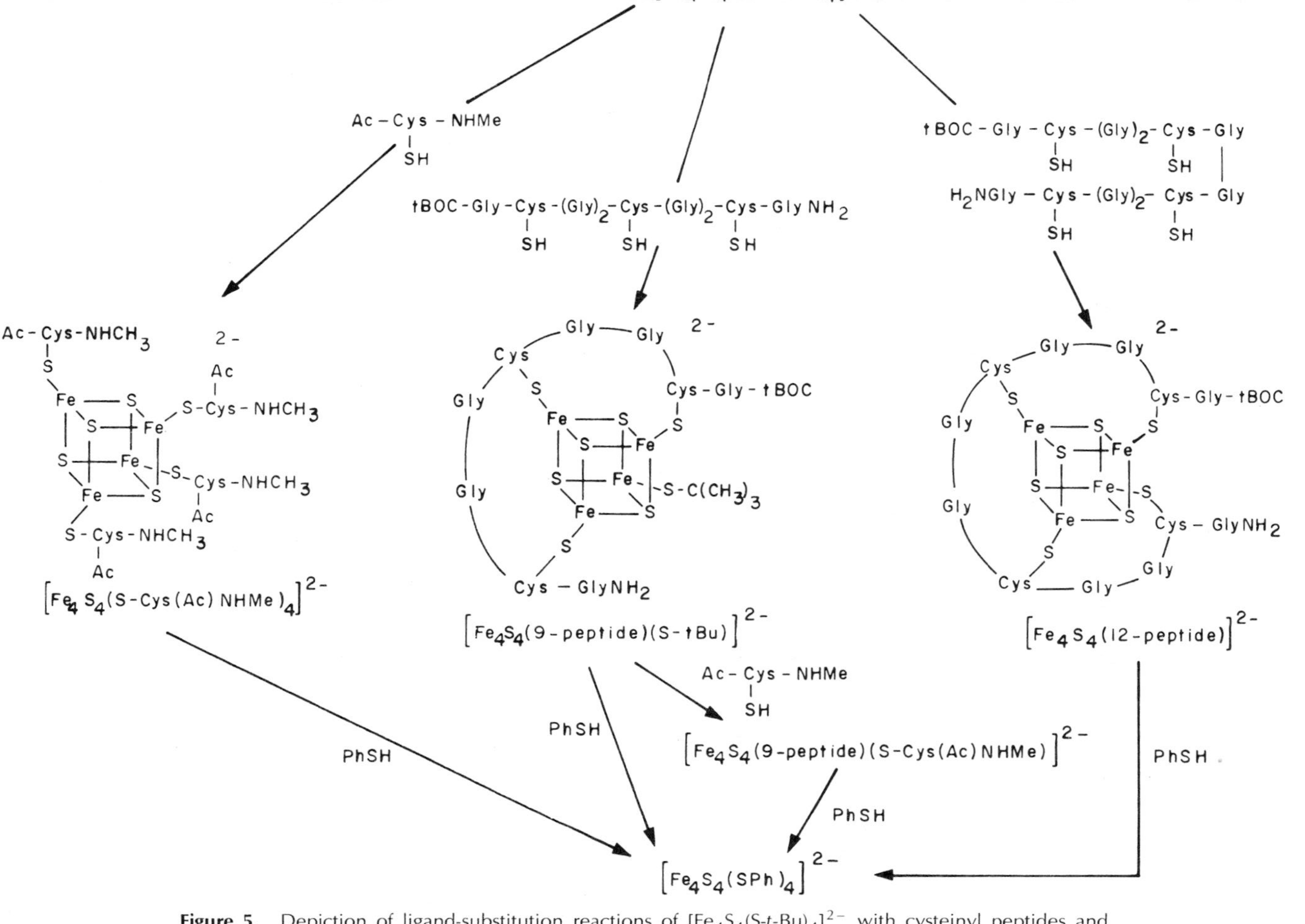

Figure 5. Depiction of ligand-substitution reactions of $[Fe_4S_4(S\text{-}t\text{-}Bu)_4]^{2-}$ with cysteinyl peptides and conversion of peptide clusters to $[Fe_4S_4(SPh)_4]^{2-}$ in DMSO solution (39).

use of the latter method is with systems in which R or R′ is aryl and the other substitutent is alkyl, owing to the red-shifted visible spectra of aryl-thiolate complexes. In all reactions considered thus far ligand substitution is not accompanied by any appreciable core degradation. However, in aqueous solution at neutral and basic pH values ligand substitution appears to be the initial event in hydrolytic destruction of analogs (52, 58) and "dissolution" of protein sites (164–166). Consequently, these reactions, when conducted in aprotic or appropriate aprotic-aqueous solvents, offer alternative routes to clusters in instances where direct synthesis from an iron salt, a sulfide source, and thiolate is difficult or where extensive product purification is required. Effective precursor clusters are those that liberate volatile thiols (e.g., R = Et, *t*-Bu), removal of which displaces the reaction toward product. The formation of peptide clusters represents a case in point. This method should prove useful with one tetracysteinyl peptide in its linear and cyclo forms (237). Spectra of the products from reactions with aqueous Fe(III) and sulfide are only somewhat suggestive of the presence of a Fe_2S_2 core unit. The reactions in Figure 6 provide an example of the preparation of a desired series of four clusters $[Fe_4X_4(YPh)_4]^{2-}$ (X, Y = S, Se) by a combination of direct synthesis and ligand substitution. The latter reactions employed

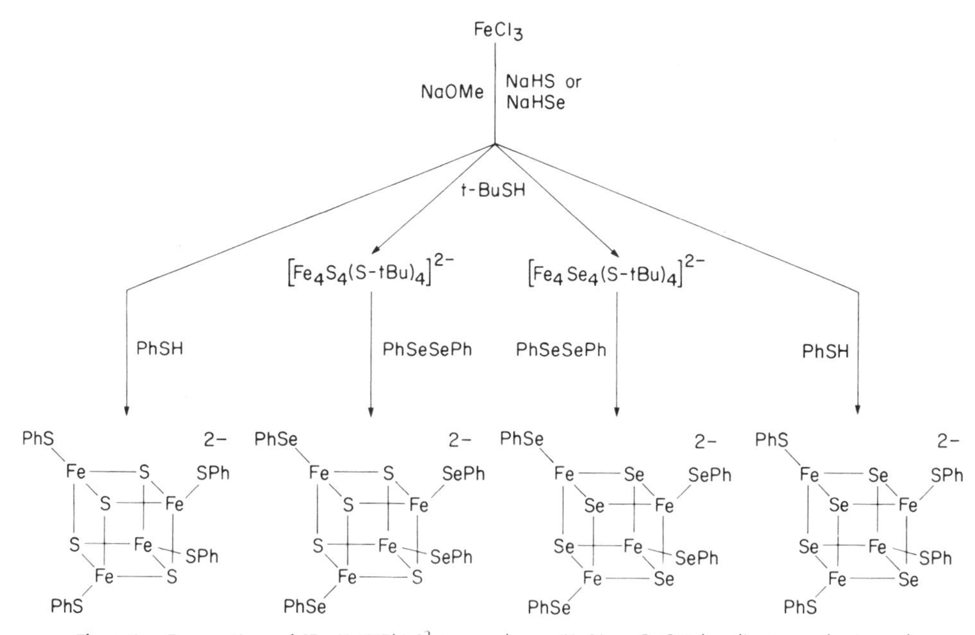

Figue 6. Preparation of $[Fe_4X_4(YPh)_4]^{2-}$ complexes (X, Y = S, Se) by direct synthesis and ligand-substitution reactions.

diphenyl diselenide instead of benzeneselenol and proceeded smoothly to give the indicated benzeneselenide clusters. Leading references to these and other applications of ligand substitution reactions are given in Tables 2 and 3.

$$[Fe_4S_4(SR)_4]^{2-} + 4EL \rightleftharpoons [Fe_4S_4(SR)_{4-n}L_n]^{2-} + nRSE \quad (32)$$

$$[Fe_2S_2(SR)_4]^{2-} + 4EL \rightleftharpoons [Fe_2S_2(SR)_{4-n}L_n]^{2-} + nRSE \quad (33)$$

Reactions (29) and (31) are themselves examples of cluster reactions with the generalized electrophile EL, as in reactions (32) and (33) (n = 1-4). Acids (including thiols), acyl halides, and anhydrides have been shown to act as electrophiles in these reactions (36, 56). The stepwise nature of these processes is demonstrated by the differential pulse polarograms for the system $[Fe_4S_4(SCH_2Ph)_4]^{2-}/CH_3COCl$, shown in Figure 7. Reactions are stoi-

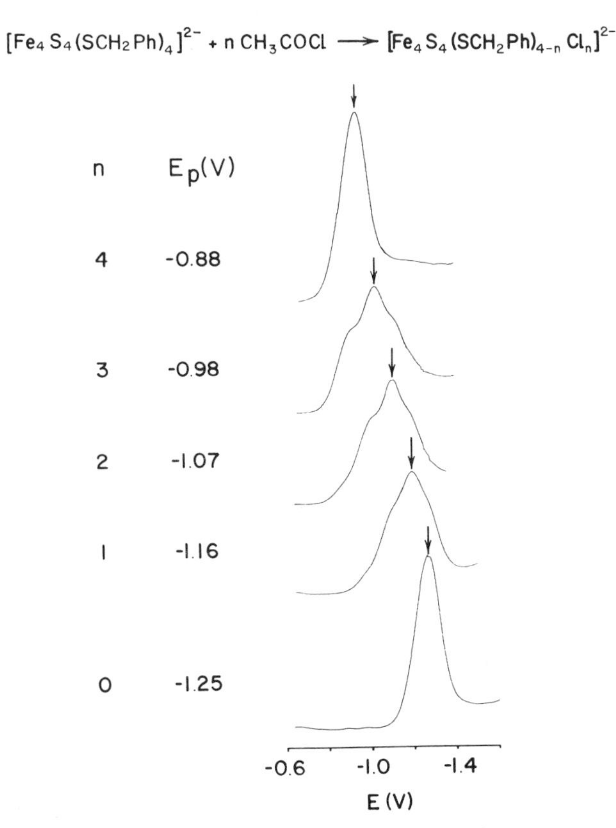

Figure 7. Differential pulse polarograms for the reaction of $[Fe_4S_4(SCH_2Ph)_4]^{2-}$ with CH_3COCl in *N*-methylpyrrolidinone solution. Arrows designate peaks whose potentials vs. SCE are given (56).

chiometric and, at least at ambient temperature, are irreversible. Increasing chloride substitution shifts potentials to less negative values, a behavior also observed with L = OAc^-, $CF_3CO_2^-$, and $CF_3SO_3^-$ (56). The overall reactions with acyl halides, depicted in Figure 8, have led to isolation of fully substituted binuclear and tetranuclear clusters. Structures of $[Fe_2S_2Cl_4]^{2-}$ (Table 5) and $[Fe_4S_4Cl_4]^{2-}$ (Table 7) reveal that replacement of thiolate produces no significant effect on the geometries of the $[4Fe\text{-}4S]^{2+}$ and $[2Fe\text{-}2S]^{2+}$ core structures. Similar results are anticipated for other clusters with nonthiolate terminal ligands. Of these, $[Fe_4S_4(OAc)_4]^{2-}$ has been generated in solution using acetic anhydride as the electrophile (56), and a salt of $[Fe_4S_4(OPh)_4]^{2-}$ has been isolated from the reaction of a thiolate cluster with phenol (238). Terminal halide ligands are more labile than thiolates, and halide clusters should find increasing utility as starting materials for the preparation of differently substituted species.

The reactions in Figure 7 with n = 1-3 equiv of acetyl chloride afford solu-

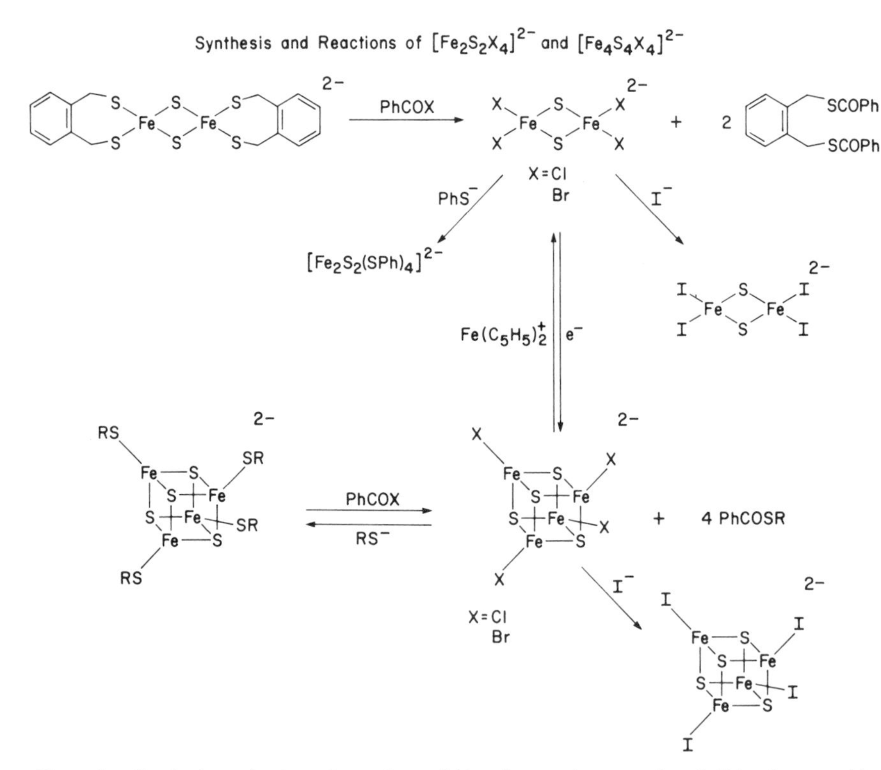

Figure 8. Synthesis and selected reactions of binuclear and tetranuclear halide clusters with X = Cl, Br, and I (36).

tions containing more than one product. Although peaks are not fully resolved, it is evident that the reactions are roughly statistical; for example, $n = 1$ eq produces $\approx 50\%$ of $[Fe_4S_4(SCH_2Ph)_3Cl]^{2-}$, and $\approx 25\%$ each of unreacted $[Fe_4S_4(SCH_2Ph)_4]^{2-}$ and $[Fe_4S_4(SCH_2Ph)_2Cl_2]^{2-}$. Reaction of $[Fe_4S_4(S\text{-}t\text{-}Bu)_4]^{2-}$ with a cysteinyl peptide, as already noted, yields multiple equilibrium constants consistent with approximately statistical behavior (57). In the thiolate-substitution equilibria produced by reaction of $[Fe_4S_4(S\text{-}t\text{-}Bu)_4]^{2-}$ and $[Fe_4S_4(S\text{-}p\text{-}C_6H_4CH_3)_4]^{2-}$, equilibrium constants approach (but are slightly less than) statistical values (43). With the finding that $K_{eq} \lesssim K_{stat}$ for core chalcogenide atom substitution in $[Fe_4X_4(SR)_4]^{2-,3-}$ clusters, the following reactivity picture emerges. Both terminal ligands and core S and Se atoms can be substituted by other groups or atoms of the same charge and (for core atoms) of not too dissimilar size in reactions affording roughly statistical product distributions. Given the long distances (6–7 Å) between terminal ligand-binding sites it is not surprising that substitution around core structures of essentially constant dimensions approaches or achieves statistical behavior. It is less obvious, however, that statistical behavior would obtain in core substitution, where heteroatom introduction must produce a definite structural change.

4.3.2 Protein Site Reactions—Core Extrusion. Occurrence of the cysteinyl peptide cluster substitution reactions in Figure 5 has led to the concept that cores of protein sites might be removed by an analogous reaction and the products identified spectroscopically in situ. Initial experiments in 1975 (239) based on reaction (34) with *Spirulina maxima* Fd_{ox} ($1Fe_2S_2$ unit) and *C. pasteurianum* Fd_{ox} ($2Fe_4S_4$ units) and benzenethiol reduced the concept to practice. Protein site cores were removed in essentially quantitative yield when the proteins, unfolded in 80% *v/v* $DMSO/H_2O$, were treated with a large excess of the thiol. The reaction products, $[Fe_2S_2(SPh)_4]^{2-}$ (λ_{max} 490 nm) and $[Fe_4S_4(SPh)_4]^{2-}$ (λ_{max} 458 nm), were readily identified and quantitated by their characteristic absorption spectra, which were previously known from measurements on the pure analogs. It was also shown that protein cores could be removed by *o*-xylyldithiol, reaction (35). The product from the algal protein is $[Fe_2S_2(S_2\text{-}o\text{-}xyl)_2]^{2-}$ (**8**), but that from the clostridial protein, while clearly a $[4Fe\text{-}4S]^{2+}$ species, is not known in detail. Reasonable formulations are oligomeric $[Fe_4S_4(S_2\text{-}o\text{-}xyl)_2]_n^{2n-}$ or, perhaps more likely, the single cluster $[Fe_4S_4(S(SH)\text{-}o\text{-}xyl)_4]^{2-}$, as shown.

$$\text{holoprotein} + \text{RSH} \xrightarrow[\text{Solvent}]{\text{Unfolding}} [Fe_nS_n(SR)_4]^{2-} + \text{apoprotein} \qquad (34)$$

$$(n = 2, 4)$$

(35) ↓ o-xyl(SH)$_2$

$[Fe_4S_4(S(SH)\text{-}o\text{-xyl})_4]^{2-}$
$[Fe_2S_2(S_2\text{-}o\text{-xyl})_2]^{2-}$ } $\xrightarrow{CF_3\text{-}C_6H_4\text{-}SH}$ $[Fe_nS_n(SR_F)_4]^{2-}$ ($n = 2, 4$) + o-xyl(SH)$_2$ (36)

Reactions of type (34) were originally described as core extrusion (239), in the common chemical context of removing a fragment from a larger structure. This term has gained currency; another laboratory has advanced the designation cluster displacement (105) for these reactions. The extrusion method has been developed rather extensively. Experimental methodology and advantages and limitations of the method have been set out in considerable detail (34, 105, 191, 240–242). Several matters deserve emphasis, and are cited with reference to the schematic representation of an extrusion reaction presented in Figure 9. [A similar depiction has been presented earlier (243).] For the reaction to proceed at a satisfactory rate (or perhaps occur at all) the protein conformation must be unfolded. While 80% DMSO is often satisfactory for small proteins, 80% *v/v* hexamethylphosphoramide (HMPA)/H_2O, introduced in an extrusion study of clostridial hydrogenase (190), is more generally applicable to both small and large proteins. The extrusion reagent must be selected not only to maximize core removal, but also to confer on the extrusion products spectroscopic differences that are sufficient to identify and quantitate one product in the presence of another. Because the distinction usually sought is between sites **1** and **3**, the analog species $[Fe_2S_2(SR)_4]^{2-}$ and $[Fe_4S_4(SR)_4]^{2-}$ must be distinguishable. Audition of numerous thiols in extrusion reactions of small Fd proteins has reinforced the utility of simple monofunctional arylthiols and *o*-xyl(SH)$_2$ (105). Extrusion of sites **1** and **3** is most reliably accomplished when they are adjusted to the $[2Fe\text{-}2S]^{2+}$ and $[4Fe\text{-}4S]^{2+}$ oxidation levels. This is particularly important in the former case because of the possibility of reaction (21), whose occurrence would result in the erroneous identification of Fe_4S_4 units. Reaction (15) presents a similar problem, but it is relatively slow and can be arrested by the presence of sufficient thiolate (30). Lastly, validity of the conclusions rests on apposite control experiments. Controls have utilized small soluble proteins such as plant, adrenal, and clostridial Fds, whose extrusion

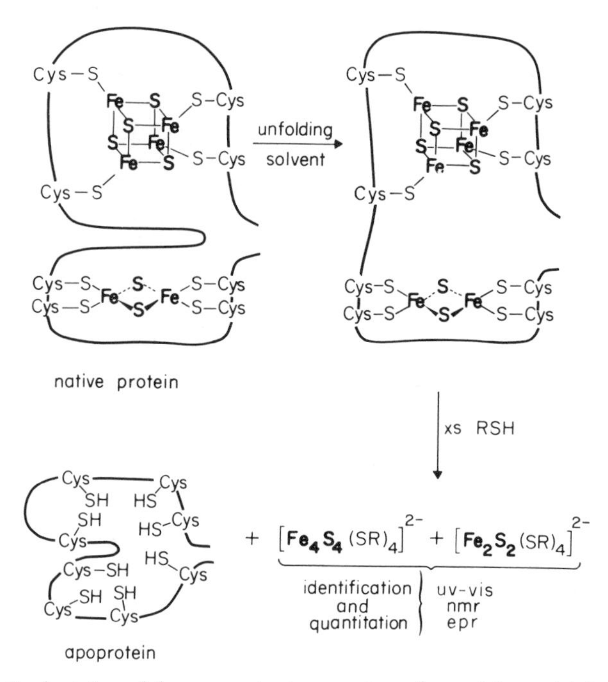

Figure 9. Schematic depiction of the core extrusion reaction of a protein containing sites **1** and **3**.

behavior under corresponding conditions is taken to parallel that of sites of the same core composition in proteins less well characterized.

The core extrusion procedure has enjoyed substantial application, as may be seen from the information assembled in Table 9. In assaying extrusion products in situ the method of choice is absorption spectrophotometry because of its simplicity and accuracy. It has been employed in a number of instances where the proteins have only Fe-S prosthetic groups (105, 190–194, 244–247), usually with benzenethiol as the extrusion reagent. A result of particular interest is the consistent detection of Fe_4S_4 units in different bacterial hydrogenases. Trimethylamine dehydrogenase is the only case where spectrophotometric analysis has been profitably applied in the presence of another chromophore (71). However, this case was advantageous in that a small peptide containing the covalently bound chromophore [later identified

Table 9 Identification of Fe-S Sites in Proteins by Core Extrusion

Protein	Molecular Weight	Fe/S (g-atoms)[b]	Other Cofactors	Method[a]	Extruded Cores	Refs.
C. pasteurianum hydrogenase						
Preparation 1	60,000	4–5	none	AS	≈1 [4Fe-4S]	190
Preparation 2	60,500	11	none	AS	3 [4Fe-4S]	191
D. gigas hydrogenase	89,500 ($\alpha\beta$)	12	none	AS	3 [4Fe-4S]	192
D. vulgaris (Miyazaki) hydrogenase	89,000 ($\alpha\beta$)	7–9	none	AS	2 [4Fe-4S]	193, 194
P. putida benzene dioxygenase						
Fd	12,300	>1	none	AS	1 [2Fe-2S]	244
Terminal oxygenase	215,000	4	none	AS	2 [2Fe-2S]	244
C. pasteurianum "paramagnetic" protein (Fd)	24,000	2	none	AS	1 [2Fe-2S]	105
A. vinelandii Fd II	24,000	2	none	AS	1 [2Fe-2S]	105
R. japonicum (bacteriod) Fd	6,740	7–8	none	AS, ICT	2 [4Fe-4S]	245
Chick renal mitochondrial Fd	11,900	—	none	ICT	(1?) [2Fe-2S]	250
B. subtilis phosphoribosyl amidotransferase	200,000 (α_4)	3–4	none	AS	≈1 [4Fe-4S]	246
Trimethylamine dehydrogenase	147,000	4	flavin	AS	1 [4Fe-4S]	71
Succinate dehydrogenase	97,000 ($\alpha\beta$)	8	flavin	ICT, NMR	2 [2Fe-2S] + 1 [4Fe-4S]	249
NADH dehydrogenase	≈700,000/flavin	16–18/25/flavin	flavin	NMR	4 [2Fe-2S] + 2 [4Fe-4S]	251
Xanthine oxidase (milk)						
Native	280,000 (α_2)	8	2 Mo, 2 flavin	NMR	4 [2Fe-2S]	34, 185
Deflavo		8	2 Mo	AS	4 [2Fe-2S]	34
Nitrogenase						
C. pasteurianum Fe protein	57,700 (α_2)	4	none	AS	1 [4Fe-4S]	190, 247
C. pasteurianum Fe-Mo protein	220,000 ($\alpha_2\beta_2$)	28 (Fe)	FeMo-co	NMR, ICT	3.4–3.9, 4 [4Fe-4S][c]	195, 196
A. vinelandii Fe-Mo protein	245,000 ($\alpha_2\beta_2$)	30 (Fe)	FeMo-co	NMR, ICT	3.7–4.0, 4 [4Fe-4S][c]	195, 196

[a] AS = visible absorption spectrophotometry; ICT = interprotein cluster transfer/EPR method; NMR = ^{19}F NMR.
[b] Values refer to Fe and S content unless noted otherwise.
[c] Some [2Fe-2S] extrusion products are observed depending on the reaction conditions (195, 196).

as a novel 6-substituted flavin (248)] could be removed and used as a control blank under extrusion conditions. Subtraction of its absorption contribution from the final extrusion spectrum resulting from the use of *p*-methoxybenzenethiol allowed identification of the extrusion product as $[Fe_4S_4(S\text{-}p\text{-}C_6H_4OCH_3)_4]^{2-}$, whose principal visible absorption band (λ_{max} 470 nm) is red-shifted from residual flavin absorption. In this way the enzyme was shown to contain $1Fe_4S_4$ unit (71).

In the general case it cannot be assumed that a protein will yield to excisement of an interfering chromophore for separate examination or, as with milk xanthine oxidase, contain a noncovalently bound chromophore that is removable with retention of other prosthetic groups. Two methods for circumventing this problem have been devised and put into practice. In one of these the extrusion reagent is *p*-trifluoromethylbenzenethiol (R_FSH), and product analysis is achieved by ^{19}F NMR spectroscopy (34), preferably with a sensitive high-field instrument. The procedure is based on reaction (34) ($R = R_F$) or sequential application of reactions (35) and (36) and the fully resolved ^{19}F resonances of R_FSH, $[Fe_2S_2(SR_F)_4]^{2-}$, and $[Fe_4S_4(SR_F)_4]^{2-}$. The latter property obtains because of the contact shifts of the two complexes, which have been synthesized and examined separately (34). Application of the method to the Fe-S flavoprotein succinate dehydrogenase (249) is shown in the upper portion of Figure 10. Here signals from the binuclear and tetranuclear analogs are observed, indicating the presence in the enzyme of Fe_2S_2 and Fe_4S_4 units in the ratio $\approx 2{:}1$. These units may be quantitated in terms of content per flavin by measurement of concentrations from spectral integration after addition of $[Fe_2S_2(SR_F)_4]^{2-}$ and $[Fe_4S_4(SR_F)_4]^{2-}$ standard solutions. The second method (241), interprotein cluster transfer, involves extrusion by reaction (35) followed by selective capture of an extruded core by its "natural" apoprotein, reaction (37). This ligand-substitution process is facilitated by increasing the aqueous content of the medium, thereby causing the reconstituted protein to refold to its native conformation, in which it is quantitatively reducible by dithionite. As shown in the EPR spectra of Figure 10, low-field resonances of the two Fd_{red} proteins are fully resolved, allowing their identification and quantitation. Application of the ^{19}F NMR and interprotein cluster transfer reactions to succinate dehydrogenase has demonstrated the presence of 1 Fe_4S_4 and 2 Fe_2S_2 units in one enzyme molecule, a

$$\left.\begin{array}{l}[Fe_4S_4(S(SH)\text{-}o\text{-xyl})_4]^{2-}\\ \\ [Fe_2S_2(S_2\text{-}o\text{-xyl})_2]^{2-}\end{array}\right\} + \left\{\begin{array}{l}\text{apoadrenal Fd}\\ \\ \text{apo } \textit{B. polymyxa}\\ \text{Fd}\end{array}\right. \xrightarrow[S_2O_4^{2-}]{\text{protein refolding}} \begin{array}{l}\text{adrenal } Fd_{red}\\ \\ \textit{B. polymyxa}\\ Fd_{red}\end{array} \qquad (37)$$

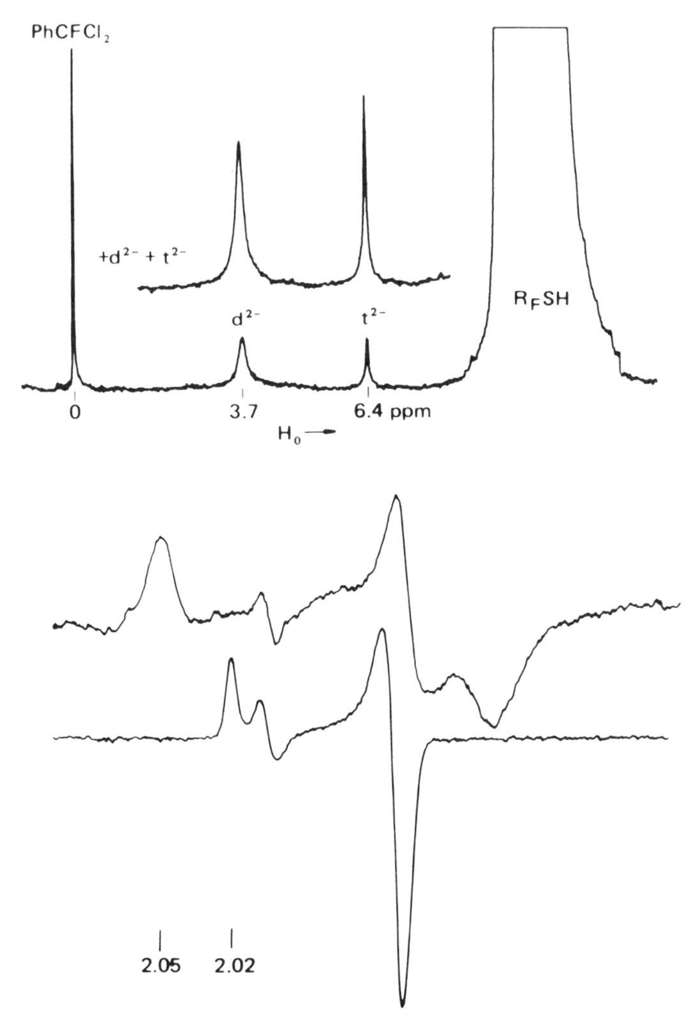

Figure 10. Extrusion of succinate dehydrogenase in 80% v/v HMPA/H_2O using ^{19}F NMR and EPR spectroscopy for product analysis (249). Top: The lower NMR spectrum is that of the reaction mixture after completion of the extrusion reaction (d^{2-} = $[Fe_2S_2(SR_F)_4]^{2-}$; t^{2-} = $[Fe_4S_4(SR_F)_4]^{2-}$), and the upper spectrum is that after addition of standard solutions of d^{2-} and t^{2-}. Bottom: EPR spectra of adrenal Fd_{red} (lower) and *B. polymyxa* Fd_{red} (upper) reconstituted in the presence of the corresponding apoproteins after extrusion, protein refolding, and dithionite reduction.

result consistent with the majority of other types of evidence. The identification of the other chromophore as a histidyl flavin was attained much earlier. Other results from application of these extrusion procedures are contained in Table 9 (34, 185, 195, 196, 245, 250, 251).

In assessing the results in Table 9, certain observations or qualifications are in order arising from the nature of the control experiments. Spectroscopic and extrusion studies of nitrogenase FeMo proteins in their semireduced (as-isolated) forms certainly point to most or all of the Fe-S content not associated with the FeMo-cofactor as being organized into Fe_4S_4 units (195–200). Such units, "P-clusters," do not have the spectroscopic properties of the conventional site **3** found in small Fds. Because such clusters are unknown in other proteins, there are obviously no rigorous control experiments for their extrusion. In dye-oxidized FeMo proteins some Fe_2S_2 cores are extruded (34). Although the conversion $Fe_4S_4 \rightarrow Fe_2S_2$ in the proteins may have been effected by oxidation, it is possible that in the semireduced forms sites with the $[2Fe\text{-}2S]^{1+}$ oxidation level are present but, that on extrusion they collapse to tetranuclear clusters, perhaps by reaction (21). This reaction has not been observed in controls (34, 191) but, as cautioned earlier (191), "No small 2-Fe Fd protein provides a suitable control of certain conceivable extrusion situations." A most unexpected problem was encountered in extrusion of the enzyme aconitase. Spectrophotometric analysis of reaction (35) showed the product to be $[Fe_2S_2(S_2\text{-}o\text{-xyl})_2]^{2-}$, leading to the conclusion that, by the extrusion criterion, the enzyme contains Fe_2S_2 unit(s) (252). Mössbauer spectral evidence is clear-cut in establishing the presence of an Fe_3S_3 unit in the enzyme (10), hence dismutation of the latter must have occurred in the extrusion experiments, which were conducted prior to knowledge of the existence of 3-Fe protein sites. Subsequently such a reaction has been observed in the extrusion of *A. vinelandii* Fd I (10). Although it could not have been known at the time, control experiments were not adequately inclusive. Lastly, note is taken of the case of hydrogenases (188). Certain of these enzymes, when (partially) oxidized, exhibit, as do 3-Fe sites, an approximately axial EPR signal at $g \approx 2.01$ which disappears on reduction. This signal is replaced in some cases by a $g \approx 1.94$-type spectrum ascribable to the $[4Fe\text{-}4S]^{1+}$ oxidation level, consistent with extrusion results, which thus far have indicated Fe_4S_4 units only. While the $g \approx 2.01$ feature has often been assigned to the $[4Fe\text{-}4S]^{3+}$ level, it is perhaps advisable not to categorize all sites in differently oxidized proteins displaying these characteristics as being of the Fe_4S_4 type.

The preceding matters notwithstanding, the prognosis for continued development and application of the extrusion method is favorable, particularly where object proteins contain $[2Fe\text{-}2S]^{2+}$ and $[4Fe\text{-}4S]^{2+}$ oxidation levels. It is anticipated that conditions will be developed for extrusion of intact Fe_3S_3

cores, thereby providing further control data necessary for extension of the method to 2-Fe, 3-Fe, and 4-Fe sites in uncharacterized proteins. Ideally, the method should be employed in a manner complementary to other physical probes of site structure. One of these is the temperature dependence and microwave power saturation characteristic of the EPR spectra of Fe-S centers in unfolded proteins (253, 254). Among results from this technique is the finding of Fe_2S_2 and Fe_4S_4 units in *Alcaligenes eutrophus* hydrogenase (255). Among the newer techniques for Fe-S site identification, the linear electric field effect (LEFE) in EPR spectroscopy (256) and MCD spectra (63, 257-259), especially of paramagnetic sites at cryogenic temperatures, hold considerable promise. One of their advantages is that proteins can be examined in the native states, thereby obviating any molecular mischief associated with the not inconsiderable perturbation of the unfolding process.

4.4 Miscellaneous Reactions of Analogs

By use of ligand-substitution processes $[Fe_4S_4(SR)_4]^{2-}$ clusters have been affixed to a cyclodextrin (260), and placed on the surfaces of a modified tin oxide electrode (59) and silica gel (60) by reaction (38). The latter system exhibited no catalytic activity for reduction of carbon monoxide, ethylene, and acetylene with dihydrogen or for oxidation of dihydrogen, among other negative indications of reactivity. Thioformimidates are reported to be formed by reaction (39) in tetramethylurea solutions, and an intermediate isocyanide-ligated cluster has been implicated (40). Catalytic reactions are described for those systems in which the cluster dianion is treated with Na dust prior to introduction of reactants. The active catalyst is claimed to be $[Fe_4S_4(SR)_4]^{4-}$, the most reduced tetranuclear cluster species detected electrochemically (30, 35, 51). Its instability, as experienced in this laboratory, calls into question its survival as a catalyst in the presence of protic reagents, but the possibility of stabilization in the form of an isocyanide adduct cannot be dismissed. Based on the reactivity features of conventional S-ligated Fe(II) complexes, the reduced clusters $[Fe_4S_4(SR)_4]^{3-,4-}$, having the $[4Fe\text{-}4S]^{1+,0}$ core oxidation

$$\text{-O-Si(CH}_2)_3\text{SH} + [\text{Fe}_4\text{S}_4(\text{S-}t\text{-Bu})_4]^{2-} \longrightarrow \text{-O-Si-(CH}_2)_3\text{S-Fe}_4\text{S}_4(\text{S-}t\text{-Bu})_3^{2-} + t\text{-BuSH} \qquad (38)$$

levels, are those that are expected to bind more strongly π-acid ligands such as isocyanides and carbon monoxide. Evidence for such binding is afforded by EPR spectra of chemically generated $[Fe_4S_4(SPh)_4]^{3-}$ in *N,N*-dimethylacetamide in the presence of carbon monoxide (261). The system appears to contain several paramagnetic components, at least one of which is a cluster–carbon monoxide adduct with the $[4Fe\text{-}4S]^{1+}$ oxidation level.

$$R'NC + RSH \xrightarrow{[Fe_4S_4(SR)_4]^{2-,4-}} R'{-}N{=}C{\begin{smallmatrix} \diagup SR \\ \diagdown H \end{smallmatrix}} \tag{39}$$

The remainder of the reactions to be described involve cluster mediation of electron-transfer processes. The reduced clusters $[Fe_4S_4(SR)_4]^{3-,4-}$ have been described as electron carriers to synthetic Mo complexes that effect reduction of dinitrogen to ammonia (262–264). Clusters were generated in situ; however, conditions in several systems [BH_4^-, $S_2O_4^{2-}$ reductant, protic media (263, 264)] are inimical to cluster stability. Evidence is lacking for survival of clusters through even one redox cycle. In other instances the reduced clusters themselves have been found to promote substrate transformation. Thus fluorenone in the presence of *n*-butyllithium and $[Fe_4S_4(SPh)_4]^{2-}$ followed by addition of water or benzenethiol is converted to bifluoren-9-yl-9, 9′-diol and/or fluoren-9-ol depending on reaction conditions (265). Formation of the latter may be represented by reaction (40), in which electrons are supplied by the reduced cluster produced by reaction of cluster dianion with *n*-BuLi. $[Fe_4S_4(SPh)_4]^{2-}$ was recovered in high yield on treatment of the

$$\text{fluorenone} + 2H^+ + 2e^- \longrightarrow \text{fluoren-9-ol (HO, H)} \tag{40}$$

mixture after reaction with benzenethiol. In the absence of the cluster 9-butylfluorenol , the expected product, was obtained in high yield. In a somewhat related system comprised of $[Fe_4S_4Cl_4]^{2-}/n\text{-BuLi}/H_2$, *cis*- and *trans*-stilbene and diphenylacetylene have been hydrogenated to 1,2-diphenylethane (266). The system $[Fe_4S_4(SPh)_4]^{3-}$/acetic acid/acetic anhydride in *N*-methylpyrrolidinone at $\approx 25°C$ has been found to reduce acetylene homogeneously to ethylene (267). The stoichiometry of the process is represented by reaction (41), but ethylene yields never exceeded $\approx 60\%$ because of a competing (unidentified) reaction which partially discharged the reducing equivalents of the cluster trianion. Systems initially containing $[Fe_4S_4(SPh)_4]^{2-}$ gave no appreciable ethylene product, nor was any significant amount of ethane found in systems based on reduced clusters. It is considered probable

$$2[Fe_4S_4(SPh)_4]^{3-} + 2H^+ + C_2H_2 \rightarrow 2[Fe_4S_4(SPh)_4]^{2-} + C_2H_4 \quad (41)$$

that a reduced acetato-substituted species, formed in a reaction analogous to reaction (32), is the principal cluster reactant. Acetylene binding would then presumably occur by displacement of acetate and must be of a fairly specific orientation inasmuch as *cis*-1,2-$C_2H_2D_2$ is the dominant stereochemical product when the reaction is conducted in the presence of a deuteron source (267). The significance of the substrate transformations described is not in their mere occurrence, for this can be achieved in other ways. Rather, it is their occurrence by virtue of electron transfer from reduced clusters. The case of acetylene reduction is particularly noteworthy, for this molecule is an alternate substrate of nitrogenase and is reduced nearly exclusively to *cis*-1,2-$C_2H_2D_2$ by the enzyme (268). Thus the results demonstrate that biologically related, reduced Fe_4S_4 units can effect reduction of at least 1 Fe-S enzyme substrate, and raise the general possibility of substrate transformation with such clusters as reaction sites in biological systems. Hydrogenase, which contains only Fe-S prosthetic groups, is the most obvious case. Taking the extrusion results (Table 9) as evidential, the system $[Fe_4S_4(SPh)_4]^{3-}$/PhSH in *N,N*-dimethylacetamide has been examined as a hydrogenase model (269). Small yields of dihydrogen ($\lesssim 30\%$), based on reaction (42), were observed. This is the first instance where H_2 has been produced in a Fe-S cluster system under thermal conditions, and as such it represents a preliminary step toward the development of a synthetic hydrogenase. Under similar conditions dihydrogen formation from the reduced double-cubane cluster $[Mo_2Fe_6S_8(SPh)_9]^{5-}$, which carries two electrons introduced at quite negative potentials, is essentially quantitative (269). This observation suggests that the doubly reduced single clusters $[Fe_4S_4(SR)_4]^{4-}$, as yet not isolated, may be more effective than $[Fe_4S_4(SR)_4]^{3-}$ as simulators of hydrogenase activity.

$$2[Fe_4S_4(SPh)_4]^{3-} + 2PhSH \rightarrow 2[Fe_4S_4(SPh)_4]^{2-} + H_2 + 2PhS^- \quad (42)$$

Lastly, demonstrations have been provided that analog clusters can replace Fds as electron carriers to biological molecules (46–48). Thus in reaction (43), conducted in aqueous solution, the water-soluble clusters $[Fe_4S_4(SCH_2CH_2OH)_4]^{2-}$ and $[Fe_4S_4(Ac\text{-}Gly_2\text{-}Cys\text{-}Gly_2\text{-}Cys\text{-}Gly_2\text{-}NH_2)_2]^{2-}$, on reduction to their trianion forms by dithionite, couple with clostridial hydrogenase. Dihydrogen production over the first hour of reaction is significant, but at parity of concentration is not as extensive as that with *S. maxima* Fd as the carrier (46). Similar results were obtained with $[Fe_4Se_4(SCH_2CH_2OH)_4]^{2-}$ (48). More recently the reverse reaction, namely dihydrogen consumption in the system *D. gigas* hydrogenase + cytochrome c_3 + Fd I or $[Fe_4S_4(SCH_2CH_2OH)_4]^{2-}/H_2$, has been achieved (47). Further, the cluster

$$S_2O_4^{2-} \xrightarrow{e^-} \left\{ \begin{array}{c} Fd_{ox/red} \\ \\ [Fe_4S_4(SR)_4]^{2-/3-} \end{array} \right\} \xrightarrow{e^-} H_2ase \left(\begin{array}{l} H^+ \\ \downarrow \\ H_2 \end{array} \right. \quad (43)$$

$[Mo_2Fe_6S_8(SCH_2CH_2OH)_9]^{3-}$, but not $[Fe_4S_4(SCH_2CH_2OH)_4]^{2-}$, was found to replace Fd in an illuminated chloroplast/Fd/hydrogenase H_2-evolving system (47). The ability of a cluster to transfer electrons in light-driven H_2 evolution otherwise has been shown for the system $[Ru(bipy)_3]^{2+}/[Fe_4S_4(SCH_2Ph)_4]^{2-}$ in aqueous *N,N*-dimethylacetamide (270). Functional substitution of Fd by synthetic clusters in enzyme systems, although at an obviously early stage of development, is the first proof that such clusters can actually execute electron transfer to enzymes. No dihydrogen has yet been produced in these systems in the absence of hydrogenase itself. However, the occurrence of reaction (42), albeit in low yield, lends credence to the possibility of developing a purely synthetic catalytic hydrogenase based on Fe-S clusters that proceeds spontaneously or is driven by illumination.

ACKNOWLEDGMENTS

Research in the field of Fe-S chemistry and biochemistry conducted in the laboratory of one of us (RHH) has been generously supported since 1972 by the National Institutes of Health. JMB is a National Science Foundation Predoctoral Fellow.

REFERENCES

1. R. H. Holm, *Acc. Chem. Res.*, **10**, 427 (1977).
2. H. Brintzinger, G. Palmer, and R. H. Sands, *Proc. Natl. Acad. Sci. USA*, **55**, 397 (1966).
3. J. F. Gibson, D. O. Hall, J. H. M. Thornley, and F. R. Whatley, *Proc. Natl. Acad. Sci. USA*, **56**, 987 (1966).
4. J. R. Herriott, L. C. Sieker, L. H. Jensen, and W. Lovenberg, *J. Mol. Biol.*, **50**, 391 (1970).
5. C. W. Carter, Jr., S. T. Freer, Ng. H. Xuong, R. A. Alden, and J. Kraut, *Cold Spring Harbor Symp. Quant. Biol.*, **36**, 381 (1972).
6. L. C. Sieker, E. Adman, and L. H. Jensen, *Nature*, **235**, 40 (1972).
7. T. Tsukihara, K. Fukuyama, H. Tahara, Y. Katsube, Y. Matsuura, N. Tanaka, M. Kakudo, K. Wada, and H. Matsubara, *J. Biochem.*, **84**, 1645 (1978).
8. K. Fukuyama, T. Hase, S. Matsumoto, T. Tsukihara, Y. Katsube, N. Tanaka, M. Kakudo, K. Wada, and H. Matsubara, *Nature*, **286**, 522 (1980).
9. C. D. Stout, D. Ghosh, V. Pattabhi, and A. H. Robbins, *J. Biol. Chem.*, **255**, 1797 (1980).
10. M. H. Emptage, T. A. Kent, B. H. Huynh, J. Rawlings, W. H. Orme-Johnson, and E. Münck, *J. Biol. Chem.*, **255**, 1793 (1980).

11. B. H. Huynh, J. J. G. Moura, I. Moura, T. A. Kent, J. LeGall, A. V. Xavier, and E. Münck, *J. Biol. Chem.*, **255,** 3242 (1980).
12. R. H. Holm, *Endeavour,* **34,** 38 (1975).
13. R. H. Holm and J. A. Ibers, in *Iron-Sulfur Proteins,* Vol. III, W. Lovenberg, ed., Academic Press, New York, 1977, Chapter 7.
14. R. W. Lane, J. A. Ibers, R. B. Frankel, G. C. Papaefthymiou, and R. H. Holm, *J. Am. Chem. Soc.*, **99,** 84 (1977).
15. P. Mascharak, R. B. Frankel, G. C. Papaefthymiou, and R. H. Holm, *J. Am. Chem. Soc.*, **103,** 6110 (1981).
16. E. J. Laskowski, R. B. Frankel, W. O. Gillum, G. C. Papaefthymiou, J. Renaud, J. A. Ibers, and R. H. Holm, *J. Am. Chem. Soc.*, **100,** 5322 (1978).
17. J. A. Ibers and R. H. Holm, *Science,* **209,** 223 (1980).
18. B. A. Averill and W. H. Orme-Johnson, *Metal Ions Biol. Syst.*, **7,** 127 (1978).
19. R. W. Lane, J. A. Ibers, R. B. Frankel, and R. H. Holm, *Proc. Natl. Acad. Sci. USA,* **72,** 2868 (1975).
20. T. Muraoka, T. Nozawa, and M. Hatano, *Chem. Lett.*, 1373 (1976); *Bioinorg. Chem.*, **8,** 45 (1978).
21. R. B. Frankel, G. C. Papaefthymiou, R. W. Lane, and R. H. Holm, *J. Phys. (Paris),* **37,** C6-165 (1976).
22. D. G. Holah and D. Coucouvanis, *J. Am. Chem. Soc.*, **97,** 6917 (1975).
23. K. S. Hagen, J. G. Reynolds, and R. H. Holm, *J. Am. Chem. Soc.*, **103,** 4054 (1981).
24. P. M. Champion and A. J. Sievers, *J. Chem. Phys.*, **66,** 1819 (1977).
25. A. Kostikas, V. Petrouleas, A. Simopoulos, D. Coucouvanis, and D. G. Holah, *Chem. Phys. Lett.*, **38,** 582 (1976); V. Petrouleas, A. Simopoulos, A. Kostikas, and D. Coucouvanis, *J. Phys. (Paris),* **37,** C6-159 (1976).
26. J. R. Anglin and A. Davison, *Inorg. Chem.*, **14,** 234 (1975).
27. G. Christou, B. Ridge, and H. N. Rydon, *J. Chem., Soc., Chem., Commun.*, 908 (1977).
28. J. J. Mayerle, R. B. Frankel, R. H. Holm, J. A. Ibers, W. D. Phillips, and J. F. Weiher, *Proc. Natl. Acad. Sci. USA,* **70,** 2429 (1973).
29. J. J. Mayerle, S. E. Denmark, B. V. DePamphilis, J. A. Ibers, and R. H. Holm, *J. Am. Chem. Soc.*, **97,** 1032 (1975).
30. J. Cambray, R. W. Lane, A. G. Wedd, R. W. Johnson, and R. H. Holm, *Inorg. Chem.*, **16,** 2565 (1977).
31. J. G. Reynolds and R. H. Holm, *Inorg. Chem.*, **19,** 3257 (1980).
32. W. O. Gillum, R. B. Frankel, S. Foner, and R. H. Holm, *Inorg. Chem.*, **15,** 1095 (1976).
33. D. Coucouvanis, D. Swenson, P. Stremple, and N. C. Baenziger, *J. Am. Chem. Soc.*, **101,** 3392 (1979).
34. G. B. Wong, D. M. Kurtz, Jr., R. H. Holm, L. E. Mortenson, and R. G. Upchurch, *J. Am. Chem. Soc.*, **101,** 3078 (1979).
35. B. V. DePamphilis, B. A. Averill, T. Herskovitz, L. Que, Jr., and R. H. Holm, *J. Am. Chem. Soc.*, **96,** 4159 (1974).
36. G. B. Wong, M. A. Bobrik, and R. H. Holm, *Inorg. Chem.*, **17,** 578 (1978).
37. B. A. Averill, T. Herskovitz, R. H. Holm, and J. A. Ibers, *J. Am. Chem. Soc.*, **95,** 3523 (1973).

38. R. H. Holm, W. D. Phillips, B. A. Averill, J. J. Mayerle, and T. Herskovitz, *J. Am. Chem. Soc.*, **96,** 2109 (1974).
39. L. Que, Jr., J. R. Anglin, M. A. Bobrik, A. Davison, and R. H. Holm, *J. Am. Chem. Soc.*, **96,** 6042 (1974).
40. A. Schwartz and E. E. van Tamelen, *J. Am. Chem. Soc.*, **99,** 3189 (1977).
41. R. H. Holm, B. A. Averill, T. Herskovitz, R. B. Frankel, H. B. Gray, O. Siiman, and F. J. Grunthaner, *J. Am. Chem. Soc.*, **96,** 2644 (1974).
42. R. B. Frankel, B. A. Averill, and R. H. Holm, *J. Phys. (Paris),* **35,** C6-107 (1974).
43. L. Que, Jr., M. A. Bobrik, J. A. Ibers, and R. H. Holm, *J. Am. Chem. Soc.*, **96,** 4168 (1974).
44. C. L. Hill, J. Renaud, R. H. Holm, and L. E. Mortenson, *J. Am. Chem. Soc.*, **99,** 2549 (1977).
45. G. Christou and C. D. Garner, *J. Chem. Soc., Dalton Trans.*, 1093 (1979).
46. M. W. W. Adams, S. G. Reeves, D. O. Hall, G. Christou, B. Ridge, and H. N. Rydon, *Biochem. Biophys. Res. Commun.*, **79,** 1184 (1977).
47. M. R'zaigui, E. C. Hatchikian, and D. Benlian, *Biochem. Biophys. Res. Commun.*, **92,** 1258 (1980).
48. M. W. W. Adams, K. K. Rao, D. O. Hall, G. Christou, and C. D. Garner, *Biochim. Biophys. Acta,* **589,** 1 (1980).
49. P. Mascharak and R. H. Holm, results to be published.
50. H. L. Carrell, J. P. Glusker, R. Job, and T. C. Bruice, *J. Am. Chem. Soc.*, **99,** 3683 (1977).
51. R. A. Henderson and A. G. Sykes, *Inorg. Chem.*, **19,** 3103 (1980).
52. R. C. Job and T. C. Bruice, *Proc. Natl. Acad. Sci. USA,* **72,** 2478 (1975).
53. R. Maskiewicz and T. C. Bruice, *J. Chem. Soc., Chem. Commun.*, 703 (1978).
54. M. A. Bobrik, L. Que, Jr., and R. H. Holm, *J. Am. Chem. Soc.*, **96,** 285 (1974).
55. G. R. Dukes and R. H. Holm, *J. Am. Chem. Soc.*, **97,** 528 (1975).
56. R. W. Johnson and R. H. Holm, *J. Am. Chem. Soc.*, **100,** 5338 (1978).
57. R. J. Burt, B. Ridge, and H. N. Rydon, *J. Chem. Soc, Dalton Trans.*, 1228 (1980).
58. T. C. Bruice, R. Maskiewicz, and R. Job, *Proc. Natl. Acad. Sci. USA,* **72,** 231 (1975).
59. R. J. Burt, G. J. Leigh, and C. J. Pickett, *J. Chem. Soc., Chem. Commun.*, 940 (1976).
60. R. G. Bowman and R. L. Burwell, Jr., *J. Am. Chem., Soc.*, **101,** 2877 (1979).
61. T. Herskovitz, B. A. Averill, R. H. Holm, J. A. Ibers, W. D. Phillips, and J. F. Weiher, *Proc. Natl. Acad. Sci. USA,* **69,** 2437 (1972).
62. R. B. Frankel, T. Herskovitz, B. A. Averill, R. H. Holm, P. J. Krusic, and W. D. Phillips, *Biochem. Biophys. Res. Commun.*, **58,** 974 (1974).
63. P. J. Stephens, A. J. Thomson, T. A. Keiderling, J. Rawlings, K. K. Rao, and D. O. Hall, *Proc. Natl. Acad. Sci. USA,* **75,** 5273 (1978).
64. J. G. Reynolds, E. J. Laskowski, and R. H. Holm, *J. Am. Chem. Soc.*, **100,** 5315 (1978).
65. G. C. Papaefthymiou, R. B. Frankel, S. Foner, E. J. Laskowski, and R. H. Holm, *J. Phys. (Paris),* **41,** C1-493 (1980).
66. R. W. Lane, A. G. Wedd, W. O. Gillum, E. J. Laskowski, R. H. Holm, R. B. Frankel, and G. C. Papaefthymiou, *J. Am. Chem. Soc.*, **99,** 2350 (1977).

67. E. J. Laskowski, J. G. Reynolds, R. B. Frankel, S. Foner, G. C. Papaefthymiou, and R. H. Holm, *J. Am. Chem. Soc.*, **101,** 6562 (1979).

68. M. A. Bobrik, E. J. Laskowski, R. W. Johnson, W. O. Gillum, J. M. Berg, K. O. Hodgson, and R. H. Holm, *Inorg. Chem.*, **17,** 1402 (1978).

69. G. Christou, B. Ridge, and H. N. Rydon, *J. Chem. Soc., Dalton Trans.*, 1423 (1978).

70. J. G. Reynolds, C. L. Coyle, and R. H. Holm, *J. Am. Chem. Soc.*, **102,** 4350 (1980).

71. C. L. Hill, D. J. Steenkamp, R. H. Holm, and T. P. Singer, *Proc. Natl. Acad. Sci. USA*, **74,** 547 (1977).

72. R. Cammack and G. Christou, in *Hydrogenases: Their Catalytic Activity, Structure, and Function,* H. G. Schlegel and K. Schneider, eds., Erich Goltze KG, Göttingen, 1978, pp. 45-56.

73. J. G. Reynolds and R. H. Holm, *Inorg. Chem.*, **20,** 1873 (1981).

74. S.-P. W. Tang, T. G. Spiro, C. Antanaitis, T. H. Moss, R. H. Holm, T. Herskovitz, and L. E. Mortenson, *Biochem. Biophys. Res. Commun.*, **62,** 1 (1975).

75. E. T. Lode and M. J. Coon, *J. Biol. Chem.*, **246,** 791 (1971).

76. J. Peisach, W. E. Blumberg, E. T. Lode, and M. J. Coon, *J. Biol. Chem.*, **246,** 5877 (1971).

77. M. Bruschi, I. Moura, J. LeGall, A. V. Xavier, and L. C. Sieker, *Biochem. Biophys. Res. Commun.*, **90,** 596 (1979).

78. I. Moura, A. V. Xavier, R. Cammack, M. Bruschi, and J. LeGall, *Biochim. Biophys. Acta*, **533,** 156 (1978).

79. I. Moura, B. H. Huynh, R. P. Hausinger, J. LeGall, A. V. Xavier, and E. Münck, *J. Biol. Chem.*, **255,** 2493 (1980).

80. K. D. Watenpaugh, T. N. Margulis, L. C. Sieker, and L. H. Jensen, *J. Mol. Biol.*, **122,** 175 (1978).

81. K. D. Watenpaugh, L. C. Sieker, and L. H. Jensen, *J. Mol. Biol.*, **131,** 509 (1979).

82. K. D. Watenpaugh, L. C. Sieker, and L. H. Jensen, *J. Mol. Biol.*, **138,** 615 (1980).

83. M. Pierrot, R. Haser, M. Frey, M. Bruschi, J. LeGall, L. C. Sieker, and L. H. Jensen, *J. Mol. Biol.*, **107,** 179 (1976).

84. E. T. Adman, L. C. Sieker, L. H. Jensen, M. Bruschi, and J. LeGall, *J. Mol. Biol.*, **112,** 113 (1977); L. H. Jensen, private communication (December 1980).

85. B. Bunker and E. A. Stern, *Biophys. J.*, **19,** 253 (1977).

86. R. G. Shulman, P. Eisenberger, B. K. Teo, B. M. Kincaid, and G. S. Brown, *J. Mol. Biol.*, **124,** 305 (1978).

87. D. Coucouvanis, D. Swenson, N. C. Baenziger, D. G. Holah, A. Kostikas, A. Simopoulos, and V. Petrouleas, *J. Am. Chem. Soc.*, **98,** 5721 (1976).

88. K. D. Watenpaugh, L. C. Sieker, J. R. Herriott, and L. H. Jensen, *Cold Spring Harbor Symp. Quant. Biol.*, **36,** 359 (1972).

89. K. D. Watenpaugh, L. C. Sieker, J. R. Herriott, and L. H. Jensen, *Acta Crystallogr.*, **B29,** 943 (1973).

90. L. H. Jensen, in *Iron-Sulfur Proteins,* Vol. II, W. Lovenberg, ed., Academic Press, New York, 1973, Chapter 4.

91. R. A. Bair and W. A. Goddard, III, *J. Am. Chem. Soc.*, **100,** 5669 (1978).

92. D. Swenson, N. C. Baenziger, and D. Coucouvanis, *J. Am. Chem. Soc.*, **100,** 1932 (1978).

93. R. H. Sands and W. R. Dunham, *Q. Rev. Biophys.*, **7,** 443 (1975), and references therein.

94. K. Ogawa, T. Tsukihara, H. Tahara, Y. Katsube, Y. Matsuura, N. Tanaka, M. Kakudo, K. Wada, and H. Matsubara, *J. Biochem.*, **81,** 529 (1977).
95. A. Kunita, M. Koshibe, Y. Nishikawa, K. Fukuyama, T. Tsukihara, Y. Katsube, Y. Matsuura, N. Tanaka, M. Kakudo, T. Hase, and H. Matsubara, *J. Biochem.*, **84,** 989 (1978).
96. H. Kagamiyama, K. K. Rao, D. O. Hall, R. Cammack, and H. Matsubara, *Biochem. J.*, **145,** 121 (1975).
97. K. S. Hagen and R. E. Palermo, unpublished results.
98. M. A. Bobrik, K. O. Hodgson, and R. H. Holm, *Inorg. Chem.*, **16,** 1851 (1977).
99. P. Müller and W. Bronger, *Z. Naturforsch.*, **34b,** 1264 (1979).
100. R. D. Shannon, *Acta Crystallogr.*, **A32,** 751 (1976).
101. J. G. Norman, Jr., P. B. Ryan, and L. Noodleman, *J. Am. Chem. Soc.*, **102,** 4279 (1980).
102. B.-K. Teo, R. G. Shulman, G. S. Brown, and A. E. Meixner, *J. Am. Chem. Soc.*, **101,** 5624 (1979).
103. D. C. Yoch and D. I. Arnon, *J. Biol. Chem.*, **247,** 4514 (1972).
104. Y. Shethna, *Biochim. Biophys. Acta,* **205,** 58 (1970).
105. B. A. Averill, J. R. Bale, and W. H. Orme-Johnson, *J. Am. Chem. Soc.*, **100,** 3034 (1978).
106. C. D. Stout, *Nature,* **279,** 83 (1979).
107. C. D. Stout, personal communication; D. Ghosh, W. Furey, Jr., S. O'Donnell, and C. D. Stout, *J. Biol. Chem.*, **256,** 4185 (1981).
108. M. Bruschi, E. C. Hatchikian, J. LeGall, J. J. G. Moura, and A. V. Xavier, *Biochim. Biophys. Acta,* **449,** 275 (1976).
109. D. C. Yoch, R. P. Carithers, and D. I. Arnon, *J. Biol. Chem.*, **252,** 7453 (1977).
110. M. G. Yates, M. J. O'Donnell, D. J. Lowe, and H. Bothe, *Eur. J. Biochem.*, **85,** 291 (1978).
111. C.-H. Wei and L. F. Dahl, *Inorg. Chem.*, **4,** 493 (1965).
112. L. Markó, B. Markó-Monostory, T. Madach, and H. Vahrenkamp, *Angew. Chem. Int. Ed. Eng.*, **19,** 226 (1980).
113. R. Meij, J. van der Helm, D. J. Stufkens, and K. Vrieze, *J. Chem. Soc., Chem. Commun.*, 506 (1978).
114. K. Gerst and C. E. Nordman, *Abstr. Am. Crystallogr., Assn.* (Summer Meeting), 1974, p. 225, abstr. E2.
115. K. A. Rubinson and G. Palmer, *J. Am. Chem. Soc.*, **94,** 8375 (1972).
116. K. S. Hagen, J. M. Berg, and R. H. Holm, *Inorg. Chim. Acta,* **45,** L17 (1980).
117. L. Pazdernik, F. Brisse, and R. Rivest, *Acta Crystallogr.*, **B33,** 1780 (1977).
118. B. Menzebach and P. Bleckmann, *J. Organomet. Chem.*, **91,** 291 (1975).
119. H.-J. Jacobsen and B. Krebs, *J. Organomet. Chem.*, **136,** 333 (1977).
120. S. Pollitz, F. Zettler, D. Forst, and H. Hess, *Z. Naturforsch.*, **31b,** 897 (1976).
121. J. A. Tiethof, J. K. Stalick, and D. W. Meek, *Inorg. Chem.*, **12,** 1170 (1973).
122. C. J. DeRanter and M. Rolies, *Cryst. Struct. Commun.*, **6,** 399 (1977).
123. E. M. McPartlin and N. C. Stephenson, *Acta Crystallogr.*, **B25,** 1659 (1969).
124. K. Anzenhofer and J. J. DeBoer, *Recl. Trav. Chim. Pays-Bas,* **88,** 286 (1969).
125. A. B. Blake and L. R. Fraser, *J. Chem. Soc., Dalton Trans.*, 193 (1975).
126. E. M. Holt, S. L. Holt, W. F. Tucker, R. O. Asplund, and K. J. Watson, *J. Am. Chem. Soc.*, **96,** 2621 (1941).

127. R. V. Thundathil, E. M. Holt, S. L. Holt, and K. J. Watson, *J. Am. Chem. Soc.*, **99**, 1818 (1977).

128. C. Giacovazzo, F. Scordari, and S. Menchetti, *Acta Crystallogr.*, **B31**, 2171 (1975).

129. See, for example, *Eur. J. Biochem.*, **93**, 427 (1979).

130. K. Dus, H. DeKlerk, K. Sletten, and R. G. Bartsch, *Biochim. Biophys. Acta*, **140**, 291 (1967).

131. W. V. Sweeney, J. C. Rabinowitz, and D. C. Yoch, *J. Biol. Chem.*, **250**, 7842 (1975).

132. D. C. Yoch and R. P. Carithers, *Microbiol. Rev.*, **43**, 384 (1979).

133. W. V. Sweeney and J. C. Rabinowitz, *Ann. Rev. Biochem.*, **49**, 139 (1980).

134. G. Strahs and J. Kraut, *J. Mol. Biol.*, **35**, 503 (1968).

135. C. W. Carter, Jr., J. Kraut, S. T. Freer, N.-H. Xuong, R. A. Alden, and R. G. Bartsch, *J. Biol. Chem.*, **249**, 4212 (1974).

136. C. W. Carter, Jr., J. Kraut, S. T. Freer, and R. A. Alden, *J. Biol. Chem.*, **249**, 6339 (1974).

137. S. T. Freer, R. A. Alden, C. W. Carter, Jr., and J. Kraut, *J. Biol. Chem.*, **250**, 46 (1975).

138. C. W. Carter, Jr., in *Iron-Sulfur Proteins*, Vol. III, W. Lovenberg, ed., Academic Press, New York, 1977, Chapter 6.

139. E. T. Adman, L. C. Sieker, and L. H. Jensen, *J. Biol. Chem.*, **248**, 3987 (1973).

140. L. H. Jensen, *Ann. Rev. Biochem.*, **43**, 461 (1974).

141. E. T. Adman, L. C. Sieker, and L. H. Jensen, *Acta Crystallogr.*, **A31**, 534 (1975).

142. E. T. Adman, L. C. Sieker, and L. H. Jensen, *J. Biol., Chem.*, **251**, 3801 (1976).

143. C. D. Stout, *J. Biol. Chem.*, **254**, 3598 (1979).

144. E. Adman, K. D. Watenpaugh, and L. H. Jensen, *Proc. Natl. Acad. Sci. USA*, **72**, 4854 (1975).

145. C. W. Carter, Jr., *J. Biol. Chem.*, **252**, 7802 (1977).

146. C. W. Carter, Jr., J. Kraut, S. T. Freer, R. A. Alden, L. C. Sieker, E. Adman, and L. H. Jensen, *Proc. Natl. Acad. Sci. USA* **69**, 3526 (1972).

147. J. M. Berg, K. O. Hodgson, and R. H. Holm, *J. Am. Chem., Soc.*, **101**, 4586 (1979).

148. C. Y. Yang, K. H. Johnson, R. H. Holm, and J. G. Norman, Jr., *J. Am. Chem. Soc.*, **97**, 6596 (1975).

149. P. J. M. Geurts, J. W. Gosselink, A. van der Avoird, E. J. Baerends, and J. G. Snijders, *Chem. Phys.*, **46**, 133 (1980).

150. B.-K. Teo and J. C. Calabrese, *Inorg. Chem.*, **15**, 2467 (1976).

151. T. C. W. Mak and F.-C. Mok, *J. Cryst. Mol. Struct.*, **8**, 183 (1978); C. D. Garner, in *Transition Metal Clusters*, B. F. G. Johnson, ed., Wiley, New York, 1980, Chapter 4.

152. R. A. Shunn, C. J. Fritchie, Jr., and C. T. Prewitt, *Inorg. Chem.*, **5**, 892 (1966).

153. C. H. Wei, G. R. Wilkes, P. M. Treichel, and L. F. Dahl, *Inorg. Chem.*, **5**, 900 (1966).

154. H. Vahrenkamp, *Angew. Chem. Int. Ed. Engl.*, **14**, 322 (1975).

155. Trinh-Toan, W. P. Fehlhammer, and L. F. Dahl, *J. Am. Chem. Soc.*, **99**, 402 (1977).

156. Trinh-Toan, B. K. Teo, J. A. Ferguson, T. J. Meyer, and L. F. Dahl, *J. Am. Chem. Soc.*, **99**, 408 (1977).

157. R. S. Gall, C. T.-W. Chu, and L. F. Dahl, *J. Am. Chem. Soc.*, **96**, 4019 (1974).

158. I. Bernal, B. R. Davis, M. L. Good, and S. Chandra, *J. Coord. Chem.*, **2**, 61 (1972).

159. C. J. Fritchie, Jr., *Acta Crystallogr.*, **B31**, 802 (1975).

160. S. C. Nyburg, *Acta Crystallogr.*, **B30,** 251 (1974).

161. S. R. Wilson and J. C. Huffman, *J. Org. Chem.*, **45,** 560 (1980).

162. G. Christou, B. Ridge, and H. N. Rydon, *J. Chem. Soc., Chem. Commun.*, 20 (1979).

163. Y. Sugiura, K. Ishizu, and T. Kimura, *Biochem. Biophys. Res. Commun.*, **60,** 334 (1974).

164. R. Maskiewicz, T. C. Bruice, and R. G. Bartsch, *Biochem. Biophys. Res. Commun.*, **65,** 407 (1975).

165. R. Maskiewicz and T. C. Bruice, *Biochemistry,* **16,** 3024 (1977).

166. R. Maskiewicz and T. C. Bruice, *Proc. Natl. Acad. Sci. USA,* **74,** 5231 (1977).

167. E. F. Wallace and J. C. Rabinowitz, *Arch. Biochem. Biophys.*, **146,** 400 (1971).

168. T. Kimura, Y. Nagata, and J. Tsurugi, *J. Biol. Chem.*, **246,** 5140 (1971); S. Arakawa, R. D. Bach, and T. Kimura, *J. Am. Chem. Soc.*, **102,** 6847 (1980).

169. J.-S. Hong and J. C. Rabinowitz, *J. Biol. Chem.*, **245,** 6574 (1970).

170. T. Kimura and S. Nakamura, *Biochemistry,* **10,** 4517 (1971).

171. D. Petering, J. A. Fee, and G. Palmer, *J. Biol. Chem.*, **246,** 643 (1971).

172. J. Rabinowitz, *Meth. Enzymol.*, **24B,** 431 (1972).

173. K. K. Rao, R. Cammack, D. O. Hall, and C. E. Johnson, *Biochem. J.*, **122,** 257 (1971).

174. K. Suhara, K. Kanayama, S. Takemori, and M. Katagiri, *Biochim. Biophys. Acta,* **336,** 309 (1974).

175. J. Fritz, R. Anderson, J. Fee, G. Palmer, R. H. Sands, J. C. M. Tsibris, I. C. Gunsalus, W. H. Orme-Johnson, and H. Beinert, *Biochim. Biophys. Acta,* **253,** 110 (1971).

176. J. A. Fee and G. Palmer, *Biochim. Biophys. Acta,* **245,** 175 (1971).

177. J. A. Fee, S. G. Mayhew, and G. Palmer, *Biochim. Biophys. Acta,* **245,** 196 (1971).

178. K. Mukai, J. J. Huang, and T. Kimura, *Biochem. Biophys. Res. Commun.*, **50,** 105 (1973); *Biochim. Biophys. Acta,* **336,** 427 (1974).

179. J. C. M. Tsibris, M. J. Namtvedt, and I. C. Gunsalus, *Biochem. Biophys. Res. Commun.*, **30,** 323 (1968).

180. W. H. Orme-Johnson, R. E. Hansen, H. Beinert, J. C. M. Tsibris, R. C. Bartholomaus, and I. C. Gunsalus, *Proc. Natl. Acad. Sci. USA,* **60,** 368 (1968).

181. S. W. May and J.-Y. Kuo, *Biochemistry,* **17,** 3333 (1978).

182. Y. Sugiura, K. Ishizu, and T. Kimura, *Biochemistry,* **14,** 97 (1975).

183. G. Palmer, in *The Enzymes,* Vol. XII, Part B, 3rd ed., P. D. Boyer, ed., Academic Press, New York, 1975, Chapter 1.

184. D. J. Steenkamp, T. P. Singer, and H. Beinert, *Biochem. J.*, **169,** 361 (1978).

185. D. M. Kurtz, Jr., G. B. Wong, and R. H. Holm, *J. Am. Chem. Soc.*, **100,** 6777 (1978).

186. J. S. Olson, D. P. Ballou, G. Palmer, and V. Massey, *J. Biol. Chem.*, **249,** 4363 (1974).

187. R. C. Bray, in *The Enzymes,* Vol. XII, Part B, 3rd ed., P. D. Boyer, ed., Academic Press, New York, 1975, Chapter 6.

188. M. W. W. Adams, L. E. Mortenson, and J.-S. Chen, *Biochim. Biophys. Acta,* **594,** 105 (1981).

189. R. R. Eady and B. E. Smith, in *A Treatise on Dinitrogen Fixation,* Sections I and II, R. W. F. Hardy, F. Bottomley, and R. C. Burns, eds., Wiley-Interscience, New York, 1979, Chapter 2 (Section II).

190. D. L. Erbes, R. H. Burris, and W. H. Orme-Johnson, *Proc. Natl. Acad. Sci. USA,* **72,** 4795 (1975).

191. W. O. Gillum, L. E. Mortenson, J.-S. Chen, and R. H. Holm, *J. Am. Chem. Soc.*, **99**, 584 (1977).

192. E. C. Hatchikian, M. Bruschi, and J. LeGall, *Biochem. Biophys. Res. Commun.*, **82**, 451 (1978).

193. I. Okura, K.-I. Nakamura, and T. Keii, *J. Mol. Catal.*, **4**, 453 (1978).

194. I. Okura, K.-I. Nakamura, and S. Nakamura, *J. Mol. Catal.*, **6**, 307, 311 (1979).

195. D. M. Kurtz, Jr., R. S. McMillan, B. K. Burgess, L. E. Mortenson, and R. H. Holm, *Proc. Natl. Acad. Sci. USA*, **76**, 4986 (1979).

196. W. H. Orme-Johnson and E. Münck, in *Molybdenum and Molybdenum-Containing Enzymes*, M. P. Coughlan, ed., Pergamon, New York, 1980, Chapter 13.

197. R. Zimmermann, E. Münck, W. J. Brill, V. K. Shah, M. T. Henzl, J. Rawlings, and W. H. Orme-Johnson, *Biochim Biophys. Acta*, **537**, 185 (1978).

198. B.-H. Huynh, M. T. Henzl, J. A. Christner, R. Zimmermann, W. H. Orme-Johnson, and E. Münck, *Biochim. Biophys. Acta*, **623**, 124 (1980).

199. L. C. Davis, M. T. Henzl, R. H. Burris, and W. H. Orme-Johnson, *Biochemistry*, **18**, 4860 (1979).

200. B. E. Smith, M. J. O'Donnell, G. Lang, and K. Spartalian, *Biochem. J.*, **191**, 449 (1980).

201. J. R. Lancaster, J. M. Vega, H. Kamin, N. R. Orme-Johnson, W. H. Orme-Johnson, R. J. Krueger, and L. M. Siegel, *J. Biol. Chem.*, **254**, 1268 (1979).

202. L. M. Siegel, in *Mechanisms of Oxidizing Enzymes*, T. P. Singer and R. N. Ondarza, eds., Elsevier-North Holland, Amsterdam, 1978, pp. 201–214.

203. J. A. Christner, E. Münck, P. A. Janick, and L. M. Siegel, *J. Biol. Chem.*, in press.

204. R. Cammack, K. K. Rao, C. P. Bargeron, K. G. Hutson, P. W. Andrew, and L. J. Rogers, *Biochem. J.*, **168**, 205 (1977).

205. G. S. Wilson, J. C. M. Tsibris, and I. C. Gunsalus, *J. Biol. Chem.*, **248**, 6059 (1973).

206. J. J. Huang and T. Kimura, *Biochemistry*, **12**, 406 (1973).

207. P. D. J. Weitzman, I. R. Kennedy, and R. A. Caldwell, *FEBS Lett.*, **17**, 241 (1971).

208. T. Ikeda, K. Toriyama, and M. Senda, *Bull. Chem. Soc. Jpn.* **52**, 1937 (1979).

209. T. Kakutani, K. Toriyama, T. Ikeda, and M. Senda, *Bull. Chem. Soc. Jpn.*, **53**, 947 (1980).

210. E. T. Lode, C. L. Murray, and J. C. Rabinowitz, *J. Biol. Chem.*, **251**, 1683 (1976).

211. N. A. Stombaugh, J. E. Sundquist, R. H. Burris, and W. H. Orme-Johnson, *Biochemistry*, **15**, 2633 (1976).

212. R. N. Mullinger, R. Cammack, K. K. Rao, D. O. Hall, D. P. E. Dickson, C. E. Johnson, J. D. Rush, and A. Simopoulos, *Biochem. J.*, **151**, 75 (1975).

213. R. J. Kassner and W. Yang, *Biochem. J.*, **133**, 283 (1973); *J. Am. Chem. Soc.*, **99**, 4351 (1977).

214. R. Cammack, *Biochem. Biophys. Res. Commun.*, **54**, 548 (1973).

215. W. D. Phillips, C. C. McDonald, N. A. Stombaugh, and W. H. Orme-Johnson, *Proc. Natl. Acad. Sci. USA*, **71**, 140 (1974).

216. M. Poe, W. D. Phillips, C. C. McDonald, and W. Lovenberg, *Proc. Natl. Acad. Sci. USA*, **65**, 797 (1970).

217. M. Poe, W. D. Phillips, C. C. McDonald, and W. H. Orme-Johnson, *Biochem. Biophys. Res. Commun.*, **42**, 705 (1971).

218. E. L. Packer, W. V. Sweeney, J. C. Rabinowitz, H. Sternlicht, and E. N. Shaw, *J. Biol. Chem.*, **252,** 2245 (1977).

219. E. L. Packer, H. Sternlicht, E. T. Lode, and J. C. Rabinowitz, *J. Biol. Chem.*, **250,** 2062 (1975).

220. N. Sutin, in *Inorganic Biochemistry,* Vol. 2, G. Eichhorn, ed., Elsevier, Amsterdam, 1973, Chapter 19.

221. L. E. Bennett, *Prog. Inorg. Chem.*, **18,** 1 (1973).

222. L. E. Bennett, in *Iron-Sulfur Proteins,* Vol. III, W. Lovenberg, ed., Academic Press, New York, 1973, Chapter 9.

223. C. A. Jacks, L. E. Bennett, W. N. Raymond, and W. Lovenberg, *Proc. Natl. Acad. Sci, USA,* **71,** 1118 (1974).

224. J. Rawlings, S. Wherland, and H. B. Gray, *J. Am. Chem. Soc.*, **99,** 1968 (1977).

225. F. A. Armstrong, R. A. Henderson, M. G. Segal, and A. G. Sykes, *J. Chem. Soc., Chem. Commun.*, 1102 (1978).

226. F. A. Armstrong and A. G. Sykes, *J. Am. Chem. Soc.*, **100,** 7710 (1978).

227. I. A. Mizrahi, F. E. Wood, and M. A. Cusanovich, *Biochemistry,* **15,** 343 (1976).

228. J. Rawlings, S. Wherland, and H. B. Gray, *J. Am. Chem., Soc.*, **98,** 2177 (1976).

229. D. Cummins and H. B. Gray, *J. Am. Chem. Soc.*, **99,** 5158 (1977).

230. B. A. Feinberg and W. V. Johnson, *Biochem. Biophys. Res. Commun.*, **93,** 100 (1980).

231. R. A. Holwerda, D. B. Knaff, H. B. Gray, J. D. Clemmer, R. Crowley, J. M. Smith, and A. G. Mauk, *J. Am. Chem. Soc.*, **102,** 1142 (1980).

232. I. A. Mizrahi, T. E. Meyer, and M. A. Cusanovich, *Biochemistry,* **19,** 4727 (1980).

233. A. G. Mauk, R. A. Scott, and H. B. Gray, *J. Am. Chem. Soc.*, **102,** 4360 (1980).

234. F. A. Armstrong, R. A. Henderson, and A. G. Sykes, *J. Am. Chem. Soc.*, **102,** 6545 (1980).

235. S. Wherland and H. B. Gray, in *Biological Aspects of Inorganic Chemistry,* A. W. Addison, W. R. Cullen, D. Dolphin, and B. R. James, eds., Wiley-Interscience, New York, 1977, pp. 289–368.

236. J.-S. Hong and J. C. Rabinowitz, *J. Biol. Chem.*, **245,** 6582 (1970).

237. N. Ohta, C. Kawasaki, M. Maeda, S. Tani, and K. Kawasaki, *Chem. Pharm. Bull.*, **27,** 2968 (1979).

238. W. E. Cleland and B. A. Averill, *Inorg. Chim. Acta,* **56,** L9 (1981).

239. L. Que, Jr., R. H. Holm, and L. E. Mortenson, *J. Am. Chem. Soc.*, **97,** 463 (1975).

240. R. H. Holm, in *Biological Aspects of Inorganic Chemistry,* A. W. Addison, W. R. Cullen, D. Dolphin, and B. R. James, eds., Wiley-Interscience, New York, 1977, pp. 71–111.

241. W. H. Orme-Johnson and R. H. Holm, *Meth. Enzymol.*, **53,** 268 (1978).

242. L. E. Mortenson and W. O. Gillum, *Meth. Enzymol.*, **69,** 779 (1980).

243. W. H. Orme-Johnson, L. C. Davis, M. T. Henzl, B. A. Averill, N. R. Orme-Johnson, E. Münck, and R. Zimmermann, in *Recent Developments in Nitrogen Fixation,* W. Newton, J. R. Postgate, and C. Rodriguez-Barrueco, eds., Academic Press, New York, 1977, pp. 131–178.

244. S. E. Crutcher and P. J. Geary, *Biochem. J.*, **177,** 393 (1979).

245. K. R. Carter, J. Rawlings, W. H. Orme-Johnson, R. R. Becker, and H. J. Evans, *J. Biol. Chem.*, **255,** 4213 (1980).

246. B. A. Averill, A. Dwivedi, P. Debrunner, S. J. Vollmer, J. Y. Wong, and R. L. Switzer, *J. Biol. Chem.*, **255,** 6007 (1980).

247. W. H. Orme-Johnson and L. C. Davis, in *Iron-Sulfur Proteins,* Vol. III, W. Lovenberg, ed., Academic Press, New York, 1977, Chapter 2.

248. D. J. Steenkamp, W. McIntire, and W. C. Kenney, *J. Biol. Chem.*, **253,** 2818 (1978).

249. C. J. Coles, R. H. Holm, D. M. Kurtz, Jr., W. H. Orme-Johnson, J. Rawlings, T. P. Singer, and G. B. Wong, *Proc. Natl. Acad. Sci. USA,* **76,** 3805 (1979).

250. P. S. Yoon, J. Rawlings, W. H. Orme-Johnson, and H. F. DeLuca, *Biochemistry,* **19,** 2172 (1980).

251. C. Paech, J. G. Reynolds, T. P. Singer, and R. H. Holm, *J. Biol. Chem.*, **256,** 3167 (1981).

252. D. M. Kurtz, R. H. Holm, F. J. Ruzicka, H. Beinert, C. J. Coles, and T. P. Singer, *J. Biol. Chem.*, **254,** 4967 (1979).

253. R. Cammack, *Biochem. Soc. Trans.*, **3,** 482 (1975).

254. H. Rupp, K. K. Rao, D. O. Hall, and R. Cammack, *Biochim. Biophys. Acta,* **537,** 255 (1978).

255. K. Schneider, R. Cammack, H. G. Schlegel, and D. O. Hall, *Biochim. Biophys. Acta,* **578,** 445 (1979).

256. J. Peisach, N. R. Orme-Johnson, W. B. Mims, and W. H. Orme-Johnson, *J. Biol. Chem.*, **252,** 5643 (1977).

257. P. J. Stephens, A. J. Thomson, J. B. R. Dunn, T. A. Keiderling, J. Rawlings, K. K. Rao, and D. O. Hall, *Biochemistry,* **17,** 4770 (1978).

258. A. J. Thomson, R. Cammack, D. O. Hall, K. K. Rao, B. Briat, J. C. Rivoal, and J. Badoz, *Biochim. Biophys. Acta,* **493,** 132 (1977).

259. M. K. Johnson, A. J. Thomson, A. E. Robinson, K. K. Rao, and D. O. Hall, *Biochim. Biophys. Acta,* **667,** 433 (1981).

260. B. Siegel, *J. Inorg. Nucl. Chem.*, **41,** 609 (1979).

261. B. A. Averill and W. H. Orme-Johnson, *J. Am. Chem. Soc.*, **100,** 5234 (1978).

262. E. E. van Tamelen, J. A. Gladysz, and C. R. Brûlet, *J. Am. Chem. Soc.*, **96,** 3020 (1974).

263. K. Tano and G. N. Schrauzer, *J. Am. Chem. Soc.*, **97,** 5404 (1975).

264. G. N. Schrauzer, P. R. Robinson, E. L. Moorehead, and T. M. Vickrey, *J. Am. Chem. Soc.*, **97,** 7069 (1975); **98,** 2815 (1976).

265. H. Inoue, N. Fujimoto, and E. Imoto, *J. Chem. Soc., Chem.Commun.*, 412 (1977).

266. H. Inoue and M. Suzuki, *J. Chem. Soc., Chem. Commun.*, 817 (1980).

267. R. S. McMillan, J. Renaud, J. G. Reynolds, and R. H. Holm, *J. Inorg. Biochem.*, **11,** 213 (1979).

268. R. W. F. Hardy, R. D. Holsten, E. K. Jackson, and R. C. Burns, *Plant Physiol.*, **43,** 1185 (1968).

269. G. Christou, R. V. Hageman, and R. H. Holm, *J. Am. Chem. Soc.*, **102,** 7600 (1980).

270. Y. Okuno and O. Yonemitsu, *Chem. Lett.*, 959 (1980).

CHAPTER 2

Iron-Sulfur Proteins: The Problem of Determining Cluster Type

WILLIAM H. ORME-JOHNSON

Department of Chemistry
Massachusetts Institute of Technology
Cambridge, Massachusetts

NANETTE R. ORME-JOHNSON

Mary Ingraham Bunting Institute
Radcliffe College
Cambridge, Massachusetts

CONTENTS

1 INTRODUCTION

Since our last summary of the properties of Fe-S proteins (1), there have appeared extensive discussions of the chemistry of synthetic Fe-S model systems with comparisons to Fe-S proteins (2, 3). Recent compilations of information on bacterial Fe-S proteins (4) and on proteins containing [4Fe-4S] centers (5), as well as a series of volumes devoted to these proteins (6), also attest to the continuing interest in this area. Given this interest we thought it appropriate to discuss the question of finding suitable methodology for the determination of cluster type in Fe-S proteins, an endeavor that must precede detailed studies of chemical mechanism. We hope to give a concise account of (1) the quality of present structural evidence on the six* recognized types of Fe-S centers, (2) the strengths and weaknesses of present noncrystallographic methods for structure elucidation, and (3) some of the results reported on the analysis of Fe-S cluster type in proteins, particularly enzymes. We wish to comment critically on certain cases that we believe will show that, while one should not automatically assume the presence of one of the known Fe-S clusters in a newly studied protein, it is not at all easy to convincingly establish the existence of novel types of centers, given the variety of properties manifested by presently known Fe-S systems. Our hope is to convey a sense of the challenge and interest of this area, in which spectroscopic, synthetic, and crystallographic approaches have been of comparable importance.

Sketches of the forms of four of the Fe-S clusters found in proteins are given in Figure 1. These are referred to as 1-Fe, [2Fe-2S], [3Fe-3S], and [4Fe-4S] centers, respectively.*

This is not a comprehensive set of the types of Fe-S clusters found in proteins. For example, the complete structures of the M and P centers (8) of the nitrogenase Mo-Fe protein are unknown (though the P centers can be made

*These six types include the four containing 1, 2, 3, or 4 Fe atoms in a cluster with documented structure and the two of unknown structure in the Mo-Fe protein of nitrogenase.

*The latest recommended IUB-IUPAC nomenclature is described in reference 7. We refer to the oxidation states of clusters calculated as the net charge on the core structure (i.e., minus the charge on the protein-donated ligands). Thus the oxidation states of the centers in Figure 1, as they are now presumed to be present in proteins, are, in parentheses, 1-Fe, (2+, 3+); [2Fe-2S], (1+, 2+); [3Fe-3S], (2+, 3+); [4Fe-4S], (1+, 2+, 3+). These charges are computed using ferric or ferrous ions and sulfide dianions as the components of the centers. The assignment given here for the [3Fe-3S] center is discussed below; the assignment for the other centers comes from comparisons of the physical properties of model compounds and proteins, summarized in references 2 and 3. The nomenclature for the charge states as used in older references such as references (2) or (3) may be arrived at by subtracting one charge for each thiolate (exo) ligand. Thus [4Fe-4S] clusters were said to traverse the (−3, −2, −1) set of states.

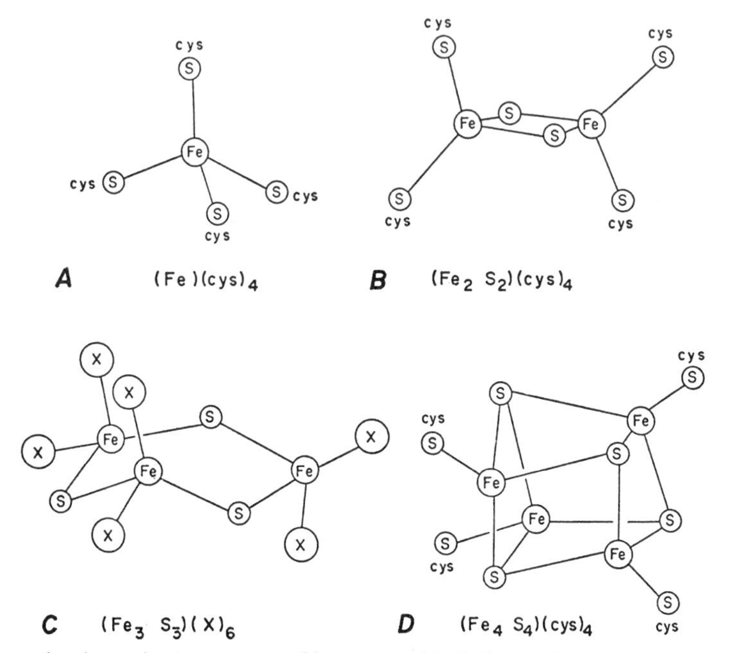

Figure 1. Approximate geometry of four types of Fe-S clusters that occur in proteins.

to yield [4Fe-4S] cores). As discussed in this chapter, these centers probably differ significantly from those shown in Figure 1. A précis of distinguishing features of FeS centers is given next. Following this we consider FeS clusters contained in proteins, in decreasing order of the accuracy with which their structures are now known, and then we discuss methods for identifying and quantitating these centers in proteins.

2 PROPERTIES OF Fe-S CENTERS IN PROTEINS AND MODEL SYSTEMS

In Table 1 we list properties of proteins, from which the presence of Fe-S centers may be inferred. Not shown is the presence of labile sulfur or sulfide displaceable by acid, normally found in quantities equimolar to Fe (1). The values listed in Table 1 are those found for rubredoxin, Spinach ferredoxin, *Desulfovibrio* ferredoxin, and *B. stearothermophilus* and *polymyxa* ferredoxins along with *Chromatium* high-potential Fe-S proteins, representing well-studied 1-Fe, [2Fe-2S], [3Fe-3S], and [4Fe-4S] proteins.

Table 1 Typical Properties of Fe-S Centers in Proteins and Model Systems

Measurement	Core Oxidation State	Cluster Type: $Fe(RS)_4$	$Fe_2S_2(RS)_4$	$Fe_3S_3L_6$	$Fe_4S_4(RS)_4$
Cysteine: Fe	—	4:1	4:2	probably 5:3	4:4
Absorbance[a] maxima, nm (absorbance/mol Fe)	3+	380 (10.8), 480 (8.8)	n.a.	305 (7.7), 415 (5.2), 453 (4.4)	325 (8.1), 375 (5.0), 450 (4.6)
	2+	311 (10.8), 333 (6.3)	325 (6.4), 420 (4.8), 465 (4.9)	425 (3.2)	305 (4.9) 390 (3.8)
	1+	n.a.	shoulder at 550 (1.1)	n.a.	unfeatured
EPR[a] values of g_1, g_2, g_3 (temperature)[b]	3+	4.3,[c] 9[c] (<20°K)	n.a.	1.97, 2.00, 2.02 (<20°K)	2.04, 2.04, 2.12 (<20°K)
	2+	n.a.	n.a.	n.a.	n.a.
	1+	n.a.	1.89, 1.95, 2.05 (<77°K)	n.a.	1.88, 1.92, 2.06 (<20°K)
Mössbauer[g] ΔE_q, δ; mm/s at 4.2°K	3+	0.50, 0.32	n.a.	(0.54, 0.27, (77°K)	(1.03, 0.40), (0.88, 0.29)[d]
	2+	3.25, 0.70	0.65, 0.27	(1.47, 0.46), (0.47, 0.30)[e]	
	1+	n.a.	(0.64, 0.25), (3.00, 0.54)[d]	n.a.	(1.32, 0.50),[f] (1.89, 0.58)[d]
Raman[h] frequency, cm^{-1}, attributed to Fe-S	3+	365, 314, 126	n.a.	n.a.	n.a.
	2+	n.a.	350	n.a.	n.a.
	1+	n.a.	n.a.	n.a.	n.a.
Average[i] Fe-S bond length (standard error) from EXAFS [model compound data]	3+	2.256 (16) [2.279 (13)]	n.a.	n.a.	n.a.
	2+	2.32 (2) [2.340 (13)]	2.227 (15) [2.234 (15)]	n.a.	2.249 (16)
	1+	n.a.	2.241 (28)	n.a.	2.262 (14)

[a]The data were obtained from references 9 and 1.
[b]Temperature range for optimum observation.
[c]These features are from the middle and lower Kramers doublets.
[d]Two subsites in 1:1 ratio.
[e]Two subsites in 2:1 ratio.
[f]Four subsites: $\Delta E_q = 1.50, 1.20, 1.10, 0.66$; all with $\delta = 0.42$.
[g]Isomer shifts (mm/sec) referred to metallic Fe at 300°K (9–13).
[h]The data were obtained from references 14–16.
[i]The data were obtained from references 17–19.

3 EVIDENCE FOR Fe-S CLUSTER STRUCTURE

3.1 1-Fe Clusters

Although the physiological role of rubredoxin is imperfectly understood (6), the protein is itself a model of tetrahedral coordination of Fe by S, a feature found in other Fe-S proteins. The study of rubredoxins has been intensive, furnishing, among other things, an interesting example of the use of non-crystallographic information in the refinement of a structure. Sequence data had led to the suggestion of four coordination of Fe by cysteine (20), and spectroscopic features of the protein were reproduced by Fe coordinated to a peptide containing four cysteine residues (21). The crystal structure of the *C. pasteurianum* rubredoxin determined with diffraction data to 1.5 Å showed a nearly tetrahedral Fe complex, with the single peculiarity that one bond was shortened to about 2.0 Å (22). This feature and the intermediate value of the parameter *E/D* obtained from EPR spectra of rubredoxins were suggested (23, 24) to arise from *entasis*, or constraint of the cluster geometry by the protein. X-Ray scattering studies using synchrotron radiation were carried out on the *C. pasteurianum* (25) and *Peptococcus aerogenes* rubredoxins (26). An EXAFS analysis was made assuming that the needed phase shifts could be obtained from studies of S-coordinated Fe compounds of known structure. The fine structure can be simulated satisfactorily only if the average Fe–S distance was the same as the *average* of the crystallographic distances in the protein, within ±0.03 Å, while the probable *maximum* difference between the longest and shortest bonds derived from the same analysis was 0.06 (25) to 0.1 Å (26). Subsequent refinement of the crystal structure using 1.2 Å data (27) verified this latter conclusion, removing the necessity for the postulate of entasis in the protein. A later EXAFS study (17) comparing salts of the analog $[Fe((SCH_2)_2C_6H_4)_2]^{1-/2-}$ with rubredoxins provided convincing evidence that, to better than 0.05 Å, the analogs and proteins have identical Fe–S distances, and that neither compound has suffered distortions away from T_d symmetry beyond that limit. Similar comparisons of the Mössbauer properties of these compounds (28), as well as the tetrakisthiophenolate Fe (II) anion (29), reinforce the general impression of the identity of the ligand environment in rubredoxins, these models, and the structure shown in Figure 1*A*. Refinements of the X-ray structure by conformational energy calculations (30), as well as a series of theoretical studies (31–35) aimed at reproducing electronic and Mössbauer spectra, starting with the (then) current best estimates of ligand geometries, provide no cause to doubt the above conclusion, nor does a second structure, of a rubredoxin from *Desulfovibrio vulgaris,* with data at 2 Å (36).

Inspection of the spectroscopic parameters listed in Table 1 would suggest that there is little difficulty in recognizing this type of center. The optically active absorbance near 500 nm (37) and the EPR characteristic of an $S = 5/2$ manifold given by the oxidized protein indeed serve to distinguish this from other known Fe-S systems. However, these properties may be conferred by a certain general kind of electronic environment of Fe, and it can be imagined that mononuclear centers with ligands or ligand arrangements other than that depicted in Figure 1*A* might have similar spectroscopic properties. A case in point is 3,4-protocatechuate dioxygenase, which appears to contain isolated Fe centers yielding EPR and absorbance characteristics similar to those of rubredoxins (38). The magnitude of the zero field splitting, deduced from the temperature dependence of the EPR signals of the ferric enzyme, was used to infer the presence of rubredoxinlike centers (24). If valid, this conclusion would be particularly interesting, since substrates cause substantial spectral perturbations, suggesting substrate binding to the chromophore. A study of the Mössbauer effect of the enzyme (38) led to the conclusion that the Fe atoms were *not* tetrahedrally coordinated by thiolate, based on a clear difference in the saturation magnetic field of the center in the enzyme (−520 kG) and in rubredoxins (−330 kG) and on the observation that the Mössbauer isomer shift exhibited by the reduced enzyme was double that of reduced rubredoxin. The observed Mössbauer parameters were characteristic of N_2 and/or O_2 coordination. Subsequent resonance Raman work strongly suggests the presence of tyrosinate ligands in this (and other) dioxygenases (39). On the other hand, the protein "desulforedoxin" from *D. gigas*, which has EPR and electronic characteristics similar to, but distinct from, those of rubredoxins, has been proposed to possess a geometric variant of the rubredoxin site, and indeed exhibits 4 cysteines per Fe atom, which is displaced on the addition of 4 equivalents of mercurial (40). Assuming that crystallographic or EXAFS criteria are less easily applied, one concludes that if the chemical, absorbance, EPR, Mössbauer, *and* Raman criteria are satisfied, then one may infer the presence of the structure of Figure 1*A*. No enzyme has yet satisfied these criteria.

3.2 4-Fe Clusters

The structure depicted in Figure 1*C* is thought to be widely distributed among electron-transfer proteins and enzymes. The more easily obtainable presumptive clues to the presence of this type of center include finding 4 or more atoms of Fe and of labile S per molecule of protein, a rather plain electronic spectrum with a peak or shoulder near 400 nm ($A_m \cong 4000$ cm^2/gm atom Fe) in samples in the nonparamagnetic (EPR-silent) oxidation state (1), and the classic slightly anisotropic low temperature ($T < 30K$) EPR signal* near $g = 2.00$ with $g_{av} < 2$ in samples exposed to denaturing solvents (e.g., 80% dimethyl sulfoxide) and dithionite (42). These criteria may be difficult to apply when more than one type of Fe-S center or other chromophores are present, so an examination of the variety of proposed detection methods is in order. To set the stage for these considerations, an outline is given of recent structural and chemical findings about tetranuclear clusters, particularly in model systems and the simpler protein cases.

The structures of the Fe-S clusters in the oxidized form of *Peptococcus aerogenes* ferredoxin has been refined by Adman, Sieker, and Jensen (43) with a 2 Å data set, and the standard deviations of the Fe and S atom positions are ≈0.05 Å. At this confidence level the clusters were identical to that found in the reduced form of the cluster in *Chromatium* high-potential iron protein, also determined from 2 Å data [(44, 45) and references therein]. Carter and co-workers (45) reported that a difference map calculated with data from reduced and oxidized specimens of the high-potential protein showed that the cluster expands about 0.16 Å along the normal to two opposite faces of the cube during reduction of the center from the 3+ to 2+ net core charge state. This expansion is at the limit of detection with the data utilized, but similar comparisons have been made (46, 47) of the synthetic tetranuclear clusters in the 2+ and 1+ net core charge states, corresponding to the reduced and superreduced HIPIP (or oxidized and reduced ferredoxin) states. These structural data are at least an order of magnitude more precise than the corresponding protein figures, and Holm and his co-workers were able to document an expansion of 0.08 Å along one of the quasi-fourfold axes, passing from a squat compressed box to an elongated box shape (47). As they point out, the Fe and labile S atoms lie at the vertices of nearly perfect interpenetrating tetrahedra, with the 6 Fe–Fe edge distances being nearly identical at 2.736 Å and the (bridging) S–S distances belonging to 2 slightly different sets, 2 at 3.592 Å and 4 at 3.650 Å. On reduction the Fe–Fe distances are unchanged, while 4 of the S–S distances are 3.592 Å and 2 of the S–S distances swell to 3.690 Å. This means that during reduction of the *protein* the bonded interactions between the 4 cysteine ligands and the Fe atoms need not change, since the Fe–Fe distances in the unconstrained model remain constant. The more important interactions for the center in ferredoxins would seem to be those between the protein and the *bridging S atoms*. Unfortunately, we do not have structural information for the reduced ferredoxin, nor do we have in hand a stable model derivative corresponding

*Details of the EPR of Fe-S proteins are given in reference 41. The g_{av} is the arithmetic mean of the apparent g-values at the extrema and crossing point of the first derivative presentation as normally recorded.

to the *oxidized* HIPIP (3+ cluster charge) state. The HIPIP data themselves do suggest a marginal (twice the standard deviation) increase in the Fe–Fe distance on reduction, from 2.73 Å to 2.81 Å, though this cannot yet be regarded with the same confidence as the model compound data. The latter do imply (46) that a substantial contribution to the inability of native HIPIP to be reduced beyond the 2+ cluster charge state is restraint on the expansion of the cluster, particularly the bridging S atom tetrahedron, imposed by the protein shell around the cluster. Carter (48) pointed out that what appear to be 4 H–bonds to cysteine S and 2 to labile S are detected in HIPIP, while in the ferredoxins each cluster enjoys 5 H–bonds to cysteine S and 4 to the bridging S atoms. One may argue about whether N–H–S hydrogen bonds are likely to be very strong or whether they are even present in the cases cited [about half the bonds posited here have N–S distances of 3.7–4.0 Å, substantially longer than the 3.25–3.55 Å seen in small molecules (49)]. Nonetheless, such dipoles should help to stabilize the increased negative charge on a reduced cluster. The real significance of the fewer "H-bonds" in HIPIP may be the correspondingly more numerous and tighter contacts with hydrophobic residues, leading to constraints on cluster expansion and preventing reduction beyond the 2+ state. This leaves open the question of whether the ferredoxins enjoy the property of reversible oxidation to the 3+ state. Though they may be oxidizable beyond the 2+ state (50), this does not seem to occur in typical enzymatic reactions of these proteins (1).

Holm and co-workers set out to systematically explore the chemistry of the tetrameric clusters. Important findings were that (1) the thiolate ligands exchange readily with exogenous thiolate (51), (2) the thiolates can be replaced by halide ions (52), and (3) the thiolates can be sequentially reacted with electrophiles of the form YX in aprotic solvents to yield $Fe_4S_4(SR)_{4-n}X_n$ and YSR (53). Where X = Cl, OAc, all four species, n = 1–4, could be generated, while for X = CF_3CO_2 and CF_3SO_3 the n = 1 species was found. These substitutions raised the redox potential of the 2+/3+ couple by ≈100 mV per added ligand, and the Cl- and OAc-ligated species were themselves found to undergo facile substitution by isonitriles. Reduced (1+ net core charge) clusters reacted far more rapidly (minutes vs. 10^2 minutes) than did oxidized clusters. Importantly, the substitution of 1 PhS^- by 1 OAc^- ligand decreased the absorbance near 400 nm by ≈25%. Differences of this magnitude between the molar absorbances of centers in proteins are often observed, often in the direction of lower absorbance, and the replacement of cysteine ligation by glutamate or aspartate, for example, which might not be otherwise detected, has become a plausible explanation since these model studies. The general mechanistic implications of substitution by these more easily displaceable ligands, for active site purposes, is obvious but as yet unexplored.

3.3 2-Fe Clusters

Unlike the tetranuclear cluster, for which no correct model was proposed prior to crystallographic studies, the essentials of the structures in binuclear clusters were deduced from chemical and spectroscopic data and "the principle of minimum astonishment" (R. H. Sands, personal communication) long before X-ray studies became feasible (1, 41). The 2 Fe and 2 sulfide atoms in a colored protein were found to yield one spin equivalent on integration of the EPR spectrum of reduced specimens. This led two groups to the concept of the binuclear cluster (54, 55). (Such evidence, coupled with the uptake of $1e^-$ equivalent per pair of Fe atoms (56) still creates a strong presumption in favor of the structure of Figure 1*B*.) Consideration of the simplest geometry incorporating four cysteine and two sulfide ligands (the disruption of the center by mercurials requires 8 equivalents) (54), as well as the desire to couple a ferric and a ferrous ion antiferromagnetically to yield the observed EPR (55), led to detailed proposals for the structure. That both iron nuclei and both sulfide nuclei couple to the electron spin moment was shown definitively from the observation of broadening of the EPR by ^{57}Fe, ^{33}S, and ^{77}Se (the latter a replacement for the labile sulfide). The spectral detail achieved with Fe (58) and Se (57) was such that simulation of the broadened spectra was possible only on the assumption that two labeled nuclei interact with the electron spin. The antiferromagnetic coupling of the Fe atoms, pairing the spins of the ferric and ferrous atoms of the [2Fe-2S] cluster via polarization of the bridging S atom electrons (59), is the source of several distinctive spectroscopic features of the center, in addition to the *g*-values of the EPR.

The success of the antiferromagnetically coupled binuclear cluster hypothesis in explaining the properties of the proteins was eventually consumated in a synthetic model with the proposed structure, known in crystalline form in the 2+ net core charge state (60). The synthesis of the model compound apparently depends on the addition of sulfide to bis-*o*-xylyldithiolato-Fe(II), with coupling and oxidation of two centers to give the binuclear Fe(III) complex. The mixed valence and binuclear Fe(II) complexes are unknown except by inference from electrochemical data (2). The dithiol can be subsequently replaced by monothiols to yield an unstable tetrathiomonotholato binuclear Fe(III) complex. The synthesis establishes a key point, in addition to the feasibility of the structure and formal oxidations states previously proposed, namely that a rational alteration of a thiolate ligand structure (replacement of monothiol with a dithiol with restricted coordination bite) leads to an alternative product. The implication for the role of the protein in the selection of the center type formed is plain.

Recently four reports have appeared of structural work on binuclear ferredoxins. *Azotobacter vinelandii* ferredoxin I, reported to contain [4Fe-4S] and

[2Fe-2S] centers (61) turns out instead to contain [4Fe-4S] and [3Fe-3S] clusters (see below). Preliminary data have appeared giving unit cell dimensions for crystals of *Aphanothece sacrum* (62) and *Halobacterium* of the Dead Sea (63) ferredoxins. With a 2.8 Å data set the chain in *Spirulina platensis* ferredoxin has been traced and the attachment of the Fe center to cysteines 41, 46, 49, and 79 determined (64). This general arrangement of ligands had been deduced from the sequence of equisetum ferredoxin, which contains only 4 cysteines (65). At this level of precision the correspondence of the electron density map to the postulated core structure is gratifying but not yet a critical test of the hypothesis. In particular, there is strong EPR evidence (66) for at least a close contact to one or more nitrogenous ligands in the plant ferredoxin cluster site. Analysis of the EXAFS measurements (19) shows a congruence of the average distances in the protein and model center, but does not speak to the question of non-S ligands. Further refinement of the structure by Kakudo, Tsukihara, Matsubara, and their co-workers (64) should settle these questions.

3.4 3-Fe Clusters

The sketch shown in Figure 1*C* represents a twisted-boat structure currently thought (C. D. Stout, personal communication) to best fit the electron-density map calculated from 2 Å single and multiple isomorphous replacement data from *A. vinelandii* ferredoxin I, a protein originally isolated by Shethna (67) and Yoch (68). The model used by Stout and his co-workers for the fit is the well-characterized compound $Sn_3S_3(CH_3)_6$ (69); the protein also contains a well-defined [4Fe-4S] center. The ligands to the [3Fe-3S] are, provisionally, five cysteines and a carboxylate (Glu 18). These straightforward and interesting crystallographic results were preceded by a period of confusion to which structural, spectroscopic, and chemical approaches all contributed. The concept of a 3-Fe center was in fact arrived at from *spectroscopic* measurements somewhat in advance of the structural study, and as a cautionary tale as well as an exemplification of the cardinal characteristics of this type of center, the story is worth at least a brief summary.

Originally, Sweeney and his co-workers proposed that two [HIPIP-type] [4Fe-4S] centers using the 2+ and 3+ core charge states were present in *Azotobacter* ferredoxin I, albeit with midpoint potentials of +340 and −420 mV (70). This assignment rested on the finding of EPR in *oxidized* states of the center, since of the [4Fe-4S] and [2Fe-2S] clusters, only the former had been observed to enjoy the 3+ charge state, at least in proteins. This could be at least casually rationalized, since the 2+ state of the [2Fe-2S] center formally contains 2 Fe(III) ions. The proposal of 2 [4Fe-4S] centers was supported by a cluster extrusion experiment of Howard and his co-workers (71), who found an absorbance maximum at 458 nm, identical to that to synthetic $[Fe_4S_4(PhS)_4]^{2-}$, in the presence of denaturants and excess thiophenol. Their published data stopped at 500 nm. These authors also partially sequenced the protein and, with some imaginative arranging of the chain, found 2 sets of 4 cysteines in positions analogous to those in authentic two-center [4Fe-4S] proteins. The extrusion experiments were repeated and extended by Averill and colleagues (72), who examined the absorbance spectra quantitatively under a variety of conditions and found evidence for only 1 [4Fe-4S] center. The spectral region beyond 500 nm also appeared to be in flux for at least several hours. Averill and co-workers concluded that the balance of the Fe was either in nonstandard forms or in a difficultly accessible region of the protein. (Both conclusions later proved to be warranted.) On this background Stout and his co-workers (61) reported a preliminary crystallographic analysis, based on 4 Å data, and suggested the presence of a [4Fe-4S] and a [2Fe-2S] center. The fit to the former cluster seemed much more satisfactory than the fit to the latter (C. D. Stout, personal communication). With these last findings in mind, Emptage and his colleagues measured the Mössbauer effect of this ferredoxin (73). The concept of the experiment was simple: A sample prepared at a potential near zero volts should yield a diamagnetic spectrum from the high-potential center, superimposed on a paramagnetic (i.e., magnetically perturbed) spectrum of the low-potential (EPR active) center. Because the integral of the Mössbauer effect of Fe atoms in proteins seems to be independent of environment, at least to $\pm 5\%$, at liquid helium temperatures (11, 74), assignment of a [2Fe-2S] cluster to one center and a [4Fe-4S] center to another, would be simple, from the study of a single spectrum. At the same time they explored the cluster-transfer analysis of the metal center composition; by capturing the 2-Fe and 4-Fe centers in appropriate apo ferredoxins, refolding, and quantitating the refolded, reduced ferredoxins by EPR, they could estimate the amounts of these centers. The Mössbauer experiment (73) did indicate that the [4Fe-4S] center was the high-potential site. However, the other center gave a Mössbauer effect integrating to 3-Fe atoms per molecule, and the cluster-transfer experiments yielded 1 to 1.5 [2Fe-2S] clusters per molecule. The three Fe atoms in the low-potential site divided into two classes with opposite signs of the hyperfine coupling instants, strongly suggesting antiferromagnetic coupling of these Fe atoms. On reduction the center became EPR silent, but the Mössbauer effect showed the presence of integer (nonzero) spin. These data required Münck and his co-workers to consider the concept of a center containing 3 Fe atoms. This was done with extreme reluctance, and only because the Mössbauer experiments forced reconsideration of the evidence. The suggestions that such a center be tried in the structural studies led Stout and his colleagues, with the higher resolution data then available, to test various possibilities along these lines, with the results indicated at the beginning of this section (75). Ultimately the Fe content of the protein appears to be 7 Fe, in reality rather difficult to distinguish from 8 (W. V. Sweeney, personal communication). One conclusion from this is that the finding of EPR near $g = 2$ from oxidized states of a center with fewer than 4 Fe atoms creates a suspicion of the presence of [3Fe-3S] centers, and that the Mössbauer effect can be definitive in this regard. In fact, examination of Mössbauer spectra from *D. gigas* ferredoxin II (76), which contains 3 Fe atoms per monomer (12 per protein) shows it to be a pure case of [3Fe-3S] centers (9). Subsequent examination of spectra of glutamate synthase of *A. vinelandii* and bovine mitochondrial aconitase [previously thought on the basis of recent extrusion experiments to have been definitively established as a [2Fe-2S] case (77)] shows that they contain [3Fe-3S] centers (73), since they have the characteristic pair of Mössbauer doublets in a 2:1 ratio, along with EPR in the oxidized state. Unfolding of [3Fe-3S] proteins in the presence of *o*-xylyldithiol appears to lead to the formation of [2Fe-2S] centers. Aside from crystallographic or combined EPR and Mössbauer studies, the only other criterion for [3Fe-3S] centers known yet is the unique form of the dependence of the EPR linear electric field effect (66) on the magnetic field. This is presently being evaluated (Orme-Johnson, Rendina, Emptage, Peisach, and Mims, in preparation), and may distinguish the oxidized forms of [3Fe-3S] and [4Fe-4S] centers.

Finally, the best estimate from the Mössbauer results (73, 78) is that in the paramagnetic oxidized state all the Fe atoms in a [3Fe-3S] cluster are ferric, and thus the net core charge in this state is 3+. Also, according to sequence comparisons (79), the ferredoxin of *P. ovalis* must contain a [3Fe-3S] center.

4 APPROACHES TO THE IDENTIFICATION AND QUANTITATION OF Fe-S CENTERS

From the preceding sections one may conclude that the noncrystallographic characterization of Fe-S clusters in proteins is fraught with many difficulties. This is the correct conclusion to draw. Even the simple ferredoxins may present individualized problems of analysis. More complex substances are no easier to deal with, since environmental factors, including the presence of other prosthetic groups, can affect nearly all the distinguishing physical characteristics of the intact protein. Following is an outline of the chief approaches that seem profitable now.

4.1 Gross Composition

Precise determinations of nonheme Fe, labile S, and cysteine content require extraordinary care. The major problems with these determinations have been outlined (80). Clearly, if 2 nonheme Fe atoms and 2 labile S atoms are found, tri- and tetranuclear centers are not present, and if fewer than 5 cysteines are present, a 3-Fe center seems even less likely. Commonly, however, substoichiometric amounts of these constituents are found, the molecular weights are inaccurate, and the protein determinations off, in the early going. Hong and Rabinowitz gave an example of the correct solution to these problems (81).

4.2 Absorbance Spectrum

If no other centers are present and the non-Fe-S components can be removed or compensated for, one can often recognize the rather more featured [2Fe-2S] centers in the 2+ oxidation state with maxima near 330, 420, and 465 nm, and shoulder at 550 nm, as well as the [4Fe-4S] clusters in the 3+ oxidation state, with maxima near 325, 375, and 450 nm. The spectra of [4Fe-4S] clusters in the 2+ and 1+ states, as well as the corresponding spectra of the [3Fe-3S] systems, are relatively unfeatured. One of the earliest identifications of Fe-S centers in an enzyme was made by Rajagopalan and Handler (82), who subtracted the FAD contribution from the spectrum of xanthine oxidase and detected the [2Fe-2S] centers, which they likened to the centers in spinach ferredoxin. As to intensities, for all known centers in the 2+ oxidation state, the extinction coefficient near 400 nm is in the range of 4000–5000 cm^2/mole (Table 1).

4.3 EPR Spectrum

4.3.1 Low-Temperature EPR Spectrum. The 1+ and 3+ oxidation states of the [4Fe–4S] clusters are paramagnetic. A table of representative data is given in reference 1 as well as in Table 1. In all but one known case the observation of EPR with $g_{av} < 2$ under reducing conditions is a property of either [2Fe-2S] or [4Fe-4S] centers in the 1+ state. The exception, not understood at present, is "center 2" of xanthine oxidase, for which $g_{av} > 2$ (83), although the finding of 1 spin/2 Fe and other evidence (82, 84) make it nearly certain that only [2Fe-2S] centers are present. One immediately suspects the presence of additional ligands to center 2, of course. In all instances investigated so far, low-temperature EPR with $g_{av} > 2$ from oxidized states arises from 3+ states of [3Fe-3S] or [4Fe-4S] proteins. A subsidiary clue may be provided by integration of the EPR spectrum (85). If a defined oxidation state can be reached quantitatively [this has not been achieved in, e.g., the 2-[4Fe-4S] ferredoxin of *Chromatium* (86)], comparison of spins to iron content can be made, where under optimum conditions both numbers are precise to $\pm 5\%$. A technique that may deserve wider trial is selective leaching of the Fe atoms in the presence of partial denaturants combined with EPR studies of intermediate states. For example, Salerno and co-workers (87) found that 6 of the 8 Fe atoms of soluble beef heart succinate dehydrogenase appeared as a reduced, paramagnetic, NO–Fe–cysteine complex in the presence of cysteine, nitrite, and dithionite, during which time an EPR signal due to a reduced [2Fe-2S] center was unaltered. This showed that this signal was not due to a reduced [4Fe-4S] cluster, also thought (88) to be present in succinate dehydrogenase.

It is worth noting that subclasses of Fe-S proteins can be discerned from the characteristics of their low-temperature EPR. For example, the [2Fe-2S] proteins are easily classifiable (1) into a chloroplast type (EPR as in Table 1); hydroxylase type (89), with axial EPR at $g = 2.02$ and 1.94, observable to 100°K; and an intermediate type of unknown function, from *azotobacter, clostridium,* and mitochondria (90), with $g = 2.01$, 1.96, and 1.94, easily observable up to 200°K.

4.3.2 EPR Spectrum of Solvent-Perturbed Proteins. An important difficulty with the direct EPR method is that the protein has a profound influence on the rate and extent of reduction (or oxidation) of the center. Also, two or more paramagnets in a single protein molecule may interact magnetically, giving complex spectra (87, 91–93). At least for qualitative purposes, these complications may sometimes be circumvented by removing or lessening the influence of the protein by reversible denaturation (42). Buffer diluted to 80% DMSO in the presence of excess $Na_2S_2O_4$ has often been used for this purpose; the low-temperature spectrum of [4Fe-4S] clusters in these circumstances is nearly axial, much like that of native *B. polymyxa* ferredoxin, whereas [2Fe-2S] centers yield a narrow, nearly isotropic resonance centered on $g = 2$ (94). If no complications intervene, this measurement can often make the initial distinction between [4Fe-4S] centers, [2Fe-2S] and [3Fe-3S] systems. Three difficulties arise, however. First, because the potentials of [2Fe-2S] and [4Fe-4S] centers in aqueous systems are quite negative compared to the dithionite couple (95), the amplitude of the resulting EPR signals tends to be small and somewhat irreproducible. Second, clusters seem to be considerably less stable in the reduced state (72), and in at least one case do not show up under these conditions (87). Third, unfolding the protein without cluster destruction may require considerable tinkering with the solvent (72).

4.3.3 EPR Relaxation and Resonance Raman. On further analysis (see the discussion in ref. 96), it may become possible to systematize the analysis of EPR relaxation characteristics and some Raman lines in terms of excited magnetic states (97–101). At present, however, the subject clearly needs further study. The 1+ oxidation states seem to have *J* values* inversely proportional to their EPR *g*-value spread, while for the 3+ oxidation state a direct relationship seems to hold (103). The general guideline, that [2Fe-2S] resonances may be seen at higher temperatures (around 77°K) than [3Fe-3S] and [4Fe-4S] centers, which are best studied below 35°K, is still operationally useful but not clear-cut enough for identification purposes (1, 41, 100).

4.3.4 ^{57}Fe Hyperfine Structure in the EPR Spectrum of Fe-S Clusters. The early EPR studies on the composition of [2Fe-2S] clusters relied on the 1:2:1 characteristic splitting of a line arising from a spin coupled to *two* $I = \frac{1}{2}$ nuclei (57). In [4Fe-4S] centers broadening is often observed in ^{57}Fe-enriched samples (e.g., ref. 104) but the larger linewidths of these resonances make observation of the phenomenon difficult. However, in a tour de force of the method, ^{57}Fe-enriched yeast mitochondria were examined at various temperatures and under various oxidation conditions, and assignments were made based on the observed splittings (105). Unfortunately, the requirement for ^{57}Fe enrichment limits the usefulness of this procedure.

4.4 Displacement and Quantitation of [2Fe-2S] and [4Fe-4S] Clusters

If an Fe-S protein could be unfolded without destruction of the clusters, the ligand-exchange technique might afford a method of identification and quantitation by conversion of the clusters to standard substances with convenient spectroscopic properties. The methodology works if the protein either contains only known cluster types or if interference from other substances can be eliminated. The methods of cluster identification employed to date include the following: (1) spectrophotometric determination of the cluster thiophenolate in the presence of excess thiophenol and denaturant (e.g., 80% DMSO) requires 10 nmol or more of protein complexed cluster for which the binuclear and tetranuclear complexes give significantly different spectra (64,

*J measures the strength of the antiferromagnetic coupling of the metal atoms in a cluster. Values of J may be inferred from susceptibility (102) and Raman (98) measurements (but see Chapter 11) and from EPR lineshape analysis (96).

106–110); (2) complexation with p-$CF_3C_6H_4SH$ in denaturing solvents and subsequent determination of the ^{19}F NMR spectrum at −15°C. Under these conditions slow exchange prevails, allowing quantitation (using at least 100 nmol of cluster) of the binuclear and tetranuclear centers (84); and (3) transfer of centers displaced with *o*-xylyldithiol in solvent onto either apo adrenodoxin, specific for [2Fe-2S] centers, or apo ferredoxin from *B. polymyxa*, specific [4Fe-4S] centers (3, 110). The last method is easily applied at the 3 nmol level, with ≈90% recovery in test cases. Each of these methods has provided much sound quantitative information, but each has been documented to fail utterly in the task of distinguishing [2Fe-2S] and [3Fe-3S] centers, as indicated previously. Methods (1) and (2) probably cannot be made to yield this distinction. Method (3) might work if a suitable acceptor protein could be found. There is some hope of this, obviously. It is also apparent that a wide variety of solvents and thiols can play a role in displacing and stabilizing clusters (72), so there is still considerable room for development of this aspect of the cluster displacement methodology, particularly in the handling of [3Fe-3S] clusters.

4.5 Mössbauer Spectroscopy

Simply stated, the disadvantage of this approach is the low natural abundance (2.2%) of ^{57}Fe and the restricted ratio of spectral range to linewidth (≈10). The overwhelming *advantages* are that the technique (111) sees all the Fe regardless of the oxidation state, magnetic and nonmagnetic species can be distinguished, coupling constants and their signs can be determined so that antiferromagnetism can be detected, and the molar γ-ray resonant absorbance of all Fe atoms at low temperature seems to be about the same (74). The hallmarks of 1-Fe, [2Fe-2S], and [3Fe-3S] centers are particularly easy to spot. Since the [4Fe-4S] centers are those most easily determined by EPR methods, the combination of the two represents the present optimum initial approach in a new situation. Briefly, rubredoxin centers are recognized from their large quadrupole splitting (2+ state) and low hyperfine (saturation) field in the 3+ state; binuclear clusters in the 1+ state show two doublets of 1:1 intensity at high temperatures, and at low temperatures show coupling constants of opposite sign; trinuclear clusters behave similarly to binuclear ones, but are magnetic in the 3+ state and have a 2:1 intensity ratio for the two subspecies. The 4 Fe centers are less easily distinguished, but at present appear to possess four distinct but quite similar subsites in the 2+ states, while addition or deletion of an electron to yield the 1+ or 3+ paramagnetic states yields 2 classes of Fe atoms, that is, a pair that is unchanged and a pair that has gained or lost the spin density (12–13, 102). Enrichment of samples in

^{57}Fe is highly desirable but, with modern equipment, by no means obligatory (9) provided that micromoles of the centers are available.

4.6 Linear Electric Field Effect

This refers to the perturbation of EPR g values by applied electric fields. When no center of symmetry is present, as in [4Fe-4S] centers, uncompensated ligand shifts in an electric field lead to shifts in the value of g. When a centrosymmetric structure such as an idealized [2Fe-2S] center is measured, no shift should be found. In practice (66) [4Fe-4S] ferredoxins give large shifts, and [2Fe-2S] centers yield much smaller effects. The dependence of the effect on g, which measures the electric polarizability and symmetry of the center viewed from different directions, is characteristic of such centers. The method is sensitive (30 nmol of centers needed), but suffers from the rarity of spectrometers capable of the measurements

4.7 MCD (see also Chapter 10)

The optical rotation induced by applied magnetic fields, unlike natural circular dichroism (CD), is mostly a property of the chromophore and is not very strongly affected by extrinsic influences (i.e., protein folding). Measurements on Fe-S proteins are expected to yield spectra characteristic of the center type (112). The effects are quite small, but two findings in this developing field seem useful: [4Fe-4S] centers in the 2+ state have bands in the region of 1100–1250 nm that are absent from [2Fe-2S] systems (113), and measurements at very low temperatures promise to give largely enhanced spectra of paramagnetic (1+ and 3+) ferredoxin states, due to the enhanced population difference in the Kramers sublevels of the ground state (114). These measurements can at the moment be made in only a very few laboratories.

4.8 Some Other Possible Methods

Direct NMR and CD methods seem useless for the identification of centers because of their large extrinsic components. By the same token, once clusters are identified these methods become of great use in studying the environments of the centers (115–116). The Raman vibrations of the *exo* and *endo* Fe–S bonds (Table 1) have not yet proven useful, although they might become so (see Chapter 11). The EXAFS measurement also is more useful for refinement of known structures than finding new ones (17, see also chapter 9). Use of any of these methods in multicenter cases is hopeless as things now stand.

5. THE METAL CENTERS IN NITROGENASE

The nitrogenase system is now commented on inasmuch as it appears to contain 2 novel types of Fe-S centers that are different in important respects from those discussed previously. The analysis given here rests on chemical, EPR, and Mössbauer measurements.

Nitrogenase activity depends on the presence of 2 proteins (Table 2). The Fe protein of nitrogenase contains a single [4Fe-4S] center (80). Considerable uncertainty exists about its role, beyond that of electron transfer to the Mo-Fe protein. The Mo-Fe proteins of *A. vinelandii, C. pasteurianum,* and *K. pneumoniae* have been examined in detail, and they have 28–33 Fe and 2 Mo atoms in a molecule (Table 2), with somewhat smaller amounts of labile S than Fe. Their Mössbauer spectra are so nearly identical that the Fe clusters in the three enzymes are very probably identical in all important details (8). The cluster-transfer technique accounts for 16 of the Fe atoms as 4 [4Fe-4S] cores (117–118). This finding has been confirmed with the ^{19}F NMR method (119). Only trace amounts of EPR arising from either 3+ or 1+ states of the [4Fe-4S] centers are seen in normal forms of the enzyme during turnover (120–121), but in CO-inhibited states substantial amounts of either form can be elicited (122). The native enzyme does exhibit EPR from an $S = 3/2$ center (120), which arises (76) from the "Mo-Fe cofactor," an extractable, nonprotein complex of approximate composition $Fe_{6\text{-}8}S_4Mo$ (other ligands unknown) (123), which complements apoproteins in extracts from certain bacterial mutants to produce active MoFe proteins (124). Reversible oxidative titrations of *A. vinelandii* MoFe proteins show (125) that the MoFe cofactor (EPR active) components yield $2e^-$ and the other components yield $4e^-$. There is a transient appearance of a $g_{av} > 2$ ferredoxin-type EPR signal during the titration (Emptage, Rawlings, and Orme-Johnson, unpublished). A detailed Mössbauer analysis of the *A. vinelandii* (125–126) and *C. pasteurianum* (8) proteins showed that the 4 [4Fe-4S] centers were antiferromagnetically coupled cluster structures, and that on oxidation by $1e^-$ each they became high spin ($S > 3/2$), with such anisotropic g-values that they are in effect EPR silent. Though these same centers yield [4Fe-4S] cores on displacement, they nonetheless must be ligated differently than any previously observed system, and we have called these new centers *P* clusters (8). The Mo-Fe cofactor does not decompose in the presence of excess thiols, nor does it yield any observable clusters (76). Thus it does not seem to contain any of the substructures of the 1-, 2-, 3-, and 4-Fe centers, at least as far as the feature of tetrahedral (labile) S ligation is concerned. Since Mössbauer spectra of the $S = 3/2$ state, taken in polarizing magnetic fields, reveal that at least 6 of the Fe atoms in each center are antiferromagnetically coupled, we feel that a second novel type of Fe-S cluster has been identified (76). These we call *M* centers (8).

Table 2 Characteristics of Selected Complex Fe-S Proteins for Which a Cluster Type Has Been Proposed

Enzyme (Source)	Molecular Weight (Subunit Composition)	EPR Characterization	Number per Molecule: Fe	Acid-Labiles	Other Prosthetic Groups	Type(s) of Fe-S Center(s)[a]	Refs.
Aconitase							
(Beef heart mitochondria)	66–97,000	+	1–3	3	—	Fe_2S_2 (*a, b*) 1 Fe_3S_3 (*c*)	77,127–129 73
Benzene-dioxygenase							
(*Pseudomonas putida*)	215,000	+	4	4	—	2 Fe_2S_2 (*a*)	130–131
Benzoate 1,2-dioxygenase							
(*Pseudomonas arvilla* C-1)	275,000	+	10	9	—	Fe_2S_2 (*a*)	132
ETF dehydrogenase							
(Beef heart mitochondria)	?	+	4	4	1 FAD	1 Fe_4S_4 (*d*)	133–136
	(66,000; α_n)		per 66, 000 MW				
Ferredoxin							
(*Azotobacter vinelandii*)	—	+	7–8	7–8	—	1 Fe_3S_3 and 1 Fe_4S_4 (*a, c*)	73 68
(*Thermus thermophylis*)	9,000	+	6–7	6–7	—	1 Fe_2S_2 and 1 Fe_4S_4 (*e, f*)	138
Glutamate synthase							
(*Escherichia coli, Azotobacter vinelandii*)	800,000 (132,000; 51,000; $\alpha_4\beta_4$)	+	10	8	2 (FAD + FMN)	1 Fe_3S_3 and 1 or 2 Fe_4S_4 (*c, d, e*)	139–140

Glutamine phosphoribosylpyro-phosphate amidotransferase							
(*Bacillus subtilis*)	200,000 (50,000; α_4)	—	3–4	3	—	1 Fe_4S_4 (*a, c*)	101
Hydrogenase							
(*Chromatium*)	98,000 (50,000; α_2)	+	4	4	—	—	141
(*Clostridium pasteurianum*)	60,000	+	4	4	—	1 Fe_4S_4 (*a, d*)	106, 142
	60,500	+	12	12	—	3 Fe_4S_4 (*a*)	109,143
(*Desulfovibrio gigas*)	90,000 (62,000; 26,000; $\alpha\beta$)	+	12	12	—	3 Fe_4S_4 (*a*)	144
(*Desulfovibrio vulgaris*)	45,000	—	1	—	—	—	145
	60,000 (30,000; α_2)	—	4	3	—	—	146
	89,000 (59,000; 29,000; $\alpha\beta$)	—	9–10	7–8	—	Fe_4S_4 (*a*)	147–149
	50,000	—	12	12	—	—	150
(*Megasphaera elsdenii*)	50,000	—	10–13	11–12	—	—	151
Nitrite Reductase							
(Spinach)	61,000	+	3	2	0.5 siroheme	1 Fe_4S_4 (*d, g*)	152–155
(*Cucurbita pepo*)	61–63,000	+	2	—	siroheme	—	156–157
(*Chlorella fusca*)	63,000	—	2	+	—	—	158
Nitrogenase Fe-protein							
(*Azotobacter vinelandii*)	64,000 (33,000; α_2)	+	3	3	—	—	159–160
(*Clostridium pasteurianum*)	56,000 (27,500; α_2)	+	3–4		—	1 Fe_4S_4 (*a, b, e*)	80, 159 161

Table 2 *(Continued)*

Enzyme (Source)	Molecular Weight (Subunit Composition)	EPR Charac-terization	Number per Molecule			Type(s) of Fe-S Center(s)[a]	Refs.
			Fe	Acid-Labiles	Other Prosthetic Groups		
(*Klebsiella pneumoniae*	66,800 (34,600; α_2)	+	4	4	—	—	80
Nitrogenase Mo-Fe protein							
(*Azotobacter vinelandii*)	214–270,000 ($\alpha_2\beta_2$)	+	24–33	24	2 Mo	3 Fe_4S_4 (*a, b, e*)	117, 124, 160, 162, 163
(*Clostridium pasteurianum*)	200–220,000 (51,000; 60,000; $\alpha_2\beta_8$)	+	12–24	8–24	2 Mo	3 Fe_4S_4 (*a, b, e*)	117, 159, 163–165
(*Klebsiella pneumoniae*)	218, 000 (51,000; 60,000; $\alpha_2\beta_2$)	+	33	—	2 Mo	—	159, 166–167
(*Rhizobium japonicum*)	180–200,000 (51–58,000; 54–61,000; $\alpha_2\beta_2$)	+	29	26	1 Mo	—	80, 166, 168
Succinate dehydrogenase							
(Beef heart mitochondria)	—	+	8	8	1 flavin	2 Fe_2S_2 and 1 Fe_4S_4 (*b, e, g*)	87–88, 137, 169–171
(*Canadida utilis*)	—	+	—	—	—	2 Fe_2S_2 and 1 Fe_4S_4 (*h*)	105

Sulfite reductase							
(*Escherichia coli*)	680,000	+	20–21	14–15	4 FMN	Fe_4S_4 (*d, g*)	172–174
	60,000;				4 FAD		
	54,000; $\alpha_4\beta_4$)				3-4 siroheme		
(Spinach)	63–69,000	+	5	3	1 siroheme	Fe_4S_4 (*d, g*)	175
Trimethylamine dehydrogenase	147,000	+	4	4	1 flavin	1 Fe_4S_4 (*a*)	176–177
Xanthine oxidase		+	8		2 Mo,	4 Fe_2S_2 (*b*)	83, 84
					2 FAD		178–180

[a]The following methods were used to determine the type(s) of Fe-S center(s):

(*a*) Displaced Fe-S cluster is identified from its absorption spectrum; see text.
(*b*) Displaced Fe-S cluster is identified from its ^{19}F NMR spectrum; see text.
(*c*) Cluster is identified from its Mössbauer spectrum; see text.
(*d*) Cluster is identified from the ratio of chemically identified Fe (and labile S) to number of spins calculated from the integrated EPR spectrum; see text.
(*e*) Displaced Fe-S cluster is transferred to an acceptor apoprotein and identified from its low-temperature EPR spectrum; see text.
(*f*) Cluster is identified from the exchange coupling constant (*J* value) calculated from the temperature dependence of the EPR linewidth and integrated intensity; see text.
(*g*) Cluster is identified from the lineshape of the low-temperature EPR spectrum of the solvent perturbed protein; see text.
(*h*) Cluster is identified from ^{57}Fe hyperfine structure of the low-temperature EPR spectrum.

6 PROGRESS IN THE ANALYSIS OF COMPLEX Fe-S PROTEINS

In Table 2 we list Fe-S proteins for which cluster analyses have been attempted, with an indication of the proposed cluster compositions. Since this endeavor is relatively new and since we have discussed the apparent limitations of the available methods, we do not comment on the reliability of the analyses.

7 CONCLUDING REMARKS

With the exception of the rubredoxinlike centers, which so far seem to be restricted to microorganisms, all types of Fe-S centers occur ubiquitously. With the possible exception of hydrogenase, the direct participation of such centers in reactions other than outer-sphere electron transfer is unproven. For aconitase and glutamine-PRPP amidotransferase electron transfer may not be part of the reaction mechanism at all unless a *virtual* oxidation occurs. These and related questions may become more accessible as one moves toward an accurate picture of the cluster core structure and ligand environment in Fe-S proteins.

REFERENCES

1. W. H. Orme-Johnson, *Ann. Rev. Biochem.*, **42,** 159, (1973).
2. R. H. Holm, J. A. Ibers, in *Iron-Sulfur Proteins,* Vol. 3, W. Lovenberg, ed., Academic, New York, 1977, p. 443.
3. B. A. Averill and W. H. Orme-Johnson, in *Metal Ions in Biological Systems,* Vol. 3, H. Sigel, ed., Marcel Dekker, New York, 1978, p. 206.
4. D. C. Yoch and R. P. Carithers, *Microbiol. Rev.*, **43,** 384 (1979).
5. W. V. Sweeney and J. C. Ravinowitz, *Ann. Rev. Biochem.*, **49,** 139 (1980),
6. W. Lovenberg, ed., *Iron-Sulfur Proteins,* Vols. 1–3, Academic, New York, 1977.
7. IUB-IUPAC, "Nomenclature of Iron-Sulfur Proteins," *Eur. J. Biochem,* **93,** 427 (1979).
8. B. H. Huynh, M. T. Henzl, J. A. Christner, R. Zimmerman, W. H. Orme-Johnson, and E. Münck, *Biochim. Biophys. Acta,* **623,** 124 (1980).
9. B. H. Huynh, J. J. G. Moura, I. Moura, T. A. Kent, J. LeGall, A. B. Xavier, and E. Münck, *J. Biol. Chem.*, **225,** 3242 (1980).
10. W. R. Dunham, A. J. Bearden, I. T. Salmeen, G. Palmer, R. H. Sands, W. H. Orme-Johnson, and H. Beinert, *Biochim. BioPhys. Acta,* **253,** 134 (1971).
11. C. Schulz and P. G. Debrunner, *J. Phys.*, **37(C6),** 153 (1976).
12. P. Middleton, D. P. E. Dickson, C. E. Johnson, and J. D. Rush, *Eur. J. Biochem.*, **88,** 135 (1978).

13. P. Middleton, D. P. E. Dickson, C. E. Johnson, and J. D. Rush, *Eur. J. Biochem.*, **104,** 289, (1980).
14. T. V. Long, II, T. M. Loehr, J. R. Allkins, and W. Lovenberg, *J. Am. Chem. Soc.*, **93,** 1810 (1971).
15. S-P. W. Tang, T. G. Spiro, K. Mukai, and T. Kimura, *Biochem. Biophys. Res. Comm.*, **53,** 869 (1973).
16. S-P. W. Tang, T. G. Spiro, C. Antanaitis, T. H. Moss, R. H. Holm, T. Herskovitz, and L. E. Mortenson, *Biochem. Biophys. Res. Commmun.*, **62,** 1 (1975).
17. B. Bunker and E. A. Stern, *Biophys. J.*, **19,** 253 (1977).
18. R. G. Shulman, P. Eisenberger, B-K. Teo, B. M. Kincaid, and G. S. Brown, *J. Mol. Biol.*, **124,** 305 (1978).
19. B-K. Teo, R. G. Shulman, G. S. Brown, and A. E. Meixner, *J. Am. Chem. Soc.*, **101,** 5624 (1979).
20. H. Backmeyer, L. H. Piette, K. Yasunobu, and H. R. Whiteley, *Proc. Natl. Acad. Sci. USA,* **57,** 122 (1967).
21. A. Davison and E. C. Switkes, *Inorg. Chem.*, **4,** 837 (1971).
22. K. Watenpaugh, L. C. Sieker, J. R. Herriott, and L. H. Jensen, *Acta Crystallogr., Sect. B.*, **29,** 243 (1973).
23. J. Peisach, W. E. Blumberg, E. T. Lode, and M. J. Coon, *J. Biol. Chem.*, **246,** 5877, 1971).
24. W. E. Blumberg and J. Peisach, *Ann. NY Acad. Sci.*, **222,** 539 (1973).
25. D. E. Sayers, E. A. Stern, and J. A. Herriott, *J. Chem. Phys.*, **64,** 427 (1976).
26. R. G. Shulman, P. Eisenberger, W. E. Blumberg, and N. A. Stombaugh, *Proc. Natl. Acad. Sci. USA,* **72,** 4003 (1975).
27. K. Watenpaugh, L. C. Sieker, and L. H. Jensen, *J. Mol. Biol.*, **131,** 509 (1979).
28. R. B. Frankel, G. C. Papaefthymiou, R. W. Lane, and R. H. Holm, *J. Phys.*, **37(C6),** 165 (1976).
29. A. Kostikias, V. Petrouleas, A. Simpoulos, D. Coucouvannis, and D. G. Holah, *Chem. Phys. Lett.*, **38,** 582 (1976).
30. D. Rasse, P. K. Warme, and H. Scheraga, *Proc. Natl. Acad. Sci. USA,* **71,** 3736 (1974).
31. G. H. Loew, M. Chadwick, and D. A. Steinberg, *Theor. Chim. Acta (Berlin),* **33,** 125 (1974).
32. J. G. Norman, Jr. and S. C. Jackels, *J. Am. Chem. Soc.*, **97,** 3833 (1975).
33. R. A. Bair, and W. A. Goddard, III, *J. Am. Chem. Soc.*, **99,** 3505 (1977).
34. L. Eisenstein and D. R. Franceschetti, *Chem. Phys. Lett.*, **50,** 167 (1977).
35. R. A. Bair and W. A. Goddard, III, *J. Am. Chem. Soc.*, **100,** 5669 (1978).
36. E. T. Adman, L. C. Sieker, L. H. Jensen, M. Bruschi, and J. LeGall, *J. Mol. Biol.*, **112,** 113 (1977).
37. W. A. Eaton and W. Lovenberg, in *Iron-Sulfur Proteins,* Vol. 2, W. Lovenberg, ed., Academic, New York, 1973, p. 131.
38. L. Que, Jr., J. D. Lipscomb, R. Zimmerman, E. Münck, N. R. Orme-Johnson, and W. H. Orme-Johnson, *Biochim. Biophys. Acta,* **452,** 320 (1976).
39. L. Que, Jr., R. H. Heistand, II, R. Mayer, and A. L. Roe, *Biochemistry,* **19,** 2588 (1980).
40. I. Moura, B. H. Huynh, R. P. Hausinger, J. LeGall, A. V. Xavier, and E. Münck, *J. Biol. Chem.*, **255,** 2493 (1980).

41. W. H. Orme-Johnson and R. H. Sands, in *Iron-Sulfur proteins*, Vol. 2, W. Lovenberg, ed., Academic, New York, 1973, p. 194
42. R. Cammack, *Biochem. Soc. Trans.*, **3,** 482 (1975).
43. E. T. Adman, L. C. Sieker, and L. J. Jensen, *J. Biol. Chem.*, **251,** 3801 (1976).
44. C. W. Carter, Jr., J. Kraut, T. Freer, and R. A. Alden, *J. Biol. Chem.*, **249,** 6339 (1974).
45. C. W. Carter, Jr., *J. Biol. Chem.*, **252,** 7802 (1977).
46. E. J. Laskowski, R. B. Frankel, W. O. Gillum, G. C. Papaefthymiou, J. Renaud, J. A. Ibers, and R. H. Holm, *J. Am. Chem. Soc.*, **100,** 5322 (1978).
47. J. B. Berg, K. O. Hodgson, and R. H. Holm, *J. Am. Chem. Soc.*, **101,** 4586 (1979).
48. C. W. Carter, Jr., in *Iron-Sulfur Proteins*, Vol. 3, W. Lovenberg, ed., New York, 1977, p. 157.
49. J. Donohue, *J. Mol. Biol.*, **45,** 231 (1969).
50. R. W. Sweeney, A. J. Bearden, and J. C. Rabinowitz, *Biochem. Biophys. Res. Comm.*, **59,** 188 (1974).
51. L. Que, Jr., M. A. Bobrik, J. A. Ibers, and R. H. Holm, *J. Am. Chem. Soc.*, **96,** 4168 (1974).
52. G. B. Wong, M. A. Bobrik, and R. H. Holm, *Inorg. Chem.*, **17,** 578 (1978).
53. R. W. Johnson and R. H. Holm, *J. Am. Chem. Soc.*, **100,** 5338 (1978).
54. H. Brintzinger, G. Palmer, and R. H. Sands, *Proc. Natl. Acad. Sci. USA*, **55,** 397 (1966).
55. J. F. Gibson, D. O. Hall, J. H. M. Thornley, and F. R. Whatley, *Proc. Natl. Acad. Sci. USA*, **56,** 987 (1966).
56. W. H. Orme-Johnson and H. Beinert, *J. Biol. Chem.*, **244,** 6143 (1969).
57. W. H. Orme-Johnson, R. E. Hansen, H. Beinert, J. C. M. Tsibris, R. C. Bartholomaus, and I. C. Gunsalus, *Proc. Natl. Acad. Sci. USA*, **60,** 368 (1968).
58. J. C. M. Tsibris, R. L. Tsai, I. C. Gunsalus, W. H. Orme-Johnson, R. E. Hansen, and H. Beinert, *Proc. Natl. Acad. Sci. USA*, **59,** 959 (1968).
59. J. G. Norman, Jr., B. P. Ryan, and L. Noodleman, *J. Am. Chem. Soc.*, **102,** 4279 (1980).
60. J. J. Mayerle, S. E. Denmark, B. V. DePamphilis, J. A. Ibers, and R. H. Holm, *J. Am. Chem. Soc.*, **97,** 1032 (1975).
61. C. D. Stout, *Nature*, **279,** 83 (1979).
62. A. Kunita, M. Koshibe, Y. Nishikawa, K. Fukuyama, T. Tsukihara, Y. Katsube, Y. Matsuura, N. Tanaka, M. Kakudo, T. Hase, and H. Matsubara, *J. Biochem.*, **84,** 989 (1978).
63. J. L. Sussman, P. Zipori, M. Harel, A. Yonath, and M. M. Werber, *J. Mol. Biol.*, **134,** 375 (1979).
64. T. Tsukihara, K. Fukuyama, H. Tahara, Y. Katsube, Y. Matsuura, N. Tanaka, M. Kakudo, K. Wada, and H. Matsubara, *J. Biochem.*, **84,** 1645 (1978).
65. S. T. Aggarwall, K. K. Rao, and H. Matsubara, *J. Biochem. (Tokyo)*, **69,** 601 (1971).
66. J. Peisach, N. R. Orme-Johnson, W. B. Mims, and W. H. Orme-Johnson, *J. Biol. Chem.*, **252,** 5643 (1977).
67. Y. I. Shetna, *Biochim. Biophys. Acta*, **205,** 58 (1970).
68. D. C. Yoch and E. I. Arnon, *J. Biol. Chem.*, **247,** 4514 (1972).
69. B. Menzebach and P. Bleckmann, *J. Organomet. Chem.*, **91,** 291 (1975).
70. W. V. Sweeney, J. C. Rabinowitz, and D. C. Yoch, *J. Biol. Chem.*, **250,** 7842 (1975).

71. J. B. Howard, T. Lorsbach, and L. Que, *Biochem. Biophys. Res. Commun.*, **70,** 582 (1976).

72. B. A. Averill, J. R. Bale, and W. H. Orme-Johnson, *J. Am. Chem. Soc.*, **100,** 3034 (1978).

73. M. H. Emptage, T. A. Kent, B. H. Huynh, J. Rawlings, W. H. Orme-Johnson, and E. Münck, *J. Biol. Chem.*, **255,** 1793, (1980).

74. A. Dwivedi, T. Pederson, and P. G. Debrunner, *J. Phys.*, **40,** 531 (1979).

75. C. D. Stout, D. Ghosh, V. Pattabhi, and A. Robbins, *J. Biol. Chem.*, **255,** 1797 (1980).

76. J. Rawlings, V. K. Shah, J. R. Chishell, W. J. Brill, R. Zimmermann, E. Münck, and W. H. Orme-Johnson, *J. Biol. Chem.*, **253,** 1001 (1978).

77. D. M. Kurtz, R. H. Holm, F. J. Ruzicka, H. Beinert, C. J. Coles, and T. P. Singer, *J. Biol. Chem.*, **254,** 4967 (1979).

78. M. Bruschi, E. C. Hatchikian, J. LeGall, J. J. G. Moura, and A. V. Xavier, *Biochim. Biophys. Acta,* **449,** 275 (1976).

79. T. Hase, S. Wakabayashi, H. Matsubara, D. Ohmori, and K. Suzuki, *FEBS Lett.*, **91,** 315 (1978).

80. W. H. Orme-Johnson and L. C. Davis, in *Iron-Sulfur Proteins,* Vol. 3, W. Lovenberg, ed., Academic, New York, 1977, p. 16.

81. J. S. Hong and J. C. Rabinowitz, *J. Biol. Chem.*, **245,** 4982 (1970).

82. K. V. Rajagopalan and P. Handler, *J. Biol. Chem.*, **239,** 1509 (1964).

83. D. J. Lowe, R. M. Lynden-Bell, and R. C. Bray, *Biochem. J.*, **130,** 239 (1972).

84. G. B. Wong, D. M. Kurtz, Jr., R. H. Holm, L. E. Mortenson, and R. C. Upchurch, *J. Am. Chem. Soc.*, **101,** 3078 (1979).

85. R. Aasa and T. Vänngård, *J. Magn. Reson.*, **19,** 308 (1975).

86. N. A. Stombaugh, J. E. Sundquist, R. H. Burris, and W. H. Orme-Johnson, *Biochemistry,* **15,** 2633 (1976).

87. J. C. Salerno, T. Ohnishi, J. Lim, and T. E. King, *Biochem. Biophys. Res. Commun.*, **73,** 833 (1976).

88. C. J. Coles, R. H. Holm, D. M. Kurtz, Jr., W. H. Orme-Johnson, J. Rawlings, T. P. Singer, and G. B. Wong, *Proc. Natl. Acad. Sci. USA,* **76,** 3805 (1979).

89. P. S. Yoon, J. Rawlings, W. H. Orme-Johnson, and H. F. DeLuca, *Biochemistry,* **19,** 2172 (1980).

90. D. Bäckstöm, M. Lorusso, K. Anderson, and A. Ehrenberg, *Biochim. Biophys. Acta,* **502,** 276 (1978).

91. R. Mathews, S. Charlton, R. H. Sands, and G. Palmer, *J. Biol. Chem.*, **249,** 4326 (1974).

92. F. J. Ruzicka, H. Beinert, K. L. Schepler, W. R. Dunham, and R. H. Sands. *Proc. Natl. Acad. Sci. USA,* **72,** 2886 (1975).

93. D. J. Lowe and R. C. Bray, *Biochem. J.*, **169,** 471 (1978).

94. R. Cammack, *Biochem. Biophys. Res. Commun.*, **54,** 548 (1973).

95. C. L. Hill, J. Renaud, R. H. Holm, and L. E. Mortenson, *J. Am. Chem. Soc.*, **99,** 2549 (1977).

96. J. P. Gayda, P. Bertrand, A. Deville, C. More, G. Roger, J. F. Gibson, and R. Cammack, *Biochim. Biophys. Acta,* **581,** 15 (1979).

97. F. Adar, H. Blum, J. S. Leigh, T. Ohnishi, J. Salerno, and T. Kimura, *FEBS Lett.*, **84,** 214 (1977).

98. H. Blum, F. Adar, J. C. Salerno, and J. S. Leigh, *Biochem. Biophys. Res. Commun.*, **77,** 650 (1977).

99. J. C. Salerno, T. Ohnishi, H. Blum, and J. S. Leigh, *Biochim. Biophys. Acta,* **494,** 191 (1977).

100. H. Rupp, K. K. Rao, D. O. Hall, and R. Cammack, *Biochim. Biophys. Acta,* **537,** 255 (1978).

101. B. A. Averill, A. Dwivedi, P. Debrunner, S. J., Vollmer, J. Y. Wong, and R. L. Switzer, *J. Biol. Chem.,* **255,** 6007 (1980).

102. B. C. Antanaitis and T. H. Moss, *Biochim. Biophys. Acta,* **405,** 262 (1975).

103. H. Blum, J. C. Salerno, P. R. Rich, and T. Ohnishi, *Biochim. Biophys. Acta,* **548,** 139 (1979).

104. N. A. Stombaugh, R. H. Burris, and W. H. Orme-Johnson, *J. Biol. Chem.,* **248,** 7951 (1973).

105. S. P. J. Albracht and J. Subramanian, *Biochim. Biophys. Acta,* **462,** 36 (1977).

106. D. L. Erbes, R. H. Burris, and W. H. Orme-Johnson, *Proc. Natl. Acad. Sci. USA,* **72,** 4795 (1975).

107. J. R. Bale, Ph.D. Dissertation, University of Wisconsin, Madison, Wisconsin (1974).

108. L. Que, Jr., R. H. Holm, and L. E. Mortenson, *J. Am. Chem. Soc.,* **97,** 463 (1975).

109. W. O. Gillum, L. E. Mortenson, J-S Chen, and R. H. Holm, *J. Am. Chem. Soc.,* **99,** 584 (1977).

110. W. H. Orme-Johnson and R. H. Holm, in *Methods in Enzymology,* Vol. 53, S. Fleisher and L. Packer, eds., Academic, New York, 1978, p. 268.

111. E. Münck, in *Methods in Enzymology,* Vol. 54, S. Fleisher and L. Packer, eds., Academic, New York, 1978, p. 346.

112. J. C. Sutherland, I. Salmeen, A. S. K. Sun, and M. P. Klein, *Biochim. Biophys. Acta,* **263,** 550 (1972).

113. P. J. Stephens, A. J. Thomson, T. A. Keiderling, J. Rawlings, K. K. Rao, and D. O. Hall, *Proc. Natl. Acad. Sci. USA,* **75,** 5273 (1978).

114. A. J. Thomson, R. Cammack, D. O. Hall, K. K. Rao, R. Briat, J. C. Rivoal, and J. Badoz, *Biochim. Biophys. Acta,* **493,** 132 (1977).

115. W. D. Phillips and M. D. Poe, in *Iron-Sulfur Proteins,* Vol. 2, W. Lovenberg, ed., Academic, New York, 1973, p. 255.

116. E. L. Packer, J. C. Rabinowitz, and H. Sternlicht, *J. Biol. Chem.,* **253,** 7722 (1978).

117. W. H. Orme-Johnson and E. Münck, in *Molybdenum and Molybdenum-Containing Enzymes,* M. Coughlan, ed., Pergamon, New York, 1980, p. 427.

118. W. H. Orme-Johnson, N. R. Orme-Johnson, C. Touton, M. Emptage, M. Henzl, J. Rawlings, K. Jacobson, J. P. Smith, W. B. Mims, B. H. Huynh, E. Münck, and G. S. Jacob, in *Molybdenum Chemistry of Biological Significance* (Proceedings of the International Symposium on Molybdenum Chemistry of Biological Significance, April 10–13, 1979, Lake Biwa, Japan), W. E. Newton and S. Otsuka, eds., Plenum, New York, 1980, p. 85.

119. D. M. Kurtz, Jr., R. S. McMillan, B. K. Burgess, L. E. Mortenson, and R. H. Holm, *Proc. Natl. Acad. Sci. USA,* **76,** 4986 (1979).

120. W. H. Orme-Johnson, W. D. Hamilton, T. Ljones, M-Y. W. Tso, R. H. Burris, V. K. Shah, and W. J. Brill, *Proc. Natl. Acad. Sci. USA,* **69,** 3142 (1972).

121. R. N. F. Thorneley, J. Chatt, R. R. Eady, D. J. Lowe, M. J. O'Donnell, J. R. Postgate, R.

L. Richards, and B. E. Smith, in *Nitrogen Fixation,* University Park, W. E. Newton and W. H. Orme-Johnson, eds., Baltimore, 1978, p. 171.

122. L. C. Davis, M. T. Henzl, R. H. Burris, and W. H. Orme-Johnson, *Biochemistry,* **18,** 4860 (1979).

123. B. E. Smith, in *Molybdenum Chemistry of Biological Significance* (Proceedings of International Symposium on Molybdenum Chemistry of Biological Significance, April 10–13, Lake Biwa, Japan, W. E. Newton and S. Otsuka, eds., Plenum, New York, 1980, p. 179.

124. V. K. Shah and W. J. Brill, *Proc. Natl. Acad. Sci. USA,* **74,** 3243 (1977).

125. R. Zimmermann, E. Münck, W. J. Brill, V. K. Shah, M. T. Henzl, J. Rawlings, and W. H. Orme-Johnson, *Biochim. Biophys. Acta,* **537,** 185 (1978).

126. B. H. Huynh, E. Münck, and W. H. Orme-Johnson, *Biochim. Biophys. Acta,* **527,** 192 (1979).

127. J. J. Villafranca and A. S. Mildvan, *J. Biol. Chem.,* **246,** 772 (1971).

128. O. Gawron, M. C. Kennedy, and R. A. Rauner, *Biochem. J.,* **143,** 717 (1974).

129. F. J. Ruzicka and H. Beinert, *J. Biol. Chem.,* **253,** 2514 (1978).

130. B. A. Axcell and P. J. Geary, *Biochem. J.,* **146,** 173 (1975).

131. S. E. Crutcher and P. J. Geary, *Biochem. J.,* **177,** 393 (1979).

132. M. Yamaguchi and H. Fujisawa, *J. Biol. Chem.,* **255,** 5058 (1980).

133. R. W. Miller and V. Massey, *J. Biol. Chem.,* **240,** 1453 (1965).

134. V. Aleman, P. Handler, G. Palmer, and H. Beinert, *J. Biol. Chem.,* **243,** 2560 (1968).

135. F. J. Ruzicka and H. Beinert, *Biochem. Biophys. Res. Commun.,* **66,** 622 (1975).

136. F. J. Ruzicka and H. Beinert, *J. Biol. Chem.,* **252,** 8440 (1977).

137. C. J. Lusty, J. M. Machinist, and T. P. Singer, *J. Biol. Chem.,* **240,** 1804 (1965).

138, T. Ohnishi, H. Blum, S. Sato, K. Nakazawa, K. Hon-nami, and T. Oshima, *J. Biol. Chem.,* **255,** 345 (1980).

139. R. E. Miller and E. R. Stadtman, *J. Biol. Chem.,* **247,** 7407 (1972).

140. A. Rendina, Ph.D. Thesis, University of Wisconsin, Madison, Wisconsin (1980).

141. P. H. Gitlitz and A. I. Krasna, *Biochemistry,* **14,** 2561 (1975).

142. G. Nakos and L. Mortenson, *Biochim. Biophys. Acta,* **227,** 576 (1971).

143. J. S. Chen and L. E. Mortenson, *Biochim. Biophys. Acta,* **371,** 283 (1974).

144. E. L. Hatchikian, M. Bruschi, and J. LeGall, *Biochem. Biophys. Res. Commun.,* **82,** 451 (1978).

145. R. H. Haschke and L. L. Campbell, *J. Bacteriol.,* **165,** 249 (1971).

146. J. LeGall, D. V. Dervartanian, E. Spilker, J-P. Lee, and H. D. Peck, Jr., *Biochim. Biophys. Acta,* **234,** 525 (1971).

147. T. Yagi, *J. Biochem.,* **68,** 649 (1970).

148. T. Yagi, K. Kimura, H. Dadoji, F. Sakai, S. Tamura, and H. Inokuchi, *J. Biochem.,* **79,** 61 (1976).

149. I. Okura, K. I. Nakamura, and T. Keii, *J. Mol. Catal.,* **4,** 453 (1978).

150. H. M. Van der Westen, S. G. Hayhew, and C. Veeger, *FEBS Lett.,* **86,** 122 (1978).

151. C. Van Dijk, S. J. Mayhew, H. J. Grande, and C. Veeger, *Eur. J. Biochem.,* **102,** 317 (1979).

152. M. J. Murphy, L. M. Siegel, S. R. Tove, and H. Kamin, *Proc. Natl. Acad. Sci. USA,* **71,** 612 (1974).

153. P. J. Aparicio, D. B. Knaff, and R. Malkin, *Arch. Biochem. Biophys.*, **169**, 102 (1975).

154. J. M. Vega and H. Kamin, *J. Biol. Chem.*, **252**, 896 (1977).

155. J. R. Lancaster, J. M. Vega, H. Kamin, N. R. Orme-Johnson, W. H. Orme-Johnson, R. J. Krueger, and L. M. Siegel, *J. Biol. Chem.*, **254**, 1268 (1979).

156. D. P. Hucklesby, D. M. James, M. J. Barwell, and E. J. Hewitt, *Phytochemistry*, **15**, 599 (1976).

157. R. Cammack, D. P. Hucklesby, and E. J. Hewitt, *Biochem. J.*, **171**, 519 (1978).

158. W. G. Zumft, *Biochim. Biophys. Acta*, **276**, 363 (1972).

159. R. R. Eady, B. E. Smith, K. A. Cook, and J. R. Postgate, *Biochem. J.*, **128**, 655 (1972).

160. D. Kleiner and C. H. Clen, *Arch. Microbiol.*, **98**, 93 (1974).

161. M-Y. W. Tso, T. Ljones, and R. H. Burris, *Biochim. Biophys. Acta*, **267**, 600 (1972).

162. R. C. Burns, R. D. Holsten, and R. W. F. Hardy, *Biochem. Biophys. Res. Commun.*, **39**, 90 (1970).

163. W. H. Orme-Johnson, E. Münck, R. Zimmerman, W. J. Brill, V. K. Shah, J. Rawlings, M. T. Henzl, B. A. Averill, and N. R. Orme-Johnson, in *Mechanisms of Oxidizing Enzymes* (Proceedings of International Congress on Mechanisms of Oxidizing Enzymes, LaPaz, Mexico, 1977), T. P. Singer and R. N. Ondarza, eds., Elsevier, New York, 1978, p. 165.

164. T. C. Huang, W. G. Zumft, and L. E. Mortenson, *J. Bacteriol.*, **113**, 884 (1973).

165. M-Y. W. Tso, *Arch. Microbiol.*, **99**, 71 (1974).

166. C. Kennedy, R. R. Eady, E. Kondorosi, and D. K. Rekosh, *Biochem. J.*, **155**, 383 (1976).

167. B. E. Smith, R. N. F. Thorneley, M. G. Yates, R. R. Eady, and J. R. Postgate, *Proc. Int. Symp. Nitrogen Fixation, 1st.*, **1**, 150 (1976).

168. D. W. Israel, R. L. Howard, H. J. Evans, and A. S. Russell, *J. Biol. Chem.*, **249**, 500 (1974).

169. T. E. King, *Biochem. Biophys. Res. Commun.*, **16**, 511 (1964).

170. K. A. Davis and Y. Hatefi, *Biochemistry*, **10**, 2509 (1971).

171. T. Ohnishi, J. C. Salerno, D. B. Winter, J. Lim, C. A. Yu, L. Yu, and T. E. King, *J. Biol. Chem.*, **251**, 2094 (1976).

172. L. M. Siegel, M. J. Murphy, and H. Kamin, *J. Biol. Chem.*, **248**, 251 (1973).

173. L. M. Siegel and P. S. Davis, *J. Biol. Chem.*, **249**, 1587 (1974).

174. L. M. Siegel, in *Mechanisms of Oxidizing Enzymes,* (Proceedings of International Symposium of Mechanisms of Oxidizing Enzymes, LaPaz, Mexico, 1977). T. P. Singer and R. N. Ondarza, eds., Elsevier, New York, 1978, p. 201.

175. R. J. Krueger, Ph.D. Dissertation, Duke University, Durham, North Carolina (1979).

176. C. L. Hill, D. J. Steenkamp, R. H. Holm, and T. P. Singer, *Proc. Natl., Acad. Sci. USA*, **74**, 547 (1977).

177. D. J. Steenkamp, T. P. Singer, and H. Beinert, *Biochem. J.*, **169**, 361 (1978).

178. W. H. Orme-Johnson and H. Beinert, *Biochem. Biophys. Res. Commun.*, **36**, 337 (1969).

179. V. Massey, in *Iron-Sulfur Proteins,* Vol. 1, W. Lovenberg, ed., Academic, New York, 1973, p. 301.

180. R. C. Bray, in *Enzymes,* Vol. 12, 3rd ed., P. Boyer, ed., Academic, New York, 1975, p. 299.

CHAPTER 3
Iron-Sulfur Protein Crystallography

CHARLES DAVID STOUT

Department of Crystallography
University of Pittsburgh
Pittsburgh, Pennsylvania

CONTENTS

1 INTRODUCTION

This chapter reviews the present knowledge on the structures of Fe-S proteins and their chromophores as derived by X-ray crystallographic studies. Although the list of solved Fe-S protein structures is short, these structures have served as benchmarks in understanding the structure and function of the Fe-S prosthetic groups in a large number of proteins. These structures have also given important insights into the role of the polypeptide in influencing the properties of Fe-S clusters.

Although considerable advances have been made in protein crystallography, a protein structure determination remains a major undertaking. Fe-S proteins, by the nature of the chromophore, tend to be relatively more labile than most other types of proteins successfully studied by X-ray diffraction methods. As a result the Fe-S protein structures solved to date are all low-molecular-weight electron-transport proteins. This, and the limitation that crystallography provides the only readily available method for visualizing molecular structure, has lent considerable emphasis to the assumption of generality in the results obtained. In view of the diverse biochemical properties of the Fe-S proteins and their ubiquity in biological systems, however, it seems clear that more structural information is needed.

Both the structure and properties of Fe-S active sites are influenced by the polypeptide environment. It is well established that reduction potentials of [4Fe-4S] clusters may range over 0.7 V. More remarkably, some proteins, such as *D. gigas* ferredoxin, may change the very covalent structure of the Fe-S center. Certainly, the conformation of a center can change, as in the ox-

Table 1 Crystallography of Fe-S Proteins

Protein	Source	Comment	Refs.
Rubredoxin	*C. pasteurianum*	structure at 1.2 Å	3–9
	D. vulgaris	structure at 2.0 Å	10, 11
	D. gigas	diffraction pattern	10
Ferredoxin (chloroplast)	*S. platensis*	structure at 2.5 Å	13–15
	A. sacrum	structure at 5 Å	16
	Halobacterium	diffraction pattern	17
Aconitase	pig heart	diffraction pattern	37, 38
High-potential Fe protein	*C. vinosum*	structure at 2.0 Å	20–27
Ferredoxin (Fe-S protein III)	*A. vinelandii*	structure at 2.5 Å	28–31
Ferredoxin (clostridial)	*P. aerogenes*	structure at 2.0 Å	23,32–36
MoFe protein (component I) of nitrogenase	*C. pasteurianum*, *A. vinelandii*	diffraction pattern	2

idation states of *Chromatium* HIPIP. Chemical data for aconitase and *Azotobacter* ferredoxin reveal that [3Fe-3S] clusters are particularly labile, emphasizing that for these centers the cluster structure is strongly dependent on protein conformation.

Iron-sulfur proteins whose crystal structures have been solved or are under study are tabulated in Table 1. Only "simple" Fe-S proteins have been studied in detail by crystallography (1), with the important exception of the MoFe protein of nitrogenase, for which suitable crystals and diffraction patterns have been obtained (2). No crystal structure studies of an Fe-S/flavin, Fe-S/Mo/flavin, or Fe-S/siroheme protein have been reported.

Review articles dealing with the structures of rubredoxin, clostridial ferredoxin and HIPIP are already in the literature (39–43).

2 EXPERIMENTAL ASPECTS OF Fe-S PROTEIN CRYSTALLOGRAPHY

Two of the most time-consuming steps in a protein crystallographic study are the growth of good quality crystals and the preparation of heavy atom derivatives. As an incentive to those pursuing new structures of Fe-S proteins, results of successful studies are tabulated in this section. An additional aspect to the study of these molecules is the use of anomalous scattering data, which can be used to locate the Fe (and S) atoms and derive phase information. Principles of protein crystallographic methods are treated by Jensen (39, 41) and Blundell and Johnson (44).

2.1 Crystallization

Because of the lability of their chromophores to oxygen, Fe-S proteins can be difficult samples to crystallize. Once suitable conditions are found, however, crystals of Fe-S proteins usually diffract well due to the ordering effect of the metal ions on the structure. Crystallization conditions and crystal data are summarized in Table 2 for 21 crystal forms of Fe-S proteins. Of these, 13 were found suitable for X-ray diffraction studies. Except for aconitase, nitrogenase and xanthine oxidase, all the proteins listed in Table 2 have low molecular weights, are acidic, and have been crystallized at high ionic strength. Except in the case of the *Halobacterium* 2-Fe ferredoxin, the salt used has invariably been $(NH_4)_2SO_4$. Except for rubredoxin (no acid-labile S) the pH values used range only from 7.0 to 8.5, and except for nitrogenase and reduced HIPIP no effort has been made to exclude molecular O_2. Straightforward batch, dialysis, or vapor diffusion techniques have been used to induce nucleation. It seems that, given this rather limited range of conditions used

Table 2 Crystal Data for Fe-S Proteins

Protein	Crystal Data and Crystallization[a]
Rubredoxin *C. pasteurianum* MW 6100 daltons Reference 3	Rhombohedral crystal form, space group *R3*, a = 38.77 Å, α = 112.37°, (Z = 1), V_m = 2.15 Å^3/dalton. Rhombs 0.2 × 0.2 × 0.2 mm. A protein solution was adjusted to pH 4.0 and made 0.75 saturated in $(NH_4)_2SO_4$ at 23°C. Crystals grew within a week.
Rubredoxin *D. vulgaris* MW 6000 daltons References 10, 11	Monoclinic crystal form, space group $P2_1$, a = 19.993, b = 41.505, c = 24.404 Å, β = 107.6°, Z = 1, V_m = 1.68 Å^3/dalton. Crystal size 0.5 × 0.5 × 0.3 mm. Protein (10 mg) was dissolved in 1 ml of 0.1 *M* sodium citrate or acetate buffer, pH 4.0, and solid $(NH_4)_2SO_4$ was added to 100 μl aliquots without stirring to about 25% *w/v* concentration. Crystals grew in sealed containers at 27°C in three days.
Rubredoxin *D. gigas* MW 6000 daltons Reference 10	Monoclinic crystal form, space group $P2_1$, a = 19.75, b = 41.70, c = 24.38 Å, β = 108.1°, Z = 1, V_m = 1.6 Å^3/dalton. Crystal size 0.5 × 0.5 × 0.3 mm. Crystals grown as for *D. vulgaris* protein.
Ferredoxin (chloroplast) *S. platensis* MW 10,890 daltons Reference 13	Orthorhombic crystal form, space group $C222_1$, a = 62.32, b = 28.51, c = 108.09 Å, Z = 1, V_m = 2.19Å^3/dalton. Crystals stabilized in 3.5 *M* $(NH_4)_2SO_4$, pH 7.4.
Ferredoxin (chloroplast) *A. sacrum* MW 10,480 daltons Reference 16	Tetragonal crystal form, space group $P4_3$, a = 92.2, c = 47.6 Å, Z = 4, V_m = 2.42 Å^3/dalton. Rhombic dodecahedra 0.8 × 0.8 × 0.4 mm. A 1-3% protein solution was dialyzed for 15 days at 4°C in 75% saturated $(NH_4)_2SO_4$ solution containing 0.7 *M* NaCl and 0.1 *M* tris·HCl, pH 7.5.
Ferredoxin (2-Fe) *Halobacterium* MW 14,000 daltons Reference 17	Hexagonal crystal form, space group $P6_322$, a = 60.6, c = 127.8 Å, Z = 1, V_m = 2.42 Å^3/dalton. Hexagonal plates 0.5 × 0.5 × 0.5 mm. Samples (1 ml) of 12.5-25 mg/ml protein solution in 4 *M* NaCl, pH 7.0, were dialyzed against 4 *M* phosphate buffer, pH 7.0, filtered, and allowed to stand for one week.
Aconitase (pI 8.5 and 8.1 isozymes) MW 65,000 daltons References 37, 38	Orthorhombic crystal form, space group $P2_12_12$, a = 174.1, b = 72.0, c = 72.8 Å, Z = 1, V_m = 3.51 Å^3/dalton. Barrel-shaped prisms 0.3 × 0.2 × 0.2 mm. Partially twinned crystals were grown from amorphous precipitate obtained with

Table 2 (Continued)

Protein	Crystal Data and Crystallization[a]
	2.55-2.59 M $(NH_4)_2SO_4$, 15 mM tricarballylate (pH 7.8) solution at an enzyme concentration of 15 mg/ml.
HIPIP *C. vinosum* MW 9300 daltons References 20, 24, 25	Orthorhombic crystal form, space group $P2_12_12_1$, a = 42.70, b = 41.86, c = 38.08 Å, Z = 1, V_m = 1.83 Å^3/dalton. Rectangular prisms 0.4 × 0.4 × 0.8 mm. *Reduced crystals:* A 10 mg/ml solution of reduced protein, as isolated, was adjusted to pH 7.9 with 0.1 N NaOH in 0.67% saturated $(NH_4)_2SO_4$, and 0.75 ml aliquots were sealed in 5 ml containers. Crystals grew within 48 hr at room temperature.Crystals were transferred to a solution of 0.80% saturated $(NH_4)_2SO_4$, pH 7.9, saturated in dithiothreitol to prevent oxidation. *Oxidized crystals:* Reduced crystals were oxidized by soaking in $K_3Fe(CN)_6$ at pH 7.9, or by air. Isomorphous with reduced crystals.
Ferredoxin (Fe-S protein III) *A. vinelandii* MW 13,000 daltons References 28, 29	Tetragonal crystal form, space group $P4_32_12$, a = 55.22, c = 95.20 Å, Z = 1, V_m = 2.5 Å^3/dalton. Square bipyramids 0.4 × 0.4 × 0.8 mm. A 5-7 mg/ml protein solution containing 0.5 M tris · HCl, pH 8.5, and 1.2 M $(NH_4)_2SO_4$ was equilibrated at 22°C against 3.5 M $(NH_4)_2SO_4$ by vapor diffusion for one day and transferred to 2°C. Crystals grew within one week.
References 28, 46	Triclinic crystal form, space group $P1$, a = 46.8, b = 58.7, c = 64.3 Å, α = 105.1, β = 82.5, γ = 116.5°, Z = 4-5, V_m = 2.20-2.75 Å^3/dalton. Rhombic plates 0.5 × 0.5 × 0.05 mm. A 7-10 mg/ml protein solution containing 0.5 M potassium phosphate buffer, pH 7.4, and 1.2 M $(NH_4)_2SO_4$ was equilibrated against 3.5 M $(NH_4)_2SO_4$ by vapor diffusion at 2°C. Crystals grew within one month.
Reference 47	Orthorhombic crystal form, space group $P2_12_12_1$, a = 58, b = 56.5, c = 72.5 Å, Z = 2, V_m = 2.05 Å^3/dalton. Platelike crystals.
Iron-sulfur protein *T. thermophilis* MW 9200 daltons Reference 70	Needlelike crystals were formed in 70% saturated $(NH_4)_2SO_4$ at 4°C.
Ferredoxin *Chromatium vinosum*	Rhombic plates and hexagonal prisms were grown from 0.5 NaCl, 0.1 M Tris · HCl, pH 7.8, dialyzed

Table 2 *(Continued)*

Protein	Crystal Data and Crystallization[a]
MW 10,000 daltons References 71, 72	against 74% saturated $(NH_4)_2SO_4$, and buffered at pH 7.0 at 2°C.
Ferredoxin *P. aerogenes* MW 6000 daltons Reference 34	Orthorhombic crystal form, space group $P2_12_12_1$, $a = 30.52$, $b = 37.75$, $c = 39.37$ Å, $Z = 1$, $V_m = 1.9$ Å^3/dalton. Lathe-shaped needles 0.1 × 0.15 × 0.5 mm. Solid $(NH_4)_2SO_4$ (3.3–3.5 *M*) was added to a concentrated protein solution containing 0.7 *M* Tris·HCl, pH 7.5, to induce turbidity. Crystals grew within 1-2 days at 4°C.
Ferredoxin *Clostridial* species MW 6000 daltons References 48, 49	Crystals were grown with proteins from five species: *C. acidi-urici*—rectangular prisms and needles; *C. tetanomorphum*—rosettes; *C. pasteurianum, C. butyricum, C. cylindrosporum*—microcrystals. Solid $(NH_4)_2SO_4$ was added to a protein solution in water with stirring. Microcrystals were dissolved in water and dialyzed against 65–80% saturated $(NH_4)_2SO_4$.
Nitrogenase component I (Mo-Fe protein) *C. pasteurianum* MW 220,000 daltons Reference 2	Monoclinic crystal form, space group $P2_1$, $a = 69.2$, $b = 150.4$, $c = 123.1$ Å, $\beta = 110°$, $Z = 1$, $V_m = 2.9$ Å^3/dalton. Rectangular plates 0.5 × 0.5 × 0.2 mm.
Nitrogenase component I (Mo-Fe protein) *A. vinelandii* MW 240,000 daltons Reference 2	Monoclinic crystal form, space group $P2_1$, $a = 208.2$, $b = 72.9$, $c = 153.9$ Å, $\beta = 105°$, $Z = 2$, $V_m = 2.4$ Å^3/dalton. Rectangular plates 0.5 × 0.5 × 0.2 mm.
Reference 73	Microcrystals 0.05 mm in length. An anaerobic solution of protein in 0.025 *M* Tris·HCl, pH 7.4, and 0.25 *M* NaCl was diluted sixfold and incubated for 1 hr at 37°C to form masses of needles.
Reference 74	Prisms 0.5 × 0.15 × 0.1 mm. Microcrystals grown at low ionic strength (73) were dissolved anaerobically in 0.025 *M* Tris·HCl, pH 7.4, 0.075 *M* NaCl, and dialyzed against 0.02 *M* NaCl with concentration.
Xanthine oxidase 360,000 daltons Reference 52	Microcrystals. A 1% protein solution in 1 *M* phosphate buffer at 0°C was adjusted to a final concentration of 8% *v/v* ethanol, centrifuged, and allowed to stand at −1°C for 36 hr.

[a] Z = number of molecules per asymmetric unit; V_m = Matthews coefficient (45).

to date, crystallization of simple Fe-S proteins depends primarily on the protein used. Nitrogenase and xanthine oxidase, on the other hand, require quite different crystallization conditions. Conditions used to crystallize some 193 macromolecules have been compiled by McPherson (53). The use of polyethylene glycols (54) and the application of statistical methods (55) represent new approaches that may be useful in crystallizing complex Fe-S proteins.

2.2 Heavy Atom Derivatization

Useful heavy atom derivatives of Fe-S proteins (Table 3) have been prepared in the standard way, by soaking. The lability of S to "soft" metal ions presents an additional problem in the preparation of derivatives, however, since commonly used reagents such as Hg or Au compounds may lead to denaturation. On the other hand, one could imagine that the appropriate reagent could bind isomorphously and nondestructively to an Fe-S cluster. Heavy atoms successfully employed in studies of Fe-S proteins are listed in Table 3.

It is apparent that uranyl complexes (UO_2^{2+} or $UO_2F_5^{3-}$) make good derivatives of Fe-S proteins. Similarly, the hard metal cations Sm^{3+} and Pr^{3+} bind well to *P. aerogenes* ferredoxin (pI $\simeq$ 4), and these ions share a number of common sites with the UO_2^{2+} derivative. For HIPIP, heavy atoms were found to bind more effectively at pH 6.5 (crystals grown at pH 7.9), and for *Azotobacter* ferredoxin it was found that mildly acidic conditions markedly enhanced the binding of the Os and Rh compounds used. In the case of *P. aerogenes* ferredoxin binding occurred only at elevated concentrations of the heavy atoms (40–100 *mM*).

As for many protein crystals, Pt complexes make good derivatives, and both HIPIP and *Azotobacter* ferredoxin have a Pt binding site adjacent to methionine. Attention has been brought to the effect of the high concentration of $(NH_4)_2SO_4$ in these reactions, as Pt-ammine complexes are much less reactive (56). However, the formation of ammine complexes is an important factor in the preparation of derivatives of *Azotobacter* ferredoxin (31). In this case $[OsO_2(NH_3)_4]^{2+}$ and $[Rh(NH_3)_5Cl]^{2+}$ complexes are formed from the starting reagents, and the cations have distinctly different binding sites compared to $PtCl_4^{2-}$. In particular, the ammine complexes are able to interact at the [3Fe-3S] cluster, perhaps through NH···S (S_γ) H bonds. The presence of 12 sites in the Rh derivative suggests that $[RhCl_6]^{3-}$ in $(NH_4)_2SO_4$ may be a generally useful heavy atom reagent.

An alternative approach to the preparation of heavy atom derivatives, not yet fully explored, would take advantage of known Fe-S cluster chemistry (64, 65, 87), cluster-transfer methods (66, 67), and Se substitutions (57, and references therein). A replacement might be possible with $[(CH_3COOHg)_4C]$, a compound that has Hg atoms in positions isomorphous to Fe in [4Fe-4S] clusters (59, 60).

2.3 Anomalous Scattering

The anomalous scattering effect has a number of applications in protein crystallography, and this is particularly true for Fe-S proteins. In analogy to the Mössbauer effect, the properly chosen X-ray wavelength acts as an independent probe of the Fe centers in the protein. Iron anomalous scattering has been used to locate individual Fe atoms and [Fe-S] clusters, to establish the enantiomorph of heavy atom constellations, and to serve as an aid in MIR phase calculations.

The anomalous difference, or Bijvoet difference, is observed as a breakdown of Friedel's law for each reflection. Patterson or Fourier maps based on these differences provide an entirely independent image of the Fe structure. Anomalous scattering effects arise when the incident X-ray radiation is close to an absorption edge of the atom in question. For protein studies the standard wavelength is 1.54 Å (CuKα radiation); the K absorption edge for Fe is 1.74 Å. Under these conditions the scattering factor for Fe is described as a complex quantity:

$$f = f_0 + f' + if''$$

	Fe^{3+}	S
f_0	$23e^-$	$16e^-$
f'	$-1.1e^-$	$+0.3e^-$
f''	$+3.4e^-$	$+0.6e^-$

where f' and f'' are the real and imaginary anomalous scattering terms. It can be seen that for Fe the imaginary component is quite large compared to the combined real component. Thus the Bijvoet differences may average to a 2–3% difference in overall intensities of Friedel-related reflections in a typical data collection experiment. Although these differences are small, it is the *signs* as well as the magnitudes of the differences which carry the information. Anomalous scattering from S has to date not been observed in the Fe-S protein structures studied. However, the structure of a 5000 dalton plant protein, crambin, has recently been solved using the anomalous scattering from S in the native structure (61).

The first application of anomalous scattering in an Fe-S protein structure was in the study of HIPIP (20). Carefully replicated observations of large Bijvoet differences were used to generate a Bijvoet difference Patterson map

Table 3 Useful Heavy Atom Derivatives of Fe-S Proteins[a]

Structure	Refs.	Derivative Formed and Conditions	
Rubredoxin (*C. pasteurianum*)	3	K_2HgI_4	1 site
		$UO_2(NO_3)_2$	1 site
			0.85 saturated $(NH_4)_2SO_4$
			pH 4.0
Ferredoxin (*S. platensis*)	13	$K_3UO_2F_5$	1 site
			3.5 *M* $(NH_4)_2SO_4$
			pH 7.4
Ferredoxin (*A. sacrum*)	16	$K_3UO_2F_5$	2 sites
			0.85 saturated $(NH_4)_2SO_4$
			pH 7.5
HIPIP (*C. vinosum*)	21, 24	$K_3UO_2F_5$	2 sites
		K_2PtBr_6	3 sites
		$K_2Pt(NO_2)_4$	2 sites
			0.80 saturated $(NH_4)_2SO_4$
			pH 6.5

Ferredoxin (*P. aerogenes*)	34	$UO_2(NO_3)_2$	7 sites
		$Sm(NO_3)_3$	10 sites
		$PrCl_3$	7 sites 3.3–3.5 *M* $(NH_4)_2SO_4$ pH 7.5
Ferredoxin (*A. vinelandii*)	31	K_2PtCl_4	4 sites 3.5 *M* $(NH_4)_2SO_4$ pH 7.6
		K_2PtCl_4	7 sites 4.8 *M* K_2HPO_4/KH_2PO_4 pH 7.4
		$K_2[OsO_2(OH)_4] \cdot H_2O$	8 sites 4.5 *M* $(NH_4)_2SO_4$ pH 5.2
		$Na_3RhCl_6 \cdot 18H_2O$	12 sites 4.5 *M* $(NH_4)_2SO_4$ pH 6.4

[a]Temperatures ranged from 2–23°C, and soaking times varied from 3 days to 3 weeks. The concentrations of metal ions employed were 1–100 *mM*.

that revealed the vectors for a single, compact Fe site. An important observation made at the time was that the Bijvoet differences fell off rapidly in intensity beyond 5 Å resolution. From this it was inferred that the Fe atoms were in a roughly spherical or tetrahedral cluster (diameter 3–4 Å), as the Fourier transform of such an object (62) falls off in intensity in just such a manner. Subsequently, coefficients were derived for generating an anomalous difference Fourier (Bijvoet difference Fourier) (21). At 4.0 Å resolution, this map, based on MIR phases, showed a single tight cluster of Fe atoms, confirming the Patterson map interpretation. The map also established the proper enantiomorph of the heavy atom positions, for if the MIR phases are inverted, the Fe site occurs as a negative peak at inversion related coordinates. At 2.0 Å resolution the Bijvoet difference Fourier clearly revealed individual peaks for the 4 Fe atoms of the [4Fe-4S] cluster (22).

Fe anomalous scattering data have been used in both the 2-Fe ferredoxin structures that have been studied to date (Table 1). In the *S. platensis* protein a Bijvoet difference Fourier map at 5 Å revealed the location of the [2Fe-2S] cluster (13). At 2.8 Å the same map showed an elongated peak to which 2 Fe sites, 2.8 Å apart, were assigned (14). In the *A. sacrum* structure a Bijvoet difference Fourier map at 5 Å resolution contained a peak, significantly above the background level, for each of the four protein molecules in the asymmetric unit (16). This map also assigned the space group as $P4_3$, as no peaks were present when $P4_1$ was assumed.

A Bijvoet difference Fourier map was also used to establish the space group and heavy atom enantiomorph in *Azotobacter* ferredoxin (29). At 4.0 Å resolution the map contained two prominent peaks in the asymmetric unit (Fig. 1*a*) corresponding to the 2 Fe-S sites in the protein. In addition, the shape of these peaks is in accord with the structure of each cluster (31). At 2.5 Å resolution the individual Fe atoms in the [3Fe-3S] cluster are resolved in the anomalous scattering map (Fig. 1*b*). This is not true for the [4Fe-4S] cluster, however, since the Fe–Fe separation here is only 2.7 Å. Apparently, 2.0 Å data are required, as in the HIPIP case, to resolve the Fe atoms of a [4Fe-4S] cluster.

In a hallmark study Phillips and co-workers have explored the use of synchrotron radiation in a study of rubredoxin (75). Synchrotron radiation offers two principal advantages for protein crystallography: the X-ray beam is an order of magnitude more intense than that obtainable from conventional sealed tubes or rotating anodes, and it is tunable. Because of the ability to tune the wavelength, data may be collected very near the absorption edge of a heavy atom in the native protein or in a derivative, maximizing the anomalous scattering effect. Consequently, very large anomalous differences can be produced and recorded in a relatively short time. Moreover, these anomalous scattering data can be used to solve the phase problem.

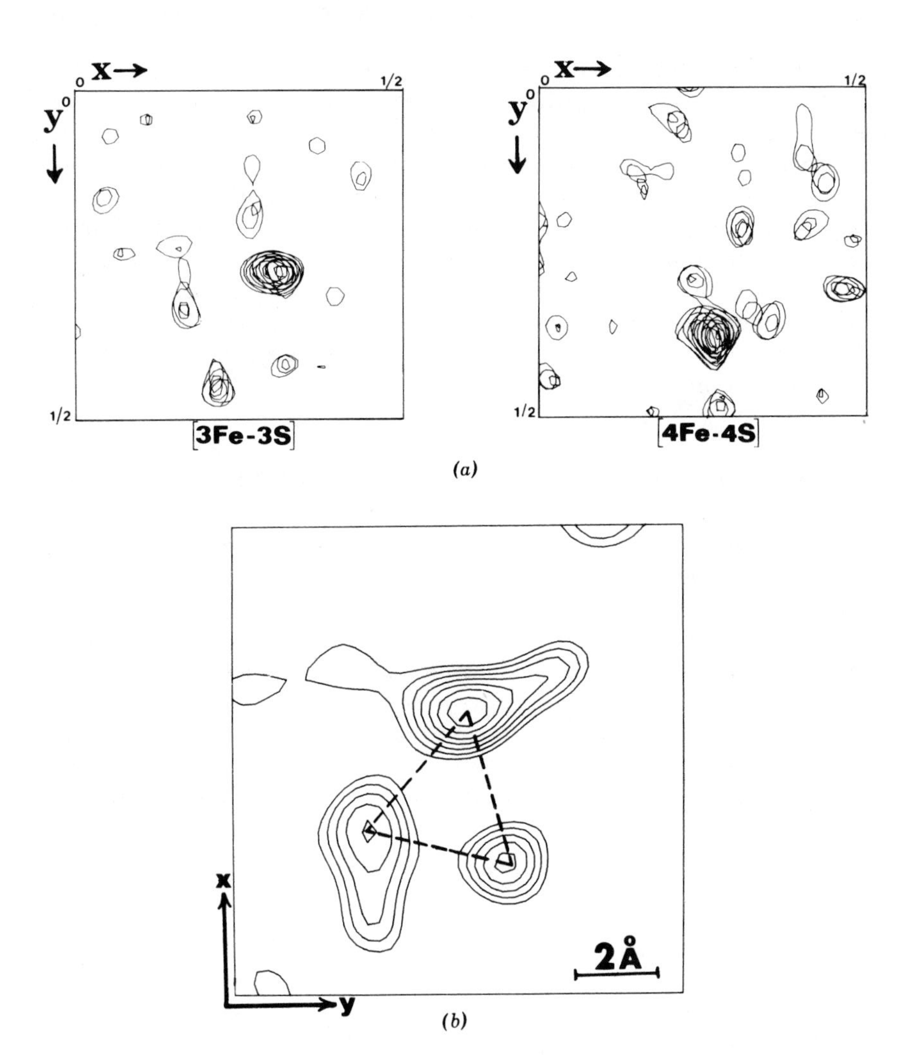

Figure 1. *(a)* Bijvoet difference Fourier map of *A. v.* ferredoxin at 4.0 Å resolution calculated with MIR phases and 969 acentric reflections. Composite sections on $z = 0.27$–0.29 for the 3-Fe center, and $z = 0.37$–0.39 for the 4-Fe center (31). *(b)* Bijvoet difference Fourier map for the 3-Fe center at 2.5 Å computed with MIR phases and 2267 acentric reflections. Section at $z = 0.28$.

In the rubredoxin study seven wavelengths were used from 1.730 Å to 1.758 Å (Fe absorption edge 1.743 Å) to record film data of an acentric projection. Changes were observed in both the real (f') and imaginary (f'') components of the Fe scattering. Difference Patterson maps were generated based on the differences in f'' only, f' only, and a combination of both. All accurately revealed the single Fe position in the protein. Changes in f' represent a change in the real scattering, analogous to an isomorphous difference. These data were combined with f'' data to derive phase angles, that agreed satisfactorily with those calculated from the refined structure (9). The results demonstrate that synchrotron X-ray radiation data is a powerful tool in the study of Fe-S proteins.

3 STRUCTURES OF Fe-S CLUSTERS IN PROTEINS

As summarized in Table 1, the structures of five classes of Fe-S proteins have been solved: rubredoxin (1 Fe), chloroplast ferredoxin (2 Fe), HIPIP (4 Fe), clostridial ferredoxin (8 Fe), and *Azotobacter* ferredoxin (7 Fe). These protein structures provide a clear image of all 4 of the presently known types of Fe-S clusters in simple Fe-S proteins, and altogether 8 independent cluster structures in 6 protein structures have been solved (Fig. 2). The cluster composition of each of the protein structures is given in Table 4. Cysteine coordination is depicted in Figure 3.

A comment on accuracy: The reader should bear in mind that protein coordinate data are derived from large crystal structures (500–1000 atoms) at

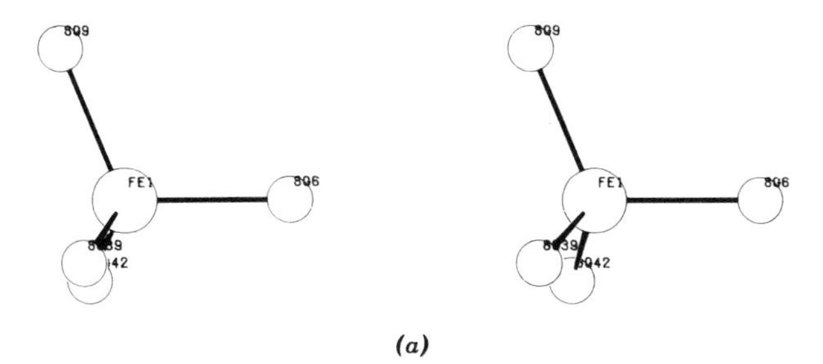

(a)

Figure 2. Structures of Fe-S clusters from protein crystallography. Coordinates in *(a)*, *(d)*, *(e)*, *(f)* from data deposited with the Protein Data Bank, Brookhaven National Laboratory, Upton, New York. Coordinates in *(b)*, *(c)*, *(g)* from the partially refined protein structures at 2.5 Å resolution. *(a)* $Fe(S\gamma)_4$ complex in *C. p.* rubredoxin (9); *(b)* $[2Fe\text{-}2S](S\gamma)_4$ cluster in *S. p.* ferredoxin (15); *(c)* $[3Fe\text{-}3S](S\gamma)_5('O_\epsilon')$ cluster in *A. v.* ferredoxin (31); *(d)* $[4Fe\text{-}4S](S\gamma)_4$ cluster in *C. v.* HIPIP (26); *(e)* $[4Fe\text{-}4S](S\gamma)_4$ cluster I in *P. a.* ferredoxin (36); *(f)* $[4Fe\text{-}4S](S\gamma)_4$ cluster II in *P. a.* ferredoxin (36); *(g)* $[4Fe\text{-}4S](S\gamma)_4$ cluster in *A. v.* ferredoxin (31).

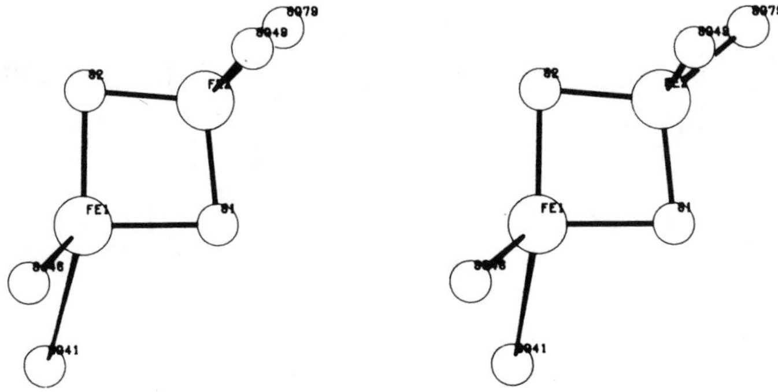

(b)

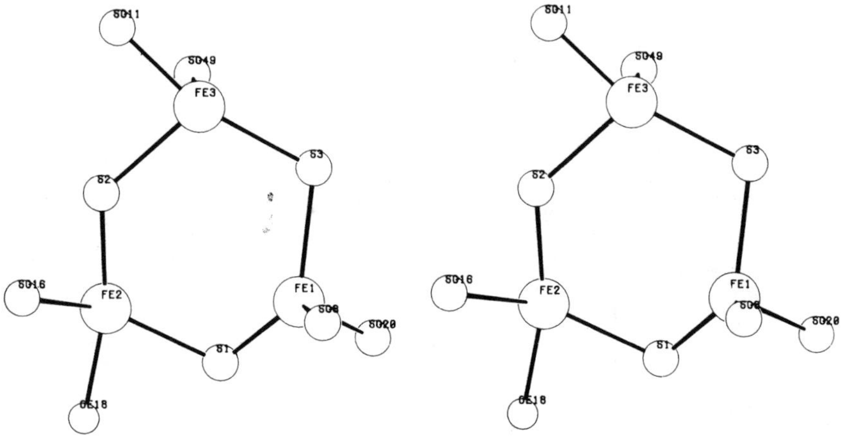

(c)

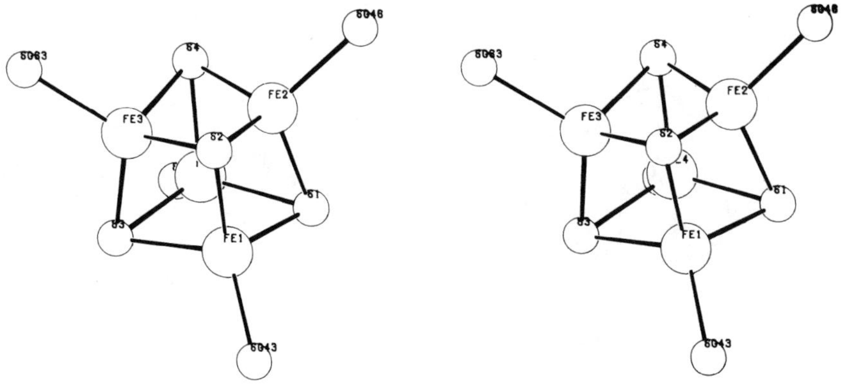

(d)

Figure 2. (Continued)

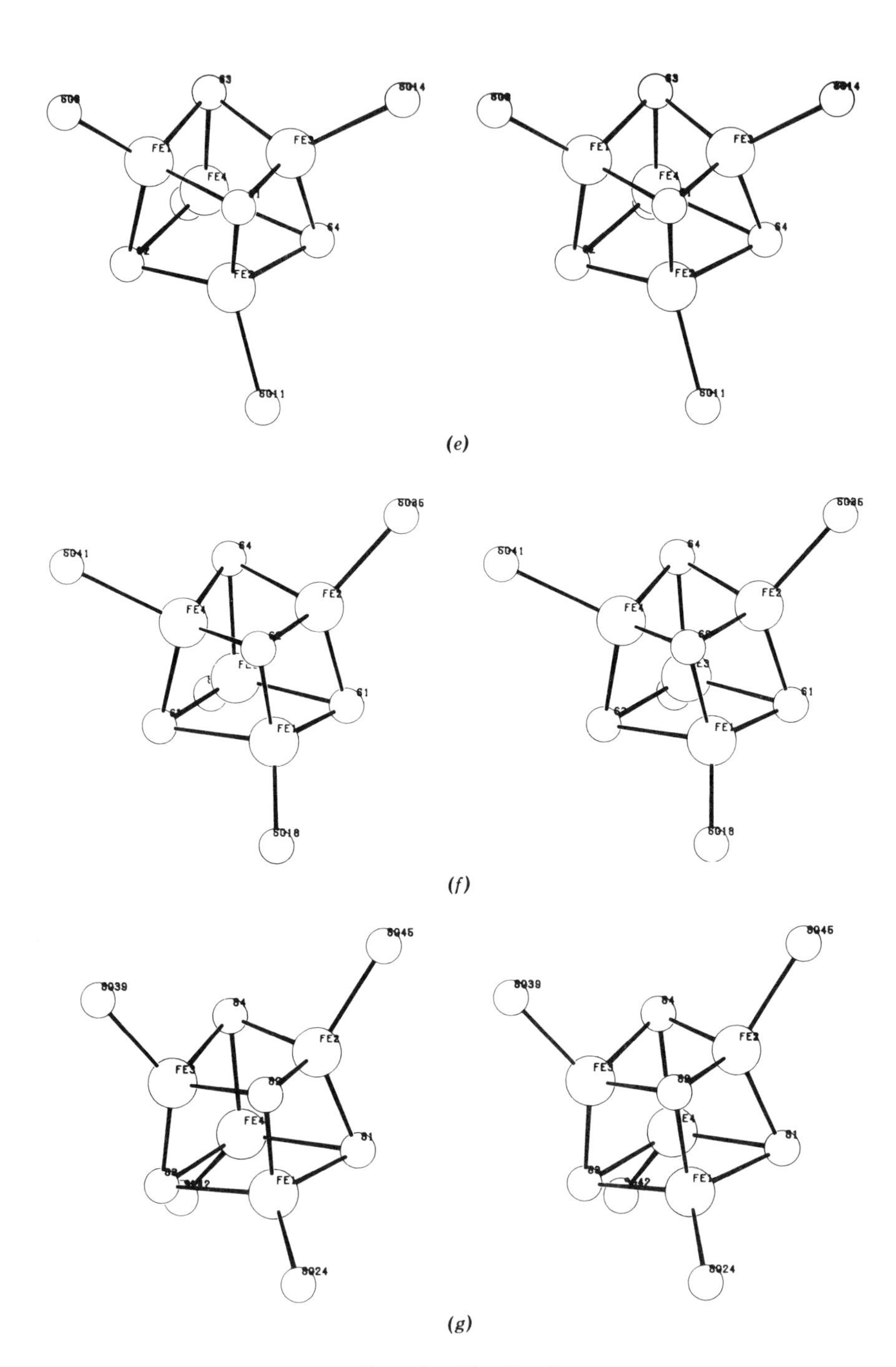

Figure 2. *(Continued)*

Table 4 Cluster Composition of Fe-S Protein Structures

Protein (Species)	Fe-S Cluster(s)	Total Fe
Rubredoxin (*C. p.*, *D. v.*)	$Fe(S_\gamma)_4$	1
Chloroplast ferredoxin (*S. p.*)	$[2Fe\text{-}2S](S_\gamma)_4$	2
HIPIP (*C. v.*)	$[4Fe\text{-}4S](S_\gamma)_4$	4
Ferredoxin (*P. a.*)	$[4Fe\text{-}4S](S_\gamma)_4$ I	8
	$[4Fe\text{-}4S](S_\gamma)_4$ II	
Ferredoxin (*A. v.*)	$[3Fe\text{-}3S](S_\gamma)_5$('$O_\epsilon$')	7
	$[4Fe\text{-}4S](S_\gamma)_4$	

less than atomic resolution (2.0 Å). Geometrical constraints for the protein components are generally assumed during refinement. In addition, protein data may suffer from systematic errors arising from absorption and crystal deterioration. However, Fe and S atoms have high electron densities, and their typical separation (2.3 Å) is resolved with 2.0 Å data. Moreover, anomalous scattering data enhances the contrast between Fe and S atoms (Fig. 1*b*). Given these factors, Fe–S bond lengths are accurate to RMS $\Delta d \approx 0.1$ Å and bond angles to 5°.

HIPIP (26) and *P. a.* ferredoxin (36) have been refined to discrepancy indices of $R = 24.7$ and 20.6%, respectively. At present the structures of *S. p.* and *A. v.* ferredoxins are not fully refined. *C. p.* (9) and *D. v.* (11) rubredoxins have been refined to $R = 12.8$ and 31.0%. The structure of the *C. p.* protein is exceptional, as the refinement has been carried out at 1.2 Å resolution, and standard deviations for the Fe–S bond lengths and angles are 0.01 and 0.5°, respectively.

3.1 $Fe(S_\gamma)_4$ Centers

The structures of the rubredoxins from *C. pasteurianum* (8, 9) and *D. vulgaris* (11) have been solved to 1.2 and 2.0 Å resolution, respectively. In both cases the individual atoms of the $Fe(S_\gamma)_4$ complex have been resolved (Fig. 2*a*). Bond distances and angles about the Fe atom in the *C. p.* protein are listed in Table 5. In this molecule there is a significant distortion from ideal tetrahedral symmetry. The bond distances vary from 2.24 to 2.33 Å, or over 6 times the standard deviation; the bond angles range from 104° to 114°, ±5° from the tetrahedral value. The distortion is apparently related to an approximate twofold axis relating the bidentate ligand peptides Cys(6)-*x*-*y*-Cys(9)-Gly and Cys(39)-*x*-*y*-Cys(42)-Gly (Fig. 3). Two of the largest angles about Fe involve adjacent S atoms in the same peptide segment, $S_\gamma(6)$ and $S_\gamma(9)$, and $S_\gamma(39)$ and $S_\gamma(42)$, while the two smallest angles involve adja-

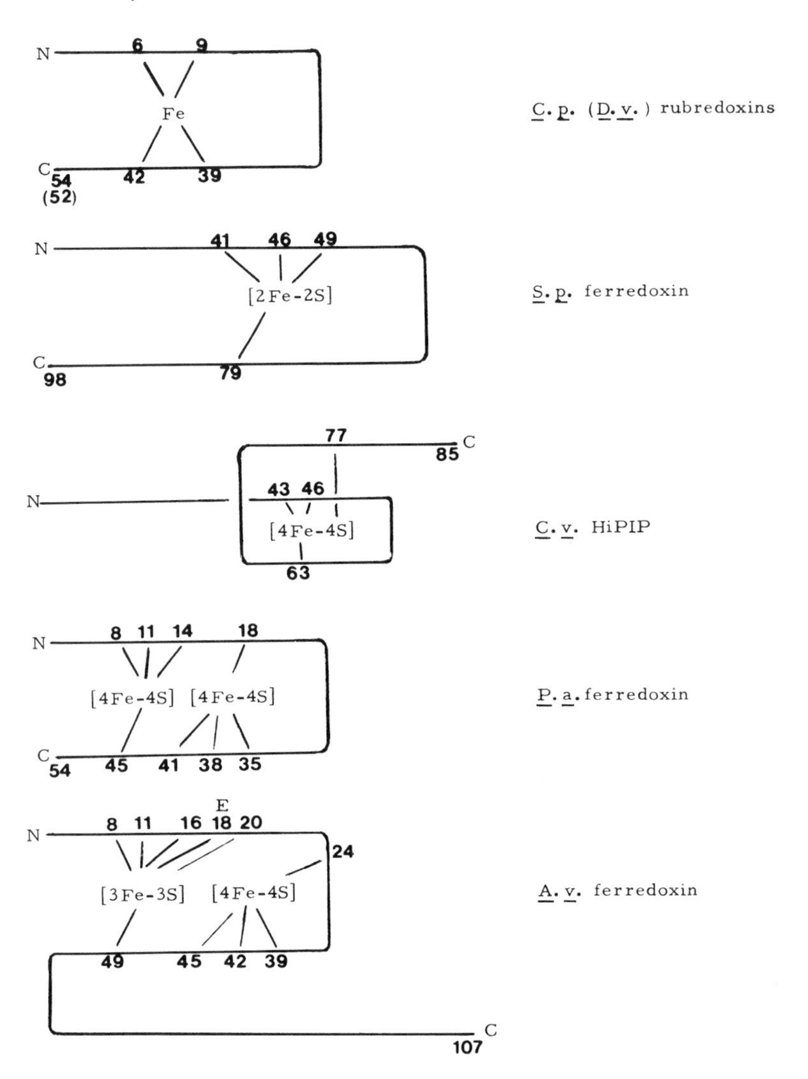

Figure 3. Residue numbers of cysteines involved in Fe-S chelation in six protein structures. The data were taken from the references given in Figure 2 and Table 1. *E*-Glutamic acid. The complete sequence analysis of *A. v.* ferredoxin (111), and the refined structure show it to have 106, not 107, amino-acids.

Table 5 Bond Lengths and Angles in $Fe(S_\gamma)_4$ Complex in *C. p.* Rubredoxin[a]

Atoms	Distance (Å)	Atoms	Angle (deg)
$Fe\text{-}S_\gamma 6$	2.333(11)	$S_\gamma 6\text{-}Fe\text{-}S_\gamma 9$	113.8(4)
$Fe\text{-}S_\gamma 9$	2.288(15)	$S_\gamma 6\text{-}Fe\text{-}S_\gamma 39$	109.0(4)
$Fe\text{-}S_\gamma 39$	2.300(15)	$S_\gamma 6\text{-}Fe\text{-}S_\gamma 42$	103.8(4)
$Fe\text{-}S_\gamma 42$	2.235(12)	$S_\gamma 9\text{-}Fe\text{-}S_\gamma 39$	103.7(4)
		$S_\gamma 9\text{-}Fe\text{-}S_\gamma 42$	114.3(5)
		$S_\gamma 39\text{-}Fe\text{-}S_\gamma 42$	112.4(5)

[a]The data were taken from references 8 and 9. Standard deviations are given in parentheses. The atom labels refer to Figure 2*a*.

cent S atoms in nonlinked peptide segments, $S_\gamma(6)$ and $S_\gamma(42)$, $S_\gamma(9)$ and $S_\gamma(39)$. Apart from this alteration of Fe-S bond angles, which must be caused by the protein matrix, the $Fe(S_\gamma)_4$ core is very similar in geometry to the synthetic analog complex, $[Fe(S_2\text{-o-xyl})_2]^{1-}$ (68, 69).

The *D. v.* rubredoxin $Fe(S_\gamma)_4$ complex also displays apparent deviation from tetrahedral geometry. In this case the $Fe\text{-}S_\gamma$ bond distances range from 2.15 to 2.35 Å (mean 2.29 Å, standard deviation 0.1 Å), and bond angles range from 102° to 124°.

3.2 [2Fe-2S] Cluster

The structure of just one protein with a [2Fe-2S] cluster, *S. platensis* ferredoxin (15), has been completed. The primary sequence of this molecule, however, is homologous with 25 other chloroplast-type ferredoxins. The Fe-S cluster (Fig. 2*b*) in its environment of 4 cysteine ligands (Fig. 3) exhibits tetrahedral coordination about each Fe atom, with the [2Fe-2S] core forming a planar 4 atom ring (Table 6). The structure of the cluster in the protein is in good agreement with the structures of the analog compounds $[Fe_2S_2(S_2\text{-}o\text{-xyl})_2]^{2-}$ and $[Fe_2S_2(S\text{-}p\text{-tolyl})_4]^{2-}$ (76, 77), and the structure deduced from spectroscopic experiments (78, and references therein).

Distances and angles for the [2Fe-2S] cluster are given in Table 7 for the partially refined structure at 2.5 Å resolution. Although the Fe-S geometry is likely to change with further refinement, one feature of the core geometry at the present stage may be significant: the shape of the 4 atom [2Fe-2S] ring. In the analog structures the Fe–Fe distance across the ring is ≈2.7 Å, indicating metal-metal interaction, while the S–S distance within the core is ≈3.5 Å, close to the van der Waals contact for these atoms. Consequently, the internal Fe–S–Fe angle is reduced to ≈75° and the core is compressed to

Table 6 Least-Squares Planes for Fe-S Clusters[a]

Cluster	Atoms Defining Plane	Deviation from Best Plane (Å)
[2Fe-2S][b]	Fe1	−0.018
	Fe2	−0.024
	S1	0.019
	S2	0.022
[3Fe-3S][c]	Fe1	−0.063
	Fe2	−0.253
	Fe3	−0.098
	S1	0.176
	S2	0.221
	S3	0.018
[4Fe-4S][c]	Fe1	0.178
	Fe2	0.190
	S1	−0.186
	S2	−0.183

[a]The atom labels refer to Figure 2.
[b]Cluster in *S. p.* ferredoxin (15). Atomic coordinates provided by Dr. T. Tsukihara.
[c]Clusters in *A. v.* ferredoxin. Similar planes are obtained for the other faces of the [4Fe-4S] cluster.

a diamond shape. In the protein the internal angles at inorganic S are 85° and 86°, and the core (in the present model) is more like a square than a diamond (Fe–Fe distance 2.72 Å, S–S distance 2.94 Å).

3.3 [4Fe-4S] Clusters

Four independent structures of [4Fe-4S]$(S_\gamma)_4$ clusters have been determined from the protein crystal structures of HIPIP, *P. a.* ferredoxin, and *A. v.* ferredoxin (Figure 2*d–g*; Table 4). For HIPIP diffraction data were collected for both oxidized and reduced crystals and the structures independently refined (HP_{ox}, HP_{red} clusters) (25). For the *P. a.* ferredoxin data were collected for the oxidized state of the protein only (Fd_{ox} I, Fd_{ox} II clusters) (36). The interatomic distance and angle data for the [4Fe-4S] clusters in HIPIP and *P. a.* ferredoxin have been summarized by Carter (42). The data for the [4Fe-4S] cluster in *A. v.* ferredoxin ($Av_{(red)}$ cluster) is added to this information in Table 8. In comparing the geometries of the four clusters and in viewing their structures in Figure 2, one overriding fact is apparent, namely that the structure of the [4Fe-4S] core is essentially the same in each of the protein environments. Nevertheless, subtle but significant differences in the stereochemistry

Table 7 Interatomic Distances and Angles in [2Fe-2S] Cluster in Partially Refined *S. p.* Ferredoxin[a]

Atoms	Distance (Å)	Atoms	Angle (deg)
Fe1-Fe2	2.72	S1-Fe1-S2	89
S1-S2	2.94	S1-Fe1-S_γ41	97
Fe1-S1	2.11	S1-Fe1-S_γ46	115
Fe1-S2	2.09	S2-Fe1-S_γ41	136
Fe2-S1	1.93	S2-Fe1-S_γ46	114
Fe2-S2	1.89	S_γ41-Fe1-S_γ46	101
Fe1-S_γ41	3.05	S1-Fe2-S2	100
Fe1-S_γ46	2.20	S1-Fe2-S_γ49	110
Fe2-S_γ49	2.38	S1-Fe2-S_γ79	116
Fe2-S_γ79	2.12	S2-Fe2-S_γ49	106
S_γ-S_γ[b]	6.29	S2-Fe2-S_γ79	122
		S_γ49-Fe2-S_γ79	101
		Fe1-S1-Fe2	85
		Fe1-S2-Fe2	86

[a] Atomic coordinates provided by Dr. T. Tsukihara, Tottori University, Japan. The atom labels refer to Figure 2*b*.
[b] Average of four distances between S_γ atoms on different Fe atoms.

of the HP_{ox}, HP_{red}, and Fd_{ox} clusters have been observed and correlated with reduction potential.

3.3.1 Structure and Geometry. The [4Fe-4S] cluster is best visualized as two interpenetrating, concentric tetrahedra of 4 Fe atoms and 4 inorganic S atoms. The Fe tetrahedron is smaller, with a mean Fe–Fe distance of close to 2.75 Å. The S tetrahedron has a mean S–S distance of about 3.55 Å. As pointed out by Carter, two limiting structures of this cluster are a perfect cube with all angles equal to 90° and a structure where each Fe has perfect tetrahedral geometry (109°), forcing the angles at S to 66° (42). In fact, a perfect cube was used as the starting model for refinement of the Fd_{ox} I and Fd_{ox} II clusters (34). In the actual structure the interior S–Fe–S angles are compressed from the tetrahedral value by 5–6°, and correspondingly the Fe–S–Fe angles are expanded to 75°. The resulting structure is therefore closer to the second limiting case, favoring tetrahedrallike Fe coordination and Fe–Fe interactions (79). It is interesting to note that the 75° angles at S are in good agreement with one of two predicted values of Pauling (80) of 74° or 133° for *spd* hybrid orbitals involving transition metals.

The geometries of these four protein-bound [4Fe-4S] clusters are in

Table 8 Interatomic Distances and Angles in [4Fe-4S] Clusters in Protein Structures[a]

Atoms	HP_{ox}	HP_{red}	Fd_{ox} I	Fd_{ox} II	$Av_{(red)}$[b]
Distances (Å)					
Fe-Fe					
Mean	2.72	2.81	2.73	2.67	2.77
Range	2.68-2.78	2.74-2.87	2.66-2.83	2.65-2.77	2.75-2.78
RMS deviation	0.04	0.04	0.06	0.07	(≈0.1)
Fe-S					
Mean	2.26	2.32	2.23	2.22	2.27
Range	2.10-2.39	2.18-2.45	1.98-2.40	1.97-2.44	2.11-2.40
RMS deviation	0.08	0.09	0.12	0.14	(≈0.2)
Fe-S_γ					
Mean	2.20	2.22	2.22	2.25	2.30
Range	2.17-2.22	2.19-2.26	1.98-2.43	1.98-2.45	2.21-2.36
RMS deviation	0.02	0.03	0.16	0.17	(≈0.2)
S-S					
Mean	3.55	3.65	3.52	3.49	3.55
Range	3.49-3.64	3.50-3.75	3.41-3.60	3.37-3.68	3.45-3.70
RMS deviation	0.06	0.11	0.07	0.12	(≈0.2)
S_γ-S_γ					
Mean	6.32	6.40	6.31	6.29	6.37
Range	6.06-6.66	6.02-6.65	5.60-6.78	5.86-6.54	5.51-6.90
RMS deviation	0.20	0.22	0.37	0.23	(≈0.4)
Angles (deg)					
Fe-S-Fe					
Mean	74	76	75	75	75
Range	72-76	72-80	—	—	73-80
RMS deviation	1.3	2.4	2	2	(≈5)
S-Fe-S					
Mean	104	104	103	104	103
Range	101-109	99-107	—	—	97-108
RMS deviation	2.4	2.6	—	—	(≈5)
S-Fe-S_γ					
Mean	115	116	115	115	114
Range	107-120	106-126	—	—	84-135
RMS deviation	4.9	5.3	3	3	(≈10)

[a]Data for HIPIP (HP) (25, 26) and *P. a.* ferredoxin (Fd) clusters I and II (36) are taken from Carter (42).

[b]Partial refinement at 2.5 Å resolution.

excellent agreement with the structures of the synthetic analogs $[Fe_4S_4(SPh)_4]^{2-/3-}$, $[Fe_4S_4(SCH_2Ph)_4]^{2-/3-}$ (81–85), and $[Fe_4S_4(S(CH_2)_2COO)_4]^{6-}$ (86). Mean Fe–Fe and S–S distances differ by no more than 0.05 and 0.12 Å, respectively, from the analog structures. Similarly, mean values of the internal angles at S and at Fe differ by no more than 2° from those of the analogs, and mean Fe–S bond distances differ by no more than 0.06 Å. It must be emphasized however, that the analog crystal structures display an order of magnitude greater precision, thereby providing a true picture of the molecular details of these compounds.

3.3.2 Redox States. Although [4Fe-4S] clusters have been found in a number of enzymes, all the clusters determined to date are from electron-transport proteins. Of these, only the structure of HIPIP has been refined in more than one oxidation state. Nevertheless, this study, in conjunction with the results obtained for oxidized *P. a.* ferredoxin crystals (36) and from the analog $Fe_4S_4(RS)_4$ complexes in solution (88, 89), has led to an important unifying description of the electronic properties of [4Fe-4S] clusters, the "three-state hypothesis" (23). According to this hypothesis, the [4Fe-4S] core can have three oxidation states, paired-spin (EPR silent), oxidized ($S = ½$, $g = 2.01$), and reduced ($S = ½$, $g = 1.94$); the protein selects for two of the three possible states. Consequently, the cluster may exhibit a reduction potential of +0.350 V (HIPIP) or −0.420 V (ferredoxin). The high-potential protein employs the spin-paired state/oxidized state couple, or HP_{red}/HP_{ox}. The low-potential protein involves the spin-paired state/reduced state couple, or Fd_{ox}/Fd_{red}. The spin-paired state is a dianion, as in $[Fe_4S_4(SPh)_4]^{2-}$ (89). HIPIP in partially denaturing solutions of DMSO/H_2O can be "super-reduced" to the third (Fd_{red}) state consistent with the hypothesis (90).

3.3.3 Cluster Asymmetry. A distinct distortion of the [4Fe-4S] cluster was observed as a result of the HIPIP refinement, which is consistent with the results of the crystal structures of a number of synthetic analog compounds (89, 91). For the HP_{red} cluster there are two distinct sets of Fe–S distances, eight long bonds, which average 2.36 Å, and four parallel, short bonds, which average 2.25 Å (23, 27). In the analog spin-paired cluster, $[Fe_4S_4(SPh)_4]^{2-}$, these sets of average distances are 2.296 and 2.267 Å, respectively (85). In $[Fe_4S_4(SCH_2Ph)_4]^{2-}$ the same average distances are 2.310 and 2.239 Å (82). A similar asymmetry of bond lengths is seen in both the Fd_{ox} I and Fd_{ox} II clusters. Therefore each of the clusters undergoes a compression, or tetragonal distortion, which lowers the symmetry from tetrahedral (T_d, $\bar{4}3m$) to tetragonal (D_{2d}, $\bar{4}2m$). The $\bar{4}$ axis is parallel to the four shorter bonds (Fig. 4). Because of its presence in $Fe_4S_4(RS)_4$ compounds as well as in proteins,

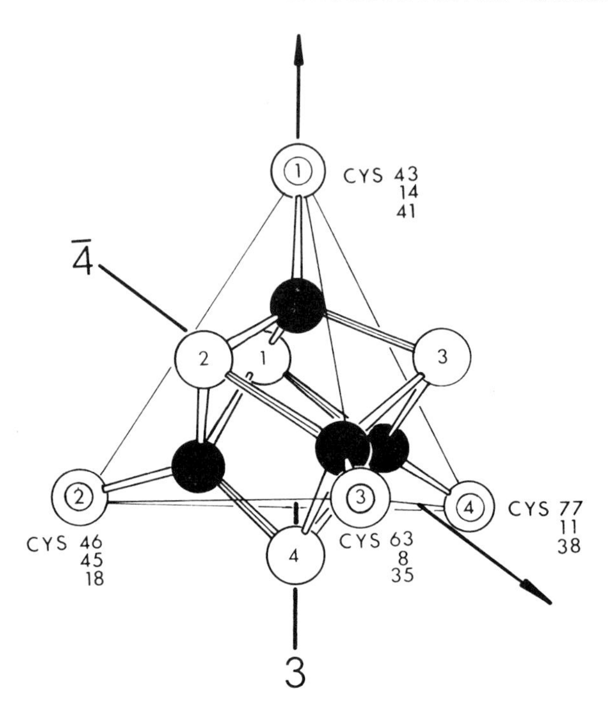

Figure 4. Unique orientation for $[4Fe\text{-}4S](S\gamma)_4$ cluster (27). Tetragonal distortion ($\bar{4}$ axis) is approximately parallel with the Fe_1–S_3 direction. Trigonal distortion (threefold axis) passes through cysteine 43 Sγ of HIPIP (or cysteines 14 or 41 of *P. a.* ferredoxin). Molecular point group is C_s with an effective mirror plane passing through the Fe-1, S-3, Fe-2, and S-4 atoms [Fe atoms (solid circles) are numbered in relation to Sγ atom labels]. [Reprinted with permission from C. W. Carter, *J. Biol. Chem.*, **252,** 7802–7811 (1977).]

tetragonal distortion must be regarded as an intrinsic property of these clusters, and it has been ascribed to Jahn-Teller effects (79, 89).

A further asymmetry of the HP_{red} $[4Fe\text{-}4S](S_\gamma)_4$ structure has been pointed out by Carter (27). Interatomic S_γ–S_γ distances in HP_{red} fall into two classes, three with an average separation of 6.24 Å and three with an average separation of 6.55 Å. The same phenomenon is observed for the analog clusters, where the respective distances in $[Fe_4S_4(SPh)_4]^{2-}$ are 6.27 and 6.50 Å (85); in $[Fe_4S_4(SCH_2Ph)_4]^{2-}$ these distances are 6.32 and 6.51 Å (82). If one combines the threefold symmetry axis produced by this trigonal distortion of S_γ ligands with the $\bar{4}$ axis generated by tetragonal compression of the core, a unique cluster orientation is obtained (Fig. 4). The molecular point group of the 12 atom $[4Fe\text{-}4S](S_\gamma)_4$ complex is thus further reduced to C_s (mirror) symmetry (27).

Properties of protein-bound [4Fe-4S] clusters can be correlated with the unique cluster orientation, suggesting that the protein matrix selects one of the two diastereomers produced by the effective cluster mirror plane. One such property is redox potential, with the apparent correlation being the interaction of aromatic residues with the cluster (27). Another is the conformational change that takes place on oxidation of the cluster. Difference Fourier maps (25) with coefficients HP_{ox}-HP_{red} show negative difference electron density between the Fe and S atoms that form the 4 short Fe–S bonds and strong positive density along the other 8 Fe–S bonds. With reference to the refined HP_{ox} and HP_{red} coordinates, the cluster, on oxidation, *contracts* 0.16 Å along the Fe-1–S-2 direction (average of 4 Fe–S bonds, Fig. 4) and *contracts* 0.08 Å along the Fe-1–S-1 direction (average of 4 Fe–S bonds). Both directions intersect the effective mirror plane. The 4 Fe–S bonds parallel with the mirror plane and the $\bar{4}$ axis *expand* 0.03 Å on oxidation (average of 4 bonds in the Fe-1–S-3 direction). Thus contraction of the protein-bound cluster does not take place along the 4 bonds already shortened by tetragonal distortion.

Protein crystallographic data for the Fd_{red} state of a [4Fe-4S] cluster are not available. The structures of $[Fe_4S_4(SPh)_4]^{2-/3-}$ and $[Fe_4S_4(SCH_2Ph)_4]^{2-/3-}$ have been studied, however (89, 91). As mentioned above, $[Fe_4S_4(SPh)_4]^{2-}$ and $[Fe_4S_4(SCH_2Ph)_4]^{2-}$ have compressed D_{2d} symmetry. On reduction the $[Fe_4S_4(SPh)_4]^{3-}$ anion *expands* 0.08 Å along the $\bar{4}$ axis and exhibits elongated D_{2d} symmetry. The core symmetry intermediate between these two structures, T_d, can be viewed as a transition state for electron transfer in Fd_{ox}/Fd_{red}. Further, this "core dimensional change" accompanying reduction would be permitted by the Fd polypeptide but precluded in HIPIP, suggesting a reason for the absence of the $HP_{s\text{-}red}$ state under physiological conditions (89).

The picture then is one of a spin-paired $[Fe_4S_4(RS)_4]^{2-}$ cluster (RS = organothiol or protein cysteine S_γ) having intrinsic, compressed D_{2d} symmetry and capable of undergoing $1e^-$ oxidation or reduction. Trigonal distortion of the RS ligands further reduces the symmetry of the 12-atom complex to C_s. On oxidation to the $[Fe_4S_4(RS)_4]^{1-}$ state (HP_{ox} data) the cluster contracts along the 8 Fe–S bonds not parallel to the $\bar{4}$ axis. In the $[Fe_4S_4(RS)_4]^{3-}$ reduced state ($[Fe_4S_4(SPh)_4]^{3-}$ data corresponding to Fd_{red}) the cluster expands along the 4 Fe–S bonds parallel to the $\bar{4}$ axis, preserving D_{2d} symmetry of the core.

3.3.4 A. v. Cluster. At the present time detailed comparisons cannot be made involving the [4Fe-4S] cluster in the *A. v.* ferredoxin structure with the refined and analog cluster structures. However, it is interesting to note that, although the *A. v.* cluster is high potential (reduction potential +0.320 V), it resides in a sequence homologous with *P. a.* ferredoxin (31). For the present it has been assumed that the [4Fe-4S] cluster is in the reduced (spin-paired) state (Table 7) because of the presence of the oxidized low-potential 3-Fe center in the same molecule. At the same time the cluster may be subject to oxidization of the X-ray beam, as is the HP_{red} cluster (25).

3.4 [3Fe-3S] Cluster

To date the only [3Fe-3S] cluster studied by crystallographic techniques has been the one in *A. v.* ferredoxin (30, 31), although accumulating biochemical and spectroscopic data clearly indicate that this type of [Fe-S] cluster may be as widespread in occurrence as 1-, 2-, and 4-Fe clusters. Entirely independent evidence for the existence of 3-Fe centers in several proteins comes from Mössbauer and EPR spectroscopy (18, 63).

The presence of 3-Fe centers raises some intriguing questions about their role in several biochemical reactions. For instance, the Mössbauer (63) and EPR evidence for a 3-Fe center in aconitase suggests that the cluster serves a regulatory function, activating the enzyme on reduction (50). Data for the *D. gigas* ferredoxin clearly demonstrate that the Fd II oligomer contains 3-Fe centers (18), whereas the Fd I oligomer contains 4-Fe clusters, in spite of the fact that the identical polypeptide is present in each case (19). The [3Fe-3S] cluster therefore may represent an intermediate in the formation of [4Fe-4S] clusters, or vice versa, and a given polypeptide may adopt two alternative conformations in incorporating the two types of clusters. In either case conformational change is implied. By its nature one would expect the open six-membered ring of a [3Fe-3S] cluster (6 Fe–S bonds) to display greater conformational freedom than the [4Fe-4S] cluster (12 Fe–S bonds). At the same time the protein might be expected to exert a greater influence on a [3Fe-3S] cluster (5 or 6 ligands) than on a [4Fe-4S] cluster (4 ligands). Although these characteristics could make [3Fe-3S] clusters suitable for certain enzymatic reactions, they do not explain why some redox proteins require a 3-Fe center when the reduction potentials of certain [4Fe-4S] and [2Fe-2S] clusters are similar.

3.4.1 Structure and Geometry. The structure of the [3Fe-3S] cluster in *A. v.* ferredoxin is shown in Fig. 2*c*. Interatomic distance data are summarized in Table 9. The coordinates of the cluster are derived from fitting $2F_o - F_c$ Fourier maps at 2.0 Å resolution. The protein is being refined by the method of Konnert and Hendrickson (92) and the overall residual R is 25%. Because of the remaining errors in the protein model and the incomplete nature of the refinement, the coordinates are estimated to be no more accurate than $\Delta d \cong 0.1$ Å. An example of the electron density is shown in

Table 9 Interatomic Distances and Angles in the [3Fe-3S] Cluster in *A. v.* Ferredoxin[a]

Atoms	Distances (Å) and Angles (deg)
Fe–Fe	
Mean	4.00 Å
Range	3.94–4.05
RMS deviation	(≈0.1)
Fe–S	
Mean	2.34
Range	2.24–2.44
RMS deviation	(≈0.2)
Fe–S_γ	
Mean	2.28
Range	2.02–2.49
RMS deviation	(≈0.2)
Fe–'O_ϵ'	2.03
RMS deviation	(≈0.2)
S–S	
Mean	4.05
Range	3.95–4.12
RMS deviation	(≈0.2)
S_γ–S_γ('O_ϵ')[b]	
Mean	6.25
Range	5.17–7.64
RMS deviation	(≈0.4)
Fe–S–Fe	
Mean	117 deg
Range	113–121
RMS deviation	(≈5)
S–Fe–S	
Mean	119
Range	115–122
RMS deviation	(≈5)
Fe1–S1–S3–S_γ8–S_γ20 (six angles)	
Mean	110
Range	82–142
RMS deviation	(≈10)
Fe2–S1–S2–S_γ16–'O_ϵ' (six angles)	
Mean	112
Range	77–160
RMS deviation	(≈10)
Fe3–S2–S3–S_γ11–S_γ49 (six angles)	
Mean	103
Range	83–147
RMS deviation	(≈10)

[a]Partial refinement at 2.5 Å resolution. The atom labels refer to Figure 2*c*.

[b]Twelve distances involving ligand atoms on different Fe atom centers.

Figure 5 (31). Independent X-ray data for the structure of the cluster is provided by anomalous scattering (Fig. 1*b*). The coordinates of the 3 Fe atoms are therefore firmly established. Positions of the 3 inorganic S atoms are derived from fitting of the cluster density, and the model is in agreement with the analytical data for equimolar Fe and inorganic S (93). Positions of the ligand atoms are derived from fitting the chemical sequence to the electron density at 2.5 Å resolution. It is anticipated that details of the structure will change as refinement proceeds.

The geometry data in Table 9 show that, although the [3Fe-3S] cluster displays essentially standard Fe–S bond lengths, the Fe–Fe distances at ≈4.0 Å are fully 1.3 Å greater than in [2Fe-2S] (Table 7) and [4Fe-4S] (Table 8) clusters. The implication of this arrangement is that there is little direct interaction of Fe orbitals. The average S–S distance at 4.05 Å is only ≈0.5 Å larger than in the [2Fe-2S] and [4Fe-4S] clusters. Interestingly, the average S_γ–S_γ and S_γ–"O_ϵ" distance, 6.25 Å, is quite similar to the S_γ–S_γ distance in the other Fe-S proteins; that is, the chelating "bite-distance" from the ligands of the protein to a 2-Fe, 3-Fe, or 4-Fe cluster is about the same.

The bond angles about each of the 3 Fe atoms average to approximately tetrahedral values (Table 9). Distortions, however, are quite large and are not likely to arise from errors in the model alone. In comparing the [3Fe-3S] cluster with the [2Fe-2S] and [4Fe-4S] clusters, a marked difference is seen for the internal angles of the core, which in the new cluster are essentially 120° at both Fe and inorganic S. Thus the angles at Fe are considerably expanded from the ≈104° value and are closer to the external S-Fe-S_γ angles of ≈115° seen in the other clusters (Tables 7 and 8). The angles at S are expanded fully 40° from the 75° values observed in [2Fe-2S] and [4Fe-4S] core

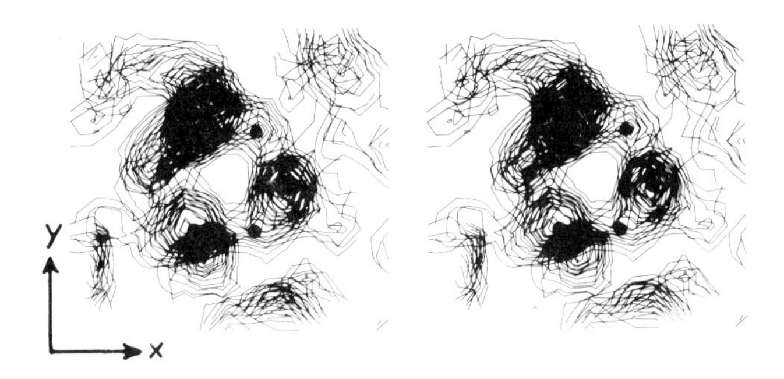

Figure 5. Electron density for the [3Fe-3S] cluster at 2.0 Å resolution computed with 6249 reflections and phases from the partially refined *A. v.* protein structure (R = 25%). Coefficients of the Fourier map $2|F_0| - |F_c|$; the Fe, S, and $S\gamma$ atoms were omitted from the phase calculation. Note: the view is not the same as in Figures 1*b*, 2*c*, or 6.

structures. This feature of the structure accommodates trinuclear association of tetrahedral Fe to form the symmetrical and nearly planar ring of the core.

Is it possible to describe the geometry of the [3Fe-3S] cluster in a way that is also chemically meaningful? As mentioned, the predicted most favored angles for bonds involving *spd* hybrid orbitals are 73° and 134° (80). Since the 73° value is in good agreement with the observed angles at inorganic S in [2Fe-2S] and [4Fe-4S] clusters, one might expect the 134° angle to be an alternative possibility for S in Fe-S clusters. If one uses this angle to construct a six atom ring with 104° angles at Fe, the ring can be made perfectly planar with Fe–Fe and S–S distances of 4.2 and 3.6 Å, respectively. This suggests that a [3Fe-3S] cluster could be planar, and a least squares plane through the [3Fe-3S] core has no atoms more than 0.25 Å from the plane (Table 6). However, the angles at S are not 134°, but closer to 120°.

An alternative limiting case for the [3Fe-3S] structure is one in which every atom has tetrahedral geometry. Precedents for this are the crystal structures of $[Sn_3S_3](CH_3)_6$ (94) and $[Fe_4(SPh)_{10}]^{2-}$ (95). The latter compound has an adamantanelike structure of four fused six-membered rings, each containing 3 Fe and 3 bridging RS groups. In this structure the mean Fe–Fe distance is 3.93 Å, and the angles about Fe range from 98° to 117°. Because it is a fused ring system, each $Fe_3(RS)_3$ cycle adopts a chair comformation.

With the Sn compound an interesting analogy can be made. By referring to Table 10 and Figure 6 it can be seen that both clusters have quite similar geometries. The mean Sn–Sn, Sn–S, Sn–C, S–S, and C–C distances are 3.78, 2.42, 2.17, 3.90, and 6.05 Å, respectively; angles about Sn and S within the "core" are 103° and 108°, respectively; and the remaining angles about Sn range from 108° to 118°. The conformation of the six-membered $[Sn_3S_3]$ ring is a twist-boat in which 1 of the S atoms (S2) lies within the plane of the Sn atoms, the other 2 being puckered above and below (Table 10). In an analogous manner, 1 S atom of the [3Fe-3S] core (S3) is markedly within the plane of the 3 Fe atoms (Table 10). When the two structures are superposed with respect to this S atom, the S_γ atoms and methyl groups become positioned about the ring in a similar manner (Figs. 6 and 2*c*). A twist-boat model therefore accounts for a fundamental feature of the ligand structure in the [3Fe-3S] cluster. Nevertheless, the [3Fe-3S] core appears to be essentially planar and symmetrical.

3.4.2 Redox States. EPR experiments with *A. v.* ferredoxin have established that both Fe-S centers are paramagnetic in their oxidized states, with reduction potentials E_m of -0.424 and $+0.320$ (96). Mössbauer spectra of the protein in its air-isolated state display characteristic spectra for a spin-paired [4Fe-4S] cluster, while the paramagnetism associated with the $g = 2.01$ signal of the isolated state arises from the low-potential center (63). The

Table 10 A Comparison of Least-Squares Planes for [3Fe-3S](S_γ)$_5$('O_ϵ') and [Sn_3S_3](CH_3)$_6$ Structures[a]

Fe-S Cluster		Sn-S Compound (ref. 94)	
Atom	Deviation (Å)	Atom	Deviation (Å)
Fe1[b]	−0.010	Sn3[a]	−0.0011
Fe2[b]	0.006	Sn1[a]	0.0007
Fe3[b]	−0.010	Sn2[a]	−0.0011
S3[b]	0.014	S2[a]	0.0015
S1	0.355	S3	1.4428
S2	0.429	S1	−1.4282
S_γ11	−0.029	C3	−1.0953
'O_ϵ'	−0.059	C2	1.2948
S_γ8	−2.35	C5	−1.9910
S_γ16	−1.74	C1	−1.2703
S_γ20	1.56	C6	1.0932
S_γ49	2.45	C4	2.0045

[a]The atom labels refer to Figures 2*c* and 6.
[b]Used to define the least-squares plane.

high-potential center is therefore the [4Fe-4S] cluster. Mössbauer spectra of the fully reduced, EPR-silent protein demonstrate that the low-potential center contains 3 Fe atoms (63). Mössbauer studies of the very similar 3-Fe center in *D. gigas* ferredoxin II confirm these results (18), and lead to a description of the oxidized cluster as containing 3 Fe^{3+}, which on reduction become 2 $Fe^{2.5+}$/1 Fe^{3+} (97). Protein crystallography of the *A. v.* protein to date has been done entirely with the molecule in its air-isolated state, that is, the oxidized state of the [3Fe-3S] cluster.

3.4.3 Conformational Change. Apart from the intuitive notion that [3Fe-3S] clusters should be conformationally active, several experimental results suggest this. Displacement reactions using organic thiols and denaturing solvents lead to cluster decomposition and rearrangement in two proteins. For *A. v.* ferredoxin different reaction conditions lead to differing yields of [2Fe-2S] and [4Fe-4S] clusters (67); for aconitase [2Fe-2S] clusters are preferentially released (51) from the presumed single [3Fe-3S] center that is present (63). Although nonphysiological, these reactions may be related to the potential "introconversion" between 4-Fe and 3-Fe states of *D. gigas* ferredoxin. They also may be related to rearrangements observed in the synthesis of Fe-S cluster analogs, for example, the dimerization of [2Fe-2S]

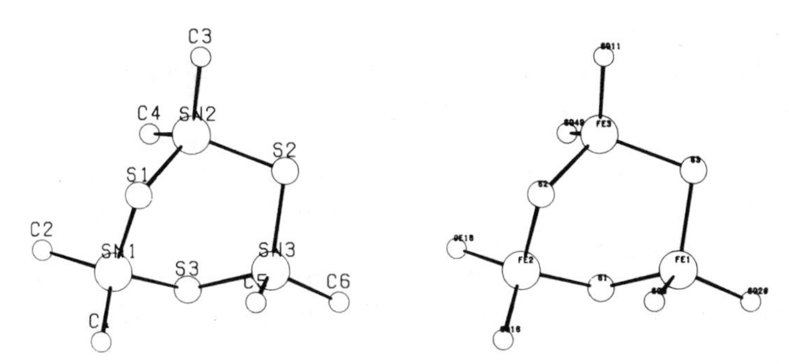

Figure 6. Coordinates of the structure of $[Sn_3S_3](CH_3)_6$ (94) shown in the same view as the $[3Fe\text{-}3S](S\gamma)_5('O_\epsilon')$ cluster (Fig. 2c) and density map (Fig. 1*b*). The corresponding atoms of the Fe-S cluster are labeled on the right side of the stereofigure; labels on the left side refer to the Sn compound and Table 9. Note that the Sn structure is considerably more puckered at S-1(Sn) and S-3(Sn) than the Fe-S cluster, resulting in a twisting of the ligand atoms on Sn-1 [compared to those on Fe-2 (Fig. 2c)].

clusters (65) and the formation of $[Fe_4S_4(SPh)_4]^{2-}$ from $[Fe_4(SPh)_{10}]^{2-}$ (58, 95). It is observed that in a heavy atom derivatives of *A. v.* ferredoxin, two compounds, $[OsO_2(NH_3)_4]^{2+}$ and $[Rh(NH_3)_5Cl]^{2+}$, are able to bind at and displace atoms of the [3Fe-3S] cluster (31).

3.5 General Remarks

Active site Fe-S structures containing 1, 2, 3, or 4 Fe atoms have been elucidated. The 1-Fe, 2-Fe and 3-Fe centers can adopt two oxidation states, while 4-Fe centers display two of three oxidation states. A fourth oxidation state for 4-Fe centers has been demonstrated in the P clusters of nitrogenase component I (98). A fifth type of Fe-S cluster, in association with Mo, must be present in the MoFe cofactor (99).

In the known structures the Fe in $Fe(S_\gamma)_4$, [2Fe-2S], [3Fe-3S], and [4Fe-4S] clusters is four-coordinate with tetrahedral or distorted tetrahedral geometry. The Fe–S and Fe–S_γ bond lengths are essentially 2.3 Å. In [2Fe-2S] and [4Fe-4S] clusters the Fe–Fe (2.7 Å) and S–S (3.5 Å) distances are similar. However, the [4Fe-4S] cluster is not geometrically a "dimer" of [2Fe-2S] clusters, as seen from the least squares planes (Table 6), since the faces of the [4Fe-4S] cluster are puckered. The Fe in the [3Fe-3S] cluster has distorted tetrahedral coordination, but the core is essentially planar. In this cluster the Fe–Fe distances, and to a lesser extent the S–S distances, are significantly larger than in other Fe-S clusters.

The formation of a variety of Fe-S structures must be due in large part to the chemical versatility of S. Inorganic S is two-coordinate in [2Fe-2S] clusters with an internal angle of 75°, but three-coordinate with the same angle in [4Fe-4S] clusters. In the [3Fe-3S] cluster, S is again two coordinate, but the internal angle is 120°.

Each known type of [Fe-S] cluster has distinctive symmetry. As pointed out by Peisach and colleagues (100), the symmetry of $Fe(S_{\gamma})_4$ and $[4Fe\text{-}4S](S_{\gamma})_4$ centers is tetrahedral to a good first approximation, while the $[2Fe\text{-}2S](S_{\gamma})_4$ center is centrosymmetric. However, the ideal symmetry of [4Fe-4S] clusters and $[Fe_4S_4(RS)_4]$ compounds is reduced by tetragonal distortion along a cluster $\bar{4}$ axis. Lasar Raman spectra of [4Fe-4S] clusters reveal additional symmetrical but nontetrahedral vibrations (101). For a $[3Fe\text{-}3S](S_{\gamma})_6$ type of cluster the idealized symmetry could be either twofold (pure twist-boat conformation) or threefold (planar core, S angles 134°). Substitution of the S_{γ} ligands or protein-induced distortion would be expected to lower this symmetry, but in no case could a [3Fe-3S] cluster be centrosymmetric. Recently, an MCD study of aconitase has found spectral features for a 3-Fe center unlike those observed for 2-Fe or 4-Fe centers (102), indicating a unique configuration for the cluster in the solution state.

4 Fe-S PROTEINS AS LIGANDS

Three of the four types of Fe-S centers discussed above incorporate the same number of ligands from the polypeptide, that is, the four cysteinyl sidechains required for $Fe(S_{\gamma})_4$, [2Fe-2S], and [4Fe-4S] clusters. The protein, while providing these ligands, must not only select for the type of Fe-S center formed, but also modulate the properties of the site. It is not surprising then that distinctive patterns are seen for the cysteine-containing peptides of all classes of Fe-S proteins. This appears to be especially true in the formation of [3Fe-3S] clusters, which require more than four protein ligands. A definitive primary sequence for simple Fe-S proteins containing [3Fe-3S] centers is apparently a hexapeptide, and in the particular case of *Azotobacter* ferredoxin, glutamic acid may act as an independent ligand to iron. The involvement of negatively charged oxyligands in the formation of Fe-S centers, if a general phenomenon, suggests that a variety of Fe-S centers may be present in other proteins.

In this section attention is focused on the cysteine-containing peptides. Details and comparisons involving the full structures of rubredoxin, *P. a.* ferredoxin, and HIPIP have been given by the respective authors (Table 1) and reviewed (39–43). Complete details of the *S. p.* and *A. v.* ferredoxin structures have not yet been described. Although each of these molecules provides fascinating insights into protein stereochemistry, the functional aspects of the overall structures are often obscure because of a lack of complementary physiological information. It has generally been assumed that structural features of an active site apply to other proteins containing the same type of active site ([Fe-S] cluster).

4.1 Cysteinyl Peptides

Three-dimensional structures are depicted in Figure 7 for five Fe-S proteins. The sixth known structure, *C. pasteurianum* rubredoxin, is similar to *D. vulgaris* rubredoxin. Although representing four types of Fe-S centers, these structures are not representative of Fe-S proteins as a whole. All function in electron-transfer reactions and have molecular weights in the range of 5000–13,000 daltons.

The cysteine coordination is summarized with respect to the primary structures in Figure 3. Tabulated primary sequence data (103) show that each of the crystal structures represents a general class of homologous sequences, that is, rubredoxins, chloroplast ferredoxins, high-potential Fe proteins, clostridial ferredoxins, and *Azotobacter*-type (7-Fe) ferredoxins. Within each of these classes the number and distribution of cysteine residues in the pri-

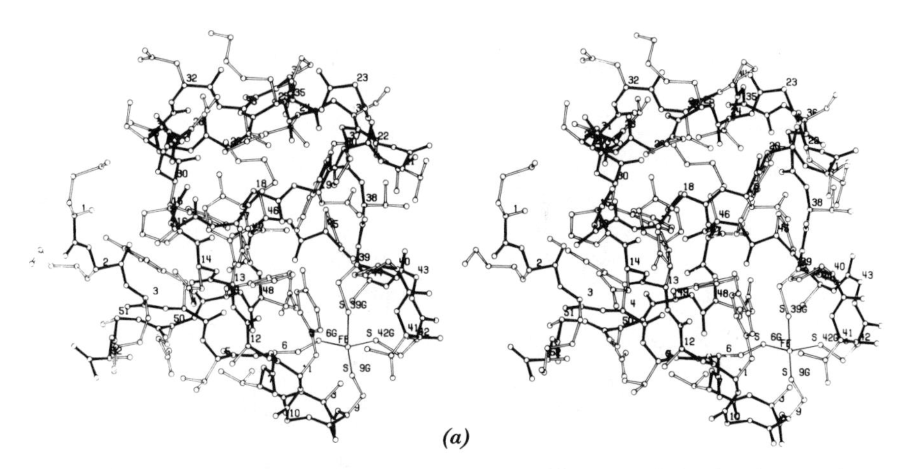

Figure 7. The crystal structures of five Fe-S proteins. *(a)* Atomic coordinates of *D. v.* rubredoxin (11). [Reprinted with permission from E. T. Adman, *J. Mol. Biol.*, **112,** 113–120 (1977) (Fig. 4).] *(b)* Alpha-carbon, cysteine, Fe, and S coordinates of *S. p.* ferredoxin (15). The stereofigures were provided by T. Tsukihara. *(c)* Schematic drawing of *C. v.* HIPIP structure (24). Arrows denote tertiary main chain to main chain H bonds within the cluster binding cavity. Stippled regions—α-helical segments; shaded regions—hairpin turns; unmarked segments—extended conformation. [Reprinted with permision from C. W. Carter, in *Iron-Sulfur Proteins,* Vol. III, W. Lovenberg, ed., Academic, New York, 1977, pp. 157–204, (Fig. 6).] *(d)* Schematic drawing of *P. a.* ferredoxin (34). [Redrawn with permission from E. T. Adman, *J. Biol. Chem.*, **248,** 3987–3996 (1973) (Fig. 6).] *(e)* Schematic drawing of *A. v.* ferredoxin (31) showing the non-cysteinyl ligand as 'OXO'.

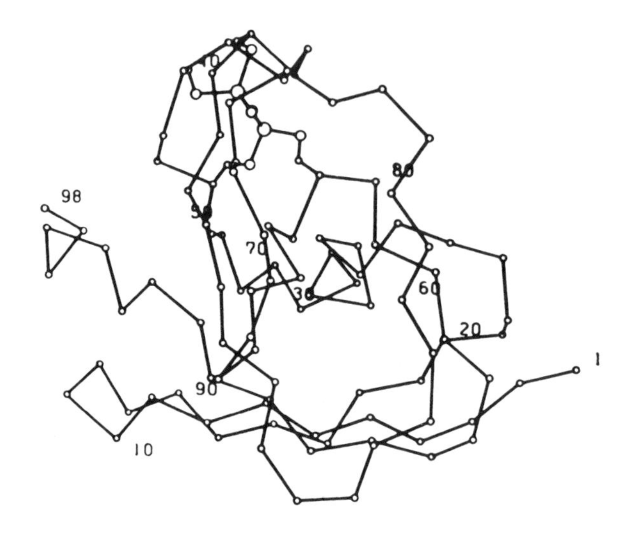

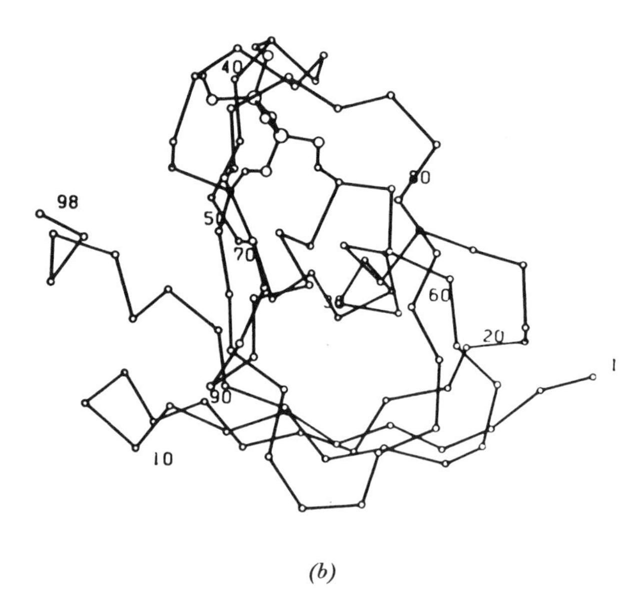

(b)

Figure 7. *(Continued)*

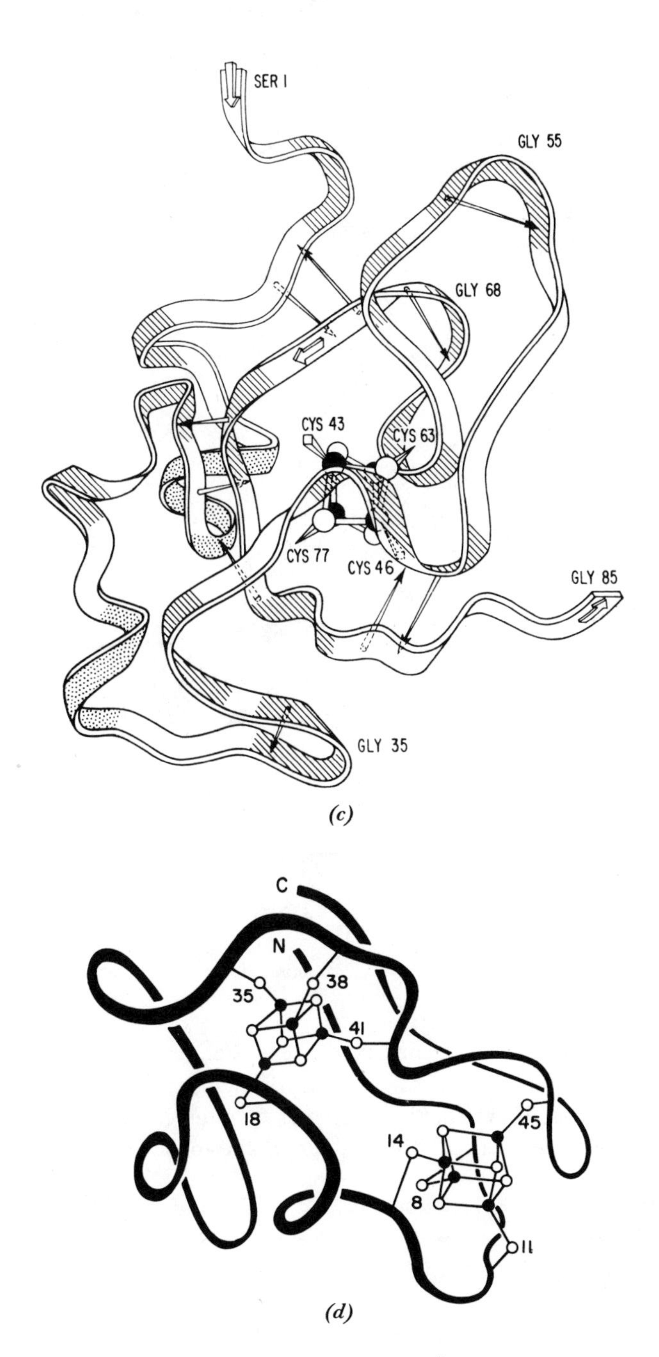

Figure 7. *(Continued)*

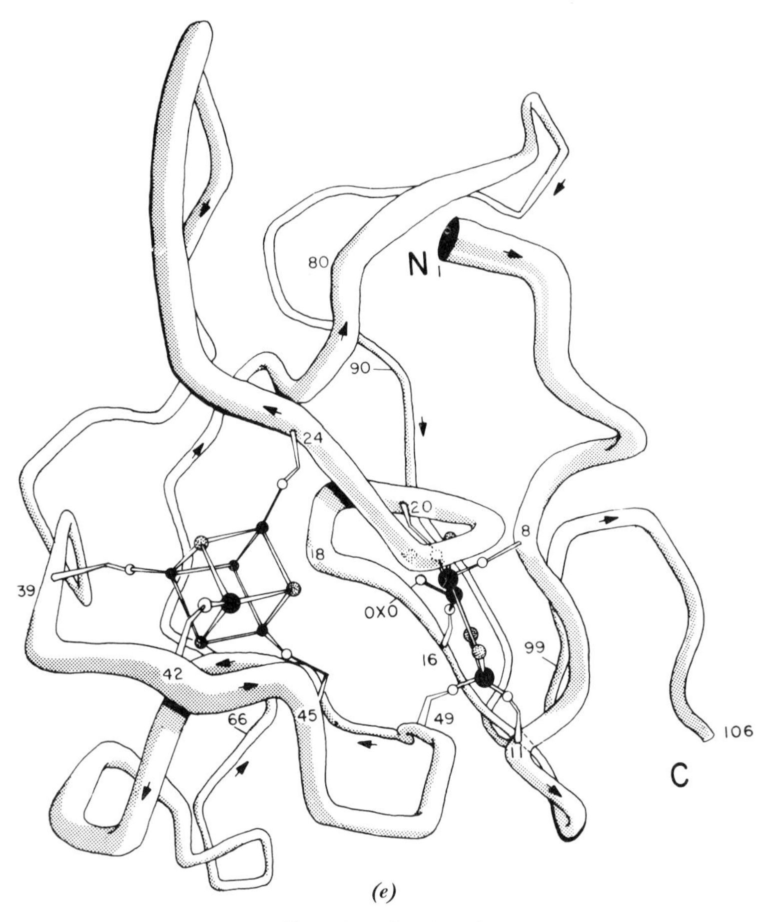

(e)

Figure 7. *(Continued)*

mary structures are strongly conserved. This fact, coupled with the X-ray structures, allows one to draw some general conclusions about the coordination of Fe-S centers. If cysteine is denoted as C, intervening amino acids as x, and the first cysteine is counted as residue n, the following 'rules' apply.

Peptide Size	Sequence	Coordination Rule
n, $n + 1$	CC	same Fe
n, $n + 2$	CxC	never occurs
n, $n + 3$	CxxC	same Fe or same cluster
n, $n + 4$	CxxxC	different clusters
n, $n + 5$	CxxxxC	same Fe or same cluster

The only example of the n, $n + 1$ rule is desulforedoxin, which has cysteines at positions 9, 12, 28, and 29 (104), and which has been established by Mössbauer experiments to contain a $Fe(S_\gamma)_4$ site (105). Although no Fe-S protein yet sequenced contains a CxC peptide, an exception may occur when one of the cysteines is replaced by glutamic acid, as in *Azotobacter* ferredoxin, and possibly by aspartic acid, as inferred from sequence data (see below).

4.1.1 NH···S H Bonds. The n, $n + 3$ rule has been observed for *P. a.* ferredoxin, rubredoxin, and HIPIP, and reasons for this preferred sequence are described by Adman and co-workers (35). CxxC-containing peptide sequences are listed in Table 11, including those for the partially refined structures of *S. p.* and *A. v.* ferredoxins. A basis of this preferred cysteine distribution is the formation of hydrogen bonds between peptide N and S (NH···S). All such hydrogen bonds in the refined structures have been tabulated (35, 42), and four types are observed (Table 11).

Type I and II H bonds are geometrically like Type I and Type II NH···O bonds in a 3_{10} helix except that the carbonyl position is exchanged for S_γ. Types I and II differ by rotation of 180° of the peptide plane between residues

Table 11 Peptide Sequences Containing CxxC in Fe-S Protein Structures, and Classes of NH···S H Bonds

Protein	Peptide Sequence[a]
C. p. rubredoxin (106)	C_6TVC_9G
C. p. rubredoxin (106)	$C_{39}PLC_{42}G$
D. v. rubredoxin (107)	C_6TVC_9G
D. v. rubredoxin (107)	$C_{39}PVC_{42}G$
P. a. ferredoxin (108)	$C_8IAC_{11}GAC_{14}K$
P. a. ferredoxin (108)	$C_{35}IDC_{38}GSC_{41}A$
C. v. HIPIP (109)	$C_{43}ADC_{46}Q$
S. p. ferredoxin (110)	$C_{46}STC_{49}A$
A. v. ferredoxin (111)	$C_8IKC_{11}K$
	$C_{39}IDC_{42}ALC_{45}D$

Type	NH···S H Bonds[b] Donor	Acceptor
I	peptide N, residue $n + 2$	S_γ, residue n
II	peptide N, residue $n + 2$	S_γ, residue n
III	peptide N, residue n'	S_γ, residue n
IV	peptide N	inorganic S

[a]One letter nomenclature is used for amino acids. References are given in parentheses.
[b]The data were taken from Adman and co-workers (35).

$n/n + 1$. In the proteins type I and II hydrogen bonds are formed within the CxxC peptide [e.g., NH(8)···S_γ(6) of *C. p.* rubredoxin], from residues outside the CxxC peptide [e.g., NH(48)···S_γ(46) of HIPIP], and to isolated cysteines [e.g., NH(20)···S_γ(18) of *P. a.* ferredoxin]. Both rubredoxin and clostridial ferredoxin contain sequences that may be denoted $C_n x_1 x_2 C_{n+3} G$. Here C_n accepts a type I H bond and C_{n+3} a type II bond, and glycine is required to avoid steric conflict with the carbonyl of C_{n+3}. This important generalization reveals that the conformation at x_2 determines whether the two cysteines will coordinate the same or different Fe atoms (35).

Type III hydrogen bonds involve peptide nitrogens distal in the sequence, and type IV involve inorganic S atoms. In this last category no regular geometry with respect to sequence is observed, except that all the donor N atoms belong to cysteine peptides. The H bonds formed occupy the fourth position in a distorted tetrahedral arrangement about inorganic S comprised of NH and 3 Fe atoms.

S. p. ferredoxin contains one CxxC type of sequence (Table 11) which binds to each Fe atom of the [2Fe-2S] cluster. As in the 8-Fe ferredoxin, NH···S bonds are formed between NH(48)···S_γ(46) within CxxC, and also between NH(43)···S_γ(41) and NH(79)···S_γ(49) (14). *A. v.* ferredoxin contains three CxxC sequences (Table 11), two at the [4Fe-4S] cluster, homologous with the clostridial ferredoxin sequences, and one at the [3Fe-3S] cluster in which the cysteines coordinate separate Fe atoms. Apparent NH···S bonds involve NH(44)···S_γ(42), NH(45)···S(1), and N_ϵ(Q52)···S_γ(11); additional interactions must be confirmed with refined coordinates.

To summarize, it is clear that CxxC sequences are favored in the formation of Fe-S centers, but they are not sufficient to determine the cluster type, since each of the five protein structures embodies such a sequence (Figure 3).

The n, $n + 4$ coordination rule applies only to the 8-Fe and 7-Fe ferredoxins where these peptides link cysteine-rich segments with isolated cysteines (Figure 3). In *P. a.* ferredoxin these sequences are $C_{14}KPEC_{18}$ and $C_{41}ASVC_{45}$; in *A. v.* ferredoxin they are $C_{20}PVDC_{24}$ and $C_{45}EPEC_{49}$.* The basis of this rule may simply be conservation of an ancestral ferredoxin gene.

The n, $n + 5$ rule is based only on the observation of the C_{11}–C_{16} loop in *A. v.* ferredoxin, which coordinates 2 Fe atoms of the [3Fe-3S] cluster, and the C_{41}–C_{46} loop in *S. p.* ferredoxin, which binds the same Fe of the [2Fe-2S] cluster. These two examples indicate that once there are four intervening residues, the peptide has considerable potential freedom as a ligand. Coincidentally, it is interesting to note that of the five cysteinyl peptides discussed here, only the first, third, and fifth participate in forming Fe-S centers.

*The C_{16}···C_{20} peptide is discussed below.

4.2 Coordination of [4Fe-4S] Clusters

Of four independent [4Fe-4S] clusters studied by protein crystallography (Fig. 2, *d–g*), two are high potential (HIPIP and *A. v.*) and two are low potential (*P. a.* 8-Fe ferredoxin). As seen in Figure 3, all are formed by at least one CxxC type of sequence; three of the four clusters in fact are formed by a CxxCxxC triplet of cysteines, consistent with repetition of the n, $n + 3$ coordination rule.

For the 8-Fe and 7-Fe ferredoxins, which are closely homologous in residues 1–50, the [4Fe-4S] clusters are formed by a triplet plus one cysteine distal in the sequence. In spite of this homology an important difference exists between the 8-Fe and 7-Fe ferredoxin [4Fe-4S] clusters. If one views along the distal cysteine S_γ–Fe bond in cluster II of *P. a.* ferredoxin [S_γ(18)–Fe(1), Fig. 2*f*], the arrangement of the cysteine triplet is clockwise. (The cysteine triplet of cluster I is also clockwise, the molecule having approximate twofold symmetry.) If one views down the corresponding bond in *A. v.* ferredoxin [S_γ(24)–Fe(1), Fig. 2*g*], the triplet is seen to be counterclockwise. The reason for this is the unique position of S_γ(24) following the C_{20}PVD turn (Fig. 7*e*), and this is a consequence of C_{20} (equivalent to C_{18} in *P. a.*) being a ligand of the [3Fe-3S] cluster. The [4Fe-4S] cluster is thus constrained to be differently oriented in its binding cavity. In particular, S_γ(39) and S_γ(45) superpose on S_γ(35) and S_γ(41) in the *P. a.* structure, but S_γ(42) (of *A. v.*) goes over the cluster rather than under because of the ligand site already occupied by S_γ(24) (Fig. 7*e*). It appears that the conformation of a 7-Fe ferredoxin is such that the presence of a [3Fe-3S] site determines that the cysteine triplet will have the opposite hand. The resulting structure presents a different environment to the [4Fe-4S] cluster, and might be a factor in selecting the high-potential redox couple ($E_m = +0.320$ V), (96) over the low-potential couple ($E_m = -0.420$ V). In a sense the *A. v.* structure represents an example of the influence of a 3Fe cluster on the properties of a neighboring cluster.

To summarize, three types of [4Fe-4S] cluster cysteine coordination patterns have been observed in the protein structures: (CxxC) + 1 + 1 in HIPIP; cysteine L(clockwise)-triplet + 1 in *P. a.* ferredoxin; and cysteine D-triplet + 1 in *A. v.* ferredoxin.

4.3 Coordination of a [3Fe-3S] Cluster

The [3Fe-3S] cluster in *A. v.* ferredoxin is bound to the protein by cysteines 8, 11, 16, 20, and 49, and a sixth ligand which is either glutamic acid 18 or an exogenous small molecule (Fig. 3) (see below). The twist boat representation of the cluster (Section 3.4) provides a good description of the ligand orientations with respect to the core, except for some distortion about Fe-2 involving 'O_ϵ' and $S_\gamma(16)$. This description is also consistent with the observation that S-3 is more in the plane of the 3 Fe atoms than are S-1 or S-2, in agreement with the conformation of the twist-boat $[Sn_3S_3](CH_3)_6$ structure (Table 10). Consequently, the planar fragment of Fe-1, S-3, and Fe-3 resides in an environment of 4 S_γ ligands (Fig. 2*c*), just as the [2Fe-2S] cluster does (Fig. 2*b*). This similarity leads one to suppose that the $[3Fe\text{-}3S](S_\gamma)_5$('$O_\epsilon$') structure is composed of a dimer fragment of $[2Fe\text{-}S](S_\gamma)_4$ into which a $Fe(S)_2(S_\gamma)$('O_ϵ') moiety has been inserted (31).

The dimer fragment is coordinated in a manner not unlike the [2Fe-2S] cluster (Fig. 8). In each the 2 Fe atoms are bound by a CxxC segment: C_{46} STC_{49} in *S. p.* ferredoxin, and C_8IKC_{11} in *A. v.* ferredoxin (Table 11). The coordination of each dimer is completed by two cysteines removed in the primary structure. In *A. v.* each of these occur in a similar peptide, $C_{20}PV$ and $C_{49}PA$.

The third Fe atom of the [3Fe-3S] cluster, Fe-2 (Fig. 2*c*), is inserted by $S_\gamma(16)$. This ligand residue belongs to a hexapeptide which appears to be a special feature of [3Fe-3S] proteins: $C_{16}VE_{18}VC_{20}P$ (31). In one sense the sequence CxExC is a triplet for [3Fe-3S] clusters, whereas a longer sequence CxxCxxC is required as a triplet for [4Fe-4S] clusters.

In Table 12 sequence data for several ferredoxins are compared with the *A. v.* primary structure. The biochemical and spectroscopic data for these proteins suggest that each contains a [3Fe-3S] cluster. In particular, this has been firmly established for *D. gigas* ferredoxin II (18). The strong homology observed for CVEVCP-type sequences reinforces the importance of this peptide, and suggests that 3-Fe centers are potentially widespread in nature. Substitutions occur, however, and apparently isoleucine or glutamic acid can replace valine. The *M. s.* sequence is interesting because it lacks C_{11}, where instead the sequence $C_8VD_{10}V$ suggests involvement of aspartic acid. In particular, the primary sequence of this protein (113) is homologous to *A. v.* ferredoxin but has eight, not nine, cysteines; the peptide CVDV is like CVEV; and coordination by the oxyligand D_{10} would involve a different Fe atom than does E_{18} *if* the structures are related. Regardless, the sequence data in Table 11 suggest that 3-Fe center coordination schemes may vary considerably. For example, in *Chromatium* ferredoxin the potential [3Fe-3S] binding site appears in the C-terminal half of the sequence (Table 12). Finally, it is interesting to note that the *D. g.* protein has two deletions such that a CxxCxxC triplet overlaps CxExC at the C_{14} position, suggesting the manner in which this molecule accommodates either a [4Fe-4S] or a [3Fe-3S] cluster (18, 19).

A comment on the role of glutamate 18: Refinement of the *Azotobacter* ferredoxin structure at 2.0 Å resolution has led to an increasingly clear image of the [3Fe-3S] cluster. In particular, electron density at the fourth coordi-

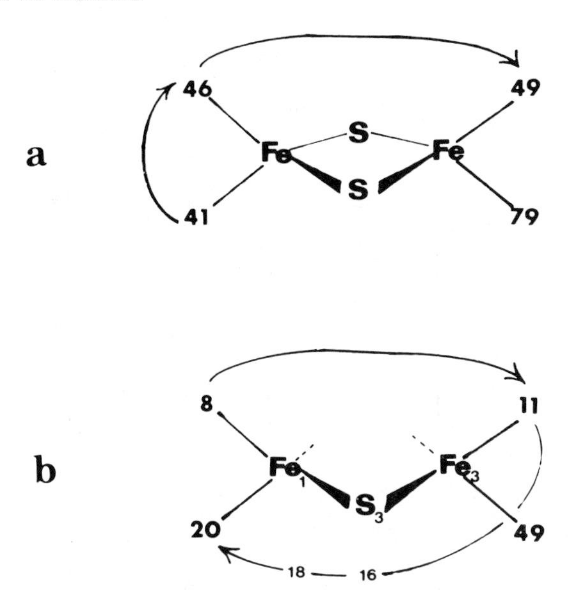

Figure 8. Cysteine coordination of two Fe-S "dimers." *(a) S. p.* ferredoxin (after Fig. 2 of ref. 14). *(b) A. v.* ferredoxin. The atom labels refer to Figure 2c.

nation position of Fe-2 has become resolved from the main chain electron density of residues 17–19, and weaker electron density for the glutamate 18 side chain has developed in the cleft (31). The peak proximal to Fe-2 has the same electron density as tightly bound water [$1.4e^-/\text{Å}^3$) and refines to $B \simeq 13\ \text{Å}^2$, the same thermal factor for the [3Fe-3S] cluster atoms. The B value of the glutamate side chain is 25–27 Å^2. Thus the crystallographic data are not inconsistent with a water (or hydroxyl) bound to Fe-2. In this model the amide NH of glutamate 18 may also donate a H bond to the O bound to Fe, and the glutamate 18 carboxyl is hydrogen-bonded to the carbonyl of alanine 75.

If the sixth ligand of the [3Fe-3S] cluster is an exogenous O atom, then the EPR and Mössbauer data argue that it could be hydroxyl in the oxidized state of the cluster. The orientation of the C_α-C_β bond of glutamate 18 is ideal for rotation of the sidechain, permitting interaction of the carboxyl group with Fe-2 or an exogenous ligand. Such a conformational change could play an important role in the reduction or activation of the cluster. Regardless, the position of glutamate 18 in the structure and its sequence conservation suggest an important role for this residue in the chemistry of the [3Fe-3S] center.

Table 12 Primary Sequence Data for Some Ferredoxins

Species				1							8			11					16		18		20			
A. v. (111)[a]				A	F	V	V	T	D	N	C	I	K	C	K	Y	T	D	C	V	E	V	C	P	V	...
Ps. o. (112)				T	F	V	V	T	D	N	C	I	K	C	K	Y	T	D	C	V	E	V	C	P	V	...
M. s. (113)				T	Y	V	I	A	E	P	*C*	*V*	*D*	*V*	K	D	K	A	C	I	E	E	C	P	V	...
T. t. (114)				T	H	V	I	C	Q	T	C	I	G	V	K	D	Q	S	C	V	E	V	C	P	V	...
D. g. (I, II) (19)				P	I	E	V	N	D	D	C	M	A	C	—	E	—	A	C	V	E	I	C	P	D	...
D. af. I (115)	A	R	K	F	Y	V	D	E	D	Q	C	I	A	C	—	E	—	S	C	V	E	I	C	P	G	...
D. af. II (115)	A	R	V	V	Y	V	D	( )	D	( )	C	I	( )	C	—	A	—	A	C	V	E	I	C	P	D	...
				33																						
C. v. (116)			...	E	P	S	L	C	T	E	C	V	E	H	Y	E	T	Q	C	V	E	V	C	P	V	...
																		S								

[a]References to sequence determinations are given in parentheses.

4.4 Factors Influencing Oxidation-Reduction Properties

Clearly, a great deal more is involved in determining the properties of a Fe-S center than coordination geometry. As discussed by Carter (42) and Adman (43), a primary factor may be NH···S hydrogen bonding (Table 11). In the *P. a.* ferredoxin the number of NH···S hydrogen bonds is almost twice that in HIPIP, which is consistent with the idea that they stabilize a more reduced cluster. However, polypeptide folding in the two structures differs dramatically. In HIPIP, amides in antiparallel β sheets participate in NH···S hydrogen bonds, whereas in ferredoxin principally the CxxCxxC triplet amides are involved.

In HIPIP (Fig. 7*c*) two portions of the molecule form an interface at the cluster binding site such that the [4Fe-4S] core is embedded in hydrophobic and aromatic residues (24). Interactions at the interface involve β-structure hydrogen bonds and close packing of hydrophobic side chains. The arrangement suggests that the protein selectively allows expansion of the [4Fe-4S] core in one direction only, as observed in the HP_{ox}-HP_{red} difference Fourier maps (25). A detailed mechanism involving change in the hydrogen bonding network on oxidation-reduction has been proposed (42).

The *P. a.* ferredoxin molecule has overall twofold symmetry (34). Antiparallel β sheet is formed in the loop between C-18 and C-35 (residues 23–25 to 28–30) and between N- and C-terminal stretches (residues 2–6 to 49–54) (Fig. 7*d*). The β-structure loop between C-18 and C-35 is homologous to a similar feature between C-46 and C-63 in HIPIP when the [4Fe-4S] clusters are placed in the same unique orientation (27). A similar antiparallel β-structure loop about the [4Fe-4S] cluster is found in the *A. v.* structure between C-24 and C-39 (Fig. 7*e*). However, the conformation of the 7-Fe ferredoxin (31) differs markedly from the 8-Fe ferredoxin due to the [3Fe-3S] cluster and concomitant *D* triplet coordination of the [4Fe-4S] cluster (Section 4.2). Consequently, the N terminus lies close to cluster II in clostridial ferredoxin, but is 20 Å away from the [4Fe-4S] cluster in the 7-Fe ferredoxin.

Rubredoxin and *S. p.* ferredoxin are similar in that each forms a shallow cavity for the Fe-S site (Fig. 7*a*, 7*b*) fairly near the surface of the molecule. Each also contains some degree of pleated sheet: 14 of 54 residues in rubredoxin (8), 21 of 98 in *S. p.* ferredoxin (15). The *S. p.* ferredoxin structure is reminiscent of other electron-transfer proteins, including flavodoxin, plastocyanin, and azurin (see ref. 43), where a protein core of β secondary structure forms a binding site at one end of the molecule.

If there is a common structural motif for the Fe-S proteins, it is a lack of secondary structure. The largest of the structures, *A. v.* ferredoxin, exhibits essentially no secondary structure at all. HIPIP, although containing two short α helices and a small segment of three-stranded pleated sheet, has 31 of

85 residues involved in hairpin turn conformations (24). A number of hairpin turns are found in the 8-Fe and 7-Fe ferredoxin structures, where the protein folding is dominated by the metal chelation. In these two molecules the sequence CPV/A is found four times, and in each instance it turns the chain away from an Fe-S cluster. In rubredoxin a 3_{10} helical H bonding pattern has been described for four turns in the molecule (8).

At least three factors have been discussed in terms of protein environmental effects on the reduction potential of Fe-S sites: proximity of aromatic sidechains, distribution of sequence invariant amino acids, and relative solvent exposure.

Special attention has been focused on the role of tyrosine 19 in HIPIP and tyrosines 2 and 28 in *P. a.* ferredoxin, as each lies adjacent to a [4Fe-4S] cluster. In HIPIP Y19 has been implicated in an electron-transfer mechanism, and its interaction with the cluster represents one aspect in protein discrimination of two cluster diastereomers (27). In *Azotobacter* ferredoxin two phenylalanines, F25 and F55, lie adjacent to opposite sides of the high-potential [4Fe-4S] cluster (31), while the aromatic homologous to Y2 of *P. a.* ferredoxin, F2, is ~20 Å removed from the [4Fe-4S] cluster. Interestingly, 7-Fe ferredoxins from *M. s.* and *T. t.* (Table 12) each have isoleucine at position 25 and tyrosine at position 55. There are no aromatic residues in close proximity to the [3Fe-3S] cluster.

Correlations are observed between the distribution of charged or sequence conserved amino acids and proximity to the active sites. In *C. p.* and *D. v.* rubredoxins a clustering of negatively charged residues, away from $Fe(S_\gamma)_4$, may be a recognition site (43). In five rubredoxin sequences residues 47 and 48, in the vicinity of the $Fe(S_\gamma)_4$ site, are found to be acidic amino acids. In chloroplast ferredoxin 22 amino acids, invariant in 26 sequences, are markedly distributed near the [2Fe-2S] cluster in the structure of the *S. p.* protein (15), but only 2 of these are charged residues. In *P. a.* ferredoxin, fully 33 of 54 amino acids are conserved or semiconserved in relation to five clostridial ferredoxin sequences (34); the majority occur within the cluster binding segments; and one invariant residue (D37) forms a salt bridge to the N terminus, perhaps stabilizing the interaction of Y2 and cluster II.

The structural importance of 12 conserved aromatic and hydrophobic residues in the [4Fe-4S] cluster binding segments of three bacterial HIPIPs has been noted (25). Electron-transfer kinetic studies of HIPIPs from four species indicate that the net negative charge on the Fe-S cluster is delocalized onto the surface of the protein through NH···S H bonds to carbonyl O atoms (117). The charge is also partially neutralized by sequence conserved lysine, arginine, or histidine residues.

A striking pattern is seen in the local environment of the [3Fe-3S] cluster in *A. v.* ferredoxin. Here the carbonyls of residues 8 and 9 are directed *to-*

ward the cluster [to Fe(2) and Fe(3), respectively] and six basic residues (K10, K12, K98, K100, H103, and R106) are arranged about the Fe-S site. This observation appears consistent with the Mössbauer electronic model for 3-Fe centers in which the oxidized cluster contains 3 Fe^{3+} (97). In other words, a reduced [3Fe-3S] cluster would bear a formal charge of −4, one charge more negative than reduced [4Fe-4S] ferredoxin clusters, and two charges more negative than reduced HIPIP clusters. Carbonyl groups may also play a role in delocalizing the negative charge of 3-Fe centers.

The high-potential [4Fe-4S] cluster in the *A. v.* protein has only one adjacent basic residue (H35), and like the HIPIP cluster it is surrounded by aliphatic and aromatic amino acids. At the same time, the carbonyls of the cysteine peptides are directed away from this cluster, a feature observed in *S. p.* and *P. a.* ferredoxins (14, 43).

A third factor influencing redox potential is solvent exclusion from the Fe-S site. In aqueous media the potential of $[Fe_4S_4(SPh)_4]^{2-/3-}$ is shifted to appreciably more negative values (88). Pulsed EPR experiments with 2-Fe, 4-Fe, and 8-Fe ferredoxins and HIPIPs show that protons in close proximity to the low-potential Fe-S centers can be exchanged in D_2O, but no such exchange is observed in HIPIPs (100). In rough structural terms the high-potential [4Fe-4S] center of HIPIP is buried, while the low-potential [3Fe-3S] center and [4Fe-4S] centers in *P. a.* ferredoxin are relatively exposed to the solvent.

To summarize, Fe-S proteins are seen to modulate the properties of an Fe-S cluster in several fundamental ways. Exclusion of the cluster from solvent by the polypeptide raises the redox potential. H bonding from amide NH to both inorganic S and S_γ stabilizes negative charge, and favors two of three possible redox states of [4Fe-4S] clusters. NH···C=O H bonds and adjacent basic residues promote delocalization of negative cluster charge. Charge distribution in the protein appears important in stabilizing [3Fe-3S] clusters. The polypeptide provides a stereospecific binding cavity and accommodates inherently distorted (D_{2d} symmetry) [4Fe-4S] clusters. At the same time the protein architecture promotes conformational change of an intrinsically asymmetric [4Fe-4S] cluster in HIPIP. Geometrical distortions of the [3Fe-3S] cluster in *Azotobacter* ferredoxin suggest a similar ability of the protein to control conformational change at this Fe-S site.

ACKNOWLEDGMENTS

The author's research has been supported by NIH grant GM-25672 and USDA grant 79-58-2424.

REFERENCES

1. International Union of Biochemistry, *Biochem. J.*, **181**, 513-516 (1979).
2. M. S. Weininger and L. E. Mortenson, *Airlie Symposium on Fe in Proteins*, Abstract VII-9, April 1980.
3. J. R. Herriott, L. C. Sieker, and L. H. Jensen, *J. Mol. Biol.*, **50**, 391-406 (1970).
4. J. R. Herriott, K. D. Watenpaugh, L. C. Sieker, and L. H. Jensen, *J. Mol. Biol.*, **80**, 423-432 (1973).
5. K. D. Watenpaugh, L. C. Sieker, J. R. Herriott, and L. H. Jensen, *Acta Crystallogr.*, **B29**, 943-956 (1973).
6. K. D. Watenpaugh, L. C. Sieker, J. R. Herriott, and L. H. Jensen, *Cold Spring Harbor Symp. Quant. Biol.*, **36**, 359-367 (1971).
7. K. D. Watenpaugh, T. N. Margulis, L. C. Sieker, and L. H. Jensen, *J. Mol. Biol.*, **122**, 175-190 (1978).
8. K. D. Watenpaugh, L. C. Sieker, and L. H. Jensen, *J. Mol. Biol.*, **131**, 509-522 (1979).
9. K. D. Watenpaugh, L. C. Sieker, and L. H. Jensen, *J. Mol. Biol.*, **138**, 615-633 (1980).
10. M. Pierrot, R. Haser, M. Frey, M. Bruschi, J. LeGall, L. C. Sieker, and L. H. Jensen, *J. Mol. Biol.*, **107**, 179-182 (1976).
11. E. T. Adman, L. C. Sieker, L. H. Jensen, M. Bruschi, and J. LeGall, *J. Mol. Biol.*, **112**, 113-120 (1977).
12. M. Bruschi, I. Moura, J. LeGall, A. V. Xavier, and L. C. Sieker, *Biochem. Biophys. Res. Commun.*, **90**, 596-605 (1979).
13. K. Ogawa, T. Tsukihara, H. Tahara, Y. Katsube, Y. Matsu-ura, N. Tanaka, M. Kakudo, K. Wada, and H. Matsubara, *J. Biochem.*, **81**, 529-531 (1977).
14. T. Tsukihara, K. Fukuyama, H. Tahara, Y. Katsube, Y. Matsu-ura, N. Tanaka, M. Kakudo, K. Wada, and H. Matsubara, *J. Biochem.*, **84**, 1645-1647 (1978).
15. K. Fukuyama, T. Hase, S. Matsumoto, T. Tsukihara, Y. Katsube, N. Tanaka, M. Kakudo, K. Wada, and H. Matsubara, *Nature*, **286**, 522-524 (1980).
16. A. Kunita, M. Koshibe, Y. Nishikawa, K. Fukuyama, T. Tsukihara, Y. Katsube, Y. Matsu-ura, N. Tanaka, M. Kakudo, T. Hase, and H. Matsubara, *J. Biochem.*, **84**, 989-992 (1978).
17. J. L. Sussman, P. Zipori, M. Harel, A. Yonath, and M. M. Werber, *J. Mol. Biol.*, **134**, 375-377 (1979).
18. B. H. Huynh, J. J. G. Moura, I. Moura, T. A. Kent, J. LeGall, A. V. Xavier, and E. Münck, *J. Biol. Chem.*, **255**, 3242-3244 (1980).
19. M. Bruschi, *Biochem. Biophys. Res. Commun.*, **91**, 623-628 (1979).
20. J. Kraut, G. Strahs, and S. T. Freer, in *Structural Chemistry and Molecular Biology*, A. Rich and N. Davidson, eds., Freeman, San Francisco, 1968, pp. 55-64.
21. G. Strahs and J. Kraut, *J. Mol. Biol.*, **35**, 503-512 (1968).
22. C. W. Carter, S. T. Freer, Ng. H. Xuong, R. A. Alden, and J. Kraut, *Cold Spring Harbor Symp. Quant. Biol.*, **36**, 381-385 (1971).
23. C. W. Carter, J. Kraut, S. T. Freer, R. A. Alden, L. C. Sieker, E. Adman, and L. H. Jensen, *Proc. Natl. Acad. Sci.*, **69**, 3526-3529 (1972).
24. C. W. Carter, J. Kraut, S. T. Freer, Ng. H. Xuong, R. A. Alden, and R. G. Bartsch, *J. Biol. Chem.*, **249**, 4212-4225 (1974).

25. C. W. Carter, J. Kraut, S. T. Freer, and R. A. Alden, *J. Biol. Chem.*, **249**, 6339-6346 (1974).
26. S. T. Freer, R. A. Alden, C. W. Carter, and J. Kraut, *J. Biol. Chem.*, **250**, 46-54 (1975).
27. C. W. Carter, *J. Biol. Chem.*, **252**, 7802-7811 (1977).
28. C. D. Stout, *J. Biol. Chem.*, **254**, 3598-3599 (1979).
29. C. D. Stout, *Nature*, **279**, 83-84 (1979).
30. C. D. Stout, D. Ghosh, V. Pattabhi, and A. H. Robbins, *J. Biol. Chem.*, **255**, 1797-1800 (1980).
31. D. Ghosh, W. F. Furey, S. O'Donnell, and C. D. Stout, *J. Biol. Chem.*, **256**, 4185-4192 (1981).
32. L. C. Sieker and L. H. Jensen, *Biochem. Biophys. Res. Commun.*, **20**, 33-35 (1965).
33. L. C. Sieker, E. Adman, and L. H. Jensen, *Nature*, **235**, 40-42 (1972).
34. E. T. Adman, L. C. Sieker and L. H. Jensen, *J. Biol. Chem.*, **248**, 3987-3996 (1973).
35. E. Adman, K. D. Watenpaugh, and L. H. Jensen, *Proc. Natl. Acad. Sci.*, **72**, 4854-4858 (1975).
36. E. T. Adman, L. C. Sieker, and L. H. Jensen, *J. Biol. Chem.*, **251**, 3801-3806 (1976).
37. J. Lee, S. C. Chang, K. Hahm, A. J. Glaid, O. Gawron, B. C. Wang, C. S. Yoo, M. Sax, and J. Glusker, *J. Mol. Biol.*, **112**, 531-534 (1977).
38. C. M. Kennedy, R. Rauner, and O. Gawron, *Biochem. Biophys. Res. Commun.*, **47**, 740-745 (1972).
39. L. H. Jensen, in *Iron-Sulfur Proteins*, Vol. II, W. Lovenberg, ed., Academic, New York, 1973, pp. 163-194.
40. L. H. Jensen, *Ann. Rev. Biochem.*, **43**, 461-474 (1974).
41. L. H. Jensen, *Ann. Rev. Biophys. Bioeng.*, **3**, 81-93, (1974).
42. C. W. Carter, in *Iron-Sulfur Proteins*, Vol. III, W. Lovenberg, ed., Academic, New York, 1977, pp. 157-204.
43. E. T. Adman, *Biochim. Biophys. Acta*, **549**, 107-144 (1979).
44. T. L. Blundell and L. N. Johnson, *Protein Crystallography*, Academic, New York, 1976.
45. B. W. Matthews, *J. Mol. Biol.*, **33**, 491-497 (1968).
46. D. C. Yoch and D. I. Arnon, *J. Biol. Chem.*, **247**, 4514-4520 (1972).
47. J. Jordanov, T. Hazen, A. Cotton, and R. Swanson, private communication (1978).
48. B. B. Buchanan, W. Lovenberg, and J. C. Rabinowitz, *Proc. Natl. Acad. Sci.*, **49**, 345-355 (1963).
49. W. Lovenberg, B. B. Buchanan, and J. C. Rabinowitz, *J. Biol. Chem.*, **238**, 3899-3912 (1963).
50. F. J. Ruzicka and H. Beinert, *J. Biol. Chem.*, **253**, 2514-2517 (1978).
51. D. M. Kurtz, R. H. Holm, F. J. Ruzicka, H. Beinert, C. J. Coles, and T. P. Singer, *J. Biol. Chem.*, **254**, 4967-4969 (1979).
52. P. G. Avis, F. Bergel, and R. C. Bray, *J. Chem. Soc.*, **1955**, 1100-1105.
53. A. McPherson, *Meth. Biochem. Anal.*, **23**, 249-345 (1976).
54. A. McPherson, *J. Biol. Chem.*, **251**, 6300-6303 (1976).
55. C. W. Carter, *J. Biol. Chem.*, **254**, 12,219-12,223 (1979).
56. G. A. Petsko, D. C. Phillips, R. J. P. Williams, and I. A. Wilson, *J. Mol. Biol.*, **120**, 345-359 (1978).

57. E. Münck, P. G. Debrunner, J. C. M. Tsibris, and I. C. Gunsalus, *Biochemistry*, **11**, 855-863 (1972).
58. K. S. Hagen, J. G. Reynolds, and R. H. Holm, *J. Am. Chem. Soc.*, **103**, in press (1981).
59. K. G. Strothkamp, J. Lehmann, and S. J. Lippard, *Proc. Natl. Acad. Sci.*, **75**, 1181-1184 (1978).
60. J. J. Lipka, S. J. Lippard, and J. S. Wall, *Science*, **206**, 1419-1421 (1979).
61. W. A. Hendrickson and M. M. Teeter, *Nature*, **290**, 107-113 (1981).
62. R. H. Stanford, *Acta Crystallogr.*, **17**, 1180-1181 (1964).
63. M. H. Emptage, T. A. Kent, B. H. Huynh, J. Rawlings, W. H. Orme-Johnson, and E. Münck, *J. Biol. Chem.*, **255**, 1793-1797 (1980).
64. R. H. Holm and J. A. Ibers, in *Iron-Sulfur Proteins*, Vol. III, W. Lovenberg, ed., Academic, New York, 1977, pp. 205-281.
65. R. H. Holm, *Acc. Chem. Res.*, **10**, 427-434 (1977).
66. W. H. Orme-Johnson and R. H. Holm, *Meth. Enzym.*, **53**, 268-274 (1978).
67. B. A. Averill, J. R. Bale, and W. H. Orme-Johnson, *J. Am. Chem. Soc.*, **100**, 3034-3043 (1978).
68. R. W. Lane, J. A. Ibers, R. B. Frankel, G. C. Papaefthymiou, and R. H. Holm, *J. Am. Chem. Soc.*, **99**, 84-98 (1977).
69. D. Coucouvanis, D. Swenson, N. C. Baenziger, D. G. Holah, A. Kostikas, A. Simopoulois, and V. Petrouleas, *J. Am. Chem. Soc.*, **98**, 5721-5723 (1976).
70. T. Ohnishi, H. Blum, S. Sato, K. Nakazawa, K. Hon-nami, and T. Oshima, *J. Biol. Chem.*, **255**, 345-348 (1980).
71. R. Bachofen and D. I. Arnon, *Biochim. Biophys. Acta*, **120**, 259-265 (1966).
72. K. Melis and C. D. Stout, unpublished results (1980).
73. V. K. Shah and W. J. Brill, *Biochim. Biophys. Acta*, **305**, 445-454 (1973).
74. A. Robbins and C. D. Stout, unpublished results (1979).
75. J. C. Phillips, A. Wlodawer, J. M. Goodfellow, K. D. Watenpaugh, L. C. Sieker, L. H. Jensen, and K. O. Hodgson, *Acta Crystallogr.*, **A33**, 445-455 (1977).
76. J. J. Mayerle, R. B. Frankel, R. H. Holm, J. A. Ibers, W. D. Phillips, and J. F. Weiher, *Proc. Natl. Acad. Sci.*, **70**, 2429-2433 (1973).
77. J. J. Mayerle, S. E. Denmark, B. V. De Pamphilis, J. A. Ibers, and R. H. Holm, *J. Am. Chem. Soc.*, **97**, 1032-1045 (1975).
78. G. Palmer, in *Iron-Sulfur Proteins*, Vol. II, W. Lovenberg, ed., Academic, New York, 1973, pp. 285-325.
79. C. Y. Yang, K. H. Johnson, R. H. Holm, and J. G. Norman, *J. Am. Chem. Soc.*, **97**, 6596-6598 (1975).
80. L. Pauling, *Proc. Natl. Acad. Sci.*, **72**, 4200-4202 (1975).
81. T. Herskovitz, B. A. Averill, R. H. Holm, J. A. Ibers, W. D. Phillips, and J. F. Weiher, *Proc. Natl. Acad. Sci.*, **69**, 2437-2441 (1972).
82. B. A. Averill, T. Herskovitz, R. H. Holm, and J. A. Ibers, *J. Am. Chem. Soc.*, **95**, 3523-3534 (1973).
83. R. H. Holm, B. A. Averill, T. Herskovitz, R. B. Frankel, H. B. Gray, O. Silman, and F. J. Grunthaner, *J. Am. Chem. Soc.*, **96**, 2644-2646 (1974).
84. B. V. DePamphilis, B. A. Averill, T. Herskovitz, L. Que, and R. H. Holm, *J. Am. Chem. Soc.*, **96**, 4159-4167 (1974).

85. L. Que, M. A. Bobrile, J. A. Ibers, and R. H. Holm, *J. Am. Chem. Soc.*, **96**, 4168-4178 (1974).
86. H. L. Carrell, J. P. Glusker, R. Job, and T. C. Bruice, *J. Am. Chem. Soc.*, **99**, 3683-3690 (1977).
87. R. W. Johnson and R. H. Holm, *J. Am. Chem. Soc.*, **100**, 5338-5344 (1978).
88. C. L. Hill, J. Renaud, R. H. Holm, and L. E. Mortenson, *J. Am. Chem. Soc.*, **99**, 2549-2557 (1977).
89. E. J. Laskowski, R. B. Frankel, W. O. Gillum, G. C. Papaefthymiou, J. Renaud, J. A. Ibers, and R. H. Holm, *J. Am. Chem. Soc.*, **100**, 5322-5337 (1978).
90. R. Cammack, *Biochem. Biophys. Res. Commun.*, **54**, 548-554 (1973).
91. J. M. Berg, K. O. Hodgson, and R. H. Holm, *J. Am. Chem. Soc.*, **101**, 4586-4593 (1979).
92. J. H. Konnert, *Acta Crystallogr.*, **A32**, 614-617 (1976).
93. Y. I. Shethna, *Biochim. Biophys. Acta*, **205**, 58-62 (1970).
94. G. Menzebach and P. Bleckman, *J. Organomet. Chem.*, **91**, 291-294 (1975).
95. K. S. Hagen, J. M. Berg, and R. H. Holm, *Inorg. Chim. Acta*, **45**, L17-L18 (1980).
96. W. V. Sweeney, J. C. Rabinowitz, and D. C. Yoch, *J. Biol. Chem.*, **250**, 7842-7847 (1975).
97. T. A. Kent, B. H. Huynh, and E. Münck, *Proc. Natl. Acad. Sci.*, **77**, 6574-6576 (1980).
98. B. H. Huynh, M. T. Henzl, J. A. Christner, R. Zimmermann, W. H. Orme-Johnson, and E. Münck, *Biochim. Biophys. Acta*, **623**, 124-138 (1980).
99. W. K. Shah and W. J. Brill, *Proc. Natl. Acad. Sci.*, **74**, 3249-3253 (1977).
100. J. Peisach, N. R. Orme-Johnson, W. B. Mims, and W. H. Orme-Johnson, *J. Biol. Chem.*, **252**, 5643-5650 (1977).
101. S-P. W. Tang, T. G. Spiro, C. Antanaitis, T. H. Moss, R. H. Holm, T. Herskovitz, and L. E. Mortensen, *Biochem. Biophys. Res. Commun.*, **62**, 1-6 (1975).
102. D. Piszkiewicz, O. Gawron, and J. C. Sutherland, *Biochemistry*, in press, 1981.
103. M. O. Dayhoff, L. T. Hunt, P. J. McLaughlin, and D. D. Jones, in *Atlas of Protein Sequence and Structure*, M. O. Dayhoff, ed., National Biomedical Research Foundation, Washington, 1976.
104. I. Moura, M. Bruschi, J. LeGall, J. J. G. Moura, and A. V. Xavier, *Biochem. Biophys. Res. Commun.*, **75**, 1037-1044 (1977).
105. I. Moura, B. H. Huynh, R. P. Hausinger, J. LeGall, A. V. Xavier, and E. Münck, *J. Biol. Chem.*, **255**, 2493-2498 (1980).
106. K. F. McCarthy, Ph.D. Thesis, George Washington University, Washington, D.C., 1972.
107. M. Bruschi, *Biochim. Biophys. Acta*, **434**, 4-17 (1976).
108. J. N. Tsunoda, K. T. Yasunobu, and H. R. Whiteley, *J. Biol. Chem.*, **243**, 6262-6272 (1968).
109. K. Dus, S. Tedro, R. G. Bartsch, and M. D. Kamen, *Biochem. Biophys. Res. Commun.*, **43**, 1239-1245 (1971).
110. K. Wada, T. Hase, H. Tokunaga, and H. Matsubara, FEBS Lett., **55**, 102-104 (1975).
111. J. B. Howard, T. Lorsbach, K. Melis, and C. D. Stout, manuscript in preparation, 1981.
112. T. Hase, S. Wakabayashi, H. Matsubara, D. Ohmori, and K. Suzuki, *FEBS Lett.*, **91**, 315-319 (1978).
113. T. Hase, S. Wakabayashi, H. Matsubara, T. Imai, T. Matsumoto, and J. Tobari, *FEBS Lett.*, **103**, 224-228 (1979).

114. S. Sato, K. Nakazawa, K. Hon-nami, and T. Oshima, unpublished results communicated by T. Ohnishi (1980).

115. C. E. Hatchikian, H. E. Jones, M. Bruschi, G. Bovier-Lapierre, J. Bonicel, P. Couchoud, and N. Forget, *Biochim. Biophys. Acta,* **548,** 471–483 (1979).

116. T. Hase, H. Matsubara, and M. C. W. Evans, *J. Biochem.,* **81,** 1745–1749 (1977).

117. I. A. Mizrahi, T. E. Meyer, and M. A. Cusanovich, *Biochemistry,* **19,** 4727–4733 (1980).

CHAPTER 4

Mössbauer Studies of [3Fe-3S] Clusters and Sulfite Reductase

ECKARD MÜNCK

Department of Biochemistry
Gray Freshwater Biological Institute
University of Minnesota
Navarre, Minnesota 55392

CONTENTS

1 INTRODUCTION

Mössbauer spectroscopy has proven to be an indispensable tool for the physical characterization of Fe-S centers. When used in conjunction with EPR, this technique can yield decisive clues for the characterization of novel systems. The utility of Mössbauer spectroscopy for Fe-S proteins is intimately connected with the fact that one is dealing with spin-coupled structures. In Section 2 we discuss why this technique is so well suited for the task.

In the past five years a few new clusters have been added to the familiar prototypes, among them the P clusters (1, 2) and cofactor centers (3, 4) of nitrogenase. Most recently the discovery of [3Fe-3S] clusters (5, 6) has added new excitement to the field and stimulated research in enzymology, bioinorganic chemistry, and physics. In Section 3 we review the spectroscopic evidence for [3Fe-3S] clusters and discuss the physics of spin coupling in some detail.

Section 4 deals with very recent Mössbauer studies of *E. coli* sulfite reductase which strongly suggest a novel active site arrangement—a siroheme exchange-coupled to a [4Fe-4S] cluster.

Since all results discussed here were obtained during the past 15 months, this chapter should be viewed in the spirit of a progress report.

2 METHODOLOGY FOR ESTABLISHING THE NATURE OF CLUSTERS

The simplest type of magnetic Mössbauer spectrum results when an effective magnetic field $\mathbf{H}_{\mathbf{eff}}$ acts on the ^{57}Fe nucleus. [Mössbauer spectroscopy as applied to biological problems has been discussed extensively in articles by Moss (7), Lang (8), Debrunner (9), and Münck (10, 11).] The resulting splittings of the nuclear ground and excited states can be computed from the Hamiltonian:

$$\hat{H} = -g_n\beta_n\mathbf{I}\cdot\mathbf{H}_{\mathbf{eff}} \tag{1}$$

where $g_n\beta_n\mathbf{I}$ is the nuclear magnetic moment. An energy level diagram and some typical Mössbauer spectra are shown in Figure 1. In general, we describe the magnetic features of a Mössbauer spectrum with a spin Hamiltonian:

$$\hat{H} = \hat{H}_e + \hat{H}_{hf} \tag{2}$$

$$\hat{H}_e = D[S_z^2 - \frac{1}{3}S(S+1) + \lambda(S_x^2 - S_y^2)] + \beta\mathbf{S}\cdot\tilde{\mathbf{g}}\cdot\mathbf{H} \tag{3}$$

$$\hat{H}_{hf} = \mathbf{S}\cdot\tilde{\mathbf{A}}\cdot\mathbf{I} - g_n\beta_n\,\mathbf{I}\cdot\mathbf{H} \tag{4}$$

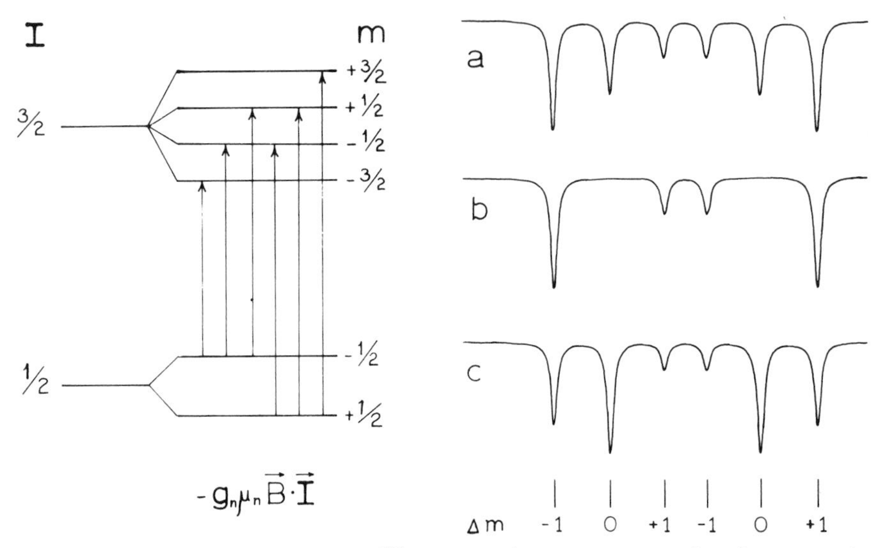

Figure 1. Nuclear Zeeman splittings of ^{57}Fe. *(a)* Mössbauer spectrum resulting from a sample with randomly distributed internal magnetic fields. Such spectra are typically observed for uniaxial systems. The internal field of each molecule in the sample is aligned parallel to *(b)* or perpendicular to *(c)* the observed Mössbauer radiation.

The zero-field splitting of the electronic system (D, λ) and the electronic Zeeman interaction are generally large compared to the hyperfine interactions, $\hat{H}_{hf}$. Thus the electronic terms of $\hat{H}_e$ define the electronic quantization axis (for a discussion see ref. 11). This allows us to write a nuclear Hamiltonian, $\hat{H}_N$, by replacing the operator **S** by an appropriately taken expectation value $\langle \mathbf{S} \rangle$:

$$\hat{H}_N = \langle \mathbf{S} \rangle \cdot \tilde{\mathbf{A}} \cdot \mathbf{I} - g_n \beta_n \mathbf{I} \cdot \mathbf{H} \tag{5}$$

$$= -g_n \beta_n \vec{\mathbf{I}} \left(-\frac{\langle \mathbf{S} \rangle \cdot \tilde{\mathbf{A}}}{g_n \beta_n} + \vec{\mathbf{H}} \right)$$

$$\hat{H}_N = -g_n \beta_n \mathbf{I} \left(\mathbf{H}_{\mathbf{int}} + \mathbf{H} \right) = -g_n \beta_n \mathbf{I} \cdot \mathbf{H}_{\mathbf{eff}} \tag{6}$$

Equation (6) connects the spin Hamiltonian with eq. (1). It can be seen that the magnetic features of the Mössbauer spectra are determined by the internal magnetic field:

$$\mathbf{H}_{\mathbf{int}} = \frac{-\langle \mathbf{S} \rangle \cdot \tilde{\mathbf{A}}}{g_n \beta_n} \tag{7}$$

It can be shown that in the absence of an applied field **H** non-Kramers systems (compounds or clusters with an even number of electrons and integer spin S) have $\langle \mathbf{S} \rangle = 0$; that is $\mathbf{H}_{\text{int}} = 0$. Consequently, in zero applied field the Mössbauer spectra of non-Kramers systems exhibit only quadrupole interactions at all temperatures. A nuclear energy level diagram and a quadrupole doublet are sketched in Figure 2. Typical non-Kramers systems are low-spin ($S = 0$) and high-spin ($S = 2$) ferrous ions, oxidized [2Fe-2S] centers, [4Fe-4S] clusters in the diamagnetic +2 state, reduced [3Fe-3S] centers, and fully reduced ferredoxin from *Clostridium pasteurianum* (both $S = 1/2$ clusters spin-couple).

Kramers systems (odd number of electrons and half-integral electronic spin), are fundamentally different. Sizable internal magnetic fields are observed even in zero applied field. (In practice one applies a weak field of a few hundred gauss, a field strong enough to decouple transferred hyperfine interactions with ligand nuclei.) With little loss of generality we can restrict the discussion to $S = 1/2$ systems; $S = 3/2$ and $S = 5/2$ systems are easily accommodated by referring to individual Kramers doublets of the multiplet. For $S = 1/2$ systems $\hat{H}_e$ of eq. (3) is simply $\beta \mathbf{S} \cdot \tilde{\mathbf{g}} \cdot \mathbf{H}$. Four arguments can be used to infer the presence of spin-coupled clusters:

1. Mössbauer spectra of proteins are usually taken on frozen solutions with a weak magnetic field applied either parallel or perpendicular to the observed γ-radiation. The spectra of the individual molecules in a frozen protein solution depend on the orientation of the molecular axes (i.e., the principal axis frame of the g tensor) relative to the directions of the applied field and the γ-beam. For an $S = 1/2$ system it follows from eq. (3) that the electronic spin is quantized along the direction $\mathbf{H}' = \tilde{\mathbf{g}} \cdot \mathbf{H}$; that is, $\langle \mathbf{S} \rangle$ has a nonzero component only along $\mathbf{H}'$. The distribution of $\mathbf{H}'$, and therefore the distributions of $\langle \mathbf{S} \rangle$ and $\mathbf{H}_{\text{int}}$, depend crucially on the electronic g values. Thus an appreciable amount of information about the magnetic features of a Mössbauer spectrum can be obtained from EPR work. This connection be-

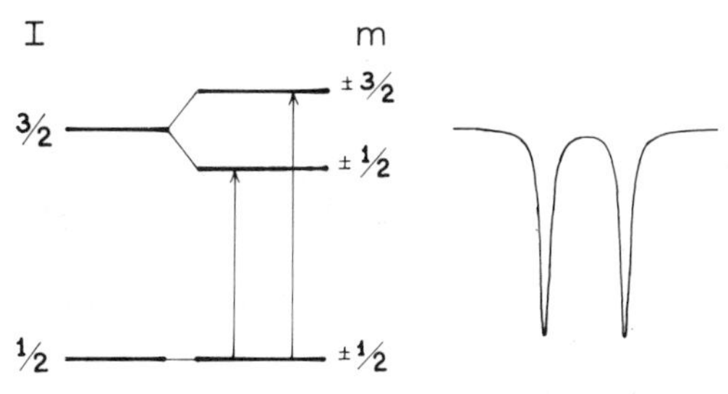

Figure 2. Quadrupole splitting in ^{57}Fe. On the right a typical spectrum, a quadrupole doublet, is shown. Compounds with integer electronic spin yield quadrupole doublets in zero applied field.

tween EPR and Mössbauer spectroscopy allows us to identify, usually unambiguously, those iron sites that are a structural component of an EPR-active center.

2. An even stronger connection between EPR and Mössbauer work can be made when the intensities of the Mössbauer absorption bands are taken into account. Since this connection has been described earlier in some detail (12), we briefly mention only the salient features. The point is best demonstrated by discussing the case of isotropic g and A tensors. For such a system $\mathbf{H}'$, $\langle \mathbf{S} \rangle$, and $\mathbf{H}_{\text{int}}$ are parallel to the applied field, independent of the molecular orientation. (*Isotropy* means that the system has no internal preferential orientation.) Thus, if the external field is applied parallel to the observed γ-radiation, each Fe nucleus in the sample experiences an internal field that is parallel to the observed γ-radiation. Since the Mössbauer radiation is magnetic dipole in character (a property of the ^{57}Fe nucleus), transitions that involve no change in the nuclear magnetic quantum numbers ($\Delta m = 0$ transitions) are forbidden (inclusion of quadrupole interactions modifies the argument slightly). In parallel field one observes a spectrum of the type shown in Figure 1*b*. If the field is applied transverse to the observed γ-radiation, the nuclear $\Delta m = 0$ lines are maximized (Fig. 1*c*).

The intensity changes observed in switching from parallel to transverse field are intimately correlated with the anisotropies of the g and A tensors; no changes are observed if $\tilde{\mathbf{g}}$ is uniaxial ($g_x = g_y = 0$). Since the intensity of an EPR signal depends on the anisotropy of the g tensor (13), a strong correlation between Mössbauer and EPR spectroscopy is indicated. (Note that a Kramers doublet with $g_x = g_y = 0$ is EPR silent.) A field dependent (parallel versus transverse) Mössbauer spectrum demands that an EPR spectrum be observed for that site. If two field dependent Mössbauer spectra and only one EPR-active species, quantitating to one spin per molecule, are observed, spin-sharing between the two iron sites is established. We apply this argument below in the discussion of [3Fe-3S] centers and sulfite reductase.

In exchange-coupled clusters the system spin S is delocalized over all Fe atoms participating in the spin-coupling. In all cases studied so far the magnetic hyperfine terms of the participating Fe sites could be written in the form $\mathbf{S} \cdot \tilde{\mathbf{A}}_i \cdot \mathbf{I}_i$, where $\mathbf{S}$ is the common spin and i denotes the Fe sites. Thus the internal fields of all subsites i are determined by the same spin expectation value, and the Mössbauer subspectra can be associated with the observed EPR signal. If all Fe sites yield distinct Mössbauer spectra, the number of Fe sites per cluster can be inferred by simply counting the number of distinct subsites. If two or more subsites are equivalent [as observed for the cofactor centers of nitrogenase (3)], the number of subsites can be deduced from the total absorption associated with the component spectra [It has been found (14) that the recoilless fraction f, which determines the integrated intensity, is quite independent of the Fe environment. For instance, the Fe sites of rubredoxin and myoglobin were found to have, to within 5%, the same f values at 4.2°K.]

3. The Fe coordination geometry in Fe-S clusters is essentially tetrahedral. The weak ligand field association with such a geometry gives rise to high-spin electronic configurations. For high-spin configurations the isotropic Fermi contact term dominates orbital and spin-dipolar contributions to the magnetic hyperfine interactions. It is well established, experimentally as well as theoretically, that the Fermi contact term is negative; that is, $A < 0$. In a spin-coupled system, however, the observable (effective) A values are referred to the system spin rather than to the spins of the constituent Fe atoms. It has been shown for [2Fe-2S] centers (15–17) and, most recently, for [3Fe-3S] clusters (18) that spin-coupling leads to a sign reversal of the effective A values for those subsites whose spin is antiparallel oriented to the system spin. Thus observation of Fe sites with positive A values gives strong evidence for a spin-coupled system.

Although the sign of H_{eff} is not measured in a conventional Mössbauer experiment (unpolarized radiation), the sign of H_{int}, hence the sign of A, is readily determined. A simple comparison [see eq. (6)] of the magnetic splittings observed in weak and strong applied fields shows whether H_{int} is parallel or antiparallel to the applied field, that is, positive or negative. Since the sign of $\langle \mathbf{S} \rangle$ is known, the sign of A is obtained from eq. (7).

While arguments (1) and (2) pertain only to Kramers systems, arguments based on the sign of A can be applied to systems with integer, nonzero electronic spin if the samples are studied in strong fields. We discuss an application in Section 3.3.

4. Mössbauer redox titrations provide a powerful way of proving the presence of spin-coupled clusters. Let us assume that we are dealing with a cluster that has half-integral electronic spin in the oxidized state and which can be reduced by a one electron step. In the oxidized state a magnetic Mössbauer spectrum is observed for each subsite. After reduction the cluster is in a non-Kramers state, and therefore only quadrupole doublets are observed. By counting the number of magnetic components converted into quadrupole doublets by the addition of one electron, the number of Fe sites per cluster is obtained.

3 [3Fe-3S] CLUSTERS

3.1 Introductory Remarks

In this section we discuss the spectroscopic and structural features of [3Fe-3S] centers which have emerged from Mössbauer and EPR studies. Although these clusters were discovered only recently (5, 6), Mössbauer work and

X-ray diffraction studies have already given us quite a detailed picture of the structure. In contrast, the function of these centers is essentially unknown. In fact, [3Fe-3S] centers are found in a variety of peculiar circumstances. In the ferredoxin from *Azotobacter vinelandii*, the protein that revealed the existence of this fundamentally new structure, the [3Fe-3S] center has a midpoint potential of $E_0' = -450$ mV; it occurs (at least in the isolated protein) together with a [4Fe-4S] center with +320 mV midpoint potential. Why? In a ferredoxin from *Thermus thermophilus* a [3Fe-3S] center occurs together with a [4Fe-4S] cluster, the latter operating at −530 mV in the +2/+1 states (J. A. Fee, E. Münck, T. A. Kent, and B. H. Huynh, unpublished).

With H. Beinert and J.-L. Dreyer, we are presently studying aconitase from beef heart mitochondria. These studies have revealed the presence of a [3Fe-3S] center. Its function, however, remains to be established; Ruzicka and Beinert (19) have suggested that the cluster may have a regulatory function. Our current Mössbauer studies suggest that conversions of the [3Fe-3S] center into other structural forms may take place when aconitase is activated in the presence of Fe and dithiothreitol; we wonder what surprises nature has in store.

In this section we focus on ferredoxin II (Fd II) from *D. gigas*, primarily because the preparations are spectroscopically pure and because only one cluster is present (20). Moreover, all [3Fe-3S] centers that have been studied so far turn out to have a similar spectroscopic signature; the pure Fd II samples, prepared by our co-workers, A. V. Xavier, J. J. G. Moura, and I. Moura from the University of Lisbon, Portugal, have given us the opportunity to discuss the spectroscopic features in some detail.

A short comment on nomenclature might be in order. We have been referring to the new centers as [3Fe-3S] clusters. Though the presence of 3 Fe sites has been established unambiguously by Mössbauer and X-ray studies, evidence for three sulfides per cluster rests primarily on the X-ray studies by Stout and co-workers (6, 21). Since the spectroscopic features of these centers are unique, we can presume that all clusters studied so far have essentially the same coordination geometry. These conclusions are also supported by the distinctive features that have emerged from protein sequences (see Chapter 3). Detailed sulfide and Fe determinations are only available for aconitase; an analysis of 14 preparations by Beinert and Dreyer (22) yielded $[S^{2-}]/[Fe] = 1.1 \pm 0.1$.

3.2 Oxidized [3Fe-3S] Centers

Here we describe, using Fd II as an example, the spectroscopic features that are observed for [3Fe-3S] centers. We take the point of view that the results are already known; this allows us to comment in somewhat more detail on certain observations that received only a cursory treatment in the original communications (5, 20).

Different Fd II preparations have systematically yielded, to within 5%, 3 Fe atoms per protein monomer. Fd II is EPR active in the oxidized form; an EPR spectrum taken at 6°K is shown in Figure 3. Quantitation against a copper-EDTA standard yields (0.93 ± 0.12) spins per 3 Fe atoms. The spectrum can be fitted very well by choosing $g_1 = 2.02$, $g_2 = 2.00$, and $g_3 = 1.97$, respectively. Note that the spectrum has a fairly long high-field tail. We suspect that this tail reflects some sort of heterogeneity, although we do not yet see a mechanism that explains such broadening around $g = 2$.

Figure 4*A* shows a Mössbauer spectrum of oxidized Fd II recorded at 77°K. Three conclusions can readily be drawn from this spectrum: (1) all 3 Fe sites appear to be equivalent (this is also observed for aconitase, the *A. vinelandii* ferredoxin, and the *Thermus thermophilus* ferredoxin, although all these proteins have distinct values for ΔE_Q); (2) there is no evidence for any Fe impurity; and (3) the isomer shift $\delta_{Fe} = 0.27$ mm/sec (relative to Fe metal) is strongly suggestive of high-spin ferric iron in a tetrahedral environment of S atoms. The last statement requires some comment. Earlier we suggested (5, 20) a tetrahedral S environment. Subsequent X-ray diffraction studies of the *A. vinelandii* protein (21) have proven this suggestion to be (almost) correct. The X-ray data reveal two sites with a tetrahedral S coordination; the coordination at the third site is found to consist of 3 S ligands

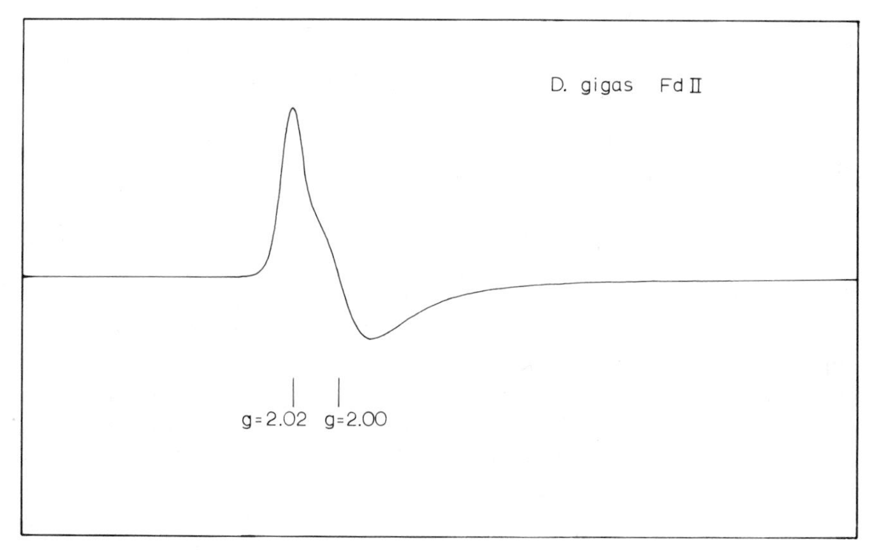

Figure 3. EPR spectrum of Fd II from *Desulfovibrio gigas* recorded at 6K. The derivative of the absorption is plotted vs. the magnetic induction.

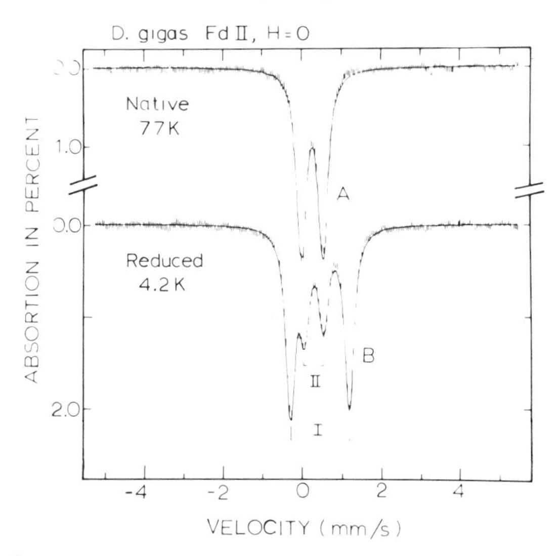

Figure 4. Mössbauer spectra of Fd II recorded in zero applied magnetic field. *(A)* Spectrum of oxidized Fd II taken at 77K. *(B)* Spectrum of reduced Fd II measured at 4.2K. The solid lines are the result of fitting quadrupole doublets to the data. The Fd II data were taken on samples containing ^{57}Fe in natural abundance (2.2%).

plus an O ligand, provided by glutamate 18. (This is a very exciting discovery by Stout and co-workers.) Since tetrahedral O environments yield isomer shifts that are ~0.1 mm/sec more positive than the corresponding S coordinations, 1 O ligand may be presumed to increase δ_{Fe} by ~0.03 mm/sec for 1 Fe site, an increase too small to be resolved and to be meaningful.

Figure 5 shows a spectrum of oxidized Fd II taken at 1.5°K. At 77°K the electronic spin relaxation is fast; consequently, the appropriately taken expectation value of the electronic spin, the thermal average, is $\langle \mathbf{S} \rangle_{th} = 0$. Therefore only quadrupole interactions are observed at 77°K. At 1.5°K, however, the spin fluctuates slowly, hence magnetic spectra are observed. The spectrum in Figure 5 reveals three distinct components; that is, the Fe sites are magnetically inequivalent. The solid curve through the data is a superposition of three spectra; the component spectra of site 1 (solid line), site 2 (dashed line), and site 3 (dotted line) are plotted above the data. We have shown earlier (5) how the spectra of site 1 and site 2 can be uniquely identified and associated with the $S = ½$ EPR signal. (A "parallel minus transverse field" difference spectrum reveals four $\Delta m = 0$ lines belonging to two field-dependent spectra; since there is only one spin per monomer present, the two spectra must be subsites in one cluster.) The third site, characterized by an unusually small magnetic hyperfine interaction, appears

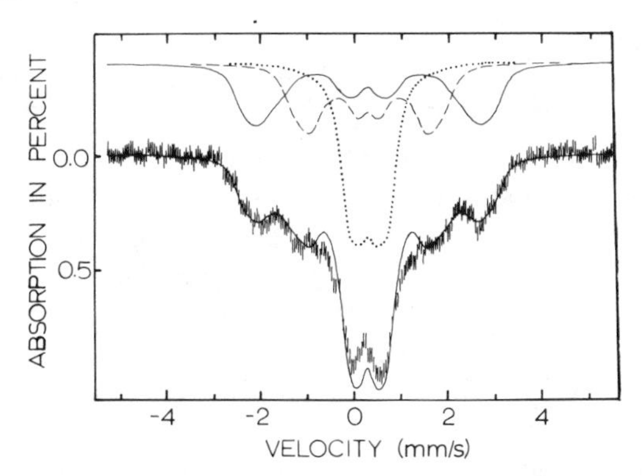

Figure 5. Mössbauer spectrum of oxidized Fd II taken at 1.5K in a field of 600 G applied parallel to the observed Mössbauer radiation. The solid line plotted over the data is the result of computing three theoretical spectra from a spin Hamiltonian. The theoretical subcomponent spectra are shown separately. For details the reader is referred to reference 20.

in Figure 5 as a broadened quadrupole doublet. Two arguments suggest that this site belongs to the cluster as well. First, since sites 1 and 2 are both high-spin ferric ($S = 5/2$) in character, the problem of the missing spin arises; two half-integer spins cannot be coupled to yield the observed $S = 1/2$ system spin; a third site with half-integral spin is required. Second, studies in strong applied fields suggest that $A > 0$ for site 3 (because of the complexity of the spectra, however, the sign of A_3 is soft). On the other hand, the high-field studies establish unambiguously that $A_1 < 0$ and $A_2 > 0$ (we recall from section 2 that $A > 0$ indicates antiparallel pairing of spins).

It is noteworthy that the spectrum in Figure 5 has rather broad features. If the magnetic hyperfine interactions were all isotropic and the sample were homogeneous, the spectrum would have considerably higher resolution. Presently, it is not clear whether anisotropies in $\tilde{A}$ or heterogeneities, or both, are the root of the rather broad spectra. This means that our knowledge of the magnetic hyperfine couplings is still somewhat crude. The best characterized A values are those obtained for the *A. vinelandii* ferredoxin. It seems that Electron Nuclear Double Resonance (ENDOR) studies are required to refine the data analyses significantly.

When we observed the first spectra of oxidized [3Fe-3S] centers, we were baffled by the observation of the small A value for site 3. Because of the isomer shifts it was clear that all three sites are rubredoxinlike and should thus exhibit rather similar *intrinsic* coupling constants. This suggested that the small A_3 would reflect details of the spin-coupling, a hint that proved to be quite fruitful.

Is it possible to uniquely distinguish spectroscopically oxidized [3Fe-3S] clusters from [4Fe-4S] centers in the +1 state, that is, the state of oxidized HIPIP? Both clusters give EPR signals with $g_{av} > 2$. The EPR signals of some of the [3Fe-3S] centers have been interpreted as "typical" HIPIP signals, indicating how similar the EPR spectra of both cluster types can be. The isomer shift of both clusters are virtually the same. Oxidized HIPIP from *Chromatium vinosum* (23) has a quadrupole splitting $\Delta E_Q = 0.80$ mm/sec (at 77°K), which varies slightly with temperature. The quadrupole splittings of oxidized 3-Fe centers are temperature independent, in accord with the high-spin ferric character of the constituent iron atoms; $\Delta E_Q = 0.53$ mm/sec for Fd II, $\Delta E_Q = 0.58$ mm/sec for the ferredoxin from *Th. thermophilus*, $\Delta E_Q = 0.63$ mm/sec for the *A. vinelandii* ferredoxin, and $\Delta E_Q = 0.71$ mm/sec for aconitase prepared from beef heart mitochondria. The overall magnetic features of the low-temperature Mössbauer spectra of [3Fe-3S] centers are not decisively dissimilar to those of HIPIP. Moreover, the magnetic spectra observed for the *Azotobacter* ferredoxin, *D. gigas* Fd II, and aconitase have pronounced differences, although they all share the features of one site characterized by a small A value. It might appear that the distinctive third site would be a reasonable identification criterion. However, a HIPIP sample with a contaminant of aggregated material [see the desulforedoxin work of Moura et al. (24)] or a high-spin ferric impurity (fast relaxing) could yield a spectrum "indicative" of a [3Fe-3S] center. Furthermore, inspection of Figure 9 in section 3.4 shows that the A values of site 3 are quite sensitive to the ratios of the exchange-coupling constants; that is, [3Fe-3S] clusters with larger A_3 values should be anticipated.

3.3 Reduced [3Fe-3S] Centers

By the addition of one electron per monomer of Fd II the spectrum shown in Figure 5 is transformed into that shown in Figure 4*B*. The latter spectrum exhibits only quadrupole doublets, characteristic of a non-Kramers system. Thus *one* electron transforms *3* Fe sites from a Kramers state into a non-Kramers state. Again, a 3-Fe center is indicated.

The spectrum of reduced Fd II shows two sharp quadrupole doublets. A least-squares fit to the data shows that doublets I and II occur in the concentration ratio 2:1; that is, 2 equivalent Fe sites contribute to doublet I. Doublet I has $\Delta E_Q = 1.47$ mm/sec and $\delta_{Fe} = 0.47$ mm/sec, while doublet II is characterized by $\Delta E_Q = 0.47$ mm/sec and $\delta_{Fe} = 0.30$ mm/sec. Within the uncertainties the isomer shift of doublet II is the same as that observed for the oxidized protein; that is, this site has remained high-spin ferric in character. At this point it is instructive to recall that oxidized rubredoxin from *C. pasteurianum* (25) has $\delta_{Fe} = 0.32$ mm/sec, while the corresponding value for the Fe^{2+} site of reduced rubredoxin is $\delta_{Fe} = 0.70$ mm/sec. Thus the δ_{Fe} value quoted above for doublet I suggests that the 2 Fe sites are roughly at the oxidation level $Fe^{2.5+}$. This suggests that the 2 Fe sites of doublet I equally share the electron that enters the complex on reduction. Since all the [3Fe-3S] centers that we have studied so far exhibit in the reduced state a spectrum such as that shown in Figure 4*B*, the two doublets probably express a characteristic structural feature of [3Fe-3S] centers. It is very tempting to speculate that the distinctive single site of doublet II corresponds to the site with the glutamate ligand.

In Figure 6*A* we have redrawn the spectrum of Figure 4*B*. Figure 6*B* shows a spectrum of the same sample recorded at the same temperature in an applied field of 600 G. The broadening of the absorption features proves that the weak applied field induces a sizable internal field ($H_{int} \approx 100$ kG); thus the ground state of reduced Fd II is paramagnetic, that is, $S \geq 1$. The Mössbauer spectra of a *diamagnetic* compound taken in zero field and in a field of 600 G are practically undistinguishable.

Figure 7 shows a Mössbauer spectrum of reduced Fd II recorded at 4.2°K in a 10 kG parallel applied field. The well-resolved magnetic features allow us to draw the following conclusions: (1) At 4.2°K the electronic spin-relaxation rate is slow compared to the nuclear precession frequencies (≈ 10 MHz). (2) Since the magnetic hyperfine fields saturate in weak applied fields, the lowest electronic spin levels are two closely spaced states. An analysis of the entire set of high-field data reveals that the 2 Fe sites of doublet I remain equivalent and therefore indistinguishable even in strong

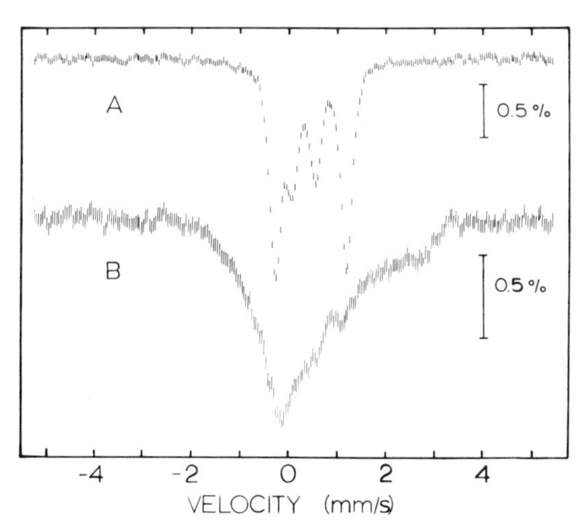

Figure 6. Mössbauer spectra of reduced Fd II taken at 4.2K in zero field *(A)* and a parallel field of 600 G *(B)*.

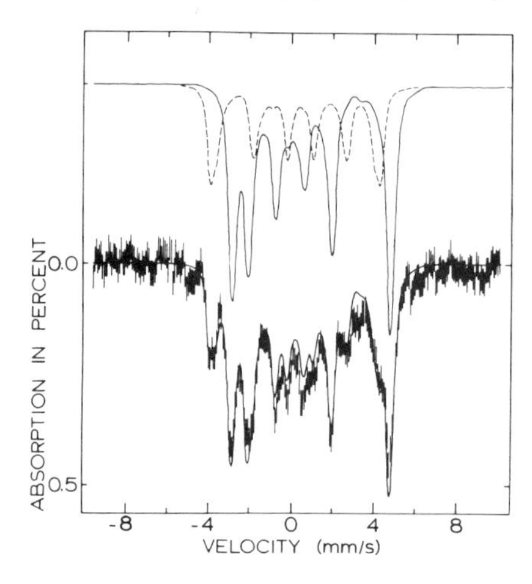

Figure 7. Mössbauer spectrum of reduced Fd II taken at 4.2K in a 10 kG field applied parallel to the observed Mössbauer radiation. The solid line is the result of theoretical calculations based on a spin Hamiltonian with $S = 2$. The decomposition of the spectrum into two subcomponents is shown as well.

fields. This observation strongly suggests that both sites are *structurally* equivalent. (Even if the electron that has entered the complex on reduction would exchange rapidly, that is, fast on the Mössbauer time scale, between the two sites, the Mössbauer spectra would be different if the two sites were inequivalent; each nucleus samples only its own site). The solid line above the data in Figure 7 shows a theoretical spectrum of site I (2 equivalent Fe sites). The dotted line shows the corresponding spectrum of site II, the site that is ferric in character. One of the most interesting results of the high-field studies is the observation of a *positive* A value for site II; this demonstrates unambiguously that site II cannot arise from an isolated Fe atom and that it must reflect an Fe site participating in spin coupling. Since the partner sites can only be the equivalent Fe atoms of Site I, a spin-coupled 3-Fe center is, once again, indicated!

The theoretical spectra in Figure 7 were computed from the spin Hamiltonian equations (2)–(4) using $S = 2$. When these spectra were computed, the electronic spin S of reduced Fd II was unknown. It turns out that the magnetic features of the Mössbauer spectra reflect essentially the properties of two closely spaced spin states. Appropriate states can be obtained by choosing a large negative zero-field splitting parameter D and an arbitrary, integer spin S; in the Fd II communication (20) we have chosen $S = 2$, primarily because our group has extensive practice with $S = 2$ systems. A very recent low-temperature magnetic CD study of Fd II by A. J. Thomson and co-workers (26) gives quite strong evidence for $S = 2$. At this point it might be appropriate to discuss some aspects of the physics of the reduced [3Fe-3S] centers.

Figure 8 shows an energy level diagram for an $S = 2$ system resulting from

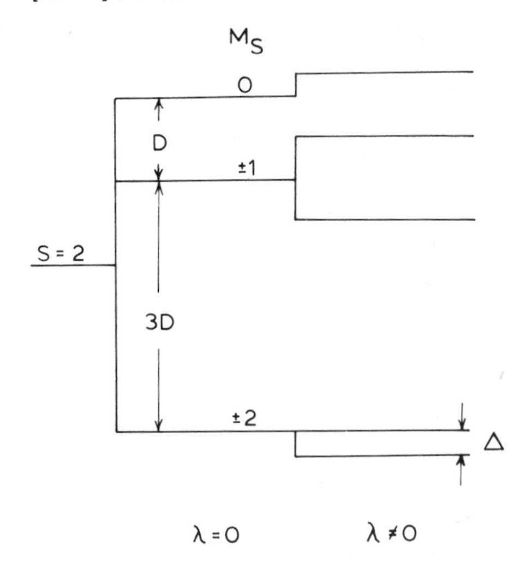

Figure 8. Energy level diagram of an $S = 2$ system resulting from a zero-field splitting term with $D < 0$. In tetragonal or trigonal symmetry the ± 2 and ± 1 sublevels are doubly degenerate. In rhombic symmetry $\lambda \neq 0$, and the two sublevels of the $\pm$ doublet have energy separation Δ.

eq. (3) for $D < 0$ and $H = 0$. For tetragonal or trigonal symmetry ($\lambda = 0$) the ground state is the $m = \pm 2$ doublet. In weak applied fields ($\beta H \ll |D|$) its magnetic properties are characterized by effective g values $g_x = g_y = 0$ and $g_z = 2g_0 S = 8$, where we have assumed that the g tensor of the system is isotropic, $g_0 = 2$. A rhombic term (λ) in the spin Hamiltonian splits the ground doublet by the amount

$$\Delta = 2|D|\left(\sqrt{1 + \frac{3}{2}\lambda^2} - 1\right)$$

which may be approximated as $\Delta \simeq 3\lambda^2 D/2$, since λ can be restricted to $0 \leq \lambda \leq 1/3$. In zero field the spin expectation values of S are zero for each of the five spin sublevels. Even a moderately weak magnetic field, when directed along the molecular z axis, can mix the two lowest levels appreciably. From a simple perturbation approach we calculate $\langle S_x \rangle \approx \langle S_y \rangle \approx 0$ and $\langle S_z \rangle = \pm 4g_0\beta H/\Delta$ where the signs refer to the upper and lower member of the doublet, respectively. Thus the internal fields are given by $H_{\text{int}}(z) = \mp 4(g_0\beta/g_n\beta_n)(A_z/\Delta)H$. From the preceding considerations we see that the two lowest spin levels are magnetically uniaxial; therefore the Mössbauer spectrum measures essentially A_z (and Δ). Furthermore, in weak applied fields the internal field is inversely proportional to Δ. For Fd II we found $\Delta = 0.35$ cm^{-1}. Similar electronic systems, albeit monoatomic, occur in rubredoxin from *C. pasteurianum* (25) and in desulforedoxin from *D. gigas* (24); these proteins have $\Delta = 2.9$ cm^{-1} and $\Delta = 0.65$ cm^{-1}, respectively.

From the preceding consideration we can readily understand the drastic difference between the spectra shown in Figure 6. Because of the small value of Δ even a weak field can induce a sizable internal field. ($\langle S_z \rangle = \pm 1.05$ for $H = 600$ G). The extreme sensitivity of the spectra to small applied fields is a distinctive feature of all reduced [3Fe-3S] clusters studied so far. These features allow us to recognize the presence of [3Fe-3S] clusters in proteins that contain other structures as well.

We have not yet published the parameter set used to compute the theoretical spectra shown in Figure 7 because these parameters are still quite tentative despite the good quality of the fits. Here we list those parameters that we consider "hard." For site I $\Delta E_Q = 1.47$ mm/sec, $\delta_{\text{Fe}} = 0.46$ mm/sec, $A_z/g_n\beta_n = -120$ kG; for site II $\Delta E_Q = -0.47$ mm/sec, $\delta_{\text{Fe}} = 0.30$ mm/sec, $A_z/g_n\beta_n = +127$ kG, and $\Delta = 0.35$ cm^{-1}. In strong applied fields, $H > 10$ kG, the $m = \pm 1$ sublevels are mixed into the ground state, yielding finite values for $\langle S_x \rangle$ and $\langle S_y \rangle$. This mixing allows us to obtain reasonable estimates for A_x and A_y.

3.4 Spin-Coupling Model for Oxidized [3Fe-3S] Clusters

As discussed in Section 3.2, the isomer shifts of the 3 Fe sites suggest a high-spin ferric ($S = 5/2$) configuration and a rubredoxinlike ligand structure is indicated. The 3 Fe sites, however, were found to exhibit drastically different magnetic hyperfine interactions. In particular, site 3 has a very small hyperfine field. For the ferredoxin from *A. vinelandii*, which yields the best resolved spectra of sites 1 and 2, the coupling constants were found to be $A_1 = -41$ MHz, $A_2 = +18$ MHz, and $|A_3| = 5$ MHz (5, 18). These observations suggest that we couple three $S = 5/2$ spins to obtain the observed ground state system spin $S = 1/2$. Since we have good reasons to believe that all Fe sites are intrinsically (roughly) the same, and rubredoxinlike, we have to explain the different A values by making the Fe sites somehow inequivalent. This can be achieved by assigning a different exchange coupling constant to each pair of Fe atoms. Thus we describe the coupling by

$$\hat{H} = J_{12}\mathbf{S}_1 \cdot \mathbf{S}_2 + J_{23}\,\mathbf{S}_2 \cdot \mathbf{S}_3 + J_{13}\,\mathbf{S}_1 \cdot \mathbf{S}_3 \tag{8}$$

where $S_1 = S_2 = S_3 = 5/2$. The system spin is formed by $\mathbf{S} = \mathbf{S}_1 + \mathbf{S}_2 + \mathbf{S}_3$.

The evaluation of the data in this model is straightforward. We are interested in an $S = 1/2$ ground state. The Hamiltonian of eq. (8) commutes with the total spin S; that is, configurations with different system spin do not mix. Fortunately, there are only two $S = 1/2$ configurations. They are obtained by coupling first S_2 and S_3 to an intermediate spin $S' = 2$ or 3, and then coupling S' to S_1 to yield $S = 1/2$. We can characterize the basis states by the kets $|S_2S_3\,(S')\,S_1;\,SM >$. Since eq. (8) mixes only the two $S = 1/2$ configurations, the most general state can be written as

$$|+\rangle = \sqrt{1-\alpha^2}\;|S_2S_3(2)S_1;\,\tfrac{1}{2}\,\tfrac{1}{2}\rangle + \alpha|S_2S_3(3)S_1;\,\tfrac{1}{2}\,\tfrac{1}{2}\rangle \tag{9}$$

Using standard techniques to evaluate the matrix elements, we can relate the A values of the coupled representation to the a values of the uncoupled system (see ref. 18):

$$A_i = 2a_i \langle + |S_{iz}| + \rangle = 2a_i \langle S_{iz} \rangle \tag{10}$$

where i enumerates the sites, and where the S_{iz} are the z components of the spin operators in the uncoupled system. If all sites are intrinsically similar, we can use the same a_i, namely that of rubredoxin, to compute the A_i in the coupled representation, the measured quantities. For fixed a_i the relevant information is then contained in the $\langle S_{iz} \rangle$. In Figure 9 the $\langle S_{iz} \rangle$ are plotted versus the mixing parameter α^2. It is only necessary to consider solutions for $0 \leq \alpha \leq 0.25$; values of α^2 outside this range correspond to a relabeling of sites. Fortunately the graph shown in Figure 9 contains all the features in which we are interested. First, experimental observations of positive and negative A values can be explained. Most interestingly, $\langle S_{3z} \rangle$ can assume very small, and even zero, values. A survey of the published data for Fe^{3+} in tetrahedral S environments shows that the A values range in magnitude from 20 to 22 MHz (see ref. 18). If we take $a_i = -20$ MHz, then $\alpha^2 = 0.01$ yields $A_1 = -45$ MHz, $A_2 = +20$ MHz, and $A_3 = +6$ MHz. Thus the simple one-parameter model reproduces the A values of the *Azotobacter* ferredoxin, $A_1 = -(41 \pm 3)$ MHz, $A_2 = +(18 \pm 3)$ MHz, and $|A_3| = (5 \pm 3)$ MHz. We see no objection to using a 10% smaller value for the a_i; this would yield an almost perfect match between the theory and the experimental data. On the other hand, the good agreement below theory and experiment might be a

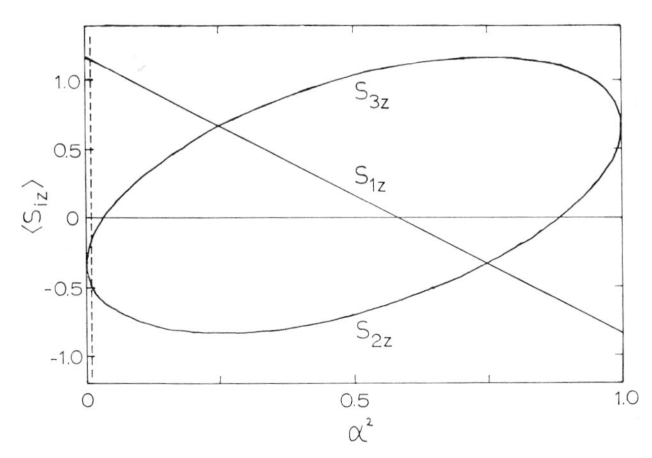

Figure 9. Expectation values $\langle S_{iz} \rangle$ for the 3 Fe sites as the function of the square of the mixing parameter α. The broken line at $\alpha^2 = 0.01$ marks the values that fit the data obtained for the *Azotobacter* ferredoxin. The parameter α can be expressed by the three coupling constants J_{12}, J_{13}, and J_{23} (18).

bit fortuitous. First, we suspect that the site with the glutamate ligand has intrinsically a somewhat large a value. Second, the above-mentioned problems of heterogeneities and anisotropies result in some sizable uncertainties for the experimental A values; the above-quoted uncertainties are "educated" estimates, not well-defined standard deviations.

The Fd II results can be fitted to the model as well. Large uncertainties, however, are associated with the experimental A values. Although $A_1 < 0$ and $A_2 > 0$ are firmly established, and a small $A_3 > 0$ is indicated, the broad magnetic spectra may suggest sizable anisotropic magnetic hyperfine interactions. Until the question of heterogeneity versus anisotropy can be resolved experimentally, we will refrain from fitting the Fd II to the model. There is no question, however, that the model reproduces the essential features of the Fd II data.

The mixing parameter α is a simple function of the three coupling constants (18). Since $\alpha^2 = 0.01$, it is instructive to explore the case $\alpha = 0$, which results when $J_{12} = J_{13}$, that is, when site 1 is equally strong exchange coupled to Sites 2 and 3. For this situation the intermediate spin S' commutes with the Hamiltonian of eq. (8), and we therefore obtain a simple analytical expression for the energies of the spin multiplets:

$$E = \tfrac{1}{2} J_{12} [S(S+1) - S'(S'+1) - S_1(S_1+1)] + \tfrac{1}{2} J_{23} [S'(S'+1) - S_2(S_2+1) - S_3(S_3+1)] \tag{11}$$

If we drop terms that shift all levels in the same way, we obtain

$$E(S_1, S') = \tfrac{1}{2} J_{12} S(S+1) + \tfrac{1}{2} (J_{23} - J_{12}) S'(S+1) \tag{12}$$

From this expression we learn that the state with $S = \tfrac{1}{2}$ and $S' = 2$ is lower in energy than the state with $S = \tfrac{1}{2}$ and $S' = 3$ when $J_{23} > J_{12}$. Furthermore, $E(\tfrac{1}{2}, 2) < E(\tfrac{3}{2}, 1)$ for $J_{12} > 0$ and $J_{12}/J_{23} > \tfrac{4}{7}$. The important result is that the coupling constants are roughly equal. A similar result (18) is found when all three exchange-coupling constants are different. This result is important in light of the findings that the application of extrusion techniques to proteins with [3Fe-3S] centers can yield quantitatively [2Fe-2S] cores. Such results could suggest that one could contemplate a 3-Fe center as consisting of a [2Fe-2S] cluster to which a third Fe atom is weakly coupled. The finding of three equally strong exchange-coupling constants argues for a fairly symmetric structure, a conclusion that is supported by the X-ray diffraction results.

The Mössbauer data have yielded only information about the relative magnitudes of the exchange-coupling constants; within a factor of 2 the three coupling constants have to be the same. Information about the absolute values has to come from techniques that are sensitive to the presence of excited state configurations. Magnetic susceptibility studies readily suggest themselves. But watch out; it follows from eq. (11) that for $0.8 < J_{12}/J_{23} < 1$ the first excited state has also $S = \frac{1}{2}$ and ($g = 2$)! MCD studies of oxidized Fd II (26) indicate that another electronic state becomes thermally populated above 100°K. Thomson and co-workers (26) suggest that the exchange coupling in oxidized Fd II may be stronger than those observed for oxidized [2Fe-2S] ferredoxins; the latter clusters have J values in the order of 200 cm^{-1}. (The authors, however, point to the need of further MCD studies at higher temperatures before definite conclusions can be drawn.)

4 SULFITE REDUCTASE: EXCHANGE COUPLING BETWEEN SIROHEME AND A [4Fe-4S] CLUSTER

4.1 Introductory Remarks

E. coli sulfite reductase and spinach nitrite reductase belong to a class of proteins that catalyze the assimilatory six-electron reduction of sulfite (nitrite) to sulfide (ammonia). Both enzymes contain a novel prosthetic group, termed *siroheme* (27), an Fe tetrahydroporphyrin of the isobacteriochlorin type with eight carboxylic acid-containing peripheral sidechains. In addition, both proteins contain an Fe-S center of the [4Fe-4S] variety. For details the reader is referred to a review article by L. M. Siegel (28).

The *E. coli* enzyme is a complex hemoflavoprotein of molecular weight 684,000 with an $\alpha_4\beta_4$ subunit structure. The holoenzyme contains 4 FAD, 4 FMN, 4 sirohemes, and 4 [4Fe-4S] clusters. Treatment of the hemoflavoprotein complex with 4 *M* urea allows the isolation of a catalytically active monomer that contains 1 siroheme and 1 [4Fe-4S] cluster. This β-monomer (here referred to as *sulfite reductase,* SiR) cannot use the natural electron donor NADH for sulfite reduction, since the flavin components are missing; it can, however, catalyze the stoichiometric reduction of sulfite to sulfide if provided with a suitable electron donor.

Examination of oxidized SiR with EPR spectroscopy reveals signals at $g_y = 6.63$, $g_x = 5.25$, and $g_z = 1.98$. These resonances are typical of high-spin ferric hemes in a slightly rhombic environment. Rueger and Siegel (29) have shown that a one-electron reduction of SiR causes the disappearance of the EPR signal. The addition of a second electron leads to the appearance of a novel EPR signal with g values at 2.53, 2.29, and 2.07 (30). On the other hand, when one-electron-reduced SiR is complexed with CN^- (or CO), the addition of a second electron yields an EPR spectrum with $g = 2.04$, 1.94, and 1.91, characteristic of reduced [4Fe-4S] clusters. Barber and Siegel (31) have assigned redox potentials to uncomplexed SiR: The first redox step has

$E_0' = -380$ mV, while the second step occurs at -450 mV. In this section we show that the siroheme and the [4Fe-4S] center function as one exchange-coupled unit.

Together with our co-workers, Lewis M. Siegel and Peter Janick at Duke University, we, Jodie A. Christner and Eckard Münck, have studied SiR in three oxidation states, complexed to ligands such as NO_2^-, CN^-, SO_3^{2-}, and NO. Here we focus entirely on oxidized sulfite reductase (32); presently the Mössbauer spectra of $2e^-$-reduced SiR and of SiR complexed to substrates and inhibitors are being evaluated. We have also studied, again with L. M. Siegel's group, a concentrated sample of oxidized spinach nitrite reductase (containing Fe in natural abundance). The spectra are virtually identical to those shown here for SiR. It should be apparent from the subsequent discussion that these observations imply that the two chromophores are exchange coupled as well.

4.2 EPR and Mössbauer Data

Figure 10 shows an EPR spectrum of oxidized ^{57}Fe-enriched SiR recorded at 7.4°K under nonsaturating conditions. The g values at 6.63, 5.24, and 1.98 fit well to a spin Hamiltonian, eq. (3), for $S = 5/2$ with $\lambda = 0.027$ and $g_0 = 2$.

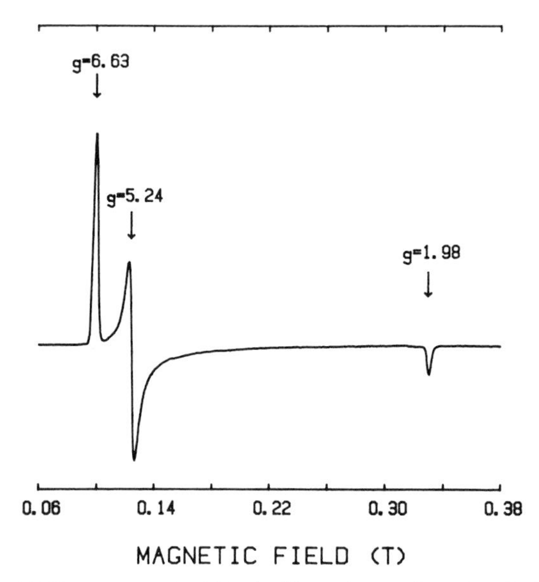

Figure 10. X-Band EPR spectrum of oxidized sulfite reductase from *E. coli*. The spectrum was recorded at 7.4K. The sample was isotopically enriched in ^{57}Fe. The Mössbauer spectra in Figures 11 and 12 were obtained from the same material.

The zero-field splitting parameter D was determined with Mössbauer and EPR spectroscopy. Both methods yield $D = +(8 \pm 1)\ \mathrm{cm}^{-1}$. With all parameters known the EPR signal was quantitated to yield a spin concentration of 0.92 spins per siroheme; that is, there is *one* spin per β-monomer. Furthermore, the spectrum in Figure 10 shows that no other EPR-active species is present. We will come back to these two important results.

Figure 11 shows a Mössbauer spectrum of oxidized SiR recorded at 190°K. One sharp quadrupole doublet is observed, showing that all Fe atoms in the sample have the same values for the quadrupole splitting ΔE_Q and the isomer shift δ_{Fe}. A least-squares fit to the data yields $\Delta E_Q = 1.00$ mm/sec and $\delta_{\mathrm{Fe}} = 0.37$ mm/sec. It is somewhat coincidental that the heme Fe and the 4 Fe sites of the [4Fe-4S] center yield the same spectrum. Concerning the siroheme the observed parameters agree with a high-spin ferric assignment.

Isomer shifts have two contributions: a chemical shift, which reflects the electronic state of the Fe site, and a temperature-dependent second-order Doppler shift. For comparison it is convenient to relate the δ_{Fe} values to 4.2°K. At 4.2°K the Fe atoms of the [4Fe-4S] center have $\delta_{\mathrm{Fe}} = 0.45$ mm/sec. This value is typical of [4Fe-4S] clusters in the diamagnetic +2 state. Typically the isomer shift increases (decreases) by 0.10–0.15 mm/sec if one electron is added (removed) from the cluster. Reduced HIPIP from *Chromatium vinosum* (23) has $\delta_{\mathrm{Fe}} = 0.44$ mm/sec; the same shift, $\delta_{\mathrm{Fe}} = 0.45$ mm/sec, has been reported for the diamagnetic [4Fe-4S] cluster of amidotransferase from *B. subtilis* (33). Thus there is no doubt that the

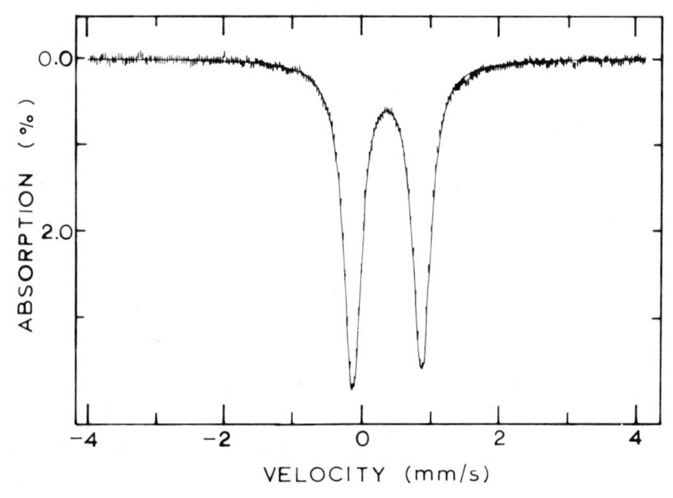

Figure 11. Mössbauer spectrum of oxidized sulfite reductase taken at 190K. The solid line is the result of least-squares fitting one quadrupole doublet to the data.

[4Fe-4S] cluster of oxidized SiR is in the +2 oxidation state, the state characterized by a ground state with $S = 0$. This assignment accords well with the EPR silence of the cluster. Moreover, the appearance of a $g = 1.94$ signal on reduction of CN^- or CO-inhibited SiR support this assignment. Thus the Mössbauer result was not surprising at all. The low-temperature spectra were surprising, however.

The general features of the low-temperature Mössbauer spectra can be predicted from the information obtained so far. For the [4Fe-4S] cluster we expect at 4.2°K the same Mössbauer spectrum as observed at 190°K, allowing, of course, for a small change in ΔE_Q and δ. The heme Fe, on the other hand should yield a spectrum exhibiting paramagnetic hyperfine structure. The shape of the heme Fe spectra can be predicted to a large extent from the observed g values (see ref. 11 for details); the only unknown is the magnetic hyperfine coupling constant A_0. Figure 12 shows a series of spectra taken at 4.2°K. The spectrum of the siroheme, with conspicuous absorption in the velocity ranges $|v| = 4$–6 mm/sec, turns out to be as anticipated. From the total splitting $|A_0| = 27$ MHz is obtained. This value is in the range typically found for high-spin ferric protoporphyrin IX compounds (see Table I of ref. 34); it is almost exactly the same as that observed for myglobin-fluoride. Thus the parameters of the siroheme, D, λ, ΔE_Q, and A_0, are in accord with the values reported for familiar hemes.

In Figure 12*A* and *B* the velocity range from -2 to $+2$ mm/sec is dominated by the contribution of the [4Fe-4S] center. The expected quadrupole doublet did not materialize; rather, the Fe sites of the cluster experience a sizable internal field, $H_{int} \approx 75$ kG. This means that the electronic environment of the 4 cluster Fe sites are characterized by $\langle \mathbf{S} \rangle \neq 0$; that is, the environment is paramagnetic! Furthermore, the intensities of the absorption lines are different in parallel and transverse applied field (compare the intensities of the absorption lines at $+0.8$ mm/sec and $+2$ mm/sec in Fig. 12*A* and *B*). As discussed in Chapter 2 this observation demands the observation of an EPR signal associated with the cluster. We recall, however, that the sample yields only *one* EPR signal, which we tacitly assumed to belong to the siroheme. We have to revise the picture now: The spin $S = 5/2$ does not only belong to the heme Fe, it is *shared* by the siroheme and the [4Fe-4S] cluster. In other words, the two chromophores are coupled.

We can cross-check this conclusion by studying the sample in strong applied fields. If the spin felt by the [4Fe-4S] cluster irons resulted from an isolated $S = 1/2$ system, $\mathbf{H}_{int} = -\langle \mathbf{S} \rangle A_0/g_n\beta_n$ would be independent of the strength of the applied field. (Similarly, the g values would be independent of the operating frequency, i.e. magnetic field, of the EPR spectrometer.) However, for $S \geq 3/2$ the applied field can mix excited states into the ground doublet, yielding a field-dependent $\langle \mathbf{S} \rangle$ and thus a field-dependent $\mathbf{H}_{int}$.

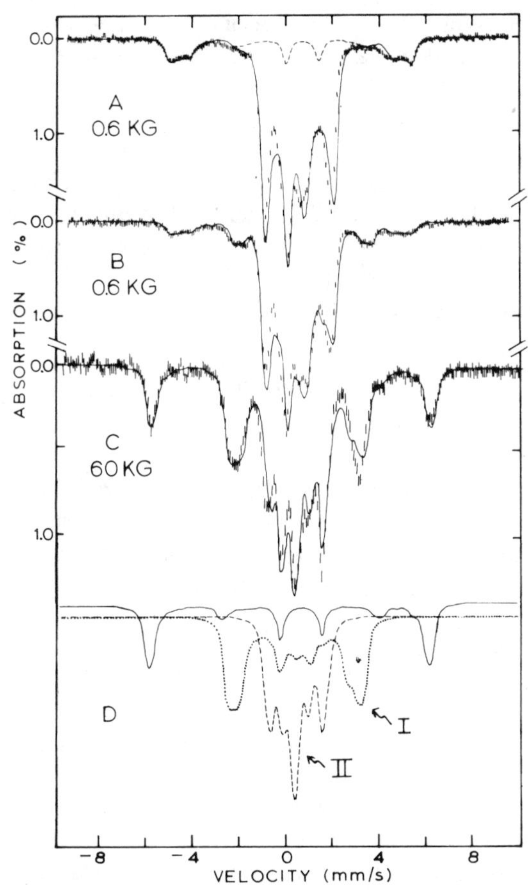

Figure 12. Mössbauer spectra of oxidized sulfite reductase recorded at 4.2K in parallel *(A, C)* and transverse *(B)* magnetic fields. The solid lines are the result of spectral simulations. In *(A)* the inner features of the siroheme spectrum are indicated by the dashed line. The spectrum in *(B)* was taken in a transverse field. The spectrum in *(D)* shows a decomposition of the 60 kG spectrum into a siroheme spectrum (solid line) and two spectra (each representing 2 Fe atoms) belonging to the [4Fe-4S] cluster. Note that the splitting of the site I spectrum increases in an applied field while that of site II decreases; the A_0 values of the sites have opposite signs. The best fits to the data are obtained if 20% of the total absorption is allotted to the siroheme spectrum, confirming independently that the Fe-S center is of the [4Fe-4S] variety.

(This field dependence is used to determine D for high-spin ferric hemes; see refs. 8 and 11.) Both the siroheme Fe and the 4 cluster Fe atoms were found to have a field-dependent $\mathbf{H}_{\mathbf{int}}$, describable with the *same* zero-field splitting parameter D. This shows that the Fe nuclei of the [4Fe-4S] cluster experience an environment with $S > 1/2$, suggesting that they share a common spin system.

An analysis of the high-field data shows that the [4Fe-4S] cluster contributes two distinct spectra of equal intensity. Site I (2 Fe sites) is characterized by $A_0 = +6.4$ MHz; site II (2 Fe sites) is found to have A_0-6.2 MHz. Note that the A_0 values have equal magnitude but opposite signs. The high-field studies also show that the A_0 of the siroheme is negative (as expected), $A_0 = -27$ MHz.

The theoretical spectra for each site i in Figure 12 were computed from the spin Hamiltonian for $S = 5/2$:

$$\hat{H} = D[S_z - \frac{35}{12} + \lambda(S_x^2 - S_y^2)] + g_0\,\beta\mathbf{H}\cdot\mathbf{S} + A_i\,\mathbf{S}\cdot\mathbf{I_i} + \text{quadrupole interactions} \qquad (13)$$

Note that the same electronic system (D, λ, $g_0 = 2$) was used for each site i. Note in particular that we have used the system spin S in the magnetic hyperfine terms (we rationalize this in Section 4.3). The best fit to the heme spectra and the [4Fe-4S] cluster spectra was found for $D = 8\ \text{cm}^{-1}$ and $\lambda = 0.027$. A decomposition of the spectrum taken in an applied field of 60 kG is shown in Figure 12*D*. The overall agreement between theory and the experiments is excellent. We have also studied (32) two samples of SiR reduced by 0.4 and 0.8 electrons per siroheme, respectively. The data clearly show that the addition of *one* electron changes 5 Fe sites from a Kramers state ($S = 5/2$) to a non-Kramers state; that is, the electron is shared by all 5 Fe sites. Furthermore, we have studied one-electron-reduced and two-electron-reduced SiR in considerable detail. These (as yet unpublished) studies lead to the same conclusion as the data discussed here: the two chromophores of SiR are coupled.

4.3 Mechanism of Coupling

We have shown in the preceding section that the two chromophores of SiR are coupled. We have also given arguments for a high-spin ferric assignment of the heme Fe, that is, $S_h = 5/2$ (the subscript refers to the heme). Moreover, we have shown that the [4Fe-4S] cluster is in the +2 oxidation state, in this state the cubane clusters have a cluster spin $S_c = 0$. Studies with substrates and inhibitors support these assignments.

It is clear that dipolar interactions of the form $\mathbf{S_h}\cdot\mathbf{J}\cdot\mathbf{S_c}$ between two paramagnetic moments are not at the root of the coupling because $S_c = 0$.

Thus exchange interactions, and therefore a covalent linkage between the two chromophores, need to be considered. Formally we can treat exchange interactions by introducing the terms $k_i\ \mathbf{S_h}\cdot\mathbf{S_i}$, where the k_i are exchange-coupling constants describing the coupling between the heme Fe and site i of the [4Fe-4S] cluster.

Our goal is twofold. First, we want to understand why magnetic hyperfine interactions are observed for the Fe sites of the Fe-S cluster. In particular, we want to understand why these interactions are of the form $A_i\mathbf{S_h}\cdot\mathbf{I_i}$, that is, why all internal fields are proportional to $\langle\mathbf{S_h}\rangle$. Second, why are the parameters observed for the siroheme essentially those of an uncoupled heme?

Rather than treating the fairly intractable problem of coupling a [4Fe-4S] center to a heme Fe, we discuss a simpler system that was suggested to us by P. G. Debrunner. We couple an oxidized, that is, diamagnetic, [2Fe-2S] cluster to the siroheme. For this system the zero-order wavefunctions are well known.

We describe the electronic Zeeman and the magnetic hyperfine interactions of the siroheme by

$$\hat{H}_h = g_0\beta\mathbf{S_h}\cdot\mathbf{H} + A_h\,\mathbf{S_h}\cdot\mathbf{I_h} \tag{14}$$

The spin sextet is described by the state vector $|S_hM\rangle$. Although the zero-field splitting is large for hemes, we can ignore this term in the subsequent discussion.

The electronic ground state of oxidized [2Fe-2S] clusters is diamagnetic, $S_c = 0$. The observed diamagnetism results from an antiparallel coupling of 2 high-spin ferric ions ($S_1 = S_2 = 5/2$). Using the successful model proposed in 1966 by Gibson and his colleagues (35), we describe the cluster by

$$\hat{H}_{\text{Fe-S}} = J\,\mathbf{S_1}\cdot\mathbf{S_2} + g_1\beta\mathbf{S_1}\cdot\mathbf{H} + g_2\beta\mathbf{S_2}\cdot\mathbf{H} + A_1\mathbf{S_1}\cdot\mathbf{I_1} + A_2\mathbf{S_2}\cdot\mathbf{I_2} \tag{15}$$

The first term describes the exchange interactions between the two high-spin ferric ions. By forming $\mathbf{S_1} + \mathbf{S_2} = \mathbf{S_c}$ a coupled representation $|S_cm\rangle$ is easily obtained. For $J > 0$ the ground state has $S_c = 0$; the first excited state, at energy J above the ground state, has $S_c = 1$. Oxidized [2Fe-2S] clusters have J values of $\sim 200\ \text{cm}^{-1}$.

To describe the coupling between the Fe-S cluster and the heme Fe, we assume that an exchange interaction acts between the heme Fe and 1 Fe site of the [2Fe-2S] cluster. Such a situation would result if the cluster were coordinated "end-on" to the heme Fe via some unspecified bridging ligand. Thus we write

$$\hat{H}_{\text{ex}} = k_1\,\mathbf{S_h}\cdot\mathbf{S_1} \tag{16}$$

The total Hamiltonian is then given by

$$\hat{H} = \hat{H}_h + \hat{H}_{\text{Fe-S}} + \hat{H}_{\text{ex}} \tag{17}$$

We solve the problem as follows: The largest term is certainly the interaction $J\ \mathbf{S}_1 \cdot \mathbf{S}_2$ describing the internal coupling of the [2Fe-2S] cluster. We assume that $\hat{H}_{ex}$ is small compared to $J\ \mathbf{S}_1 \cdot \mathbf{S}_2$ but large compared to the Zeeman and hyperfine terms. For $k_1 = 0$ the ground manifold of the uncoupled system has the state vector $|g\rangle = |S_h M;\ S_c m\rangle \equiv |5/2\ M;\ 00\rangle$. Next we treat $\hat{H}_{ex}$ by standard perturbation theory. For these calculations we take into account only the excited states of the [2Fe-2S] cluster. This is a reasonable assumption, since the relevant excited states of the heme (the spin quartets) are at least a few thousand wavenumbers above the ground manifold. Since $\hat{H}_{ex}$ is linear in S_1, the perturbation can mix only the $S_c = 1$ state into the ground state. A straightforward calculation yields for the perturbed system ($S_h = 5/2$)

$$|g\rangle' = |S_h M;\ 00\rangle - \frac{k_1}{J} \sum_m \mathbf{B}^{\mathbf{m}}(1) \cdot \mathbf{S}_{\mathbf{h}} |S_h M;\ 1m\rangle \tag{18}$$

where m labels the substates of the triplet at energy J. The vector $\mathbf{B}^{\mathbf{m}}(1)$ has components $B_q^m(1) = \langle 1m\ |S_{1q}|\ 00\rangle$, with $q = x, y, z$. Equation (18) is in a particularly useful form, since it contains S_h explicitly as an operator [see Griffith (36)]. Next we use $|g\rangle'$ to compute the matrix elements of the Zeeman and magnetic hyperfine interactions, retaining only terms linear in the perturbation. By using standard procedures of angular momentum coupling and by evaluating only matrix elements involving $\mathbf{S}_1$ or $\mathbf{S}_2$, we obtain

$$S_h M' \,|[g_0 - F(g_1 - g_2)]\beta \mathbf{H} \cdot \mathbf{S}_{\mathbf{h}} + A_h \mathbf{S}_{\mathbf{h}} \cdot \mathbf{I}_{\mathbf{h}} - F A_1\ \mathbf{S}_{\mathbf{h}} \cdot \mathbf{I}_1 + F A_2\ \mathbf{S}_{\mathbf{h}} \cdot \mathbf{I}_2\ |S_h M$$

with $F = 35\ k_1/6J$. The operator in this expression is the desired spin Hamiltonian:

$$\hat{H} = g^c \beta \mathbf{H} \cdot \mathbf{S}_{\mathbf{h}} + A_h^c\ \mathbf{S}_{\mathbf{h}} \cdot \mathbf{I}_{\mathbf{h}} + A_1\ \mathbf{S}_{\mathbf{h}} \cdot \mathbf{I}_1 + A_2 \mathbf{S}_{\mathbf{h}} \cdot \mathbf{I}_2 \tag{19}$$

The quantities in the coupled system (superscript c) can be expressed by those of the uncoupled representation:

$$g^c = g_0 - F(g_1 - g_2)$$
$$A_h^c = A_h; \qquad A_1^c = -FA_1; \qquad A_2^c = +FA_2 \tag{20}$$

These results provide us with some understanding of the experimental observations. First, we have achieved our goal of demonstrating that all internal fields are proportional to $\langle \mathbf{S}_{\mathbf{h}} \rangle$. Second, the parameters of the heme are virtually unaffected by the coupling because the heme has no low-lying excited states that are mixed by $\hat{H}_{ex}$ into the ground manifold. The g values are those typically observed for a heme. Since both Fe atoms of an [2Fe-2S] center are high-spin ferric, we can expect that $|g_1 - g_2| \lesssim 0.01 g_1$; hence the second term in g^c is negligible for $|F| < 1$. The EPR data show that the second term in the expression for g^c is very small in SiR.

In the absence of $\hat{H}_{ex}$ the ground state of the [2Fe-2S] (or [4Fe-4S]) cluster is diamagnetic. The coupling of 1 Fe site to the heme Fe perturbs the internal coupling, adding some paramagnetism into the ground state. This is expressed by eq. (18). Of particular interest are the results for A_1^c and A_2^c. First, the "observed" coupling constants A^c are seen to depend on the ratio k_1/J. Thus the hyperfine coupling constants give information about the strength of exchange coupling. Second, for $A_1 = A_2$, a reasonable assumption, we have the result that $A_1^c = -A_2^c$. When we first suspected the [4Fe-4S] cluster of SiR to be exchange coupled to the siroheme, we expected that the parameters of the Fe site where the attachment to the heme occurs would differ quite distinctly from those of the other three sites. Rather, two pairs of Fe sites with equal but opposite internal fields were observed. The model discussed here gives us a plausible explanation: The local spin $\mathbf{S}_1$ is antiparallel to $\mathbf{S}_h$ if $k_1 > 0$. (this is the significance of the minus sign in the expression for A_1^c). The strong internal coupling $J\,\mathbf{S}_1 \cdot \mathbf{S}_2$ in turn orients $\mathbf{S}_2$ antiparallel to $\mathbf{S}_1$, and thus parallel to $\mathbf{S}_h$.

The strength of the exchange coupling in SiR cannot be estimated reliably by reference to our simple system. The numerical values of F depend on the magnitudes and number of participating spins.

The discussion presented in this chapter should be taken as a simple start toward understanding some very complex phenomena.

ACKNOWLEDGMENTS

I am indebted to my colleagues who have contributed to the work described here, Drs. H. Beinert, J.-L. Dreyer, M. H. Emptage, J. A. Fee, J. LeGall, B. H. Huynh, T. A. Kent, J. J. G. and I. Moura, W. H. Orme-Johnson, L. M. Siegel, and A. V. Xavier, and to the graduate students J. A. Christner and P. A. Janick. The studies in my laboratory were supported by NSF Grant PCM 8005610 and NIH Grant GM 22701.

REFERENCES

1. R. Zimmermann, E. Münck, W. J. Brill, V. K. Shah, M. T. Henzl, J. Rawlings, and W. H. Orme-Johnson, *Biochim. Biophys. Acta,* **537,** 185 (1978).
2. B. H. Huynh, M. T. Henzl, J. A. Christner, R. Zimmermann, W. H. Orme-Johnson, and E. Münck, *Biochim. Biophys. Acta,* **623,** 124 (1980).

3. B. H. Huynh, E. Münck, and W. H. Orme-Johnson, *Biochim. Biophys. Acta*, **527,** 192 (1979).
4. E. Münck, in *Recent Chemical Applications of Mössbauer Spectroscopy*, Advances in Chemistry Series, American Chemical Society, Washington, D.C., 1981, p. 305.
5. M. H. Emptage, T. A. Kent, B. H. Huynh, J. Rawlings, W. H. Orme-Johnson, and E. Münck, *J. Biol. Chem.*, **255,** 1793 (1980).
6. C. D. Stout, D. Ghosh, V. Pattabhi, and A. Robbins, *J. Biol. Chem.*, **255,** 1797 (1980).
7. T. H. Moss, *Meth. Enzymol.* **27,** 912 (1973).
8. G. Lang, *Q. Rev. Biophys.*, **3,** 1 (1970).
9. P. Debrunner, in *Spectroscopic Approaches to Biomolecular Conformation*, D. W. Urry, ed., American Medical Association, Chicago, 1969. p. 209.
10. E. Münck, *Meth. Enzymol.*, **54,** 346 (1978).
11. E. Münck, in *The Porphyrins*, Vol. IV, D. Dolphin, ed., Academic, New York, 1978, Chapter 8, p. 379.
12. E. Münck and B. H. Huynh, in *ESR and NMR of Paramagnetic Species in Biological and Related Systems*, Series C, Vol. 52, I. Bertinin, ed., NATO Advanced Study Institutes, Reidel Publishing Company, Dordrecht, The Netherlands, 1979.
13. R. Aasa and T. Vänngard, *J. Magn. Reson.*, **19,** 308 (1975).
14. A. Dwivedi, T. Pederson, and P. G. Debrunner, *J. Phys.*, **40,** C2-531, (1979).
15. W. R. Dunham, A. J. Bearden, I. T. Salmeen, G. Palmer, R. H. Sands, W. H. Orme-Johnson, and H. Beinert, *Biochim. Biophys. Acta*, **253,** 134 (1971).
16. E. Münck, P. G. Debrunner, J. C. M. Tsibris, and I. C. Gunsalus, *Biochemistry*, **11,** 855, (1972).
17. R. H. Sands and W. R. Dunham, *Q. Rev. Biophys.*, **7,** 443, (1975).
18. T. A. Kent, B. H. Huynh, and E. Münck, *Proc. Natl. Acad. Sci.*, **77,** 6574 (1980).
19. F. J. Ruzicka and H. Beinert, *J. Biol. Chem.*, **253,** 2514 (1978).
20. B. H. Huynh, J. J. G. Moura, I. Moura, T. A. Kent, J. LeGall, A. V. Xavier, and E. Münck, *J. Biol. Chem.*, **255,** 3242 (1980).
21. D. Ghosh, W. Furey, S. O'Donnell, and C. D. Stout, *J. Biol. Chem.*, **256,** 4185 (1981).
22. T. A. Kent, J.-L. Dreyer, M. H. Emptage, I. Moura, J. J. G. Moura, B. H. Huynh, A. V. Xavier, J. LeGall, H. Beinert, W. H. Orme-Johnson, and E. Münck, in *Symposium on Interaction between Iron and Proteins in Oxygen and Electron Transport*, C. Ho, ed., Elsevier-North Holland, Amsterdam, in press (1982).
23. D. P. E. Dickson, C. E. Johnson, R. Cammack, M. C. W. Evans, D. O. Hall, and K. K. Rao, *Biochem. J.*, **139,** 105 (1974).
24. I. Moura, B. H. Huynh, J. LeGall, R. P. Hausinger, A. V. Xavier, and E. Münck, *J. Biol. Chem.*, **255,** 2493 (1980).
25. C. Schulz and P. G. Debrunner, *J. Phys. Colloque*, **37,** 154, (1976).
26. A. J. Thomson, A. E. Robinson, M. K. Johnson, J. J. G. Moura, I. Moura, A. V. Xavier, and J. LeGall, *Biochim. Biophys. Acta*, **670,** 93 (1981).
27. M. J. Murphy, L. M. Siegel, H. Kamin, and D. Rosenthal, *J. Biol. Chem.*, **248,** 2801 (1973).
28. L. M. Siegel, in *Mechanisms of Oxidizing Enzymes*, T. P. Singer and R. N. Ondarza, eds., Elsevier-North Holland, New York, 1978, p. 201.

29. D. C. Rueger and L. M. Siegel, in *Flavins and Flavoproteins,* T. P. Singer, ed., Elsevier, Amsterdam, 1976, p. 610.

30. P. A. Janick and L. M. Siegel, *Abstracts XIth Int. Conf. Biochem.*, Toronto, 1979, p. 427.

31. M. Barber and L. M. Siegel, unpublished.

32. J. A. Christner, E. Münck, P. A. Janick, and L. M. Siegel, *J. Biol. Chem.*, **256,** 2098 (1981).

33. B. A. Averill, A. Dwivedi, P. Debrunner, S. J. Vollmar, J. Y. Wong, and R. L. Switzer, *J. Biol. Chem.*, **255,** 6007 (1980).

34. E. Münck and P. M. Champion, *J. Phys.*, **35,** C6-33 (1974).

35. J. F. Gibson, D. O. Hall, J. H. M. Thornley, and F. R. Whatley, *Proc. Natl. Acad. Sci. USA,* **56,** 987 (1966).

36. J. S. Griffith, *The Theory of Transition-Metal Ions,* Cambridge University Press, Cambridge, 1971, Section 12.3.4.

CHAPTER 5

Hydrogenase and Other Iron-Sulfur Proteins From Sulfate-Reducing and Methane-Forming Bacteria

JEAN LEGALL

Department of Biochemistry, University of Georgia
Boyd Graduate Studies Research Center
Athens, Georgia

JOSÉ J. G. MOURA

Gray Freshwater Biological Institute
University of Minnesota
Centro de Química Estrutural
I.S.T., Av. Rovisco Pais
Lisbon, Portugal

HARRY D. PECK, JR.

Department of Biochemistry, University of Georgia
Boyd Graduate Studies Research Center
Athens, Georgia

ANTÓNIO V. XAVIER

Gray Freshwater Biological Institute
University of Minnesota
Centro de Química Estrutural
I.S.T., Av. Rovisco Pais
Lisbon, Portugal

CONTENTS

1 INTRODUCTION

The biological role of electron-transfer proteins justifies the amount of work devoted to the understanding of their mechanism of action (1, 2). This type of protein constitutes the best characterized group, in terms of physical and structural properties. Recent reviews have dealt with the different classes of specialized electron-transfer proteins: flavoproteins (3), cytochromes (4), and nonheme Fe proteins (5, 6). Theoretical work (7–9), as well as detailed information about their different active centers, is also available from model studies on synthetic analogs (10). An understanding of the mechanism of electron transfer is being sought by widely different techniques, ranging from pure solution chemistry to spectroscopic methods and genetics. These studies have provided evidence of a number of common characteristics that allow the development of several hypotheses, with consequent elimination of some other possible mechanisms. However, because of the complexity of the subject it is not yet always possible to differentiate between the different postulated mechanisms.

1.1 Sulfate Reduction and Methane Formation—A Model System

In the above perspective the sulfate-reducing bacteria and the methanogenic bacteria present electron-transfer systems with the following advantages: (1) Some of the electron carriers can be obtained in a soluble fraction. (2) The stability of most of the electron carriers enables full purification of the physiologically active components. (3) The precise definition of the electron donor/acceptor enables the reconstitution *in vitro* of the electron-transfer chains, which mimic the *in vivo* situation, and testing of the proposed mechanisms. (4) The presence of *intrinsic* probes in the electron-transfer components (iron, molybdenum, flavin, etc.) is important for the application of spectroscopic methods. (5) Some of these proteins are relatively simple, with well-defined active centers and low molecular weight. Their properties can be extrapolated to more complex situations using this system as a model.

Implications of the study of the metabolism of sulfate-reducing and methanogenic bacteria *per se* are summarized in Table 1.

1.2 The Organisms

Sulfate-reducing organisms are strict anaerobes that obtain energy for growth from the oxidation of a limited number of organic compounds and molecular H_2 (11). The reducing equivalents are used in the reduction of S compounds (i.e., SO_4^{2-}, SO_3^{2-}) to the level of H_2S. These organisms have been classified into two genera: *Desulfovibrio* and *Desulfotomaculum* (12,

Table 1 Relevance of Sulfate Reducers and Methanogenic Bacteria

	Sulfate Reducers	Methanogenic Bacteria
Ecological and nutritional	Production of sulfide, pollution, corrosion, metal control	Terminal step in the long degradation process of organic matter
Energetic	H_2 production	Methane formation
Taxonomic	Interrelations between species, primitive organisms (evolution)	Representative of a group of bacteria defined as "archaebacteria" with unusual properties: (1) presence of 16S rRNA (2) lipid composition (3) absence of peptidoglycans in the cell wall
Molecular biochemistry	Enzymes representative of electron transfer classes: cyt *b, d, c,* flavoproteins, Fe-S proteins. Models for more complex situations	Unusual electron carriers: F_{420} (diazaflavin), F_{430} (Ni), F_{342}. Fe-S centers in a different metabolic pathway

13). The most striking differences are the absence of cytochrome *c* type proteins and desulfoviridin in the second genus. Biebl and Pfennig (14) verified that some sulfate reducers can also use colloidal S as a "respiratory" substrate; some bacteria belonging to the genus *Desulfuromonas* cannot use sulfate but can utilize colloidal sulfur as their terminal electron acceptor. Recent reviews on S metabolism and the physiology of sulfate-reducing bacteria have been published (15, 16).

The methanogenic bacteria are a diverse group of strictly anaerobic bacteria that have the metabolic capacities of producing methane from a limited number of simple substrates (17). They are widely distributed in nature, where the bacteria play a basic and essential role in the decomposition of complex organic matter. From the physiological point of view these bacteria can be divided into two broad groups—the H_2-utilizing methanogenic bacteria, which reduce CO_2 and often formate to methane, and the acetate-utilizing methanogenic bacteria, which convert acetate to CO_2 and methane. The latter group may also produce methane from methanol and methylamines, and some can reduce CO_2 to methane with H_2 (18). Most but possibly not all contain an active hydrogenase, and the presence of this enzyme appears to be obligatory for the group of many of the H_2-utilizing methanogenic bacteria. A second classification has been developed, based on the sequence analysis of 16 S ribosomal RNA, which concludes that the methanogens, as a group, are only distantly related to other bacteria and should be placed in a third kingdom, the Archaebacteria, along with some halophiles and thermoacidophiles (19). The methane-forming bacteria are further subdivided into three orders—the *methanobacteriales*, the *methanococcales*, and the *methanomicrobiales*—based largely on analyses of 16 S RNA and cell type. All orders contain genera capable of reducing CO_2 to the methane, but the acetate-utilizing methanogenic bacteria are only found in the *methanomicrobiales*. Because of the difficulty of growing larger amounts of these organisms, the biochemistry of methanogenesis is far from being understood.

These two groups of bacteria may participate in synthrophic associations, which are essential in the decomposition of complex organic substrates. Peck (20) discussed the way that energy transfer occurs between the two systems by means of molecular H_2 (Fig. 1). The microbiology of the process is not completely understood and represents an attractive field of research. Little is known about the H_2 production and the way the interspecies H_2 transfer occurs.

Thermodynamic considerations of the reactions involved (substrate reduction and methane formation), as indicated in Table 2, show that the growth of an organism with, for example, ethanol (reaction A) is generally not

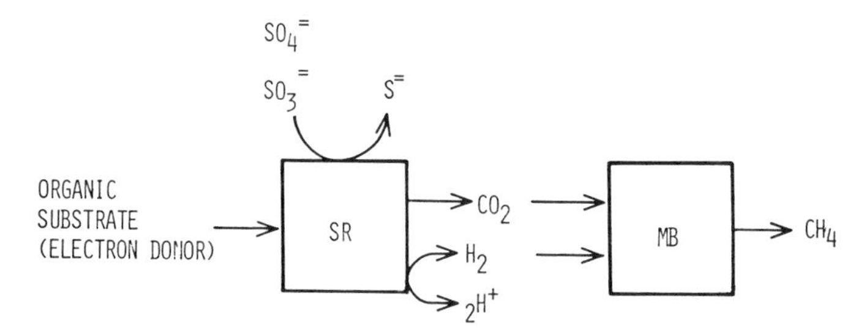

Figure 1. Schematic representation of the interrelationship between sulfate reducers (SR) and methanogenic bacteria (MB).

Table 2 Free Energy for Alcohol Oxidation and Methane Formation[a]

	Reaction	$\Delta G'_0$ (kcal)
A	$CH_3CH_2OH \longrightarrow CH_3COO^- + H^+ + 2H_2$	+1.5
B	$4H_2 + CO_2 \longrightarrow CH_4 + 2H_2O$	−33.2
A + B		−31.7

[a]Adapted from reference 231.

favorable, but the equilibrium of the overall reaction can be driven in the presence of a methane-forming organism (reaction A + B).

1.3 The Electron Carriers

Sulfate reducers and methanogenic bacteria are considered to be a group of organisms that carry out metabolic reactions that are primitive in the evolutive scale (20, 21). But primitive is not simple, and the mechanisms involved are highly sophisticated and complex. This complex electron-transfer chain, and in particular the S metabolism, has attracted considerable research in the field of structure/function relationships of the different proteins involved (20, 22).

A variety of electron carriers have been identified and characterized (Table 3): a complex cytochrome system (*d, b,* and mono- and multihemic *c* type), nonheme iron (ferredoxin, rubredoxin, hydrogenase), flavoproteins and complex electron carriers containing Mo, Co, and siroheme associated or unassociated with nonheme Fe (23–26).

A variety of electron carriers have also been isolated from methanogenic organism, but their characterization is in a preliminary state (see Table 4).

Three main reasons may be indicated for the incomplete understanding of the electron-transfer chains involved: (1) Screening of electron carriers is presently not complete. (2) The pathway of sulfate reduction is still controversial (26). (3) Important heterogeneity is observed when the different *Desulfovibrio* spp. are considered. However, a group of common carriers can be observed, and the state of purification of these components is indicated in Table 5.

The main goal of this chapter is to discuss the role of Fe-S centers in these two distinct metabolic pathways: sulfate reduction and methane formation. Structural information is used to understand the constraints that may modulate the redox potential and the physiological activity. The active centers to be discussed may be included in the known basic structures: [Rb] type, [2Fe-2S], [3Fe-3S], and [4Fe-4S] (27–29). Their association with other redox centers such as Mo and siroheme is also considered.

The discussion is organized in the following way:

1. The physiological and biochemical aspects of the role of hydrogenase are reviewed.
2. Two rubredoxin type proteins isolated from *Desulfovibrio gigas* are described. Sequence, EPR, and Mössbauer data were put together to explain the different constrains imposed by the polypeptide chain on the simplest Fe-S center.
3. As a complement to Chapter 4, establishment of the presence of the

Table 3 Selected Electron-Transfer Proteins Isolated from *Desulfovibrio* spp. Representative of Different Types of Redox Centers[a]

Protein	MW × 10^{-3} (subunits)	Redox Potential (mV)	Type of Center	Refs.
Heme Fe proteins				
Cyt b[e]	n.d.	n.d.	1 heme *b* (His, His)	232
C_{553} (D. vulgaris)	9	−100 to 0	1 heme *c* (Met, His)	22
Cyt c_3	13	−235, −235, −306, −315	4 heme *c* (His, His)	22, 176
Cyt c_3	26	negative	8 heme *c* (His, His)	22
Cyt c_7 (*Drm. acetoxidans*)	9	−107, −177, −177	3 heme *c* (His, His)	233
Nonheme Fe proteins Simple				
Rubredoxin	6	+6	$Fe(Cys)_4$ Rb	135
Desulforedoxin	7.6(2 × 3.8)	−35	Rb distorted	142
Ferredoxin II	24(4 × 6)	−130	$[3Fe\text{-}3S]^{+3(+3,+2)}$	160
Ferredoxin I	18(3 × 6)	−50	$[3Fe\text{-}3S]^{+3(+3,+2)}$	
		−455	$[4Fe\text{-}4S]^{+2(+2,+1)}$	160
Ferredoxin (*Drm. acetoxidans*)	6	n.d.	probably 2 × [4Fe-4S]	141
Complex nonheme Fe proteins				
Hydrogenase	89(62 and 26)	−30	3 × [4Fe-4S]*d*	80, 84
Mo(Fe-S) protein[b]	120		6 × $[2Fe\text{-}2S]^{+2(+2,+1)}$ Mo	180, 181
Sulfite reductase[c] (desulfoviridin)	200		siroheme + probably [4Fe-4S]	
APS reductase	220 (3 × 72, 1 × 20)		[Fe-S] + flavin 12 Fe, 12 S + 1 FAD	245

[a]All the electron carriers are referred to *D. gigas* except where otherwise indicated.
[b]See Table 12.
[c]See Table 13.
[d]Chemical analysis and extrusion experiments indicate the presence of 3 × [4Fe-4S] cores (80). However, EPR measurements (84) show the presence of "isotropic" signals in the native preparations, suggesting the presence of [3Fe-3S] cores.
[e]Also present in *D. africanus*.

Table 4 Electron-Transfer Proteins Isolated from *Methanosarcina barkeri*[a]

Protein	Active Center
Ferredoxin	$[3Fe\text{-}3S]^{+3(+3,+2)}$
Hydrogenase	[Fe-S]
P_{590}(sulfite reductase)	Siro + [Fe-S]
Cyt *b*	1 heme *b* (His, His)
F_{342}	?
F_{430}	Ni (pyrrole)
F_{420}	5-diazaflavin mononucleotide
B_{12}-protein	corrinoid

[a]Adapted from unpublished data of I. Moura, J. J. G. Moura, M. H. Santos, A. V. Xavier, J. M. Wood, and J. LeGall.

novel [3Fe-3S] core is exemplified with the *Methanosarcina barkeri* Fd, and *D. gigas* Fd is used to discuss the problem of accommodation of either a [3Fe-3S] or a [4Fe-4S] core in the same polypeptide chain.

4. The association of Mo with Fe-S centers is described in two proteins isolated from sulfate reducers, *D. gigas* and *D. africanus*.
5. Siroheme and Fe-S containing enzymes present in the sulfate reducing pathway as terminal reductases are briefly described, and an analogous enzyme isolated from *M. barkeri* is discussed.
6. The possibility of using the structural and physicochemical data accumulated by the study of the isolated components of these electron-transfer chains is presented in situations closer to the *in vivo* system.

The physiological relevance of the structural and physicochemical properties, as well as the protein compartmentation of Fe-S proteins is discussed throughout this chapter. Special attention is given to the techniques used for structural characterization of the centers, and in particular we include a short review of the most relevant NMR studies previously reported.

As a result of the specific properties of the proteins discussed, the degree of knowledge of their physicochemical properties versus their physiological activity is widely different, and this fact is reflected in the depth with which they are discussed. Another Fe-S protein, present in sulfate reducers, APS reductase (16), is not described in this chapter.

2 HYDROGENASE

Molecular H_2 has long been regarded either as an esoteric, if not trivial, product of fermentation or as a substrate for specific reductions such as sul-

Table 5 Distribution of Electron-Transfer Proteins in Sulfate-Reducing Bacteria[a]

Organism	Nonheme Fe Proteins[b]								Heme Proteins		
	FD			SIR Dissimilative							
	1 center	2 centers	Rb	DSV	DSR	SIR Assimilative	Mo Fe/S	Flavodoxin	c_3 13,000	c_3 26,000	c_{553} or Monohemic
Drm. acetoxidans	—	P	+	—	—	—	n.d.	n.d.	SC(c_7)	—	—
D. vulgaris (Hildenborough)	P	—	S	P	—	—	n.d.	S	S	+	SC
D. vulgaris (Miyazaki)	n.d.	n.d.	n.d.	P	—	n.d.	n.d.	S	SC	—	P
D. gigas	SC	—	SC	P	—	—	P	P	SC	C	—
D. desulfuricans (Norway 4)	P	n.d.	P	—	P	P	n.d.	—	SC	—	+
D. salexigens	+	n.d.	P	+	—	—	+	P	S	+	+
D. strain 9974	+	n.d.	+	—	+	+	n.d.	—	P	n.d.	P
D. africanus	P		P	+	—	—	P	—	+	n.d.	n.d.
D. desulfuricans (El Algheila Z)	+	n.d.	n.d.	+	—	—	n.d.	P	SC	n.d.	n.d.

[a] S = amino acid sequence determined; P = purified to homogeneity; + = present but not yet purified; C = crystallized; n.d. = not determined; DSV = desulfoviridin; DSR = desulforubidin.

[b] For hydrogenase see Tables 6 and 7.

fate to sulfide; however, information is accumulating that H_2 can play a unique and important role in the bioenergetics of both anaerobic and aerobic microorganisms. Molecular H_2 itself possesses several important physical and chemical properties that are significant for biological systems. It has a large heat of combustion, and is highly diffusable and freely permeable to biological membranes, but the oxidation products of H_2 (protons) are not freely permeable to biological membranes in the context of the chemiosmotic hypothesis. Thus any organism that has the enzymological capacity of oxidizing H_2 on the external surface of the cytoplasmic membrane with a net formation of protons has the potential for generating a proton gradient without the intervention of a typical Mitchell loop, that is, direct electron transfer–coupled proton translocation. Hydrogenase itself can be viewed as being able to effect proton translocation, and has the simplest proton translocating system (30). It is of extreme interest from physiological, biochemical, and evolutionary points of view (31).

Hydrogenase and H_2 are intimately involved in three areas of basic biochemical and economic interest: (1) nitrogen fixation, (2) photoproduction of H_2, and (3) complex fermentations of biomass to methane and other chemicals. In nitrogen fixation the recovery of the H_2 produced by nitrogenase is believed to decrease the energy requirements for N_2 fixation and increase legume productivity (32). The production of H_2 from chloroplasts in the presence of hydrogenase is being extensively studied (33) as one solution to the present energy crisis, and the basic physical and biochemical properties of hydrogenase figure centrally in these studies. Interspecies H_2 transfer is an essential interaction in complex anaerobic microbial associations, in particular those converting biomass to methane in diverse ecological environments (34). An understanding of hydrogenase and H_2 metabolism in these anaerobic processes is essential before complex microbial fermentations can be utilized for significant production of fuels and chemicals.

2.1 Physiological Aspects

The microbiological formation of molecular H_2 attracted the attention of early investigators simply because H_2 was a different and unexpected fermentation product. Stephenson and Strickland (35) first recognized that an enzyme, which they termed *hydrogenase*, was responsible for the biological activation of H_2 and was probably essential for the fermentative formation of H_2. These workers determined enzyme activity by measuring the utilization of H_2 in the presence of the artificial electron acceptor, methylene blue, and proposed that the enzyme catalyzed the following reversible reaction:

$$H_2 \rightleftharpoons 2H^+ + 2e^-$$

As the bacteria that they were investigating evolved H_2 from formate, through the formate hydrogenlyase reaction, it was thus implicit that the activation of hydrogen is reversible and involved in both the utilization and production of H_2. This supposition concerning the enzyme(s) involved in H_2 utilization and formation has been repeatedly questioned and remains a major problem in the metabolism of H_2.

The early literature on hydrogenase has been summarized by Gest (36) and Gray and Gest (37). In general, hydrogenase activity appeared to occur sporadically among microorganisms, and the presence of the enzyme could not be correlated absolutely with taxonomic relationships or with broad physiological types. A relationship between N_2 fixation and hydrogenase activity was recognized in that all microorganisms that fix nitrogen have hydrogenase activity. However, the converse is not true; that is, all microorganisms that have hydrogenase activity do not fix N_2.

Until 1971, when the first homogeneous preparations of hydrogenase were reported (38–40) the early enzymology was generally concerned out of necessity with crude or less purified soluble preparations and whole cells. Nevertheless some very useful information was obtained from these early studies, and many of the results obtained have since been confirmed with homogeneous preparations of hydrogenase. Farkas and co-workers (41) found that hydrogenase catalyzes an isotope-exchange reaction between D and H in which the products HDO, D_2O, and HD are formed, and Krasna and Rittenberg (42) reported that hydrogenase also catalyzes the *ortho-para* conversion of hydrogen. Studies on the "exchange reaction" and the *ortho-para* conversion of H_2 resulted in a proposal for a general mechanism of hydrogenase activity involving (42, 43) the formation of an enzyme hydride as the initial reaction of hydrogenase:

$$\text{Enzyme} + H_2 \rightleftharpoons \text{Enzyme} - H^- + H^+$$

The mechanism is based on the ability of hydrogenase to catalyze the *ortho-para* conversion of H_2 and the primary formation of HD rather than D_2, thus indicating that only 1 of the atoms of H_2 can freely exchange with the protons of water. Fe was very early implicated as a component of hydrogenase in nutrition studies by Waring and Werkman (44), and carbon monoxide was found by Kubowitz (45) to inhibit formation of hydrogen from pyruvate by intact cells of *Clostridium butyricum* in a light reversible reaction. Studies with cellfree preparation were generally inconclusive in this regard; however, they did establish a general sensitivity of hydrogenase toward inactivation by molecular oxygen (46).

Over the past 10 years there has been renewed interest in the physiology and biochemistry of hydrogenase, which has been accelerated by the energy crisis because of the potential for the photoproduction of H_2 from water. Hydrogenases have been purified to homogeneity from anaerobic bacteria, fermentative microorganisms, the sulfate reducing bacteria, the methanogenic bacteria, photosynthetic bacteria, facultative anaerobes, and aerobes, both H_2-oxidizing and H_2-fixing microorganisms. The hydrogenases appear to constitute a heterogeneous group of enzymes with regard to their molecular weights, specific activities, subunit structure, prosthetic groups, O_2 sensitivity, and electron donor or acceptor specificity, but they all have in common the presence of Fe-S centers. An additional complexity to the study of these proteins is the demonstration of two or more hydrogenases of different properties in a single organism which appear to serve different functions, H_2 uptake and H_2 production.

Significant advances have also been made in our understanding of the physiological role of hydrogenase in a number of different types of microorganisms. In fermentative bacteria it has been clearly established (47) that the function of hydrogenase is to dispose of excess reducing power by the reduction of protons via the pyridine nucleotides, as originally suggested by Gray and Gest (37). In *Clostridium klyuveri* it is only through the production of H_2 that the bacteria can obtain, from substrate phosphorylation, the energy required for growth (48). In addition to hydrogenase, molecular H_2 can be produced during N_2 fixation by nitrogenase in an irreversible reaction (49). The relationship of H_2 to N_2 fixation appears to be to reoxidize or scavenge H_2 formed by nitrogenase and then minimize the ATP requirement for N_2 fixation (50). These hydrogenases appear to be unidirectional, and have been loosely termed *uptake* hydrogenases (51). H_2 has also been recognized as an important intermediate in fermentations involving two or more microorganisms, such as the methane fermentations in which H_2 serves as the vehicle for the transfer of energy and reductant between microbial species. This process, termed *interspecies hydrogen transfer* (52), is obligatory for the growth of certain organisms recently isolated from these anaerobic environments (53, 54). Studies on the cellular localization of hydrogenase have demonstrated that a significant fraction of the total hydrogenase activity of a microorganism can be localized on the external surface of the cytoplasmic membrane (periplasmic area) (55, 56). This has led to the important concept that a proton gradient can be generated without the intervention of a typical Mitchell loop by the oxidation of H_2 on the external surface of the cytoplasma concomitant with the vectorial transfer of electrons across the membrane and their utilization by cytoplasmic reductases (57, 58).

2.2 Biochemical Aspects

In 1971 two independent studies appeared on hydrogenase from different bacterial sources utilizing enzymes that represented, at that time, the most homogeneous preparations of the hydrogenase. Nakos and Mortenson (39) isolated hydrogenase from *C. pastorianum* and found that it contained no Mo, but 4 Fe and 4 labile S g-atoms per 60,000 daltons. The enzyme showed a single peak by disk electrophoresis and was purified by use of a manometric assay which measures hydrogen evolution catalyzed by hydrogenase from reduced methyl viologen. The enzyme was inactivated by exposure to O_2 or heat, but was stable in the presence of hydrogen at 0°C for 2–3 mo. In the reduced condition (under H_2) the enzyme exhibited a $g = 1.94$ type EPR signal, characteristic of the low-potential-type Fe-S cluster. The authors noted that half of the Fe could be chelated with *o*-phenantroline. The second report, concerning the purification of hydrogenase from *D. vulgaris* (59), is discussed in a subsequent section.

Chen and Mortenson (60) concluded that their previous hydrogenase preparation, although showing a single peak in disk electrophoresis, was in fact heterogeneous based on the reactivity of the preparation with the Fe chelator, *o*-phenanthroline. They were able to modify the purification procedure by using hydroxyapatite chromatography as the final step. Hydrogenase was then purified with an additional 6.6-fold increase in specific activity. The enzyme now contained 12 Fe and 12 labile S g-atoms per 60,000 daltons arranged in 3 [4Fe-4S] clusters (60, 61). This enzyme was no longer reactive with *o*-phenanthroline and did not dissociate into subunits in the presence of sodium dodecyl sulfate and 2-mercaptoethanol. Methyl viologen, benzyl viologen, methylene blue, and ferredoxin were each suitable as an electron carrier. Under routine assay conditions (1 μmol methylviologen + $Na_2S_2O_4$) hydrogenase exhibited a specific activity of 550 μmol H_2/min · mg. The EPR spectrum exhibited g values at 2.079 and 1.892 in the reduced form, and g values at 2.009, 2.046, and 2.005 in the oxidized form. Based on the EPR spectra of the hydrogenases obtained at different redox states, a mechanism for hydrogenase was proposed in which 2 of the 3 [4Fe-4S] clusters act as redox sensors (56).

Erbes and colleagues (62) reported on hydrogenase from *C. pastorianum* using the purification procedure of Chen and Mortenson (60). They were able to purify hydrogenase to homogeneity and to the same specific activity; however, their enzymed contained 4 Fe and labile sulfide atoms per 60,000 daltons. EPR studies indicated the appearance of the low-potential type $g = 1.9$ EPR signal on reduction, while oxidation resulted in the appearance of the high-potential Fe-S signal at $g = 2.04$. Stopped-flow EPR kinetics determined that the oxidation-reduction times of the EPR-detectable components occurred within the turnover time of the enzyme. The competitive inhibitor carbon monoxide caused, in a reversible fashion, an alternation of the reduced $g = 1.9$ type signal. This observation was further explored by reaction of ^{13}CO with reduced enzyme, resulting in broadening of the reduced signal, suggesting the formation of a direct CO complex with the Fe-S system. The Fe-S cluster of the enzyme was chemically displaced and identified optically as a [4Fe-4S]-type cluster. It was concluded that the Fe-S cluster served as the site of binding of gaseous ligands.

Gillum and co-workers (61) concluded, from chemical extrusion studies of the hydrogenase of *C. pastorianum* prepared by the technique of Chen and Mortenson (60), that the enzyme contained 3 sites of the [4Fe-4S] type. This result differed somewhat from the report of Erbes and co-workers (62) on detection of a single 4-Fe site in their hydrogenase preparation from *C. pastorianum*, but agreed with the results of Chen and Mortenson (60).

Chen and Mortenson (63) have purified a unidirectional hydrogenase from the periplasmic space of *C. pastorianum*, and clearly differentiated it from the previously studied bidirectional hydrogenase found in the cytoplasm. The enzyme is quite active, has a smaller molecular weight, and is less O_2 sensitive than the bidirectional hydrogenase from the same organism. However, the unidirection enzyme was not pure, showing multiple protein bands in SDS polyacrylamide gel electrophoresis. As with the bidirectional hydrogenase, the unidirectional enzyme was strongly inhibited by carbon monoxide, suggesting that it may also be a nonheme Fe protein.

2.3. Hydrogenases of Sulfate-Reducing Bacteria

The metabolism of molecular H_2 has figured centrally in the development of our present concepts regarding the biochemistry and physiology of respiratory sulfate reduction, and the hydrogenase has been extensively studied. *Desulfovibrio vulgaris* contains a soluble hydrogenase of high specific activity which is largely localized in the periplasmic space. The purified hydrogenase is unique among hydrogenases in requiring the four-heme cytochrome c_3 (M_r = 13,000) for activity with naturally occurring electron acceptors such as ferredoxin, flavodoxin, and rubredoxin (64). Hydrogenase is produced whether these bacteria are grown on H_2 plus sulfate or on organic substrates plus sulfate; however, H_2 is not generally believed to be produced as a major fermentation product, and the role of hydrogenase, if any, in this mode of growth has been problematical (65).

The study of H_2 metabolism in *Desulfovibrio* is of special interest since (1) they constitute a very heterogeneous genus inside of which nature seems to have used all solutions to the same problem: how to make energy from the oxidation of organic matter and the reduction of S compounds; (2) they possess a very intricate electron transport system that is fairly well characterized, in particular in the case of the species *D. gigas*, as far as individual redox proteins are concerned; (3) the study of their hydrogenases is a key to a complete understanding of the respiratory reduction of S compounds since H_2 is thermodynamically situated halfway between S compounds and the organic energetic substrates, and can be either produced or utilized by the bacteria; (4) they play a crucial role in the complex bacterial associations that lead to the complete decomposition of organic matter with the ultimate formation of sulfides, carbon dioxide, and methane. In this aspect it constitutes the only known group of bacteria that can clearly participate in interspecies transfer as either H_2-producing (66) or the H_2-utilizing (53) microorganisms, and thus appear to be prime bacteria for investigating the biochemistry, bioenergetics, and physiology of H_2 metabolism and interspecies H_2 transfer in anaerobic microbial associations.

2.3.1 Localization of Hydrogenases and Variations in H_2 Metabolism of *Desulfovibrio*. As described above, the sulfate-reducing bacteria are composed of two generally recognized genera, *Desulfovibrio* and *Desulfotomaculum*, which do not appear to be closely related (65, 12). The hydrogenase of *Desulfotomaculum* is bound on the inner aspect of the cytoplasmic membrane and has a low specific activity (67, 68). Although less well studied than *Desulfovibrio*, H_2 metabolism appears to be more limited in *Desulfotomaculum* than in *Desulfovibrio*. The various species of *Desulfovibrio* contain multiple hydrogenase activities that can be periplasmic (55), membrane bound (69), or cytoplasmic, but no general pattern is evident. Heterogeneity within the genus *Desulfovibrio* is a well-established fact that has been documented by the comparative study of the primary structure of several families of proteins, such as cytochrome c_3, ferredoxin, and rubredoxin (70). This heterogeneity can even be detected within a given species, and is particularly evident as far as hydrogenase metabolism is concerned. Some of these differences among species of *Desulfovibrio* are summarized in Table 6.

The three strains of *D. vulgaris* that have been well studied so far.appear to have a completely different behavior toward molecular H_2. The strain Myazaki, in which the hydrogenases are membrane bound, has been reported to produce a burst of hydrogen (0.014 mol H_2/mol lactate) during the earliest phase of growth (71), and this H_2 is consumed in later phases. In contrast (72) the Hildenborough strain produces large amounts of hydrogen during all phases (0.5 mol H_2/mol lactate). Finally, the Marburg strain is specialized in H_2 utilization in that this strain can grow on H_2 and sulfate plus acetate. Although identical growth conditions have not been used by the authors, these differences are striking. The problem is to determine whether they can be interpreted in terms of enzyme location (see Table 7). Tsuji and Yagi (71) have interpreted their results by postulating that the H_2 that is produced during the earliest phase of growth by the "large" hydrogenase is reutilized by a second "small" hydrogenase, whose role would then be similar to the role of hydrogenase in N_2 fixation, where the enzyme scavenges H_2 produced by nitro-

Table 6 Comparison of Various Strains of *Desulfovibrio*

Strain	Particularity
D. vulgaris Hildenborough	Continuous H_2 production during sulfate reduction; external and internal hydrogenases
D. vulgaris Myazaki	Burst of H_2 production; two hydrogenases on membrane
D. vulgaris Marburg	Growth with H_2 sulfate and acetate; single perplasmic hydrogenase
D. gigas	Well-characterized redox carriers; in vitro electron transfer–coupled phosphorylation; external and internal hydrogenases
D. desulfuricans Norway 4	Structure of cytochrome c_3 known; S^0 reducer; lacks desulfoviridin
D. desulfuricans 27774	Nitrate can substitute for sulfate as terminal electron acceptor; membrane-bound hydrogenase
D. africanus	Does not produce H_2; membrane-bound hydrogenase
D. salexigens	No available data

Table 7 Localization of Hydrogenases in Various Strains of *Desulfovibrio*

Strain	Periplasmic	Membrane	Cytoplasmic
D. vulgaris			
Hildenborough	M_r 52,000	0	+
Myazaki		large M_r 180,000 small M_r 89,000	no more than 10%
Marburg	+	0	0
D. gigas	M_r 89,000	0	+
D. desulfuricans			
Norway 4		+	?
27774		+	+
D. africanus		+	?
D. salexigens	?	?	?

genase, thereby conserving ATP (73). Apparently such an explanation cannot hold for Hildenborough, since important amounts of H_2 are produced during all growth phases; the growth equation can be formulated as follows (72):

$$CH_3\text{-}CHOH\text{-}COO^- + 0.37SO_4^{2-} + 0.56H^+ \longrightarrow CO_2 + 0.98CH_3\text{-}COO^- + 0.02CH_3CH_2OH + 0.16H_2S + 0.215HS^- + 0.5H_20 + 0.48H_2$$

Apparently the presence of two types of hydrogenase is not sufficient to allow hydrogen reutilization by this strain. The strain Marburg appears to represent a different type of organism, and it is reported to contain only one type of hydrogenase, which is located in the periplasm (57).

Going back to the differences within the genus *Desulfovibrio*, it should be noted that *D. africanus* has been shown not to produce molecular H_2 in conditions where *D. gigas* and *D. vulgaris* did accumulate this gas (74). This picture is further complicated by the report that the presence of H_2 during growth on organic substrates does not enhance cell yields (75).

Thus from the existing data it is not yet possible to formulate a general pattern regarding the number, localization, or specific functions of hydrogenase found in *Desulfovibrio*. However, it is reasonably well established that *D. gigas* and *D. vulgaris*, strains Hildenborough and Myakazi, contain at least two hydrogenases (Table 7). In the case of *D. gigas* one of the hydrogenases is periplasmic and the other cytoplasmic (76). It should be pointed out that the selective elution of hydrogenase from the external surface of the cytoplasm appears to be a complex process (77, 78), and simple cell breakage can yield a membrane-bound enzyme. There appears to be considerable variability in this process among *Desulfovibrio*, and other techniques may have to be employed to localize membrane-bound hydrogenases.

2.3.2 Purification and Physical Properties of Hydrogenases from *Desulfovibrio*. LeGall and colleagues (59) reported on hydrogenase isolated from *Desulfovibrio vulgaris*, which is structurally similar to the hydrogenase of *C. pastorianum* (39) (see Table 8). The enzyme showed essentially a single peak in the disk electrophoresis and analytical centrifuge. Enzymatic reduction in the presence of H_2 yielded a typical low-potential Fe-S $g = 1.9$ EPR type signal which was, however, rather temperature insensitive, being detected as high as $-50°C$. The enzyme was found to exist in a dimer form of 60,000 daltons and a monomer form of 30,000 daltons. The dimer and monomer forms were concentration dependent, with the dimer form favored at high protein concentrations. The specific activity of the enzyme was comparable to that found by Nakos and Mortenson (39).

With the use of selective elution procedures, hydrogenases have been puri-

Table 8 Physicochemical Properties of Hydrogenases from C. *pastorianum* and Sulfate-Reducing and Methane-Forming Bacteria[a]

	Clostridium pasteurianum	*Desulfovibrio vulgaris* (Myazaki)	*Desulfovibrio vulgaris* (Hildenborough)	*Desulfovibrio gigas*	*Methanobacterium thermoautotrophicum*
MW ($\times 10^{-1}$)	60	89	52	89	185
(subunits)		(59 and 28)		(62 and 26)	(164 and 90)
Fe/mole	12	8	12	12	30–40
Sulfide/mole	12	8	12	12	n.d.
Active center	3 × [4Fe-4S]	[Fe-S]	[Fe-S]	3 × [4Fe-4S]	[Fe-S]
Specific activity (μ mol H^2 evolved/min·mg)	4000	90	10,400	91	1000 (60°C)
Physiological electron carriers	Fd, flav	cyt c_3	cyt c_3	cyt c_3	
Artificial electron carriers	BV, MV, MCB, Fd	MV	MV, BV	MV, BV, FMN	MV
O_2 sensitivity	sensitive	relatively stable	relatively stable	sensitive	sensitive

[a]Adapted from reference 235 and from unpublished results of (H. D. Peck, Jr., L. E. Mortenson, and J. LeGall).

fied from the periplasmic space of *D. gigas* and *D. vulgaris* (Hildenborough), and they are quite different as regards subunit structure molecular weight and specific activity. Van der Westen and co-workers (79) purified to homogeneity the bifunctional hydrogenase from *D. vulgaris*, and this particular hydrogenase is termed the *external* hydrogenase. The hydrogenase exhibited a high specific activity (≈ 3200) and was not irreversibly inactivated by O_2 as is the case with the hydrogenase from *C. pastorianum*, but in other properties it was similar to the clostridial hydrogenase. The enzyme contained 12 atoms each of Fe and sulfide per molecular weight of 52,000 daltons, and SDS polyacrylamide electrophoresis indicated a single polypeptide chain. It is curious that both hydrogenases are bidirectional, but the clostridial enzyme is cytoplasmic and the enzyme from *Desulfovibrio* is periplasmic. A second external hydrogenase has been purified from *D. gigas* which contains 3 [4Fe-4S] clusters, but differs in several significant properties from the external hydrogenase of *D. vulgaris*. It has a low specific activity (≈ 90), a molecular weight of 89,000 daltons, and subunits of 60,000 and 30,000 daltons (80). The presence of a high-potential-type Fe-S signal in the $g = 2$ region in an enzyme that operates at a very negative redox potential is a paradox. One explanation is that the signal is not a high-potential-type resonance, but arises from a novel 3-Fe center recently discovered by Münck and his co-workers by Mössbauer spectroscopy in a number of Fe-S containing proteins such as ferredoxin II from *D. gigas* (81), ferredoxin from *Azotobacter vinelandii* (82), and mamalian aconitase (83).

Cammack and colleagues (84), in a study of different hydrogenases presenting the same $g = 2.02$ signal, suggested several hypotheses to explain its presence. The first is that the cluster giving the $g = 2.02$ signal could have a regulatory function in controlling the activity of the active site (56); the second, that it would be a secondary electron acceptor; and the third, that it represents the higher oxidation state of a 4-Fe cluster that can undergo further reduction at or below the potential of H_2. An interesting consequence of the first hypothesis is that such a cluster could specifically bind O_2, thus protecting the H_2 activating side by converting the protein into a reversible inactivated form.

2.3.3 Relationships Between *Desulfovibrio* Hydrogenases and the Electron Carriers. All hydrogenases from *Desulfovibrio* which have been purified so far are dependent on the presence of cytochrome c_3 for activity toward the other electron carriers. Exceptions are the viologen dyes and FMN, which can react directly with the enzyme (64). This specificity for cytochrome c_3 is remarkable, since other hydrogenases such as the enzyme from *Clostridium pastorianum* can react directly with ferredoxin (60). This particuliarity has been interpreted as a physiological necessity for *Desulfovibrio* to avoid short-circuiting between hydrogenase and ferredoxin by using cytochrome c_3, which can be linked to a vectorial electron transfer necessary for the generation of ATP (57).

In contrast, *M. elsdenii* (85), *C pastorianum* (L. E. Mortenson and J. LeGall, unpublished), and *Methanosarcina barkeri* (Y. Berlier, P. Lespinat, and J. LeGall, unpublished) hydrogenases are unable to react with cytochrome c_3.

However, the demonstration that cytochrome c_3 is essential for the reduction of other electron carriers is not sufficient to prove that electrons are actually following from hydrogenase to cytochrome c_3 and then to the other carriers. In fact, MS experiments with pure hydrogenase from *D. gigas* (Y. Berlier, P. Lespinat, and J. LeGall, unpublished) have shown that the exchange reaction can be initiated by addition of either cytochrome c_3 or oxygen scavengers such as the glucose/glucose oxidase system. Since cytochrome c_3 reacts rapidly with O_2 (86), its role could be to remove any molecule of this gas which would be reversibly bound to the hydrogenase active site, preventing it from reaction with any electron carriers. The two hypotheses are represented as follows:

$$2H_2 + O_2 : \text{hydrogenase} \xrightarrow{\text{cytochrome } c_3} \text{hydrogenase} + H_2O$$

then

$$\begin{array}{lllll} & & & & \text{toward} \\ & & & \text{flavodoxin} \rightleftharpoons & \text{specific} \\ H_2 \rightleftharpoons \text{hydrogenase} & \rightleftharpoons & \text{cytochrome } c_3 \rightleftharpoons & \text{flavodoxin} \rightleftharpoons & \text{electron} \\ & & & \text{rubredoxin} \rightleftharpoons & \text{donors or} \\ & & & & \text{acceptors} \end{array}$$

or

$$\begin{array}{llll} & & & \text{toward} \\ & & \text{cytochrome } c_3 \rightleftharpoons & \text{specific} \\ H_2 \rightleftharpoons \text{hydrogenase} & \rightleftharpoons & \text{ferredoxin} \rightleftharpoons & \text{electron} \\ & & \text{flavodoxin} \rightleftharpoons & \text{donors or} \\ & & \text{rubredoxin} \rightleftharpoons & \text{acceptors} \end{array}$$

Apparently cytochrome c_3 can play a different role, as Yagi (87) has reported a fivefold stimulation of Myazaki hydrogenase activity in the presence of viologens, whereas Hatchikian and co-workers (80) noted that such a stimulation was not visible after addition of cytochrome c_3 to the purified hydrogenase from *D. gigas*. Owing to the structural differences that exist within the cytochrome c_3 family (70), this is quite plausible. In fact, such a change in function has already been noted as far as the reduction of colloidal S is concerned (88). In *D. desulfuricans* (Norway 4) a mixture of hydrogenase and cytochrome c_3 readily reduced S^0 in the presence of molecular H_2, whereas cytochrome c_3 from *D. vulgaris* Hildenborough is rapidly inhibited by sulfide, which is the reaction product. As a consequence, only the first organism can grow in the presence of colloidal S (89).

The electron carriers that have been tested so far for their reactivity toward the system hydrogenase + cytochrome c_3 are ferredoxins, flavodoxin, and rubredoxin. The larger cytochrome c_3 (M_r 26,000), which has been found in both *D. vulgaris* Hildenborough and *D. gigas* (90, 91), also reacts with hydrogenase, but has not been extensively studied. The same observation holds for the monoheme cytochromes c_{553} and $c_{553(550)}$ (93).

2.3.4 Physiological Roles of Hydrogenase in *Desulfovibrio*. *Desulfovibrio* is the only group of bacteria that can clearly participate in interspecies H_2 transfer as the H_2-producing or the H_2-utilizing microorganism, and there has been considerable interest in the bioenergetics and physiology of H_2 metabolism in these bacteria. Hatchikian and co-workers (74) first reported the accumulation of H_2 during growth on lactate plus sulfate, but there is no stoichiometry between the amount of H_2 produced and the sulfate reduced; more recently, Tsuji and Yagi (71) have reported a burst of H_2 production during the early stages of growth and a consumption of the H_2 during later stages of the growth cycle. These investigators also reported the presence of two membrane-bound hydrogenases, one of molecular weight 70,000 and another having a molecular weight of 180,000 daltons. They suggested that the lower molecular weight hydrogenase was involved in H_2 utilization and the higher molecular weight hydrogenase was involved in H_2 production. The physiological role of the low molecular weight hydrogenase was viewed as similar to the role of hydrogenase in N_2 fixation, where it scavenges H_2 produced during N_2-fixation, thereby conserving ATP (71). More recently, the formation of 0.5 mol H_2/mol lactate oxidized in the presence of sulfate has been reported (72), but this does not appear to be a universal observation. The role of H_2 in the metabolism of these microorganisms is further complicated by the reports that the presence of H_2 does not stimulate growth (75) or inhibit growth on lactate plus sulfate (71).

Badziong and Thauer (57) and Odom and Peck (unpublished) have investigated the localization of various electron-transfer proteins, dehydrogenases, and reductases in cell fractions of *Desulfovibrio*. The basic approach is to assay intact cells (or spheroplasts) and extracts for an enzymatic activity in the presence of a nonpermenant electron donor or acceptor such as a viologen dye (94). APS reductase, bisulfite reductase, thiosulfate reductase, and pyruvate dehydrogenase are exclusively localized in the cytoplasm, while fumarate reductase and lactate dehydrogenase are bound to the internal aspect of the cytoplasmic membrane (57, 76, 95). Hydrogenase and formate dehydrogenase, along with some *c*-type cytochrome, behave as external or periplasmic enzymes, but with *D. gigas* significant internal hydrogenase can be demonstrated in spheroplasts after removal of the external hydrogenase (76). This appears to be a high molecular weight hydrogenase similar to that described by Tsuji and Yagi (71). In this regard Postgate (96) noted that hydrogenase activity with certain acceptors was stimulated 3–5-fold by cell lysis.

In order to investigate the physiological role of the internal hydrogenase in *Desulfovibrio*, the amount of internal hydrogenase relative to intact cells was determined with *D. vulgaris* grown in a chemostat under conditions of increasing sulfate limitation, that is, under conditions that favor H_2 production (J. M. Odom and H. D. Peck, Jr., unpublished). Preliminary results indicate a 2–5-fold increase in internal hydrogenase at the lowest dilution rates, where sulfate is limiting. The involvement of the internal hydrogenase with the production of H_2 from organic substrates such as lactate and ethanol is also suggested by the observation that *D. vulgaris* grown on H_2 and sulfate does not contain any cytoplasmic hydrogenase (57).

The sulfate-reducing bacteria, members of the genus *Desulfovibrio*, were the first nonphotosynthetic anaerobic microorganisms in which an electron transfer–coupled phosphorylation was demonstrated. This phosphorylation was first shown with whole cells (97), and subsequently with cellfree extracts utilizing H_2 as the electron donor and sulfite (98) or fumarate (99) as the electron acceptor. Because of the requirement of ATP for sulfate reduction, growth with sulfate as electron acceptor with organic substrates requires electron transfer–coupled phosphorylation to obtain a net yield of ATP for growth (100). From the results of enzyme localization studies with *D. vulgaris* (Marburg) grown on H_2 and sulfate, a scheme for energy coupling in *Desulfovibrio* has been proposed (57) which involves vectorial electron transfer across the cell membrane with a proton gradient being produced by the direct scalar production of protons from the oxidation of H_2 on the external surface of the membrane and the consumption of protons in the cytoplasm by the reduction of sulfate. The protons produced would of course be utilized for the production of ATP by means of ATPase (101). This mechanism for the generation of a proton gradient illustrated below, is quite different from the chemisomotic model for respiratory sulfate reduction proposed by Wood (102) in that it does not involve a typical Mitchell loop.

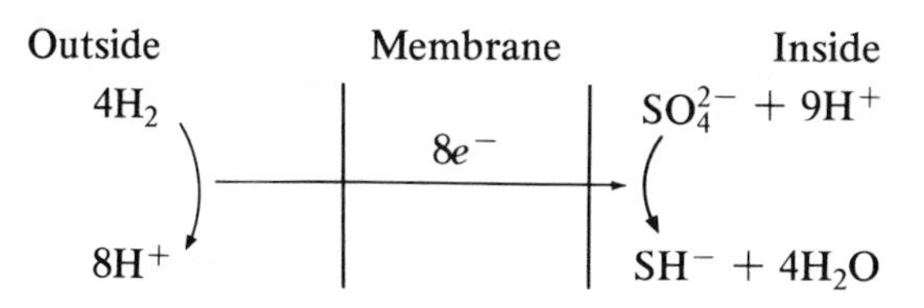

A similar scheme has also been proposed to explain the production of ATP by *D. gigas* inverted vesicles when they reduce colloidal S with molecular H_2 (103).

Rapid proton production coupled to the oxidation of H_2 and reduction of sulfite has been observed (C. L. Liu and H. D. Peck, Jr., unpublished), and the gradient is collapsed by uncoupling agents. An $H^+/2e^-$ ratio of 1.0–2.1 is observed, which suggests that protons are only produced from the oxidation of H_2. From these results it is proposed (34) that this vectorial electron transfer is the major mechanism for energy coupling in the sulfate-reducing bacteria, not only for those growing on H_2 but also for those growing on organic substrates plus sulfate. The scheme (Fig. 2) involves the internal or cytoplasmic oxidation of lactate and pyruvate to acetate, CO_2, and H_2 which, as a permeant molecule, can diffuse across the cytoplasmic membrane and be oxidized by the external hydrogenase. The electrons are subsequently transferred across the membrane to the cytoplasm, where they are used to reduce sulfate to sulfide, thus forming a proton gradient. The mechanism is very similar to interspecies H_2 transfer, and explains the unique property of *Desulfovibrio* being both an efficient H_2-producing and H_2-utilizing organism when grown in mixed culture. Consistent with this proposal, it was found that the protein composition of membranes is quite simple, containing

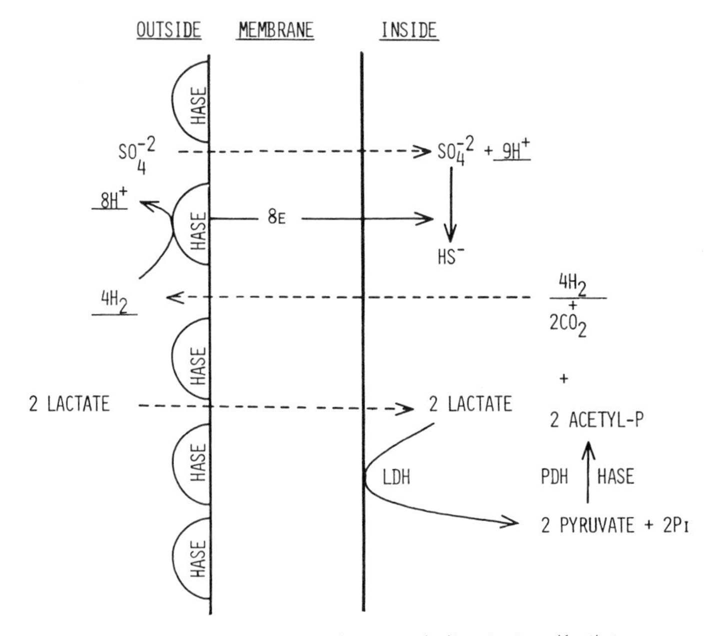

Figure 2. Bioenergetics of H_2 metabolism in *Desulfovibrio*.

only four to six major bands depending on growth conditions. In addition, to obtain lactate oxidation coupled with sulfate reduction by spheroplasts from which hydrogenase has been removed, there is an absolute requirement for externally added hydrogenase.

A major problem with the scheme involves the production of H_2 by internal hydrogenase ($E_0' = -414$ mV) with electrons derived from the oxidation of lactate to pyruvate ($E_0' = -190$ mV). However, H_2 is produced from lactate, as evidenced by interspecies H_2 transfer, and it may be that the external hydrogenase maintains the intracellular concentration of H_2 low enough to "pull" the oxidation of lactate in the direction of H_2 formation. The scheme is also supported by enzyme localization studies, and offers an explanation to the confusing observations regarding H_2 metabolism in these and related microorganisms.

Indeed, the physiological role of cytochrome c_3 (M_r 13,000) in respiratory substrate reduction remains an enigma in spite of that fact that it was the first electron-transfer protein to be isolated from the sulfate-reducing bacteria. Cytochrome c_3 is generally assumed to be unique to respiratory sulfate reduction in *Desulfovibrio,* and this conclusion is supported by the fact that it is produced in high concentrations under all growth conditions. With regard to its role as a cofactor for hydrogenase, it seems illogical that such an exceptional and sophisticated tetraheme structure evolved simply to couple hydrogenase with ferredoxin, a reaction that occurs directly with many hydrogenases. Alternatively, cytochrome c_3 may be essential for the thermodynamically unfavorable coupling of hydrogenase ($E_0' = -445$ mV) with the oxidation of lactate to pyruvate ($E_0' = -190$ mV). This mechanism (see below) implies that the 4 hemes of cytochrome c_3 (which have different redox potentials) accept 2 high-potential electrons from lactate and 2 low-potential electrons from pyruvate. Then, by a conproportionation-type mechanism, the excess of energy of the electron derived from the oxidation of pyruvate is utilized to lower the potential of the 2 electrons coming from lactate, in order to facilitate the formation. This would necessitate different electron-transfer sequences from lactate and pyruvate to cytochrome c_3:

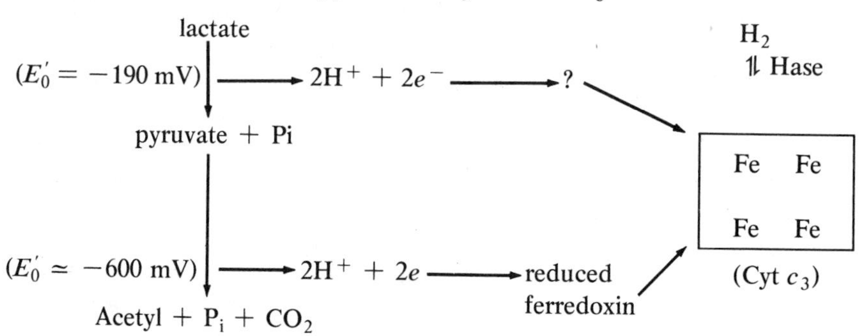

2.4 Hydrogenases of Methanogenic Bacteria

The widespread distribution of hydrogenase in both the Archaebacteria and Eubacteria, plus its involvement in energy metabolism and scalar proton translocation suggests that H_2 metabolism involving hydrogenase is important in the bioenergetics of their common ancestor. Thus hydrogenase may be considered to be a very early or "primitive" enzyme, and vectorial electron transfer associated with hydrogen oxidation as a primitive energy-coupling system.

2.4.1 Localization of Hydrogenases and Variations in the H_2 Metabolism of Methanogens. Although the presence of hydrogenase activity in extracts of methanogens has been known since the first studies on in vitro methane formation (104), it has only been recently that serious attempts have been made to purify and characterize the hydrogenase. The hydrogenase activity of the methanogens (105) is clearly internal, as activity with viologens can be increased 80-fold by cell rupture (105); however, the number of internal localizations of hydrogenase has not been resolved. Centrifugation of crude extracts commonly yields both soluble and particulate forms of hydrogenase, and multiple bands of hydrogenase activity are observed after polyacrylamide electrophoresis of the soluble protein. For example, Fuchs and co-workers (106) studied hydrogenase activity in crude extracts of *Methanobacterium thermoautotrophicum* and reported multiple bands of hydrogenase activity with molecular weights of 60,000, 120,000, and 500,000 daltons. Others have reported even higher molecular weight forms of hydrogenase and differences among the hydrogenases of various methanogens (105, 107). As H_2 and hydrogenase are fundamental in energy coupling by the methanogenic bacteria, it is of central importance that these problems be resolved. There is a good probability that some of these observations regarding hydrogenase are artifactual in that hydrogenase is known to "stick" to other proteins (108). Alternatively, hydrogenase may reside in high molecular weight complexes such as electron-transfer particles that are loosely bound to cytoplasmic membrane structures. Cell breakage would then lead to the formation of one or more soluble but high molecular weight hydrogenases. Doddema and colleagues (109) have employed cytochemical techniques to localize hydrogenase on the internal membrane system; however, their results, although highly provocative, were neither quantitative in terms of total hydrogenase nor specific in terms of the electron acceptors employed. In support of this general concept McKellar and Sprott (105) have shown that the solubilization of the particulate hydrogenase from *Methanobacterium* strain G2R by detergents yields a high molecular weight hydrogenase ($\approx 900,000$) that is identical to the soluble form of the enzyme. It is difficult to rationalize multiple hydrogenases in terms of the thermodynamics of the reduction of carbon dioxide to methane; however, hydrogenases of different electron-acceptor specificities might be required for one or more of the reductive steps, for the assimilation of carbon dioxide or for the formation of H_2 from formate in those species that grow on formate.

2.4.2 Purification and Physical Properties of Hydrogenases from Methanogenic Bacteria. As there have been no published reports of the purification and properties of a hydrogenase from the methane bacteria, the results of two unpublished studies are presented. A soluble hydrogenase has been purified from *Methanobacterium thermoautotrophicum* (L. E. Mortenson, H. D. Peck, Jr., and J. LeGall, unpublished), and its physical and biochemical properties studied. The hydrogenase is pure by gel electrophoresis, and exhibits a molecular weight of 185,000 $\pm$ 15,000 daltons. The protein is composed of four subunits of two types, with molecular weights of 64,000 and 40,000 daltons, and contains 30–40 nonheme Fe atoms; however, the types of clusters have not yet been determined. It has a specific activity measured in the presence of bovine serum albumin at 30°C of 200 μmol H_2/min·mg, but at the growth temperature of the bacterium, 60°C, the specific activity increases to 1000 μmol H_2/min·mg. The optimum activity is at pH $= 8$. As reported for the hydrogenase from *Methanobacterium* strain G2R (105), the hydrogenase is rapidly inactivated by O_2, but this inactivation can be completely reversed by dithionite or glucose plus glucose oxidase. In terms of electron acceptor specificity, the hydrogenase does not reduce cytochrome c_3 ($M_r = 13{,}000$) (Y. Berlier, P. Lespinat, and J. LeGall, unpublished), clostridial ferredoxin, or F_{420}. From its distinctive physical and biochemical properties, it appears to be the third unique hydrogenase to be found in anaerobic bacteria. However, as with all hydrogenases, it is a nonheme Fe enzyme and contains subunits of $\approx$ 60,000 molecular weight.

The *Methanosarcina barkeri* hydrogenase activity is present in the soluble and membrane fractions (I. Moura, J. J. G. Moura, M. H. Santos, A. V. Xavier, and J. LeGall, unpublished). The soluble hydrogenase was purified almost to homogeneity and found to be highly sensitive to the presence of O_2. *M. barkeri* hydrogenase can reduce ferredoxin and cytochrome *b* from the same organism, as well as utilize H_2 in the reduction of sulfite (see Section 6).

2.4.3 Relationships Between Hydrogenases and Electron Carriers. The methane bacteria contain a unique array of novel low molecular weight cofactors, and appear to lack many conventional cofactors such as FAD, FMN, and menaquinone. These new cofactors include coenzyme M, 2-mercaptoethanesulfone acid (110, 111), F_{420}, a deazaflavin (112, 113), a Ni-containing compound (115, 116), possibly a tetrapyrrole (117), F_{342} (114),

Bo, and YFC (118); their presence has been utilized as part of the rationale for separating the methanogens from procaryotes (101, 103). Four species of the H_2-utilizing *Methanobacterium* have been shown to have high levels of corrinoids, but the highest levels were found in the acetate-utilizing bacterium, *Methanosarcina barkeri*, growing on methanol (119). A B_{12} protein, required for the conversion of methanol to CH_4, has been purified from extracts of *M. barkeri* (120, J. M. Wood, J. J. G. Moura, I. Moura, M. H. Santos, A. V. Xavier, and J. LeGall, unpublished), and probably accounts for the high levels of corrinoids in this organism. The H_2-utilizing methanogens have not been shown to contain ferredoxin, flavodoxin, rubredoxin, or cytochromes (121); however, both the membranes and soluble protein exhibit EPR signals in the oxidized state characteristic of either HIPIP-type [4Fe-4S] or [3Fe-3S] centers and present, when reduced, at least two $g = 1.94$ type signals (122).

M. barkeri, which is only distantly related to the *Methanobacteria*, contains CoM (123), F_{420} (101), F_{430}, and F_{342} (J. J. G. Moura, I. Moura, M. H. Santos, J. LeGall, and A. V. Xavier, unpublished), and in addition, has a *b*-type cytochrome (124) and a ferredoxin that contains a single [3Fe-3S] cluster (J. J. G. Moura, I. Moura, B. H. Huynh, P. Hausinger, J. B. Howard, A. V. Xavier, and J. LeGall, unpublished) (see Table 4). This ferredoxin is reduced by hydrogenase. Partially purified hydrogenases from *Methanobacteria* reduce FMN and FAD but not cytochrome c_3 or clostridial-type ferredoxin.

F_{420} is the only one of the unique cofactors that has been shown to be redox active and is reduced in crude extracts by H_2 and formate (125, 126). The two formate dehydrogenases found in *Methanococcus vannielii* appear to be specific for F_{420} and coupled to $NADP^+$ by means of an F_{420} linked $NADP^+$ reductase (127). The specificity of the hydrogenase from the *Methanobacteria* for F_{420} is less clear. In general, crude preparations and the higher molecular weight forms of hydrogenase reduce F_{420}, but the lower molecular weight forms are inactive toward this cofactor (107, 128). Together, these observations suggest that a protein in addition to hydrogenase may be required for F_{420} reduction.

2.4.4 The Physiological Role of Hydrogenase in Methanogenic Bacteria. Hydrogenase is present in all known methanogenic bacteria, and in the H_2-utilizing methanogens it must be assumed to play an essential role in energy coupling. The proposal of Thauer and co-workers (129) for the generation of a proton gradient by vectorial electron transfer is extremely attractive and can be outlined as follows:

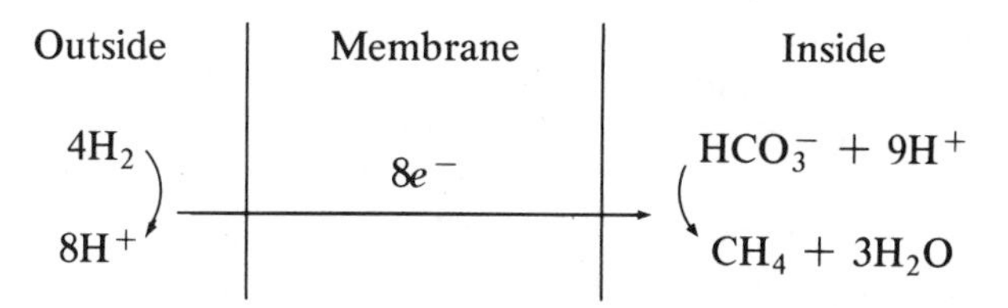

Eight protons are produced from the oxidation of 4 H_2 molecules on one side of a membrane impermeable to protons except through an ATPase, and the eight electrons are transferred across the membrane, where they are utilized for the reduction of HCO_3^- to CH_4 with the consumption of nine protons. The net effect is the formation of a proton gradient equivalent to four molecules of ATP assuming an H^+/ATP ratio of 2, and suggests that the efficiency of energy coupling approaches 100%. In view of the facts that methane formation does not require substrate amounts of ATP and protons are probably not required for the transport of CO_2 or CH_4 across the membrane, this projected yield of ATP does not appear to be consistent with the low growth yields of methanogens (130), and is barely consistent with the overall thermodynamics of methane formation ($\Delta G_0' = -32.4$ kcal). Alternatively, mechanistic restrictions may limit vectorial electron transfer to less than four of the reductive steps in methane formation. This explanation would be consistent with the observations of Gunsalus and Wolfe, indicating that a hydrogenase specific for F_{420} is an integral component of methyl Coenzyme M reductase from *Methanobacterium thermoautotrophicum* (107), and with the different reactivities of hydrogenase preparations toward F_{420}.

The evidence for vectorial electron transfer is based on the quantitative localization of enzymes and electron-transfer components relative to the cytoplasmic membrane, and on the demonstration of the formation of a proton gradient coupled to the specific redox system. The evidence in this regard from the methane bacteria is not yet compelling. As hydrogenase and formate dehydrogenase reductase are clearly internal enzymes (105, 131), vectorial electron transfer would appear to be eliminated as the mechanism for energy coupling in these bacteria; however, H_2-utilizing methanogens contain an internal membrane system that has been proposed as the site of energy coupling (132).

Doddema and co-workers (109) have presented some extremely interesting evidence concerning the putative role of hydrogenase and the inner membrane system in the bioenergetics of *M. thermoautotrophicum.* Through the use of cytochemical techniques, hydrogenase was clearly localized on the exterior aspect of the inner membranes, as indicated by the reduction of $Fe(CN)_6^{3-}$ and tetrazolium salts. However, some caution must be exercised in the interpretation of these results. $Fe(CN)_6^{3-}$ and tetrazolium salts are not specific electron acceptors for hydrogenase, and the possibility exists that these compounds were only interacting with electron-transfer components at the surface of the membrane. In this regard McKellar and Sprott (105) have reported that the hydrogenase from *Methanobacterium* strain G2R does not reduce $Fe(CN)_6^{3-}$. Membrane vesicles were prepared which appeared to have a mixed orientation and to contain only 20% of the total hydrogenase activity. H_2-dependent ATP synthesis was demonstrated, although the nature of the electron acceptor was not clear, and the generation of a proton gradient and membrane potential were shown by means of fluorescent dyes. Most interesting, the presence of ADP/ATP translocase has been shown with these vesicle preparations (133). Although more extensive and quantitative studies on this membrane system are required, it is now possible to propose a scheme for the role of hydrogenase in energy generation in the H_2-utilizing methanogens (Fig. 3).

ADP and ATP would exchange across the internal membrane as in mitochondria, and a proton gradient would be generated by vectorial electron transfer coupled to the oxidation of H_2. ATPase has been cytochemically localized on the internal membranes, but there are no reports on the specific localization of the methane-forming system; however, under certain condi-

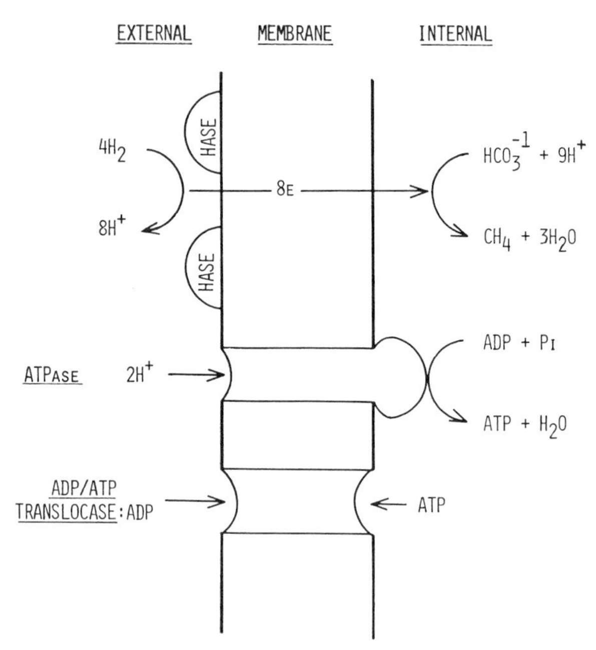

Figure 3. Bioenergetics of H_2 metabolism in methanogens.

tions of cell breakage the methane-forming system appears to be localized in a membrane fraction (134). Although a scheme such as this accounts for many of the observations on the bioenergetics of methane, it should be regarded as highly tentative.

3 RUBREDOXIN-TYPE PROTEINS

Rubredoxins are the simplest Fe-S proteins. The active center is characterized by the absence of labile S and the presence of 1 Fe atom linked in a tetrahedral arrangement to the S atoms of four cysteinyl residues. Rubredoxins have been isolated from sulfate- and S-reducing organisms: *D. gigas* (135), *D. vulgaris* (136), *D. salexigens* (137), *D. desulfuricans* (Norway 4) (138), *D. desulfuricans* (El Algheila Z) and *D. desulfuricans* 27774 (I. Moura, J. J. G. Moura, A. V. Xavier, J. LeGall, and H. D. Peck, Jr., unpublished), *D. desulfuricans* (Berre S) (140), *Drm. acetoxidans* (141), and *D. africanus* (139). A comparison of the amino acid composition is presented in Table 9. The overall distribution is quite similar, but differences can be observed, in particular when charged amino acid residues are considered. Physicochemical and some relevant biological data are shown in Table 10. A variation of the basic rubredoxin type of core is present in desulforedoxin isolated from *D. gigas* (142). Desulforedoxin (Dx) is a relatively small nonheme Fe protein of molecular weight 7900 daltons, and contains two identical subunits, 2 Fe atoms and no labile S. The amino acid sequence is known (143), and each subunit has four cysteine residues. Studies on the sulphydryl content of apo- and holoprotein indicate that the four cysteines participate in the maintainance of the Fe atom (144).

3.1 Spectroscopic Data

A comparison of visible and UV absorption spectra of rubredoxin and desulforedoxin shows similarities between the two chromophores (145). However, the spectrum of desulforedoxin is not the superposition of two rubredoxin spectra, as is the case of the 2-Fe rubredoxin isolated from *Pseudomonas oleovorans* (146). The oxidized form of rubredoxin is high-spin ferric (e.g., *D. gigas* Rb has a magnetic moment of 5.73 BM in the oxidized state (145)). All the Rb isolated from sulfate and S-reducing bacteria (see Table 10 and Fig. 4) show an EPR spectrum that does not differ from that reported for *C. pastorianum* rubredoxin (147). They exhibit resonances at $g = 4.3$ and $g = 9.4$ (below 20°K). These resonances can be interpreted (148) as arising from the transitions within the ground and first excited doublet state (middle Kramers doublet) of the high-spin ferric center ($S = 5/2$) with a

Table 9 Amino Acid Composition of Rubredoxins Isolated from Sulfate- and S-Reducing Organisms

	D. gigas	D. salexigens	D. desulfuricans (Berre S)	Drm. acetoxidans	D. vulgaris	D. africanus	D. desulfuricans (Norway 4)
Lys	6	3	4	2	4	4	5
His	0	0	0	0	0	0	0
Arg	0	0	0	0	0	0	0
Trp	1	n.d.	1	n.d.	1	3	n.d.
Asp	8	8	7	8	7	9	13
Thr	2	2	2	2	3	1	4
Ser	2	1	2	2	2	2	0
Glu	4	7	8	4	3	5	5
Pro	5	5	6–7	5	6	6	5
Gly	5	6	6	6	6	5	7
Ala	4	3	6	4	4	2	5
Cys	4	4	4	4	4	4	4
Val	3	2	5	4	5	6	6
Met	1	1	1	2–3	1	1	1
Ile	2	0	2	2	0	1	0
Leu	1	2–3	0	1	1	0	1
Tyr	3	2	3	3	3	3	4
Phe	2	2	3	2	2	2	2
Total	52	48–49	60–61	51–52	52	54	62
Ref.	49	137	140	141	150	166	138

Table 10 Physicochemical Data on Rubredoxin-Type Proteins Isolated from Sulfate- and S-reducing Organisms

	E'_0 (mV)	EPR g Values	Molar Extinction Coefficients per Fe atom (at indicated wavelengths) $\times 10^{-3}$	Magnetic Moment μ_{eff}		$K_M(M)$[b]	Refs.
				Oxidized Form	Reduced Form		
Rubredoxins							
D. gigas	+6	4.3 (also 9.4 below 20°K) (oxidized form)	8.5(376), 7.0(493)	5.73	4.96	6.2×10^{-6}	145, 153, 178
D. salexigens	−31	*a*	7.8(376), 6.3(493)	n.d.		n.d.	137, 178
D. vulgaris	n.d.	*a*	n.d.	n.d.		5.3×10^{-5}	153
D. desulfuricans (Norway 4)	n.d.	n.d.	8.1(376), 6.9(490)	n.d.		n.d.	138
D. africanus	n.d.	n.d.	13.4(378), 11.5(490)	n.d.		n.d.	166
Drm. acetoxidans	−46	*a*	7.1(380), 8.8(490)	n.d.		n.d.	141, 178
Desulforedoxin							
D. gigas	−35	7.7(5.7), 4.1, 1.8 (oxidized form)	7.8(370), 4.6(507)	5.7	4.90	n.d.	145, 178

[a] Identical to *D. gigas* rubredoxin.
[b] K_M determined for the NADH-H^+ rubredoxin oxidoreductase from *D. gigas*.

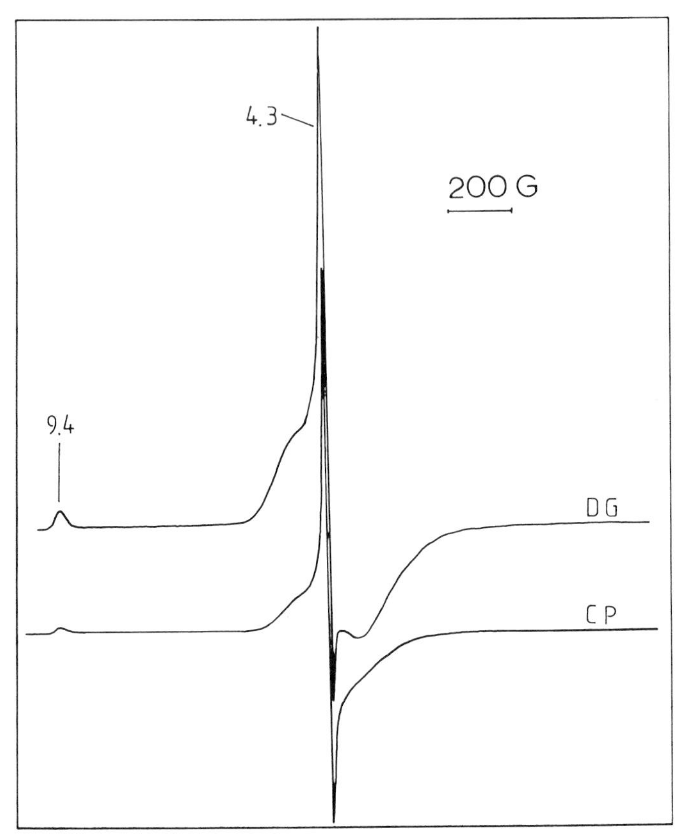

Figure 4. Comparison of the EPR spectra of rubredoxin from *D. gigas* and *C. pastorianum* (adapted from ref. 144). Experimental conditions: temperature 12°K; microwave power 20 mW; frequency 9.25 GHz; modulation amplitude 1 mT; gain 100.

higher degree of rhombic distortion ($E/D = 0.28$). As the temperature is lowered the $g = 4.3$ signal decreases in intensity as the middle Kramers doublet becomes depopulated and the $g = 9.4$ originating from the ground state increases. The Dx EPR spectrum indicates that the Fe is high-spin ferric (144, 145). Native Dx shows a nearly axial EPR signal ($E/D = 0.08$) with principal g features at 7.7 and 5.7, and broad components at 4.1 and 1.8 (Fig. 5). The observed g values calculated for a high-spin ferric ion, associated with three Kramers doublets with $E/D = 0.08$, are shown in the insert of Figure 5. The resonances at 4.1, 7.7, and 1.8 were attributed to the ground state doublet ($\pm\frac{1}{2}$) and the resonance at $g = 5.7$ to the middle doublet ($\pm\frac{3}{2}$). When the temperature is decreased (Fig. 5), the signal at 7.7 in-

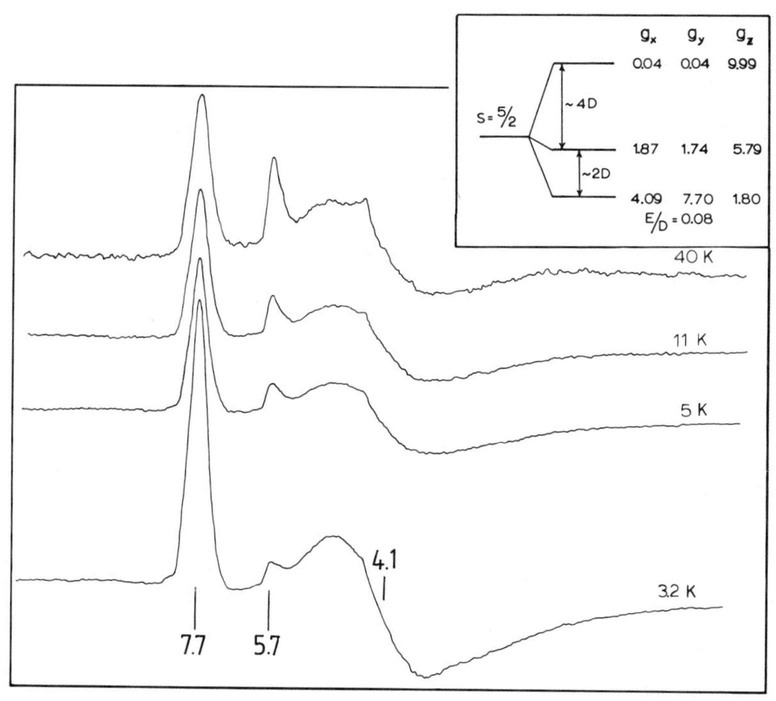

Figure 5. EPR spectra of *D. gigas* desulforedoxin at the indicated temperatures and the following power and gains: *(a, b)* 20 W, 6.3×10^3; *(c)* 20 W, 10^4; *(d)* 0.2 mW, 2.0×10^4. Other instrument settings: frequency 9.25 GHz, modulation amplitude 10 G (144).

creases while the signal at $g = 5.7$ decreases. The zero field splitting was estimated to be 2 cm^{-1} (144).

The comparison of the zero field splitting parameters of Dx and Rb in the oxidized form indicates geometrical differences between the structures of the active centers. The saturation fields calculated for Dx (−385 kG) and Rb (−410 kG) from Mössbauer spectroscopic measurements also indicate more covalent binding in the first protein. In the reduced form (EPR silent) the Mössbauer isomer shift is typical of ferrous ion ($S = 2$) with tetrahedral S coordination. Moura and co-workers (144) also showed that differences were observed in the orbital ground state of Dx and Rb by Mössbauer arguments.

3.1.1 EPR Quantitations. At low salt concentrations two surprising results are observed in the EPR and Mössbauer spectra of oxidized Dx (144): (1) the EPR signal quantifies only up to ≈ 0.5 spin/Fe atom, and (2) the Mössbauer spectra show, at low temperature (4.2°K), a quadrupole doublet with $\Delta E_Q = 0.75$ mm/sec that may account for up to 50% of the total intensity. Furthermore, in the magnetic spectrum, the isomeric shift ($\delta = 0.25$ mm/sec) is the same for the doublet and the magnetic component, suggesting that both spectra have their origin in the chromophore of *D. gigas* Dx.

This conclusion was confirmed by three observations (144): (1) when high magnetic fields are applied ($H > 10$ kG), the two components give the same Mössbauer spectrum, (2) samples with different proportions of the two components have the same ratio (intensity of the optical spectrum per Fe content); (3) the Mössbauer spectrum of the reduced form of Dx shows a single quadrupole doublet. Increasing the ionic strength decreases the amount of fast-relaxing material, with concomitant increase of the magnetic component and increased number of spins quantified by EPR measurements up to ≈ 1 spin/Fe atom. The aggregated material could not be detected by EPR due to the fast-relaxing conditions. Correct spin quantitation can only be obtained by changing the environment of the EPR-active spins so that they become detectable by EPR. In this case the increase in ionic strength changes the relaxation properties of the center. This problem may be the origin of low EPR quantitation observed in several other cases.

3.2 The Active Center

The differences in the electronic structure of the chromophores of Dx and Rb observed by Mössbauer and EPR measurements were interpreted as the result of different stereochemical constraints imposed by the polypeptide chain on the simple Fe core (144). The primary structure of *D. gigas* Rb (149), *D. vulgaris* Rb (141), and *C. pastorianum* Rb (151) are known, showing that 31 amino acid residues are unchanged when the three sequences are considered. Remarkably conservative positions are observed for cysteine, proline, and aromatic residues. Also, the polypeptide chains in the proximity of the Fe binding side are homologous. The binding peptides have sequences of the type

-Cys(6)-X-X-Cys(9)-Gly-

and

-Cys(39)-X-X-Cys(42)-Gly-

The degree of homology between the sequence of *D. gigas* Dx and Rb is small (143). Only in the N-terminal part, two cysteine residues can be recognized in an arrangement of the rubredoxin type

-Cys(9)-X-X-Cys(12)-Gly-

while the other two cysteines are adjacent to each other:

-Cys(28)-Cys(29)-Gly-

The 4 cysteines are bound to the Fe (144). The comparison of the sequence data gives an indication that the two adjacent cysteine residues may be responsible for spectroscopic differences from those of rubredoxin, by different constraints in the active center. It might also be important to observe that the dimer structure of Dx can be accommodated in two ways, as indicated in Figure 6. The X-ray data are awaited with great interest.

3.3 Physiological Activity

Odom and co-workers (153) have shown that the NADH/H^+ rubredoxin oxidoreductase from *D. gigas* is highly specific for the rubredoxin isolated from other organisms (*D. vulgaris* and *C. pastorianum*, see Table 10). Since the main dissimilarities seen in the primary sequences (*D. gigas, D. vulgaris,* and *C. pastorianum*) as well as the X-ray structures available [*C. pastorianum* (154) and *D. vulgaris* (155)] are the number and positions of charged residues, this may explain the high specificity observed.

Rubredoxin-type proteins are the Fe-S proteins for which more physicochemical measurements have been done. However, their physiological role is not understood. The value of the redox potential associated with this center is quite positive and difficult to place in the scheme so far worked out for the sulfate-reducing organisms. When the range of redox potentials offered by the redox carriers known so far is compiled (see Table 11), rubredoxin-type proteins represent an extreme value. The span, controlled by the H_2O/H_2 couple and the S compounds H_2S/S^{n-} couple (156), has narrow limits and this may be the reason for the absence of Cu- and Mn- containing enzymes in these systems.

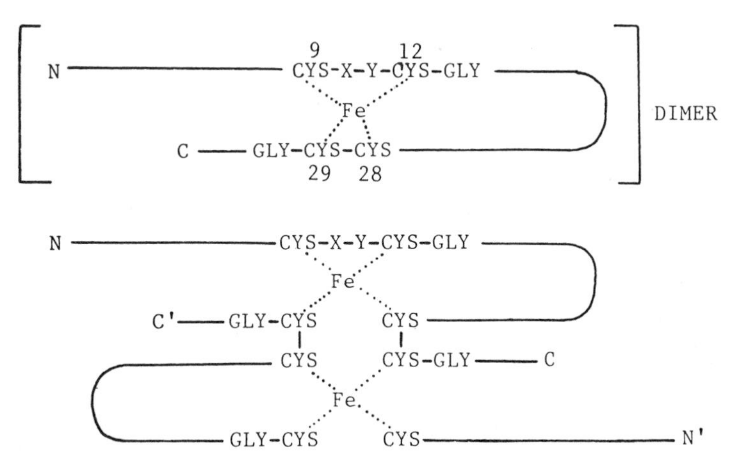

Figure 6. Two possible structures of *D. gigas* desulforedoxin. (Adapted from ref. 152.)

Table 11 Redox Potentials in S Metabolism

E (V)							
−0.5	−0.4	−0.3	−0.2	−0.1	0	0.1	0.2
	[Fe/S]		[Fe/S]				
	Mo			Cyt *b*			
			Cyt *c*				
		Cyt c_3					
	Flavodoxin		Flavodoxin				
					Rb		
	H_2O/H_2				H_2S/S^{n-}		

4 FERREDOXINS

Ferredoxins are present in all the *Desulfovibrio* spp. examined as well as in *Desulfuromonas acetoxidans* and *Methanosarcina barkeri.* No ferredoxin was found in *Methanobacterium thermoautotrophicum* (L. E. Mortenson, H. D. Peck, Jr., and J. LeGall, unpublished). A comparison of the amino acid compositions is shown in Table 12.

The best characterized of the ferredoxins indicated are those of *D. gigas* and *M. barkeri.* In this section we use the spectroscopic data of *M. barkeri* ferredoxin to illustrate how the presence of the novel [3Fe-3S] center can be proved. *D. gigas* ferredoxins are discussed in terms of the structural and physiological relevance of the [4Fe-4S] and [3Fe-3S] centers.

4.1 *M. Barkeri* Ferredoxin

The analysis of *M. barkeri* Fd properties by EPR and Mössbauer spectroscopy clearly enables the identification of the type of cluster present. The method of analysis is quite similar to the one followed for *A. vinelandii* Fd (82) and *D. gigas* Fd II (81).

The analysis of the active core is a good example where the use of unenriched ^{57}Fe protein, containing $\sim 2\ \mu$mol Fe/0.3 ml sample, is sufficient to obtain the necessary Mössbauer and EPR data that can provide an unambiguous identification of the Fe-S core.

Iron quantitation can be misleading in the identification of the core (29, 157, 158).

In the oxidized state *M. barkeri* Fd exhibits a fairly isotropic signal centered around $g = 2.01$ (Fig. 7). A very small $g = 1.94$ type signal is observed on reduction. The EPR redox transition is associated with a

Table 12 Amino Acid Composition of Ferredoxins Isolated from Sulfate and S-Reducing Bacteria

	D. gigas	*D. desulfuricans* (Norway 4) Fd I	*D. desulfuricans* (Norway 4) FD II	*D. desulfuricans* (Berre S)	*D. africanus* FD I	*D. africanus* FD II	*D. africanus* FD III	*Drm. acetoxidans*
Lys	1	2	2	2	3	1	3	1
His	0	0	0	1	1	1	0	0
Arg	1	0	0	0	1	1	0	0
Trp	0	n.d.	n.d.	0	n.d.	n.d.	n.d.	n.d.
Asp	11	5	6	10	5	6	8	7
Thr	0	3	0	3	1	2	4	4
Ser	3	3	3	2	2	4	2	4
Glu	9	11	11	11	13–14	8	14	6
Pro	4	3	2	3	3	3	0	2
Gly	1	2	5	6	2	3	4	5
Ala	6	7	6	2	6	5	3	6
Cys(1/2)	6	6	9	6	5–6	4	6	8
Val	5	3	1	5	7	5	9	3
Met	2	2	4	0	2	1	1	0
Ile	5	5	2	4	3	4	3	3
Leu	1	1	1	2	0	0	3	0
Tyr	0	0	0	0	2	1	2	2
Phe	1	1		0	2	2	0	0
Total	56	54	52	57	58	51	62	51
Ref.	160	138	168	236	166	166	167	141

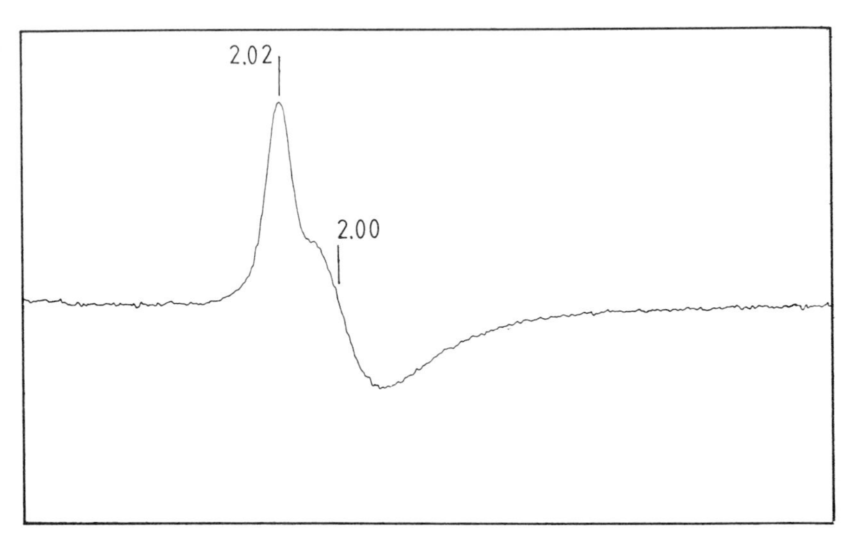

Figure 7. EPR spectra of *M. barkeri* ferredoxin in the native (oxidized) state. Instrument settings: microwave power 10 mW; frequency 9.215 GHz; modulation amplitude 46; gain 1.25×10^4; temperature 8°K. [I. Moura, J. J. G. Moura, B. H. Huynh, R. P. Hausinger, J. B. Howard, A. V. Xavier, and J. LeGall, unpublished.]

negative redox potential in *M. barkeri* Fd, *A. vinelandii* Fd, and *D. gigas* Fd II (see Table 13). Temperature and power studies indicate that the material is homogeneous and only one EPR-active species is present.

At 4.2°K in zero applied field the Mössbauer spectrum of the one-electron dithionite-reduced *M. barkeri* Fd shows two quadrupole doublets (I. Moura, J. J. G. Moura, B. H. Huynh, R. Hausinger, J. B. Howard, M. H. Santos, A. V. Xavier, and J. LeGall, unpublished). The spectrum closely resembles that of reduced *D. gigas* Fd II (Fig. 8; compare with Fig. , Chapter 4). The observed Mössbauer parameters for doublet I and doublet II (see Table 13) are indistinguishable from those of *D. gigas* Fd II. When the Mössbauer spectrum of the reduced sample is recorded at 4.2°K with a field of 600 G applied parallel to the γ beam, both doublets are substantially broadened due to induced magnetic hyperfine interactions (159). Similar behavior was observed for *D. gigas* Fd II and *A. vinelandii* Fd (81, 82, 150), proving that this ferredoxin contains [3Fe-3S] centers.

4.2 *D. Gigas* Ferredoxin

D. gigas Fd is isolated in different oligomeric forms of the same basic unit of $M_r = 6000$. Ferredoxin II (Fd II) and Ferredoxin I (Fd I) are two of the oligomeric forms in which the ferredoxin of this organism is isolated (160).

Table 13 **Physicochemical Properties of [3Fe-3S] Center Ferredoxins**

	D. gigas Fd II	*M. barkeri*	*A. vinelandii*
Cluster type	$[3Fe\text{-}3S]^{+3(+3,+2)}$	$[3Fe\text{-}3S]^{+3(+3,+2)}$	$[3Fe\text{-}3S]^{+3(+3,+2)}$ $[4Fe\text{-}4S]^{+2(+2,+1)}$
Number of cysteines	6	≃5	9
Redox potential (mV)	−130	negative	−420 +350
Optical properties (oxidized form)			
Absorption spectrum	23.1(305), 15.7(405), 13.3	16(400)	[a]
$E_m M(\lambda_{max}$, nm) per cluster	(453 shoulder) negative bands at	n.d.	
CD (λ, nm)	317, 423 positive bands at 474, 580		
EPR			
g values	2.02(2.02, 2.00, 1.97 from simulation)	2.02	2.01
Mössbauer at 4.2°K			
Isomer shifts (mm/sec)	0.27 ± 0.03 (oxidized)	n.d.	0.27 ± 0.04[b]
	0.46 ± 0.02 (reduced doublet I)	0.46 ± 0.04	~0.47
	0.30 ± 0.02 (reduced doublet II)	0.30 ± 0.06	~0.29
Quadrupole splittings (mm/sec)	0.54 ± 0.03 (oxidized)	n.d.	0.63 ± 0.05
	1.47 ± 0.03 (reduced doublet I)	1.40 ± 0.04	~1.45
	0.47 ± 0.02 (reduced doublet II)	0.40 ± 0.06	~0.40
Ref.	81	[c]	82

[a]The optical properties reflect a superposition of the two types of centers.
[b]The data indicated refer only to the [3Fe-3S] core present.
[c]I. Moura, J. J. G. Moura, B. H. Huynh, R. P. Hausinger, J. B. Howard, A. V. Xavier, and J. LeGall, unpublished results.

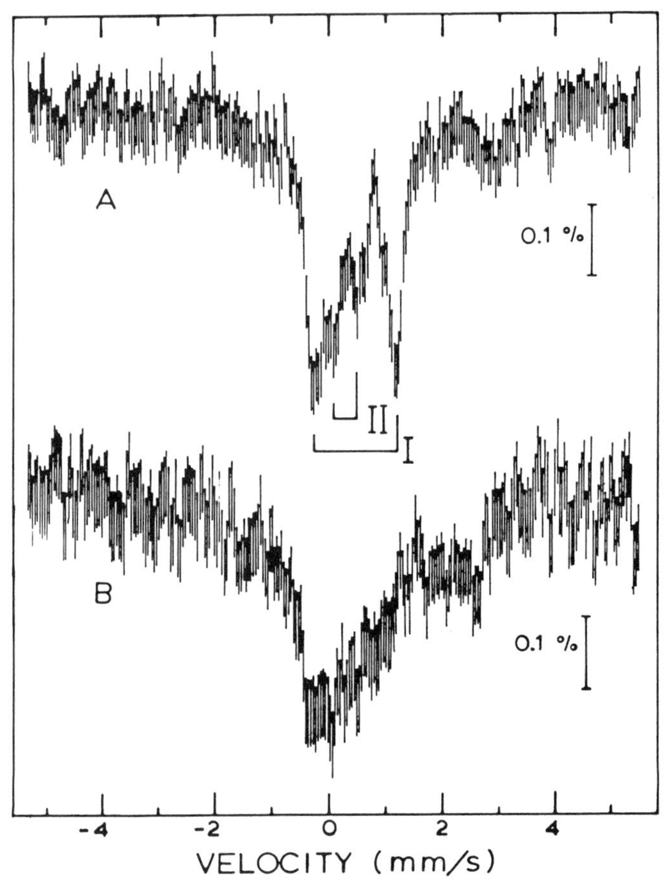

Figure 8. Mössbauer spectra of *M. barkeri* ferredoxin at 4.2°K. *(A)* Zero-field spectrum. The experimental data on doublets I and II are summarized in Table 13. *(B)* 600 G field applied parallel to the γ radiation. [I. Moura, J. J. G. Moura, B. H. Huynh, R. P. Hausinger, J. B. Howard, A. V. Xavier, and J. LeGall, unpublished.]

Although the primary sequence of the basic unit of both oligomers has been verified to be identical (161), the two proteins differ in optical properties, physiological activities, and oxidation-reduction potentials (160, 162, 163).

The active center was fully characterized in *D. gigas* Fd II by EPR and Mössbauer spectroscopy, and the presence of a [3Fe-3S] center was firmly established (162, 81) (for details see Chapter 4). In the oxidized form Fd I also shows an "isotropic" type signal centered around $g = 2.0$. However, the amount of this signal varies from only 10 to 30% in the sample preparations examined so far (81, 162, 29). On reduction an intense EPR signal is obtained with g values at 2.07, 1.94, and 1.92 (29, 162). The intensity of this reduced signal accounts for most of the missing intensities observed in the native preparation. At 4.2°K this form has a magnetic Mössbauer spectrum almost superposable with the spectrum obtained for reduced [4Fe-4S] clusters. On oxidation the $g = 1.94$ signal disappears and the Mössbauer spectrum modifies to a quadrupole pattern (29, 81). High-field studies proved that this pattern originates from diamagnetic sites, strongly implying the presence of [4Fe-4S] cores in Fd I, in addition to the [3Fe-3S] cores.

Recent MCD studies of Fd II have also indicated another method of identification and characterization of this cluster (164). These results are discussed in Chapter 10.

4.2.1 Ligands of the Active Center. Stout and colleagues (165) solved the structure of the 7 Fe ferredoxin from *A. vinelandii* and showed the way the [3Fe-3S] and [4Fe-4S] HIPIP-type center were ligated to the polypeptide chain. Cys 8, 11, 16, 20, and 49 bind the [3Fe-3S] core, implying the participation of the two moieties of the polypeptide chain in the binding. Cys 24, 39, 42, and 45 bind the 4-Fe center. A sixth ligand is required to bind the 3-Fe center, and this might be provided by glutamic 18 or by an exogenous molecule (water or hydroxyl) (see Chapter 3). Thus Fe-S centers can be bound by ligands other than S.

The sequence Cys(16)-Val-Glu(18)-Val-Cys(20)-Pro- provides at least two of the ligands to [3Fe-3S] cluster. Similar sequences are present in *D. gigas* Fd II (161), *Ps. ovalis, Chromatium* protein, *T. thermophilus* Fd, *M. segmatis* Fd (165), *D. africanus* Fd II and III (166, 167), *D. desulfuricans* (Norway 4) Fd I and II (168), and *M. barkeri* Fd (Section 4.1). A comparison of the partial sequences of *A. vinelandii* Fd, *D. gigas* Fd II, and *D. africanus* Fd I and Fd II is shown in Figure 9. Like *D. gigas* Fd I, an intense isotropic signal is present in the EPR spectra of the oxidized ferredoxins from *D. desulfuricans* (Norway 4), together with a $g = 1.94$ signal observed for the reduced form (I. Moura, J. J. G. Moura, A. V. Xavier, R. Cammack, and J. LeGall, unpublished).

Amino acid sequence studies of *M. barkeri* [3Fe-3S] Fd indicates that position 18 is occupied by an aspartic residue and that a glutamic residue is present in the next position. Although glutamic might not be the sixth ligand of the [3Fe-3S], the presence of an acidic residue between Cys(16) and Cys(20) could play an important role in the stabilization of this type of core.

4.2.2 The Accommodation of [3Fe-3S] and [4Fe-4S] by the Same Polypeptide Chain. After reconstitution of the *D. gigas* Fd II apoprotein with Fe and sulfide under reducing conditions, an EPR-silent sample is obtained. After reduction with dithionite an EPR signal is observed ($g_z = 2.07$, $g_y =$

			10					15					20	
AV	CYS	ILE	LYS	CYS	LYS	TYR	THR	ASP	CYS	VAL	GLU	VAL	CYS	PRO
DG	CYS	MET	ALA	CYS	-	-	GLU	ALA	CYS	VAL	GLU	ILE	CYS	PRO
DA I	CYS	ILE	ALA	CYS	-	-	GLU	SER	CYS	VAL	GLU	ILE	CYS	PRO
DA II	CYS	ILE	( )	CYS	-	-	ALA	ALA	CYS	VAL	GLU	ILE	CYS	PRO
DA III	( )	ASP	( )	( )	-	THR	GLY	ASP	( )	( )	GLX	VAL	( )	VAL
N I	( )	ILE	GLY	( )	-	-	GLU	ALA	( )	VAL	GLU	( )	( )	PRO
N II	( )	ASP	LYS	( )	-	ILE	GLY	SER	( )	GLU	ALA	VAL	( )	( )
M S	CYS	THR	GLY	CYS	-	-	GLY	( )	CYS	VAL	ASP	GLU	CYS	PRO

Figure 9. Partial amino acid sequences of selected ferredoxins: (AV) *A. vinelandii* Fd; (DG) *D. gigas* Fd II; (DA I, II, and III) *D. africanus* Fd I, Fd II, and Fd III; (N I and II) *D. desulfuricans* (Norway 4) Fd I and II); (MS) *M. barkeri* Fd.

1.94, and $g_x = 1.92$) similar to that of [4Fe-4S] centers at the +1 oxidation level. Mössbauer spectroscopy studies further confirm the EPR analysis. The ^{57}Fe-reconstituted Fd II sample contains only [4Fe-4S] centers with spectral features similar to the well-established [4Fe-4S] ferredoxin isolated from *Bacillus stearothermophilus* (29, and J. J. G. Moura, I. Moura, T. A. Kent, B. H. Huynh, E. Münck, A. V. Xavier, and J. LeGall, in preparation). Figure 10 compares the Mössbauer spectra obtained for the reconstituted ^{57}Fe *D. gigas* Fd II in the reduced state with the simulated spectra obtained using the data reported for *B. stearothermophilus* (169).

The Mössbauer and EPR studies described in the native preparations of the oligomeric forms of *D. gigas* Fd I and Fd II, as well as the results obtained with the reconstituted Fd II, enable us to conclude that the same polypeptide chain can accommodate [3Fe-3S] and [4Fe-4S] cores. In this respect native Fd II represents a homogeneous preparation containing only [3Fe-3S] centers. However, after reconstitution only [4Fe-4S] centers could be observed. Also, Fd I as isolated contains both [3Fe-3S] and [4Fe-4S] cores. By chromatographic criteria Fd II and Fd I are homogeneous preparations and can be readily separated (160).

The observation that the same polypeptide chain can provide ligands to both types of center raises the question of how the polypeptide chain must fold to provide the structural requirements of the binding site.

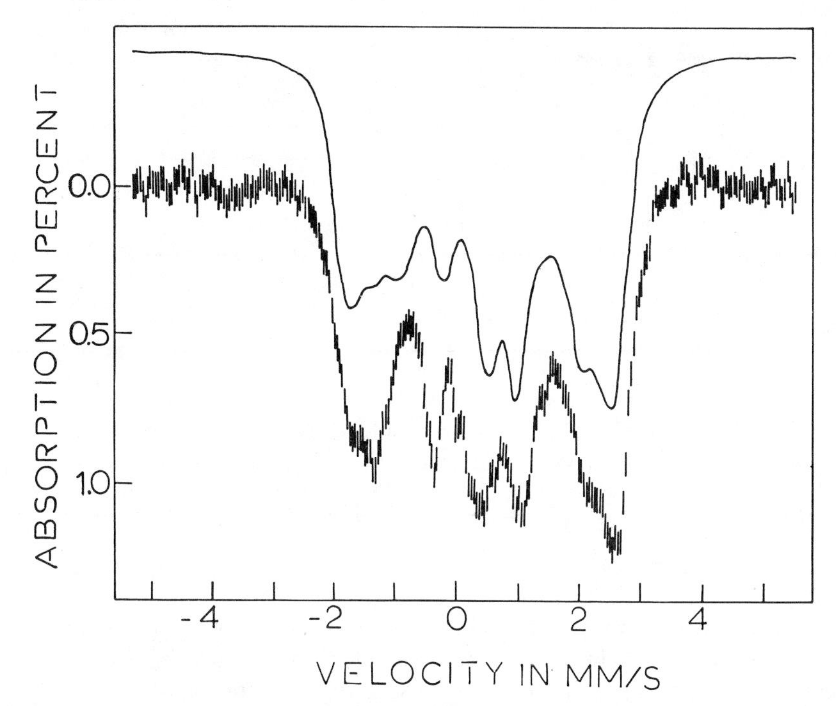

Figure 10. Mössbauer spectrum of reconstituted ^{57}Fe *D. gigas* Fd II taken at 4.2°K in a field of 600 G applied parallel to the γ radiation of observation. The upper trace is a simulated Mössbauer spectrum obtained using *B. stearothermophilus* [4Fe-4S] ferredoxin parameters (169). [I. Moura, J. J. G. Moura, B. H. Huynh, T. A. Kent, A. V. Xavier, E. Münck, and J. LeGall, unpublished.]

B. stearothermophilus Fd depicts the minimal ligand requirement (4 cysteinyl residues) for the binding of a [4Fe-4S] center (170). Four cysteinyl residues are observed in homologous positions in the *P. aerogenes* 2 × [4Fe-4S] Fd (171) in addition to the other 4 cysteinyl residues that bind the second core that is present (Fig. 11). Figure 11 also shows the ligands involved in the binding of the [3Fe-3S] and [4Fe-4S] centers in *A. vinelandii* Fd (165, and Chapter 3). Homologous conservative positions are observed for the [4Fe-4S] center. The probable 6 ligands required for the maintainance of the [3Fe-3S] center are also shown. This analysis suggests the type of binding site required for the building up of [3Fe-3S] and [4Fe-4S] centers. It is remarkable that both types of ligand requirement can be fulfilled by the amino acid sequence of *D. gigas* Fd: the cysteinyl residues (8, 11, 14, 18, 51) plus the glutamic residue (16) can ligate the [3Fe-3S] core, or the cysteinyl residues [8, 11, 14, and 51 (or optionally 41)] can maintain the [4Fe-4S] core.

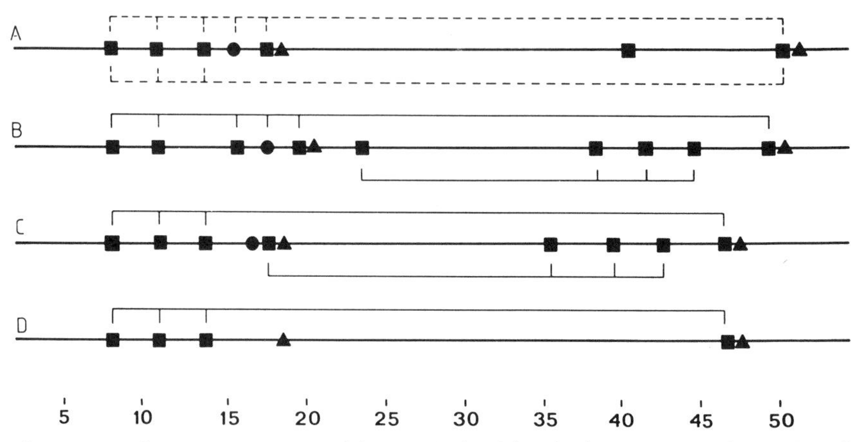

Figure 11. Schematic comparison of the cysteinyl residues in the sequences of [4Fe-4S] and [3Fe-3S] ferredoxins: (■) cysteine; (●) glutamic; (▲) proline residues. *(A) D. gigas* Fd II; *(B) A. vinelandii* Fd; *(C) P. aerogenes* Fd; *(D) B. stearothermophilus* Fd.

The evidence for the accommodation of two different types of center by the same polypeptide chain indicates that the knowledge of the primary structure cannot always be used to predict the actual type of core that is present.

It is interesting to note that [3Fe-3S] cores can be present in the following forms:

Single clusters	*D. gigas* Fd II, *M. barkeri* Fd
Associated with:	
[4Fe-4S]$^{(+2,+3)}$ cores	*A. vinelandii* Fd
or	
[4Fe-4S]$^{(+2,+1)}$ cores	*D. gigas* Fd I, *T. thermophilus* Fd

4.2.3 Physiological Relevance. Two important metabolic pathways in sulfate-reducing bacteria are the phosphoroclastic reaction (172) and the sulfite-reduction system (173). The main advantage of the *Desulfovibrio* spp. electron-transfer system is that the two pathways can be considered separately following either the evolution or consumption of H_2 (see also Fig. 1):

Hydrogen production:

$$\begin{array}{l} \text{Electron donor} \cdots \rightarrow \\ \text{(pyruvate)} \end{array} \begin{array}{c} \text{Ferredoxin} \\ \\ \text{Flavodoxin} \end{array} \rightarrow \text{Cyt } c_3 \rightarrow \text{Hase} \rightarrow \left(\begin{array}{l} H^+ \\ \\ H_2 \end{array} \right.$$

Hydrogen consumption:

$$\left.\begin{matrix} H_2 \\ H^+ \end{matrix}\right) \rightarrow \text{Hase} \rightarrow \text{Cyt } c_3 \rightarrow \begin{matrix} \text{Ferredoxin} \\ \\ \text{Flavodoxin} \end{matrix} \rightarrow \cdots \rightarrow \begin{matrix} \text{Electron} \\ \text{acceptor} \\ (SO_3^{2-}) \end{matrix}$$

Table 14 summarizes the state of knowledge about the participation of electron carriers identified in the metabolic pathways of this group of bacteria.

D. gigas Fd II and Fd I have been implicated in these electron-transfer chains (160, 164). Fd I is active in the phosphoroclastic reaction. Under the same conditions Fd II is not active. A long incubation time is necessary until the participation of Fd II can be detected. On the contrary, Fd II is more active than Fd I as an electron carrier in the sulfite reductase system. The redox potential of sulfite reduction is quite controversial, but a mean value of −110 mV has been suggested. Pyruvate can donate electrons at ≈ −600 mV. As previously discussed (29, 164), the redox potential of the [3Fe-3S] centers of Fd II better matches that of sulfite reduction, and the redox potentials of

Table 14 Electron-Carrier Proteins Identified as Active in Three Relevant Pathways in Sulfate Reducers[a]

Desulfovibrio spp.	Sulfite Reductase[b]	Thiosulfate Reductase	Pyruvate Dehydrogenase
D. gigas	ferredoxin(237) flavodoxin(238) c_3 (M_r 13,000)	ferredoxin(238) flavodoxin c_3 (M_r 13,000)[c] c_3 (M_r 26,000)	ferredoxin(237) flavodoxin(239) c_3 (M_r 13,000)
D. vulgaris (Hildenborough)	flavodoxin(240) c_3 (M_r 13,000)(247)	ferredoxin(242) c_3 (13,000)(241)	ferredoxin(172) flavodoxin(243)
D. vulgaris (Myazaki)	flavodoxin(244) c_3		flavodoxin(244) c_3
D. desulfuricans (Norway 4)	ferredoxin(168)	ferredoxin(138) c_3 (M_r 26,000)	ferredoxin(168)
D. salexigens	flavodoxin(137)	n.d.	n.d.
D. africanus	ferredoxin(166, 167)	n.d.	ferredoxin(166, 167)

[a]Superscript numerals within parentheses are references.
[b]Cytochrome c_3 (M_r 13,000) does not stimulate the sulfite reductase pathway, but its presence is necessary for the electronic transfer between hydrogenase and the other components of the electron-transfer chain.
[c]Cytochrome c_3 (M_r 26,000) is specific in this reaction ref. (232), but maximum stimulation is observed in the presence of cytochrome c_3 (M_r 13,000).

[4Fe-4S] centers have the necessary negative redox potential to mediate the electron transport from pyruvate dehydrogenase.

The observation of the $g = 2.02$ EPR signal of Fd II in the presence of *D. gigas* crude extract and pyruvate can also give information about the participation of this electron carrier in the phosphoroclastic reaction. In a time-course experiment performed in a similar way to the manometric experiments (J. J. G. Moura, A. V. Xavier, and J. LeGall, unpublished), the "isotropic" signal decreases and features at $g = 1.94$ develop. This result supports (see also Section 4.2.2) the suggestion of an interconversion between [3Fe-3S] and [4Fe-4S] cores and provides an explanation for the time lag observed before Fd II becomes active in the phosphoroclastic pathway. The interconversion process may be a general property of these centers, since in aconitase the building up of other structural forms of [Fe-S] centers has also been suggested when the enzyme is activated in the presence of Fe and dithiothreitol (Chapter 4).

The observation of different physiological activities for the two types of centers of *D. gigas* ferredoxin has been taken as indicative of their physiological relevance (160, 162, 163). It should, however, still be kept in mind that Fd II may be produced by oxidation of Fd I before or during the purification, by a process analogous to *C. pastorianum* ferredoxin (162). It should also be remembered that the $g = 2$ signal is often seen in purified 4-Fe and 8-Fe ferredoxin (157, 158). Furthermore, by treatment with ferricyanide the "$g = 2$ signal" of Fd I doubles its intensity, although that of Fd II shows little change (162).

Flavodoxin can replace ferredoxin in the indicated pathways (see Table 14). This protein has also two redox potentials available:

$$-440 \text{ mV } (\text{Flav}_{\text{red}}/\text{Flav}_{\text{semiquin}}) \qquad \text{and} \qquad -150 \text{ mV } (\text{Flav}_{\text{semiquin}}/\text{Flav}_{\text{ox}}) \tag{174}$$

The comparison of these redox potentials with the redox potentials available for the [4Fe-4S] centers in Fd I and the [3Fe-3S] centers in Fd II may explain the interreplacement between these electron carriers. This observation generates a very important question: are these electron carriers just mediators that must match the redox potentials of the donor/acceptor partners, or is there any specificity in the electron transport? To better answer this question it is important to probe the specific interactions between the electron-transfer proteins as well as the alteration of physicochemical properties of their redox centers on complex formation. Another approach to the understanding of this problem is discussed in Section 7.

The specific interaction between *D. gigas* cytochrome c_3 and Fd II was studied by NMR, showing that there is a specific binding (175). These studies also show that the rates of intramolecular electron transfer between the hemes are altered in the presence of ferredoxin. EPR measurements of the midpoint redox potential of the hemes of cytochrome c_3 show that there is also a modification of the redox potential of one of the hemes (176). Further NMR studies (I. Moura, J. J. G. Moura, M. H. Santos, A. V. Xavier, and J. LeGall, unpublished) indicate that Fd I, flavodoxin, and rubredoxin (177) also bind the cytochrome c_3 with different modifications of the NMR spectral parameters, which are also different from those induced by Fd II. The redox potential of rubredoxin is 30 mV more positive in the presence of cytochrome c_3 (178).

Recently the participation of flavodoxin isolated from five different *Desulfovibrio* strains was tested in the sulfite reduction pathways, in crude extracts of *D. gigas* devoid of the acidic electron carriers (A. R. Lino, J. J. G. Moura, A. V. Xavier, and J. LeGall, unpublished). The specific interaction of the same flavodoxins was probed by NMR spectroscopy toward the *D. vulgaris* cytochrome c_3 (M_r 13,000) (I. Moura, J. J. G. Moura, M. H. Santos, A. V. Xavier, and J. LeGall, unpublished). Regardless of the *Desulfovibrio* strain from which flavodoxin was isolated, similar results were obtained both for the stimulation of molecular H_2 consumption and the perturbation of the NMR spectral parameters induced by protein-protein interaction, indicating a lack of specificity toward the origin of the electron carrier.

5 Mo AND Fe-S CONTAINING PROTEINS

It has been shown that Mo is associated with Fe-S centers in proteins isolated from sulfate-reducing organisms (179). The *D. gigas* Mo protein contains 1 Mo atom and 12 Fe atoms per molecule, and has a molecular weight of 120,000 daltons (180, 181). No subunits or dimeric forms have been found, and no FAD was detected. Other Mo proteins associated with Fe-S centers have been isolated from *D. africanus* (182) and *D. salexigens* (J. LeGall, unpublished). The available physicochemical data are summarized in Table 15.

A detailed analysis of the Mo and Fe-S centers of the *D. gigas* Mo protein has been reported. The presence of 6 [2Fe-2S] centers was strongly suggested by optical (visible and CD) and EPR spectroscopy (191). The visible and CD spectra can be directly compared with known (2Fe-2S] proteins, since the only chromophores contributing to the visible spectrum are the Fe-S cores. The presence of [2Fe-2S] centers is also confirmed by monitoring the EPR spectrum of the reduced protein in 80% DMSO (180). Further information could be obtained by EPR measurements of the reduced enzyme. The temperature dependences shown in Figure 12 indicate two types of Fe-S centers. At 77°K a center named [Fe-S]I shows EPR features at $g_x = 1.93$, $g_y = 1.94$, and g_z

Table 15 Characteristics of the Mo Fe-S Proteins from *Desulfovibrio* Species

	D. gigas		*D. africanus*
Mo/molecule	1		5-6
Fe/molecule	12		20
MW $\times 10^{-3}$(subunit)	120		112(10 $\times$ 11.5)
Optical wavelength (nm) (molar extinction coefficients $\times 10^{-3}$)	462(58), 415(57), 320(101)		650(48.4), 410(64.0) 325, 280
EPR	([Fe-S]I) 1.899, 1.935, 2.022 ([Fe-S]II) 1.97, 2.07, 2.12, Mo 1.97		n.d.
Redox potentials (mV)	([Fe-S] IA)	−260	n.d.
	([Fe-S] IB)	−440	
	([Fe-S] II)	−285	
	(MoVI ⇌ MoV)	−450	
	(MoV ⇌ MoIV)	−530	
Refs.	180, 181		182

= 2.02. As the temperature is decreased, a signal arising from the [Fe-S]II center develops at g_x = 1.89 and g_z = 2.06 (181). The picture is even more complex, since EPR redox titrations (Fig. 12*B*) show that center [Fe-S]I differentiates into two components [Fe-S]I-A (−260 mV) and [Fe-S]I − B (−440 mV). Thus three types of Fe-S centers can be observed based on redox potential and relaxation properties. The redox properties of Fe-S centers and Mo are shown in Table 15.

The interaction between the Fe-S centers is evident in Figure 12*A*. When the temperature is lowered, the EPR spectrum shows that a splitting develops at the g_z peak of [Fe-S]I. This is consistent with a spin-spin interaction between Fe-S centers.

In this protein the Mo center does not participate in the center containing Fe and S. However, the interaction between Mo and Fe-S centers could also be probed (179, 181).

The protein isolated from *D. africanus* (182) differs from the *D. gigas* Mo protein in its optical properties and metal content (Table 15). No EPR data were published, and the Fe-S centers have not been characterized. However, based on the fact that the visible spectrum closely resembles the spectrum obtained by mixing ferrous salt with excess of sodium sulfide, it could be

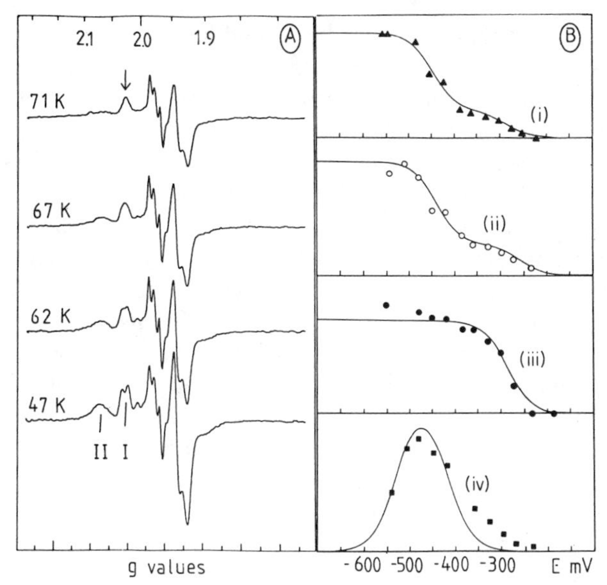

Figure 12. EPR spectra of *D. gigas* Mo [2Fe-2S] protein (adapted from ref. 181). *(A)* Temperature effect. Samples reduced with dithionite. Microwave power 1 mW; frequency 9.06 GHz; modulation amplitude 0.4 millitesla. *(B)* Redox titration curves. EPR signal intensities plotted vs. redox potential: (i) Fe-S center I measured by the height of $g = 1.94$ signal at 77°K; (ii) the same features at 24°K; (iii) Fe-S center II measured by the height of the $g = 2.06$ signal at 24°K; (iv) Mo(V) signal measured at 77°K.

speculated that the protein could function in the cell as an active Fe-S donor for the biosynthesis of Fe-S proteins.

The physiological role of the Mo proteins in the sulfate-reduction pathway is not yet understood. Activities of *D. gigas* Mo protein that were tested include electron transport between hydrogenase and bisulfite reductase, formate dehydrogenase, and xanthine oxidase (180). In each case the tests were negative.

6 Fe-S AND SIROHEME CONTAINING ENZYMES

Reduction of sulfite to sulfide is a six-electron reaction that is catalyzed by sulfite reductase. This is a central reaction in the S biocycle. There are two types of sulfite reductases: the *assimilatory,* which participate in the synthesis of S-containing biomolecules, and the *dissimilatory*, which are terminal reductases in the reduction of sulfate. Assimilatory-type reductases can carry out the complete reduction of sulfite to sulfide, but the presence of the products trithionate and thiosulfate has been suggested to occur during the function of dissimilatory enzymes (183, 184). The mechanism of dissimilatory reduction of sulfate is still quite controversial (16). Assimilatory sulfite reductases have been purified from different sources: algae, plants, and bacteria. Their physiological role and the structure of their active sites have been studied in detail (185). Chapter 4 deals in great detail with structural data on assimilatory sulfite reductases; an assimilatory enzyme was shown to be present in *D. vulgaris* (186, 187). Different dissimilatory sulfite reductases have been isolated from sulfate-reducing bacteria and have been used as taxonomic markers: desulfoviridins for *D. gigas* and *D. salexigens* (188, 189, J. J. G. Moura, I. Moura, A. V. Xavier, and J. LeGall, unpublished); desulforubidin for *D. desulfuricans* (Norway 4) and *D. desulfuricans* strain 9974 (190, J. J. G. Moura, I. Moura, A. V. Xavier, and J. LeGall, unpublished); and P_{582} for *D. desulfutomaculum* (191). Their optimal pH-activity profile is $\simeq 6.0$, suggesting that the enzymes reduce bisulfite (HSO_3^-), and they have been termed *bisulfite reductases.*

Table 16 summarizes the properties of this group of enzymes isolated from S-metabolizing bacteria and *M. barkeri.* The dissimilatory sulfite reductases have complex structures. Desulfoviridin has a molecular weight of 220,000 in a tetrameric $\alpha_2\beta_2$ structure. Desulforubidin and P_{582} were reported to have molecular weights of 225,000 and 145,000, respectively. The dissimilatory enzymes have optical spectra typical of the siroheme-containing enzymes, with bands at 590 nm, but desulfoviridin shows additional bands at lower wavelengths (680 nm).

Treatment of desulfoviridin, P-582, and the *E. coli* enzyme with acetone/HCl yields a sirohydrochlorin-containing extract (demetallized siroheme) with characteristic fluorescence. Desulforubidin was shown to have the same basic structure by demetallization of the extracted heme (192).

Siroheme is a universal choice as active center for the reduction of sulfite (and also nitrite). Assimilatory and dissimilatory sulfite reductases exhibit characteristic ferric-heme-type EPR signals around $g = 6.0$ (192–194). However, the structure of the active center of the dissimilatory sulfite reductases is not well defined in respect to the Fe quantitation. Approximately two sirohemes are bound per enzyme molecule which also contains Fe-S centers.

Under reducing conditions these enzymes show an EPR signal at $g = 1.94$, but in general no precise definition of the type of center has been made yet, although two [4Fe-4S] centers were indicated to be present in *D. gigas* desulfoviridin (194).

A sulfite reductase designated as P_{590} was isolated from *M. barkeri* (J. J. G. Moura, I. Moura, M. H. Santos, A. V. Xavier, and J. LeGall, unpublished). The spectral characterization closely resembles the assimilatory *D. vulgaris* enzyme (Table 16).

Table 16 Summary of Physical and Chemical Properties of Sulfite Reductase (SIR)[a]

	Desulfoviridin	*Desulforubidin*	P_{582}	Thiobacillus	*D. vulgaris* (Assimilative)	*M. barkeri*
MW ($\times 10^{-3}$)	200	225	194	190	27	60
Absorption bands (nm)	375, 390, 408, 580, 628	392, 545, 580, 720 (weak)	392, 582 582	390, 590 590	405, 545, 590	395, 545, 590, 700 (weak)
Fe content (total)	18	17.9	16	14	n.d.	n.d.
Nonheme Fe	16	15.9	14	12	+	+
Labile sulfide content	14	15	14	10	+	+
Siroheme	—	2	1.3	1.6	+	+
Siroporphin	2	—	—	—	—	—
Dithionite	—	+	+	+	+	+
Pyridine	—	+	+	+	+	+
CO	—	+	+	+	+	—
MW of subunits $\times 10^{-3}$	45, 50	45, 50	40, 45	40, 48	27	$\simeq 30$
Type of subunit	$\alpha_2\beta_2$	$\alpha_2\beta_2$	$\alpha_2\beta_2$	α	α_2	
Specific activity	820[b]	610	550	620		
	632[c]	410			900	90

[a]Adapted from references 186, 189–191, and from unpublished results of C. L. Liu and H. D. Peck, Jr., and I. Moura, J. J. G. Moura, M. H. Santos, A. V. Xavier, and J. LeGall.
[b]Expressed as μmol SO_3^{2-}/min·mg.
[c]μmol H_2/min·mg.

The enzymes show a pH-dependent activity profile with an optimal value at pH = 6.0–6.5. Bisulfite reductases have so far only been associated with dissimilatory-type metabolism. However, the respiratory utilization of S compounds does not correlate with our knowledge of *M. barkeri* metabolism.

Another possibility would be that *M. barkeri* sulfite reductase is utilized to gain energy from reduced S compounds in a very similar way as suggested for *T. denitrificans* (195).

7 EPR STUDIES OF MEMBRANE AND SOLUBLE FRACTIONS

The obtaining of physicochemical and structural data on purified electron carriers is a necessary task to understand their electron-transfer chains. The complex electron-transfer chains, in respiration, photosynthesis, or sulfate reactions, are electron conductors that perform in a preferential way between the electron donor and the final electron acceptor. Increasing complexity can be introduced if, after the available information on the purified components, the data are used in the interpretation of the properties of artificial binary complexes, ternary complexes, and so on . . . , that mimic the *in vivo* situation. In this regard it is particularly interesting to look at the crude soluble fractions or membrane fractions where the system is less damaged and closer to the *in vivo* condition. In this respect the modifications introduced by chemical reductants or physiological electron donors can give important information. EPR is a suitable technique for this kind of approach to the problem, as demonstrated in studies of mitocondrial preparations (196) and photosynthetic membrane systems (197). When applied to sulfate reduction and methane formation three types of information may be obtained:

1. The identification of electron carriers, whose properties have been studied in detail, when purified.
2. The electron-transfer kinetics between the redox components.
3. The role of certain components by examining extracts devoid of them (e.g., protein extracts devoid of acidic components).

Some preliminary data on the EPR study of crude protein extracts of the soluble fractions are shown in Figure 13 (J. J. G. Moura, A. V. Xavier, R. Cammack, and J. LeGall, unpublished). Relevant electron carriers can be deduced from the spectra. The region at $g = 2$ is complex because of the great variety of Fe-S centers present, as discussed. However, a comparison of the amount of isotropic signal present at $g = 2.01$ in comparison to the $g = 1.94$ signal can be obtained. Also, the reduction of cytochrome c_3, with features at $g = 3.0$ and sulfite reductase with g features around $g = 6.0$, can be followed. Kinetic data can be determined from appropriate experiments.

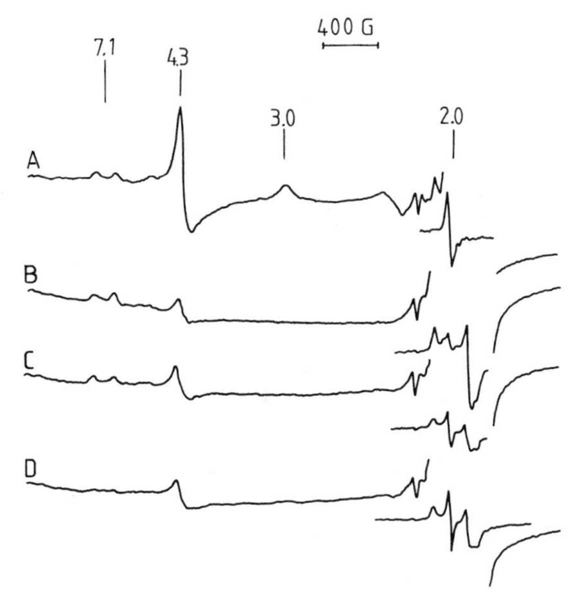

Figure 13. EPR studies on *D. gigas* crude extracts. Effect of electron donors and substrates. Spectra recorded at 20°K and modulation 10 G. The central features were recorded at modulation 1 G. *(A)* Native crude extract; *(B)* reduced with sodium dithionite; *(C)* under H_2; *(D)* in the presence of pyruvate.

By the use of low-temperature EPR spectroscopy dithionite-reducible Fe-S centers were observed in soluble and membrane-bound forms in *Methanobacterium bryantii* (198). The soluble fraction contains an isotropic-type signal typical of an oxidized Fe-S center (attributable to either HIPIP [4Fe-4S] or [3Fe-3S] centers) (Fig. 14). This could be a signal of a ferredoxin component analogous to the *M. barkeri* Fd. The membrane fraction contains a variety of signals. On reduction at least two types of $g = 1.94$ type signal were observed, in addition to radical signals. In the oxidized form a very unusual rhombic signal was observed at high temperatures, with g values significantly above 2.0. This signal was attributed to either a new type of cluster or to a transition metal such as Ni.

8 CHARACTERIZATION OF Fe-S PROTEINS BY NMR

The study of the three-dimensional structure, the determination of exchange rates, and the study of dynamic aspects of protein conformation, made more accessible by recent technological advances in the NMR spectro-

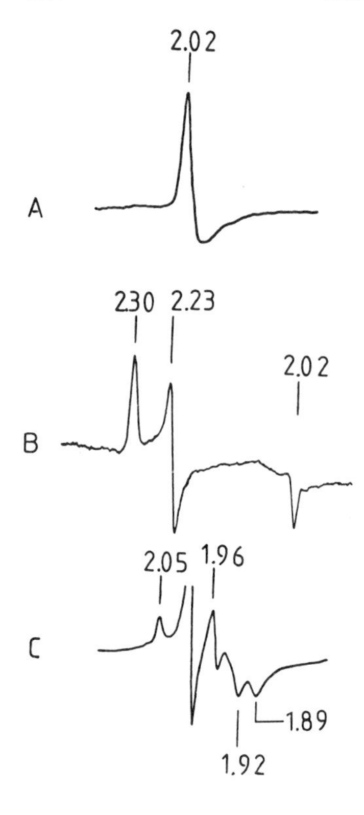

Figure 14. EPR spectra of *M. bryantii* soluble and membrane fractions (adapted from ref. 198): *(A)* soluble cell extract, oxidized state, 12°K; *(B)* oxidized membranes, 24°K; *(C)* dithionite-reduced membranes, 8°K.

scopy, yield important parameters for a better understanding of the mechanism and function of proteins. In this section the available methods are discussed and developments are reviewed.

Most of the data so far obtained by the NMR studies of Fe-S proteins are related to the presence of intrinsic paramagnetic probes. Until very recently this form of spectroscopy had been used almost exclusively to study the nature of the type of centers and their oxidation states. Very little information has been obtained about the biological properties of these centers and how they can be influenced by the rest of the protein.

Rather than undertaking an exhaustive review and citing the work in the order in which it appears in the literature (for that purpose see refs. 199–201), we indicate the scope and limitations of the use of NMR spectroscopy in the study of Fe-S proteins.

8.1 Methods

The NMR spectral parameters, chemical shifts. coupling constants, and relaxation rates are very much dependent on the environment of the nucleus, making high-resolution NMR a most useful technique for structural studies.

The full structural potential of the technique is only fulfilled with the use of paramagnetic probes, and in particular paramagnetic metal ions, since they may induce enormous perturbations on the nuclei chemical shifts (*shift probes*) and relaxation rates (*broadening probes*) which may be used to extract geometric parameters. The alteration of the spectral parameters is related to the geometric functions when the paramagnetic perturbations are the dipolar interactions between the nuclei and the unpaired electron (pseudo-contact shifts and dipolar relaxation) (202). These are the only types of interactions directly related to the geometric parameters. However, the delocalization produced by the electron spin at the resonating nucleus can induce perturbations (contact shift and scalar relaxation), which may be used to obtain information about the metal-ligand interaction (203). Furthermore, the contribution of contact shifts may make it impossible to extract the pseudo-contact contribution to the total shift, precluding the utilization of paramagnetic metal ions as molecular structure probes. Also, the increase in the relaxation rate of the nucleus under observation should obviously not be such that its resonance peak is too broad to be observed.

In some fortunate cases, for example, Fe-S proteins, the paramagnetic metal ions already exist in the native protein, and they can be used as intrinsic probes. In other cases it is sometimes possible by isomorphous replacement, or by finding in the protein a single binding site, to introduce an extrinsic probe. These requisites make some of the lanthanide ions particularly suitable as pseudo-contact shift probes, since their contact contribution is usually negligible and they have very short electron spin relaxation times (they hardly contribute to broaden the resonance peaks) (204).

Other useful intrinsic structural probes are the so-called *ring current probes* (205), for example, aromatic amino acid residues and the heme groups. The ring currents induce anisotropic shifts that again are proportional to geometrical parameters.

With most modern superconducting NMR spectrometers it is possible to obtain a good spectrum of 1 *mM* protein samples in less than 1 h. Good resolution can be obtained in some regions of the ^{1}H NMR spectra of small proteins (up to 20,000 daltons). Several NMR techniques have been developed to obtain specific answers to particular problems (206). Difference spectroscopy has been used to improve the spectral resolution [convolution (207)] and to simplify the spectra [paramagnetic difference spectroscopy (208)] considerably. Rate processes with characteristic times between 10^{-5} and 1 sec can be observed, and important conformational mobilities can be studied (206), which may be important in explaining the mechanism of protein function. Such studies have been applied to calculate the mobility of tyrosine and phenylalanine residues within the structure of cytochrome *c* (206).

With the use of different combinations of pulse methods (206), considerable spectral simplification as well as a large number of assignments can be obtained. The degree of simplification achieved was illustrated by Brown and Campbell (209) in their study of the aromatic region of the 270 MHz spectrum of hemoglobin (M_r 64,000). When a special pulse sequence (90°-τ-180°-τ-aquisition) is used, the only peaks left in the spectrum belong to histidine residues that have relatively long transverse relaxation times. The delay between the initial 90° pulse and the start of data aquisition (2τ) is a variable that allows resonances with a slow transverse relaxation rate (the sharper resonances) to be selected from the complex spectrum.

The use of pulse sequences of the type ($90_{x'}$-τ-$180_{y'}$-τ-acquisition) for resonance assignments is based on the different phase dependence of the various types of multiplets (210); for example, the value of 2τ can be chosen so that the doublets are 180° out of phase with respect to the phase of the singlets.

Simplification of the spectra is obtained by subtracting a normal spectrum from a double irradiated, nuclear Overhauser enhanced spectrum (211). The nuclear Overhauser effect (212) changes the intensity of the NMR resonance of a nuclear spin (another spectral parameter) when the NMR resonance of another nuclear spin is saturated. This enhancement can be used to determine the distance between the two nuclei, and can thus also be used as another geometric probe, as well as to assign resonances.

If a nucleus is in two different environments, its observed spectrum can be drastically altered by the exchange rate between these environments (213). It is often possible to obtain exchange-rate information from the analysis of these spectra. With the use of double irradiation it is possible to obtain information about the "transfer of magnetization" between the two environments in the form of cross-saturation (214). This results in the reduction of the cross-saturated resonance since, due to the second irradiation, the magnetization transferred is null. The cross-saturation technique can give values of exchange rates (215, 216), and is of great help in assigning resonances of a given nucleus in an interconverting state with different environments (214). The method is particularly easy to carry out when the resonance is in the crowded diamagnetic region in one state and shifted away from this region in the other. However, it is still possible to obtain cross-assignments, even when the peaks are buried in the diamagnetic region, by using difference spectroscopy.

8.2 Applications

At least one of the oxidation states of the known Fe-S cores is paramagnetic (Table 17). The presence of the paramagnetic center makes them good candidates for the application of NMR methods. However, the high magnetic moment as well as the long electron relaxation time make it difficult to ob-

Table 17 Formal Oxidation States of Fe-S Centers and Possible Redox Transitions[a]

[Rb]	[2Fe-2S]	[3Fe-3S]	[4Fe-4S][b]		
Fe^{3+} (EPR)[f]	Fe^{3+}-Fe^{3+}	3 Fe^{3+} (EPR)	3 Fe^{3+}-1 Fe^{2+} (EPR)	$HIPIP_{ox}$	
↕ ≃0 mV[c]	↕ −420 to −200 mV[d]	↕ −400 to −130 mV[e]		↕ +350 mV	
Fe^{2+}	Fe^{3+}-Fe^{3+} (EPR)	2 $Fe^{+2.5}$-1 Fe^{3+}	2 Fe^{3+}-2 Fe^{2+} (+2)	$HIPIP_{nat}$	Fd_{nat}
				↕ ~ −600 mV	↕ −450 to −280 mV[d]
			1 Fe^{3+}-3 Fe^{2+} (EPR)(+1)	$HIPIP_{superred}$	Fd_{red}

[a]The designation of formal valences is useful for the description of the oxidation level, but in practice, due to the covalency of the Fe atoms in the cluster structure, the Fe atoms cannot be distinguished. For the [3Fe-3S] core the electron added on reduction of the center is shared by 2 of the Fe atoms (81) (for details see Chapter 4). However, the spin-coupling model worked for the [3Fe-3] core, indicating that this structure is a covalently linked structure and should not be considered a [2Fe-2S] cluster weakly attached to a 3 Fe atom (159).
[b]The oxidation levels indicated for the [4Fe-4S] structure are determined by adding the formal charges on the Fe and labile S atoms.
[c]Rubredoxin-type proteins have a narrow range of redox potentials (29).
[d][2Fe-2S] and [4Fe-4S] cores have a wide range of redox potentials (29). Limits are indicated in the table.
[e]The redox potentials indicated refer to the [3Fe-3S] core of Av Fd and *D. gigas* Fd II.
[f](EPR) designates EPR-active species.

serve some of the protein resonances because of excessive line broadening. This is one of the reasons that so little information was obtained with rubredoxin-type proteins, since both oxidized ($S = 5/2$) and reduced ($S = 2$) states have a high magnetic moment (199). Indeed, the methods described above have not yet been applied in a systematic way to the study of this type of proteins.

The NMR spectroscopy of Fe-S proteins has been limited to the study of the resonances shifted by the influence of nearby intrinsic probes (characterization of Fe-S centers, their redox state, and their ligands), to the measurement of the magnetic susceptibility of the protein solutions (217), and to obtain the limits to the rates of electron exchange between different oxidation states (218).

The assignment of the contact-shifted resonances of the cysteinyl β-CH_2 protons of the 2 × [4Fe-4S] ferredoxin from *C. pastorianum* in both oxidation states gave the first evidence that the eight-cysteine residues are directly bound to the Fe-S centers (219). A confirmation of this assignment was achieved by comparison with the NMR spectra of model compounds (220, 221). The temperature dependence of the contact-shifted resonances and the magnetic susceptibility measurements of protein solutions in both oxidation states confirmed the existence of antiferromagnetic coupling between the Fe atoms (222). Magnetic susceptibility measurements by NMR have also been used to ascertain the number of unpaired spins and to elucidate the spin state of the Fe atoms of Fe-S centers (217).

These studies have been repeated for a great number of Fe-S proteins, allowing a characterization of the respective clusters as well as a classification of [4Fe-4S] ferredoxins, according to the *three state hypothesis* (223) (see Table 18). The NMR spectra have also been applied to characterize Rb [2Fe-2S], and [4Fe-4S] centers. The main conclusions may be listed as follows:

1. The strong downfield shifts are directly related to the intrinsic paramagnetism; for example, $[4Fe\text{-}4S]^{+3}$ and $[4Fe\text{-}4S]^{+1}$ clusters produce stronger downfield shifts than $[4Fe\text{-}4S]^{+2}$ states.
2. The position of the shifted resonances are temperature dependent.
3. Further assignments to the contact origin of the paramagnetic shift are based on the breadth of the lines in conjunction with conclusion 2.
4. The displaced resonances are generally assigned to the β-CH_2 protons of the cysteinyl residues that are involved in the binding of the Fe-S centers. Further support comes from the observation of similar resonances in model compounds.
5. Generally, in proteins the β-CH_2 protons cannot rotate freely and the protons are inequivalent, hence one proton intensity peaks are observed.
6. Model compounds present equivalent β-CH_2 protons, and the contact

Table 18 Magnetic Susceptibility Data on [4Fe-4S] and [3Fe-3S] Centers

	μ_{eff}		Temperature	
	Oxidized	Reduced	(°C)	Ref.
[4Fe-4S]				
B. polymyxa Fd	0.90	1.60	22	218
Chromatium HIPIP	1.84	0.84		218
[3Fe-3S]				
D. gigas Fd II[a]	1.54	2.71	25	b

[a]The values shown in this table differ from those previously reported (224). A possible explanation for this difference might be the presence of [4Fe-4S] centers in the protein previously used. The new value has been obtained with Fd II purified from three different cell batches.
[b]Our unpublished data.

shifted resonances increase the downfield chemical shift when the temperature is increased. This fact is attributed to anti-Curie paramagnetism of the $[4Fe\text{-}4S]^{+2}$ state. This paramagnetism arises from thermal population of excited spin states in the antiferromagnetic exchange-coupled system.

7. For $[4Fe\text{-}4S]^{+3}$ and $[4Fe\text{-}4S]^{+1}$ systems anti-Curie and Curie temperature dependences are observed, but only anti-Curie dependence is observed for the $[4Fe\text{-}4S]^{+2}$ state.

The conjunction of conclusions 6 and 7 and comparison with model compounds show that the polypeptide chain modulates and imposes distortions in the active core. Also, differences in the observed temperature dependence suggest that in proteins the spin distribution is asymmetric.

The observation that the NMR spectrum of half-reduced 4-Fe ferredoxins consists of a superposition of the spectra of the oxidized and the reduced forms shows that the exchange between the two states is "slow" in the NMR time scale. In contrast the NMR spectrum of half-reduced 8-Fe ferredoxins is an average spectrum indicative of "fast" electron exchange, suggesting an intermolecular cluster interaction that facilitates the exchange mechanism (218, 224).

A more sophisticated method of overcoming spectral complexity has been the selective deuteration of proteins, accomplished by growing the bacteria on deuterated amino acids; it has been used to help in the resonance assignments of bound cysteinyl and other protons of clostridial type ferredoxins (225). This study suggests that the molecular environments of the cysteinyl residues are identical, although their physical environments may differ. Alternatively, the ^{13}C NMR spectra of isotopically enriched *C. acidi-urici, C.*

pastorianum, and *P. aerogenes* ferredoxin (226) have been used to show that the two tyrosyl residues are in magnetically equivalent environments, in both the oxidized and reduced protein, and that each of them is adjacent to one of the metal clusters, as anticipated from the X-ray structure of *Peptococcus aerogenes* (227).

A very thorough ^{13}C NMR study of the same ferredoxins has further shown that the three-dimensional amino acid environments of the corresponding [4Fe-4S] clusters in each protein are similar (228).

The only published NMR spectrum of [3Fe-3S] ferredoxin is that of *D. gigas* Fd II (223) (Fig. 15). By comparison with previous assignments of ferredoxin spectra (199), the contact-shifted resonances *A, B, C,* and *D* were assigned to cysteinyl β-CH_2 protons. Their distances to the paramagnetic center, calculated from the resonance linewidths, confirm their assignment as cysteinyl protons. The temperature dependence of these resonances shows that the contact-shifted resonances have a nonuniform temperature dependence, suggesting that a completely delocalized model for the spin densities does not apply.

The proton resonance of the aromatic amino acid residue of *D. gigas* ferredoxin, PHE22 (161), was assigned by spin echo double resonance. Its resonances have the same chemical shifts in both oligomeric forms of ferredoxin and in both oxidation states, suggesting that PHE22 is not in the immediate proximity of the cluster (224).

An important characterization of the oxidation state of Fe-S centers by

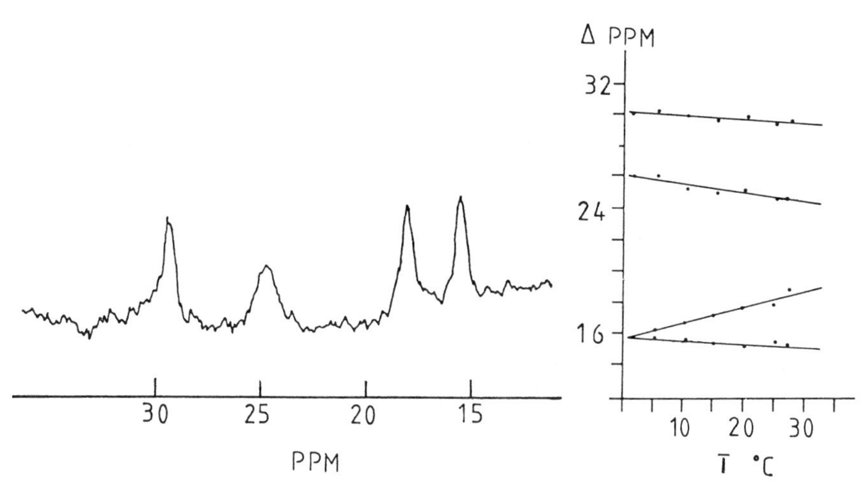

Figure 15. *(A)* Low-field region of the 270 MHz NMR spectrum of oxidized *D. gigas* Fd II at 27°C (pH ≃7.6, [Fd II] = 3 *mM* per monomer. *(B)* Temperature dependence of the low-field resonances of oxidized *D. gigas* Fd II.

NMR is the determination of the magnetic susceptibility. Although the temperature range used in NMR measurements precludes a full characterization of the spin state of the Fe-S cluster, they are very useful when used for comparison. A renewed interest in these measurements appears with the discovery of the [3Fe-3S] centers. Although both oxidized HIPIP [4Fe-4S] centers and oxidized [3Fe-3S] centers have EPR spectra with features at $g = 2$, they can be distinguished by magnetic susceptibility measurements (see Table 18), since the [4Fe-4S] center becomes diamagnetic on reduction and the [3Fe-3S] center becomes more paramagnetic (see Chapter 4).

A very elegant technique developed to identify the type of clusters present in ferredoxins uses ^{19}F NMR (229a). This method has been applied to elucidate the type of Fe-S center of succinate dehydrogenase (229b).

^{35}Cl NMR was recently used to study the weak anion binding of ferredoxin from *Halobacterium* (230).

ACKNOWLEDGMENTS

The authors would like to acknowledge the early pioneering work of Dr. A. R. Prévot on bacterial anaerobiosis, through which many modern researches have been made possible, and Monique Scandellari for her contribution in obtaining the methane bacteria proteins.

We would also like to thank I. Moura, M. H. Santos, B. H. Huynh, E. Münck, and R. Cammack for their collaboration.

This research was supported in part by grants from the CNRS Solar Energy Program, PIRDES, COMES (J. L.) contract no. DEAS09-80 ER 10499, from the Department of Energy (H. D. P.), from N.I.H. no. 25879 (A. V. X.), and the JNICT, INIC, and the Calouste Gulbenkian Foundation of Portugal.

NOTE ADDED IN PROOF

Since this review was written important experiments have introduced the concept of a dynamic relationship between the iron sulfur structures:

i) Upon prolonged treatment with excess ferricyanide, [4Fe-4S] clusters of *Cl. pastorianum* ferredoxin were converted into [3Fe-3S] clusters [1]. Similar experiments have also been carried out with *D. gigas* ferredoxin [2].

ii) Controlled reconstitution experiments of the active center of *D. gigas* FdII (containing only [3Fe-3S] cores) were performed in such a way that a reconstituted protein loaded with either only [4Fe-4S] or mixtures of [3Fe-3S] and [4Fe-4S] cores could be obtained [2].

iii) Mössbauer and EPR spectroscopies have been used to extensively study the process of cluster conversion in *D. gigas* FdII. [3Fe-3S] cores could be converted into [4Fe-4S] clusters after incubation with Fe^{2+} in the presence of dithiothreitol [2].

iv) Using 95% enriched ^{57}Fe in the incubation process, isotopic labelling of one (possibly two) subsite of the newly formed [4Fe-4S] cluster was obtained [3].

v) EPR studies of *D. gigas* cell extracts in the presence of pyruvate, have shown that the [3Fe-3S] centers of the ferredoxin are converted into [4Fe-4S] centers [4].

vi) Mössbauer spectroscopy was used to demonstrate that the [3Fe-3S] clusters of beef heart aconitase are converted into [4Fe-4S] clusters when the enzyme is activated with ferrous iron [5].

REFERENCES

[1] Thomson, A. J., Robinson, A. E., Johnson, M. K., Cammack, R., Rao, K. K., and Hall, D. O., *Biochim. Biophys. Acta*, **637**, 423–432 (1981).

[2] Moura, J. J. G., Moura, I., Kent, T. A., Lipscomb, J. D., Huynh, B. H., LeGall, J., Xavier, A. V., and Münck, E. (1982) *J. Biol. Chem.*, in press.

[3] Kent, T. A., Moura, I., Moura, J. J. G., Lipscomb, J. D., Huynh, B. H., LeGall, J., Xavier, A. V. and Münck, E. (1982) *FEBS Lett.*, in press.

[4] Moura, J. J. G., LeGall, J. and Xavier, A. V., in preparation.

[5] Kent, T. A., Dreyer, J.-L., Kennedy, M. C., Huynh, B. H., Emptage, M. H., Beinert, H. and Münck, E. (1982) *Proc. Natl. Acad. Sci.* (USA) in press.

REFERENCES

1. G. R. Moore and R. J. P. Williams, *Coord. Chem. Rev.*, **18**, 125 (1976).
2. S. Wherland and H. B. Gray, in *Biological Aspects of Inorganic Chemistry*, A. W. Addison, W. R. Gullen, D. Dolphin, and B. R. James, eds. Wiley, New York 1977, pp. 289–367.
3. S. G. Mayhew and M. L. Ludwig, in *The Enzymes*, P. D. Boyer, ed., Vol. XII, Academic, New York, 1975, pp. 57–118.
4. R. E. Dickerson and R. Timkovich, in *The Enzymes*, Vol. XII, P. D. Boyer, ed, Academic, New York, 1975, pp. 57–118.
5. D. C. Yoch and R. P. Charithers, *Microbiol. Rev.*, **43**, 384 (1979).
6. W. Lovenberg, ed., *Iron-Sulfur Proteins*, Vol. 1 (1973). Vol. 2 (1973), and Vol. 3 (1977), Academic, New York.
7. J. J. Hopfield, *Proc. Natl. Acad. Sci. USA*, **71**, 3640 (1974).
8. R. A. Marcus and N. Sutin, *Inorg. Chem.*, **14**, 213 (1975).

9. A. G. Mauk, R. A. Scott, and H. B. Gray, *J. Am. Chem. Soc.*, **102,** 4360 (1980).
10. R. H. Holm and J. A. Ibers, "Synthetic Analogues of the Active sites of Iron-Sulfur proteins, in *Iron-Sulfur Proteins,* Vol. 3, W. Lovenberg, ed., Academic, 1977.
11. J. LeGall and J. R. Postgate, in *Advances in Microbial Physiology*, Vol. 10, A. H. Rose and D. W. Tempest, eds., Academic, London, 1973 pp. 81-133.
12. J. R. Postgate and L. L. Campbell, *Bacteriol. Rev.*, **30**, 732 (1966).
13. L. L. Campbell and J. R. Postgate, *Bacteriol. Rev.*, **29,** 359 (1965).
14. M. Biebl and N. Pfennig, *Arch. Microbiol.*, **112,** 115 (1977).
15. L. M. Siegel, in *Metabolism of Sulfur Compounds,* Vol. VII, 3rd ed., D. M. Greenbey, ed., 1975, pp. 2117-286.
16. J. LeGall, D. V. DerVartanian, and H. D. Peck, Jr., *Curr. Topics Bioenerg.*, **9,** 237 (1979).
17. W. E. Balch, G. E. Fox, L. J. Magrum, G. R. Woese, and R. S. Wolfe, *Microbiol. Rev.*, **43,** 160 (1979).
18. R. S. Wolfe and I. J. Higgins, *Int. Rev. Biochem.*, **21,** 267 (1979).
19. G. E. Fox, E. Stackebraudt, R. B. Hespell, J. Gibson, J. Maniloff, T. A. Dyer, R. S. Wolfe, E. W. Balch, R. S. Tanner, L. J. Magrum, L. B. Zablen, R. Blakemove, R. Gupta, L. Boney, B. J. Lewis, D. A. Stahl, K. R. Loehrsen, K. M. Chen, and C. R. Woese, *Science,* **204,**457 (1980).
20. H. D. Peck, *Symp. Soc. Gen. Microbiol.*, **24,** 241 (1974).
21. D. O. Hall, in no. 162, *Bioinorganic Chemistry II,* K. N. Raymond, ed., Advances in Chemistry Series, American Chemical Society, Washington, D.C. 1977, p. 227.
22. J. LeGall, M. Bruschi, C. E. Hatchikian, *Proceedings of the Tenth FEBS Meeting,* Vol. 44, P. Desnuelle and A. M. Michelson, eds., North Holland-American Elsevier, 1975, p. 277.
23. C. E. Hatchikian and J. LeGall, *Biochim. Biophys. Acta,* **267,** 479 (1972).
24. J. LeGall and N. Forget, in *Biomembranes,* S. Fleischer and L. Parker eds., Methods in Enzymology, Vol. XIII, Academic, New York, 1978, p. 613.
25. J. J. G. Moura, I. Moura, M. Bruschi, J. LeGall, and A. V. Xavier, *Biochem. Biophys. Res. Commun.*, **92,** 962 (1980).
26. J. LeGall, D. V. DerVartanian, and H. D. Peck, Jr., *Curr. Topics Bioenerg.*, **9,** 237 (1979).
27. D. O. Hall, R. Cammack, and K. K. Rao, in *Iron in Biochemistry and Medicine,* A. Jacobs and M. Worwood, eds., Academic, 1974, p. 279.
28. W. Lovenberg, ed., *Iron-Sulfur Proteins,* Vols. I and II (1975) and Vol. III (1977), Academic, New York.
29. A. V. Xavier, J. J. G. Moura, and I. Moura, *Struct. Bonding,* **43,** 188 (1981).
30. R. W. Jones, *Biochem. J.*, **188,** 345 (1980).
31. J. A. Raven and F. A. Smith, *J. Theor. Biol.*, **57,** 301 (1976).
32. S. L. Albrecht, R. J. Maier, F. J. Hanus, S. A. Russell, D. W. Emerich, and R. J. Evans, *Science,* **203,** 1255 (1979).
33. D. O. Hall, *Solar Energy,* **22,** 307 (1979).
34. H. D. Peck, Jr., and M. Odom, in *Trends in the Biology of Fermentations for Fuels and Chemicals,* in press (1981).
35. M. Stephenson and L. H. Strickland, *Biochem. J. (London),* **25,** 205 (1931).

36. H. Gest, *Bacteriol. Rev.*, **18,** 43 (1954).
37. C. H. Gray and H. Gest, *Science,* **148,** 186 (1965).
38. J. LeGall, D. V. DerVartanian, E. Spilker, Jin-Po Lee, and H. D. Peck, Jr., *Biochim. Biophys. Acta,* **234,** 525 (1971).
39. G. Nakos and L. E. Mortenson, *Biochim. Biophys. Acta,* **277,** 576 (1971).
40. R. H. Haschke and L. L. Campbell, *J. Bacteriol.*, **105,** 294 (1971).
41. A. Farkas, L. Farkas, and J. Yudkin, *Proc. Roy. Soc. London, B,* **115,** 373 (1934).
42. A. I. Krasna and D. Rittenberg, *J. Am. Chem. Soc.*, **76,** 3015 (1951).
43. R. W. Spencer, L. Daniels, G. Fulton, and W. H. Orme-Johnson, *Biochemistry,* **19,** 3678 (1980).
44. W. S. Waring and C. H. Werkman, *Arch. Biochem.*, **4,** 75 (1944).
45. F. Kubowitz, *Biochem. J.*, **274,** 285 (1934).
46. W. K. Joklik, *Austr. J. Expt. Biol. Med.*, **28,** 321 (1950).
47. R. K. Thauer, G. Leimenstoll, and K. Decker, *Biochim. Biophys. Acta,* **305,** 268 (1973).
48. R. K. Thauer, K. Jungermann, H. Henninger, J. Wenning, and K. Decker, *Eur. J. Biochem.*, **4,** 173 (1968).
49. W. A. Bulen and J. R. LeConte, *Proc. Natl. Acad. Sci. USA,* **65,** 979 (1966).
50. R. O. D. Dixon, *Arch. Mikrobiol.*, **85,** 193 (1972).
51. L. E. Mortenson and J. S. Chen, in *Microbial Iron Metabolism,* J. B. Meilands, ed., Academic, New York, (1974).
52. M. P. Bryant, E. A. Wolin, M. J. Wolin, and R. S. Wolfe, *Arch. Microbiol.*, **59,** 20 (1967).
53. D. R. Boone and M. P. Bryant, *Appl. Environ. Microbiol.*, **40,** 626 (1980).
54. M. J. MacInerny, M. P. Bryant, and N. Pfennig, *Arch. Microbiol.*, **122,** 129 (1979).
55. G. R. Bell, J. LeGall, and H. D. Peck, Jr., *J. Bacteriol.*, **120,** 994 (1974).
56. J. S. Chen, in *Hydrogenases: Their Catalytic Activity, Structure and Function,* W. G. Schlegel and K. Schneider, eds. E. Goltz, Gottingen, 1978, p. 57.
57. W. Badziong and R. K. Thauer, *Arch. Microbiol.*, **125,** 167 (1980).
58. A. Kroger, *Biochim. Biophys. Acta,* **505,** 129 (1978).
59. J. LeGall, D. V. DerVartanian, E. Spilker, J. P. Lee, and H. D. Peck, Jr., *Biochim. Biophys. Acta,* **234,** 525 (1971).
60. J. Chen and L. E. Mortenson, *Biochim. Biophys. Acta,* **371,** 283 (1974).
61. W. O. Gillum, L. E. Mortenson, J. S. Chen, and R. H. Holm, *J. Am. Chem. Soc.*, **99,** 584 (1977).
62. D. L. Erbes, R. H. Burris, and W. H. Orme-Johnson, *PNAS,* **72,** 4795 (1975).
63. J. S. Chen and D. K. Blanchard, *Biochim. Biophys. Res. Commun.*, **84,** 1144 (1978).
64. G. R. Bell, J. P. Lee, H. D. Peck, Jr., and J. LeGall, *Biochimie,* **60,** 315 (1978).
65. J. R. Postgate, in *The Sulfate Reducing Bacteria,* Cambridge University Press, 1979.
66. M. P. Bryant, L. L. Campbell, C. A. Reddy, and M. R. Crabill, *Appl. Environ. Microbiol.*, **33,** 1162 (1977).
67. J. M. Akagi and L. L. Campbell, *J. Bacteriol.*, **82,** 927 (1961).
68. C. S. Buller and J. M. Akagi *J. Bacteriol.*, **88,** 440 (1964).
69. T. Yagi, A. Endo, and K. Tsuji, in *Hydrogenases: Their Catalytic Activity, Structure and Function,* H. G. Schlegel and K. Schneider, eds., C. Goltz, Gottingen, 1978, p. 107.

70. M. Bruschi and J. LeGall, in *Iron and Copper Proteins,* K. T. Yasunobu, M. F. Mower, and O. Hayaishi, eds., Plenum, New York, 1976, p. 57.
71. K. Tsuji and T. Yagi, *Arch. Microbiol.,* **125,** 35 (1980).
72. A. S. Traore, C. E. Hatchikian, J. P. Belaich, and J. LeGall, *J. Bacteriol.,* **145,** 191 (1981).
73. R. O. D. Dixon, *Nature,* **262,** 173 (1976).
74. E. C. Hatchikian, M. Chaigneau, and J. LeGall, in *Microbial Production and Utilization of Gases,* H. K. Schlegel, K. Gottschalk, and N. Pfennig, eds., Goltze, Gottingen, 1976, p. 109.
75. B. Khosrovi, R. MacPherson, and J. D. A. Miller, *Arch. Microbiol.,* **80,** 324 (1971).
76. J. M. Odom and H. D. Peck, Jr., unpublished observations.
77. S. van der Westen, G. Mayhew, and C. Veeger, *FEBS Lett.,* **7,** 35 (1980).
78. S. M. Martin, B. R. Glick, and W. G. Martin, *Can. J. Microbiol.,* **26,** 1209 (1980).
79. H. M. van der Westen, S. G. Mayhew, and C. Veeger, *FEBS Lett.,* **86,** 122, (1978).
80. E. C. Hatchikian, M. Bruschi, and J. LeGall, *Biochim. Biophys. Res. Commun.,* **82,** 451 (1978).
81. B. H. Huynh, J. J. G. Moura, I. Moura, T. A. Kent, J. LeGall, A. V. Xavier, and E. Münck, *J. Biol. Chem.,* **255,** 3242 (1980).
82. M. H. Emptage, T. A. Kent, B. H. Huynh, J. Rawlings, W. H. Orme-Johnson, and E. Münck, *J. Biol. Chem.,* **255,** 1793 (1980).
83. T. A. Kent, J. L. Dreyer, M. H. Emptage, I. Moura, J. J. G. Moura, B. H. Huynh, A. V. Xavier, J. LeGall, H. Beinert, and W. H. Orme-Johnson, in *Interactions Between Iron and Proteins in Oxygen and Electron Transfer,* H. Chen, ed., Elsevier, New York, in press (1981).
84. R. Cammack, M. V. Lalla-Maharajh, and K. Schneider, in *Interactions Between Iron and Proteins in Oxygen and Electron Transport,* C. Mo, ed., Elsevier, New York, in press (1981).
85. S. G. Mayew, C. van Dick, and E. M. van der Westen, in *Hydrogenases: Their Catalytic Activity, Structure and Function,* H. G. Schlegel, ed., E. Goltze, Gottingen, 1978, p. 125.
86. J. R. Postgate, *J. Gen. Microbiol.,* **14,** 545 (1956).
87. T. Yagi, *J. Biochem. (Tokyo),* **68,** 649 (1970).
88. G. Fauque, O. Herve, and J. LeGall, *Arch. Microbiol.,* **121,** 261 (1979).
89. M. Biebl and N. Pfennig, *Arch. Microbiol.,* **121,** 115 (1979).
90. E. C. Hatchikian, M. Bruschi, J. LeGall, and M. Dubourdieu, *Bull. Soc. Fr. Physiol. Veg.,* **15,** 381 (1969).
91. J. LeGall, M. Bruschi-Heriaud, and D. V. DerVartanian, *Biochim. Biophys. Acta,* **234,** 449 (1971).
92. J. LeGall and M. Bruschi-Heriaud, in *Structure and Function of Cytochromes,* K. Okunuki, M. D. Kamen, and K. Suzuki, eds., Univ. of Tokio Press and Univ. Park Press, 1968, p. 467.
93. M. Bruschi and J. LeGall, *Biochim. Biophys. Acta,* **271,** 48 (1972).
94. R. W. Jones and P. B. Carlond, *Biochem. J.,* **164,** 199 (1977).
95. D. Steenkamp and H. D. Peck, Jr., unpublished observations.
96. J. R. Postgate, *Bacteriol. Rev.,* **29,** 425 (1965).
97. H. D. Peck, Jr., *J. Biol. Chem.,* **235,** 2734 (1960).

98. H. D. Peck, Jr., *Biochem. Biophys. Res. Commun.*, **22,** 112 (1966).
99. L. L. Barton, J. LeGall, and H. D. Peck, Jr., *Biochem. Biophys. Res. Commun.*, **41,** 1036 (1970).
100. H. D. Peck, Jr., *Bacteriol. Rev.*, **26,** 67 (1962).
101. L. J. Guarraia and H. D. Peck, Jr., *J. Bacteriol.*, **106,** 890 (1971).
102. P. M. Wood, *FEBS Lett.*, **95,** 12 (1978).
103. G. Fauque, L. L. Barton, and J. LeGall, Excepta Medica 1980, Sulfur in Biology (CIBA Foundation Symposium 72), 1980, pp. 71–86.
104. E. A. Wolin, M. J. Wolin, and R. S. Wolfe, *J. Biol. Chem.*, 2882 (1963).
105. R. C. McKellar and G. D. Sprott, *J. Bacteriol.*, **139,** 231 (1979).
106. G. Fuchs, J. Woll, P. Scherer, and R. K. Thauer, in *Hydrogenases: Their Catalytic Structure and Function*, H. G. Schlegel, ed., Goltar, Gottingen, 1976.
107. R. P. Gunsalus and R. S. Wolfe, *J. Biol. Chem.*, **755,** 1891 (1980).
108. G. Nakos and L. E. Mortenson, *Biochemistry*, **10,** 2442 (1971).
109. H. J. Doddema, C. van der Vridt, G. D. Vogels, and M. Veenhois, *J. Bacteriol.*, **140,** 1081 (1979).
110. B. C. McBride and R. S. Wolfe, *Biochemistry*, **10,** 2317 (1971).
111. C. D. Taylor and R. S. Wolfe, *J. Biol. Chem.*, **249,** 4879 (1974).
112. P. Cheeseman, A. Tomswood, and R. S. Wolfe, *J. Bacteriol.*, **112,** 527 (1972).
113. L. D. Eirich, G. D. Vogels, and R. S. Wolfe, *Biochemistry*, **17,** 4583 (1978).
114. R. P. Gunsalus and R. S. Wolfe, *FEBS Lett.*, **3,** 191 (1978).
115. W. B. Whitman and R. S. Wolfe, *Biochem. Biophys. Res. Commun.*, **92,** 1196 (1980).
116. G. Diekert, B. Kler, and R. K. Thauer, *Arch. Microbiol.*, **124,** 103 (1980).
117. G. Diekert, H. H. Giles, R. Jaechen, and R. K. Thauer, *Arch. Microbiol.*, **128,** 256 (1980).
118. C. van der Drift and J. M. Koltjens, *Anton. Leeu. J. Microbiol.*, **46,** 106 (1980).
119. J. Krzycki and J. G. Zeikus, *Curr. Microbiol.*, **3,** 143 (1980).
120. T. C. Stadtman and B. A. Blaylock, *Fed. Proc.*, **25,** 1957 (1966).
121. J. G. Zeikus, G. Fuchs, W. Kenealy, and R. K. T. Thauer, *J. Bacteriol.*, **132,** 604 (1977).
122. J. R. Lancaster, Jr., *FEBS Lett.*, **115,** 285 (1980).
123. W. E. Balch and R. S. Wolfe, *J. Bacteriol.*, **137,** 256 (1979).
124. W. Kühn, K. Fiebics, R. Walther, and G. Gollschalk, *FEBS Lett.*, **105,** 771 (1979).
125. S. F. Tzeng, R. S. Wolfe, and M. P. Bryant, *J. Bacteriol.*, **121,** 184 (1975).
126. S. F. Tzeng, M. P. Bryant, and R. S. Wolfe, *J. Bacteriol.*, **121,** 192 (1975).
127. J. B. Jones and T. C. Stadtman, *J. Biol. Chem.*, **255,** 1049 (1980).
128. J. S. Chen and D. K. Blanchard *Microbiol. Proc.*, 144 (1980).
129. R. K. Thauer, K. Songormann, and K. Decker, *Microbiol. Rev.*, **41,** 100 (1977).
130. G. T. Taylor and S. J. Pirt, *Arch. Microbiol.*, **113,** 17 (1977).
131. M. L. Schauer and J. G. Ferry, *J. Bacteriol.*, **142,** 800 (1980).
132. J. G. Zeikus and V. G. Bowey, *Can. J. Microbiol.*, **21,** 121 (1975).
133. H. J. Doddema, C. A. Claesey, D. B. Gell, C. van der Drift, and G. D. Vogels, *Biochem. Biophys. Res. Commun.*, **95,** 1288 (1980).
134. F. D. Sauer, J. D. Erfle, and S. Mahadevan, *Biochem. J.*, **190,** 177 (1980).

135. E. J. Laishley, J. Travis, and H. D. Peck, Jr., *J. Bacteriol.*, **98,** 302 (1969).
136. M. Bruschi and J. LeGall, *Biochim. Biophys. Acta,* **263,** 279 (1971).
137. I. Moura, J. J. G. Moura, M. Bruschi, and J. LeGall, *Biochim. Biophys. Acta,* **591,** 1 (1980).
138. M. Bruschi, C. E. Hatchikian, L. A. Golovleva, and J. LeGall, *J. Bacteriol.*, **129,** 30 (1977).
139. C. E. Hatchikian, H. E. Jones, and M. Bruschi, *Biochim. Biophys. Acta,* **548,** 471 (1979).
140. D. J. Newman and J. R. Postgate, *Biochim. Biophys. Acta,* **238,** 385 (1968).
141. I. Probst, J. J. G. Moura, I. Moura, M. Bruschi, and J. LeGall, *Biochim. Biophys. Acta,* **502,** 38 (1978).
142. I. Moura, M. Bruschi, J. LeGall, J. J. G. Moura, and A. V. Xavier, *Biochem. Biophys. Res. Commun.*, **75,** 1037 (1977).
143. M. Bruschi, I. Moura, A. V. Xavier, J. LeGall, and L. C. Sieker, *Biochem. Biophys. Res. Commun.*, **90,** 596 (1979).
144. I. Moura, B. H. Huynh, R. P. Hausinger, J. LeGall, A. V. Xavier, and E. Münck, *J. Biol. Chem.*, **255,** 2498 (1980).
145. I. Moura, A. V. Xavier, R. Cammack, M. Bruschi, and J. LeGall, *Biochim. Biophys. Acta,* **533,** 156 (1978).
146. E. T. Lode and N. Y. Coon, *J. Biol. Chem.*, **246,** 791 (1971).
147. W. Lovenberg and B. E. Sobel, *Proc. Natl. Acad. Sci. USA,* **54,** 183 (1965).
148. W. E. Blumberg, in *Magnetic Resonance in Biological Systems,* A. Ehrenberg, B. C. Malsmtrom, and T. Vanngard, eds., Pergamon, Oxford, 1967, pp. 119–133.
149. M. Bruschi, *Biochim. Biophys. Acta,* **434,** 4 (1976).
150. M. Bruschi, *Biochem. Biophys. Res. Commun.*, **70,** 615 (1976).
151. K. F. MacCarthy, referenced by W. A. Eaton and W. Lovenberg, in *Iron-Sulfur Proteins,* Vol. II, W. Lovenberg, ed., Academic, 1973.
152. L. C. Sieker, M. Bruschi, J. LeGall, I. Moura, and A. V. Xavier, *Ciência Biol. (Portugal),* **5,** 145 (1980).
153. J. M. Odmon and H. D. Peck, Jr., unpublished, referenced in reference 26.
154. K. D. Watenpaugh, L. C. Sieker, J. R. Herriot, and L. H. Jensen, *Acta Crystallogr.*, **B29,** 943 (1973).
155. E. T. Adman, L. C. Sieker, M. Bruschi, and J. LeGall, *J. Mol. Biol.*, **112,** 113 (1977).
156. R. J. P. Williams, *Q. Rev.*, **24,** 331 (1970).
157. R. Cammack, "Functional Aspects of Iron-Sulfur Proteins," in *Metallaproteins: Structure, Function and Chemical Aspects,* V. Weser, ed., Thieme, Stuttgart, 1979, pp. 162–184.
158. W. H. Orme-Johnson and N. Orme-Johnson, *Ann. Rev. Biophys. Bioeng.*, in press (1980).
159. B. H. Huynh and T. A. Kent, in *Mössbauer Studies in Biomolecules,* B. V. Thosar and P. K. Iyengar, eds., Advances in Mössbauer Spectroscopy, Elsevier, Amsterdam, 1981.
160. M. Bruschi, C. E. Hatchikian, J. LeGall, J. J. G. Moura, and A. V. Xavier, *Biochim. Biophys. Acta,* **449,** 275 (1976).
161. M. Bruschi, *Biochem. Biophys. Res. Commun.*, **91,** 623 (1979).
162. R. Cammack, K. K. Rao, D. O. Hall, J. J. G. Moura, A. V. Xavier, M. Bruschi, J. LeGall, A. Deville, and J. P. Gayda, *Biochim. Biophys. Acta,* **490,** 311 (1978).

163. J. J. G. Moura, A. V. Xavier, C. E. Hatchikian, and J. LeGall, *FEBS Lett.*, **89**, 177 (1978).

164. A. J. Thomson, A. E. Robinson, M. K. Johnson, J. J. G. Moura, I. Moura, A. V. Xavier, and J. LeGall, *Biochim. Biophys. Acta*, Vol. **670**, 93 (1981).

165. C. D. Stout, D. Gosh, V. Pattabhi, and A. Robins, *J. Biol. Chem.*, **255**, 1797 (1980).

166. C. E. Hatchikian, H. E. Jones, and M. Bruschi, *Biochim. Biophys. Acta*, **548**, 471 (1979).

167. C. E. Hatchikian and M. Bruschi, *Biochim. Biophys. Acta*, **634**, 41 (1981).

168. F. Guerlesquin, M. Bruschi, G. Bovier Lapierre, and G. Fauque, *Biochim. Biophys. Acta*, **626**, 127 (1980).

169. P. Middleton, D. P. E. Dickson, C. E. Johnson, and J. D. Rush, *Eur. J. Biochem.*, **88**, 135 (1978).

170. T. Hase, N. Ohmiya, H. Matsubara, R. N. Mullinger, K. K. Rao, and D. O. Hall, *Biochem. J.*, **159**, 55 (1976).

171. K. T. Yasunobu and M. Tanaka, in *Iron-Sulfur Proteins*, Vol. 2, W. Lovenberg, ed., Pergamon, New York, 1973, p. 27.

172. J. M. Akagi, *J. Biol. Chem.*, **242**, 2478 (1967).

173. J. LeGall and N. Dragoni, *Biochem. Biophys. Res. Commun.*, **23**, 145, (1966).

174. M. Dubourdieu, J. LeGall, and V. Favaudon, *Biochim. Biophys. Acta*, **376**, 519 (1975).

175. J. J. G. Moura, A. V. Xavier, D. J. Cookson, G. R. Moore, R. J. P. Williams, M. Bruschi, and J. LeGall, *FEBS Lett.*, **81**, 275 (1977).

176. A. V. Xavier, J. J. G. Moura, J. LeGall, and D. V. DerVartanian, *Biochimie*, **61**, 689 (1979).

177. I. Moura, J. J. G. Moura, M. H. Santos, and A. V. Xavier, *Ciência Biol. (Portugal)*, **5**, 195 (1980).

178. I. Moura, J. J. G. Moura, M. H. Santos, A. V. Xavier, and J. LeGall, *FEBS Lett.*, **107**, 419 (1979).

179. J. J. G. Moura and A. V. Xavier, "Molybdenum Proteins," in *New Trends in Bioinorganic Chemistry*, R. J. P. Williams and J. J. R. Fraústo da Silva, eds., Academic, New York, 1978.

180. J. J. G. Moura, A. V. Xavier, M. Bruschi, J. LeGall, R. Cammack, and D. O. Hall, *Biochem. Biophys. Res. Commun.*, **72**, 782 (1976).

181. J. J. G. Moura, A. V. Xavier, R. Cammack, D. O. Hall, M. Bruschi, and J. LeGall, *Biochem. J.*, **173**, 419 (1978).

182. C. E. Hatchikian and M. Bruschi, *Biochem. Biophys. Res. Commun.*, **89**, 725 (1979).

183. K. Kobayashi, E. Takahashi, and M. Ishimoto, *J. Biochem.*, **72**, 879 (1972).

184. K. Kobayashi, Y. Seki, and M. Ishimoto, *J. Biochem. (Tokyo)*, **72**, 879 (1972).

185. K. Prabhakararao and D. J. D. Nicholas, *Biochem. Biophys. Acta*, **180**, 253 (1969).

186. J. P. Lee, J. LeGall, and H. D. Peck, Jr., *J. Bacteriol.*, **115**, 529 (1973).

187. H. L. Drake and J. M. Akagi, *Biochem. Biophys. Res. Commun.*, **71**, 1214 (1976).

188. Y. Seki, K. Kobayashi, and M. Ishimoto, *J. Biochem.*, **85**, 705 (1979).

189. J. P. Lee and H. D. Peck, Jr., *Biochem. Biophys. Res. Commun.*, **45**, 583 (1971).

190. J. P. Lee, J. LeGall, and H. D. Peck, Jr., *J. Bacteriol.*, **115**, 453 (1973).

191. P. A. Trudinger, *J. Bacteriol.*, **104**, 158 (1970).

192. M. J. Murphy, L. M. Siegel, H. Kamin, D. V. DerVartanian, L. P. Lee, J. LeGall, and H. D. Peck, Jr., *Biochem. Biophys. Res. Commun.*, **54**, 82 (1973).

193. C. L. Liu, D. V. DerVartanian, and H. D. Peck, Jr., *Biochem. Biophys. Res. Commun.*, **91,** 962 (1979).

194. M. H. Hall, R. H. Prince, and R. Cammack, *Biochim. Biophys. Acta,* **581,** 27 (1979).

195. M. Schedel, J. LeGall, and J. Baldensperger, *Arch. Microbiol.*, **105,** 339 (1975).

196. H. Beinert, "EPR Spectroscopy of Components of Mitochondrial Electron Transfer," in *Methods in Enzymology,* S. Fleisher and L. Packer, eds., Vol. IV, part E, Academic, New York, 1978.

197. R. Malkin and A. J. Bearden, "Bound Iron-Sulfur Centers in Photomagnetic Membranes (Higher Plants and Bacteria), in *Methods in Enzymology,* Vol. 69, Academic, New York, 1980, pp. 238–249.

198. J. R. Lancaster, Jr., *FEBS Lett.*, **115,** 285 (1980).

199. W. D. Philips and M. Poe, in *Iron Sulfur Proteins,* Vol. 2, W. Lovenberg, ed., Academic, New York, 1973, p. 255.

200. G. Palmer, in *The Enzymes,* Vol. 12, 3rd ed., P. D. Boyer, ed., Academic, New York, 1975, p. 1.

201. W. V. Sweeney and J. C. Rabinowitz, *Ann. Rev. Biochem.*, **49,** 139 (1980).

202. R. A. Dwek, R. J. P. Williams, and A. V. Xavier, in *Metal Ions in Biological Systems,* Vol. 4, H. Sigel, ed., Marcel Decker, New York, 1974, p. 61.

203. T. J. Swift, in *NMR of Paramagnetic Molecules,* G. N. LaMar and W. de W. Horrocks, Jr., eds., Academic, New York, 1973, Chapter 2.

204. C. D. Barry, J. A. Glasel, R. J. P. Williams, and A. V. Xavier, *J. Mol. Biol.*, **84,** 471 (1974).

205. L. M. Jackman and S. Sternhell, *NMR Spectroscopy in Organic Chemistry,* 2nd ed., Pergamon, Oxford, 1969.

206. I. D. Campbell and C. M. Dobson, *Meth. Biochem. Anal.*, **25,** 1 (1979).

207. I. D. Campbell, C. M. Dobson, R. J. P. Williams, and A. V. Xavier, *J. Magn. Reson.*, **11,** 172 (1973).

208. I. D. Campbell, C. M. Dobson, R. J. P. Williams, and A. V. Xavier, *Ann. NY Acad. Sci.*, **222,** 163 (1973).

209. F. F. Brown and I. D. Campbell, *FEBS Lett.*, **65,** 322 (1976).

210. I. D. Campbell, C. M. Dobson, R. J. P. Williams, and P. E. Wright, *FEBS Lett.*, **57,** 96 (1975).

211. F. M. Poulse, J. C. Hoch, and C. M. Dobson, *Biochemistry,* **19,** 2579 (1980).

212. J. H. Noggle and R. E. Schirmer, *The NOE Effect,* Academic, New York, 1971.

213. P. F. Knowles, D. Marsh, and H. W. E. Rattle, *Magnetic Resonance of Biomolecules,* Wiley, London, 1976.

214. A. G. Redfield and R. K. Gupta, *Proc. Cold Spring Harbor Symp. Quant. Biol.*, **36,** 405 (1971).

215. R. J. Gupta and A. S. Mildvan, *Meth. Enzymology,* **54,** 51 (1978).

216. S. Forsen and R. A. Hoffman, *J. Chem. Phys.*, **39,** 2892 (1976).

217. M. Poe and W. D. Phillips, *Meth. Enzymology,* **24,** 304 (1972).

218. W. D. Phillips, C. C. MacDonald, N. A. Stombaugh, and W. H. Orme-Johnson, *Proc. Natl. Acad. Sci. USA,* **71,** 140 (1974).

219. M. Poe, W. D. Phillips, C. C. McDonald, and W. Lovenberg, *Proc. Natl. Acad. Sci. USA,* **65,** 797 (1970).

220. B. A. Averill, T. Herzkovitz, R. H. Holm, and J. A. Ibers, *J. Am. Chem. Soc.*, **95**, 3523 (1973).

221. R. H. Holm, H. D. Phillips, B. A. Averill, J. J. Meyerle, and T. Herskovitz, *J. Am. Chem. Soc.*, **76**, 2109 (1974).

222. W. D. Phillips, C. C. MacDonald, and R. G. Bartsch, *Proc. Natl. Acad. Sci. USA*, **67**, 682 (1970).

223. E. Edman and L. H. Jensen, *Proc. Natl. Acad. Sci. USA*, **69**, 3526 (1972).

224. J. J. G. Moura, A. V. Xavier, M. Bruschi, and J. LeGall, *Biochim. Biophys. Acta*, **459**, 278 (1977).

225. E. L. Parker, W. V. Sweeney, J. C. Rabinowitz, H. Sternlicht, and E. N. Shaw, *J. Biol. Chem.*, **252**, 2245 (1977).

226. E. L. Parker, H. Sternlicht, and J. C. Rabinovitz, *Proc. Natl. Acad. Sci. USA*, **69**, 3278 (1972).

227. E. T. Adman, L. C. Sieker, and L. H. Jensen, *J. Biol. Chem.*, **248**, 3987 (1973).

228. E. L. Parker, J. C. Rabinovitz, and H. Sternlicht, *J. Biol. Chem.*, **253**, 7722 (1978).

229a. G. B. Wong, D. M. Kurtz, Jr., R. H. Holm, L. E. Mortenson, and R. G. Upchurch, *J. Am. Chem. Soc.*, **101**, 3078 (1979).

229b. C. J. Coles, R. H. Holm, D. M. Kurtz, Jr., W. H. Orme-Johnson, J. Rawllings, T. P. Singer, and G. B. Wong, *Proc. Natl. Acad. Sci. USA*, **76**, 3805 (1979).

230. P. Reimarsson, B. Lindman, and M. M. Werber, *Arch. Biochem. Biophys.*, **202**, 664 (1980).

231. M. P. Bryant, V. H. Varel, R. A. Frobish, and H. R. Isaacson, *Biological Potentials of Thermophilic Methanogenesis from Cattle Wastes in Microbial Energy Conservation*, H. G. Schlegel and J. Barnea, eds., 1976, pp. 347–359.

232. C. E. Hatchikian and J. LeGall, *Biochim. Biophys. Acta*, **267**, 479 (1972).

233. I. Probst, M. Bruschi, N. Pfennig, and J. LeGall, *Biochim. Biophys. Acta*, **460**, 58 (1977).

234. H. E. Jones, *Arch. Mikrob.*, **80**, 78 (1971).

235. M. W. W. Adams, L. E. Mortenson, and J. S. Chen, *Biochim. Biophys. Acta*, **594**, 105 (1981).

236. Y. A. Zubieta, K. Mason, and J. R. Postgate, *Biochem. J.*, **133**, 851 (1973).

237. J. LeGall, and N. Dragoni, *Biochem. Biophys. Res. Commun.*, **23**, 145 (1966).

238. J. LeGall and C. E. Hatchikian, *C. R. Acad. Sci.*, **264**, 2583 (1977).

239. C. E. Hatchikian and J. LeGall, *Ann. Inst. Pasteur*, **118**, 288 (1970).

240. H. L. Drake and J. M. Akagi, *J. Bacteriol.*, **132**, 132 (1977).

241. M. Ishimoto and Y. Koyama, *J. Biochem.*, **44**, 233 (1957).

242. B. H. Suh and J. M. Akagi, *J. Bacteriol.*, **99**, 219 (1956).

243. H. L. Drake and J. M. Akagi, *J. Bacteriol.*, **132**, 139 (1977).

244. K. Irie, K. Kobayashi, M. Kobayashi, and M. Ishimoto, *J. Bichem. (Tokyo)*, **73**, 313 (1973).

CHAPTER 6

Iron-Sulfur Centers In Photosynthetic Electron Transport

M. C. W. EVANS

Department of Botany and Microbiology
University College, London

CONTENTS

1 INTRODUCTION

Fe-S centers play an important role in all membrane-bound electron-transport systems that have been thoroughly investigated. Photosynthetic electron transport in both oxygenic and anoxygenic organisms involves Fe-S centers (1); however, the number of centers involved is far fewer than in mitochondria. In oxygenic organisms, the blue-green algae, eukaryotic algae, and higher plants, 5 Fe-S centers have been well characterized. One of these, the Rieske Fe-S center (2), appears to be analogous to the component in mitochondria (3, 4), and functions in the oxidative intermediary electron-transport chain. The remainder are all low-potential centers involved in the transfer of electrons from the photosystem I reaction center (5–7). A number of Fe-S centers have been identified in membrane fractions of purple photosynthetic bacteria (8, 9). Only one, the Rieske Center, has been directly implicated in photosynthetic electron transport (10); the others function in secondary electron-transport processes, for example, in succinic dehydrogenase or perhaps NAD reduction. The green photosynthetic bacteria also contain a number of membrane-associated Fe-S centers (11, 12), including a Rieske center and a low-potential center involved in the early stages of photochemical electron transport (13).

This review is mainly concerned with the properties of the Fe-S centers in oxygenic photosynthetic organisms and with a comparative account of the less well-investigated bacterial systems.

2 PHOTOSYNTHETIC ELECTRON TRANSPORT

To understand the importance of the Fe-S centers in photosynthesis, some understanding of the overall process of photosynthetic electron transport is required. Figure 1 shows a diagram of the electron-transport chain as currently understood in oxygenic photosynthesis. The overall process can be represented by the equation

$$2H_2O + 2NADP \xrightarrow{\text{Light}} 2NADPH_2 + O_2$$

The overall process involves two photochemical reactions in which the photochemical oxidation of a specialized chlorophyll molecule, the reaction center chlorophyll, is used to transfer an electron from a high-potential electron donor to a low-potential electron acceptor (14). The first photochemical system, photosystem II, operates in the potential range of +1000 mV to −600 mV (15–17). Photooxidation of the reaction center chlorophyll (P680) results in the generation of an oxidant that is utilized for water oxidation, and the reduction, through a pheophytin intermediary electron carrier (18), of a plas-

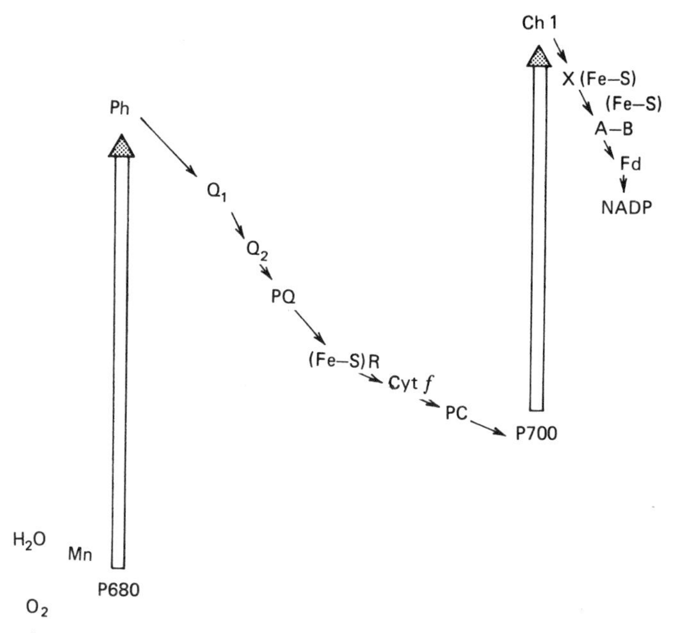

Figure 1. The noncyclic photosynthetic electron-transport chain in oxygenic photosynthetic organisms. P680, reaction center chlorophyll of photosystem II; P700, reaction center chlorophyll of photosystem I; Ph, pheophytin intermediary electron acceptor; Chl, chlorophyll intermediary electron acceptor; Q_1, Q_2, bound quinone electron acceptors; PQ, plastoquinone; Fe-S, iron-sulfur centers; Cyt *f*, cytochrome *f*; PC, plastocyanin.

toquinone electron acceptor (19). Two protein-associated quinone molecules are thought to function in direct electron transport, with a larger pool of quinone molecules required for electron transport–coupled ATP synthesis. A cytochrome b_{559} is also associated with photosystem II (20). No Fe-S centers have been identified in photosystem II. However, Fe may be associated with the quinone electron acceptors (18). The oxidation of the quinone electron acceptors is analogous to the oxidation of ubiquinone in mitochondria by the cytochrome *bc* complex II. An analogous complex, the *bf* complex, has been partially purified from chloroplasts, and contains cytochrome b_{563}, cytochrome *f*, a cytochrome b_{559} component, and the Rieske Fe-S center (21, 22). The quinone electron acceptors function as a two electron gate transferring electrons to the *bf* complex only after reduction to the PQH_2 level (23). The electrons are then transferred from the *bf* complex through the copper protein plastocyanin to the photosystem I reaction center. This reaction center functions in the potential range +400 mV to about −900 mV, transferring an electron from plastocyanin through the reaction center chlorophyll (P700) and an intermediary monomeric chlorophyll (24, 25) electron carrier to a chain of bound Fe-S centers, whose function is analogous to the quinone electron acceptors in the higher-potential photosystem II (26, 27). The electron is then transferred to NADP in a reaction involving a soluble Fe-S protein, ferredoxin, and a flavoprotein Fd NADP oxidoreductase (28). Apart from the direct flow of electrons from water to NADP, electrons may also flow in a cycle around either photosystem I or II (29).

Figure 2 shows the electron-transport system in the purple photosynthetic bacteria. This system is entirely analogous to the mitochondrial electron-transport chain between ubiquinone and cytochrome *c*, with the photosynthetic reaction center acting as the oxidant, accepting electrons from cytochrome *c*. The purple bacterial reaction center is the most completely characterized

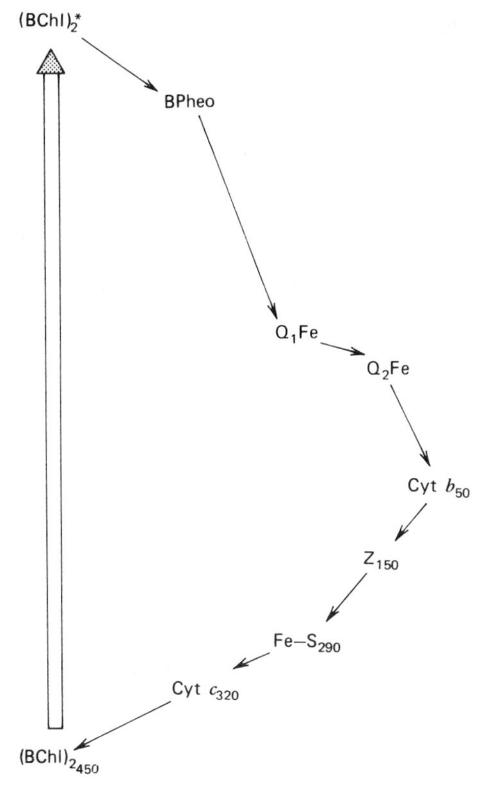

Figure 2. The electron-transport chain in purple photosynthetic bacteria. $(BChl)_2$, reaction center chlorophyll; BPheo, bacteriopheophytin intermediary electron carrier; Q_1Fe, Q_2Fe, bound quinone electron acceptors; Cyt, cytochrome; Fe-S, Rieske iron-sulfur center; Z, bound quinone electron carrier.

reaction center. It can be obtained in highly purified form suitable for optical spectroscopy, and has been used for kinetic measurements in the picosecond time range which have led to a good, though still incomplete, understanding of the mechanism of energy trapping. Absorption of a quantum of light by the special pair of reaction center bacteriochlorophyll (P890) results in the formation of an excited state and transfer of an electron from P890 through another bacteriochlorophyll (B800) to bacteriopheophytin (I) and then to a protein-bound ubiquinone-Fe (Q_1) complex that stabilizes the charge separation. The initial stages, P890 to I, take less than 10 psec and the transfer to Q_1 150 psec (30, 31, 14). The electron is then transferred from Q_1 to a secondary quinone acceptor Q_2 and through the cytochrome chain by a cyclic pathway back to the reaction center (32). Only one Fe-S center, the Rieske center, has been identified in this electron-transfer pathway (10). The photoreduction of NAD by succinate in these organisms is mediated by an energy-linked reverse electron transport (33), not by transfer of electrons through the reaction center. The mechanism by which low-potential reductants are generated from thiosulfate is unknown, as is the role in photosynthetic electron transport of the well-characterized low-potential ferredoxins and high-potential Fe proteins in these organisms. The photosynthetic pathways in the green photosynthetic bacteria are poorly defined. There is strong evidence that they have a noncyclic electron-transport system for NADP reduction (34), coupled to oxidation of sulfide and probably thiosulfate (35) quite analogous to photosystem I of higher plants, with a potential range of +220 mV to a potential below −550 mV (13), involving bound Fe-S centers, soluble ferredoxin, and a flavoprotein reductase. However, there is also evidence from other workers for the presence of a reaction center similar to that found in the purple bacteria (36, 37). Although the differences may have arisen because of the difficulties involved in working with these bacteria, it is possible that they have two reaction center types, operating either as in higher plants or independently, one for NAD reduction and one for ATP synthesis.

3 THE BOUND Fe-S CENTERS OF PHOTOSYSTEM I

This group of Fe-S centers is the most intensively studied in the photosynthetic electron-transport systems. The existence of bound Fe-S centers functioning in the primary photochemical reactions of photosystem I was discovered by Malkin and Bearden (5). It had been known from the work of Commoner and his co-workers (38) that illumination of chloroplasts, even at very low temperature (4.2K), resulted in the photooxidation of P700 as detected by EPR spectrometry. In a classical experiment Malkin and Bearden (5) showed that this low-temperature photooxidation was coupled to the photoreduction of an Fe-S center, center A (Figure 3). This reaction is essentially irreversible and temperature independent at temperatures below 77K. There is a stoichiometric transfer of electrons from P700 to center A (39, 40). P700 is measured in these experiments as EPR signal 1, a 7.5 G wide free radical attributed to the oxidation of a chlorophyll dimer (41, 42). At room temperature this signal has been identified as arising from the same component as the optical signal of P700 attributed to the photosystem I reaction center chlorophyll. A second Fe-S center, center B, is also associated with the electron-acceptor complex of photosystem I. It was first identified in preparations that had been reduced by illumination during freezing in the presence

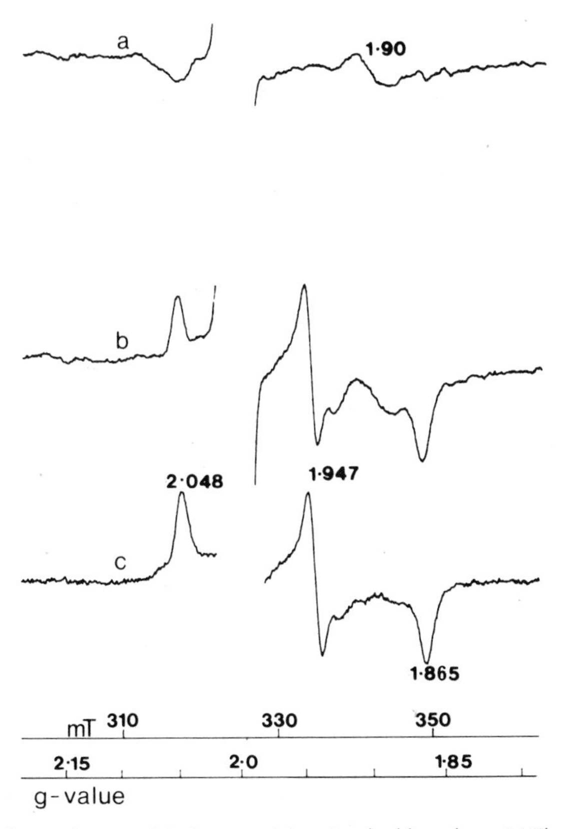

Figure 3. The photoreduction of Fe-S center A in spinach chloroplasts. *(a)* The EPR spectrum at 15K of spinach chloroplasts reduced with sodium ascorbate and frozen in the dark. *(b)* The EPR spectrum of *(a)* after illumination for 30 sec at 15K. *(c)* Spectrum *(a)* – *(b)*, showing the light-induced signal of center A. The signal at g = 1.90 in spectrum *(a)* arises from the Rieske center. The large radical at g = 2.00 of signal II and P700 is not shown.

of sodium dithionite (6). It has subsequently been shown to undergo photoreduction at cryogenic temperatures. This photoreduction is also irreversible at low temperature (Fig. 4). The two centers A and B are present in photosystem I preparations in equivalent amounts (43). When these centers were discovered, they were identified as "primary" electron acceptors of photosystem I. However, in experiments in which irreversible electron transfer from P700 to these centers is observed, a small proportion of P700 photooxidation is usually reversible on a time scale of 1–200 msec. McIntosh and co-workers obtained kinetic evidence for a component with an EPR signal at $g = 1.76$ (44). Subsequently, it was shown that this is part of the EPR spectrum of a component, X, with a broad spectrum around $g = 1.89$ (Fig. 5), which could be observed in samples prepared under extreme reducing conditions with freezing under illumination, or could be reversibly photoreduced at cryogenic temperatures in samples in which centers A and B were reduced (7, 45). There is a stoichiometric transfer of electrons from P700 to this component in this photochemical reaction (46). Despite its atypical EPR spectrum, which has led to speculation that it arises from a quinone Fe complex similar to that found in bacterial reaction centers, it now seems clear that this is also an Fe-S center (46). These 3 Fe-S centers form the electron carriers of the pri-

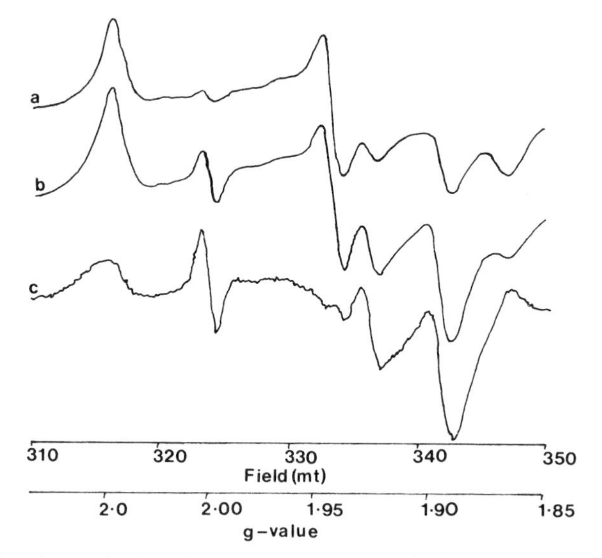

Figure 4. The photoreduction of Fe-S center B in spinach photosystem I particles. *(a)* EPR spectrum at 20K of the particles partially reduced with sodium dithionite, showing center A and some reduction of center B. *(b)* EPR spectrum of *(a)* folowing illumination for 30 sec at 20K. *(c)* Spectrum *(b)* – *(a)* showing the reduction of center B. The spectrum is distorted by the interaction with center A.

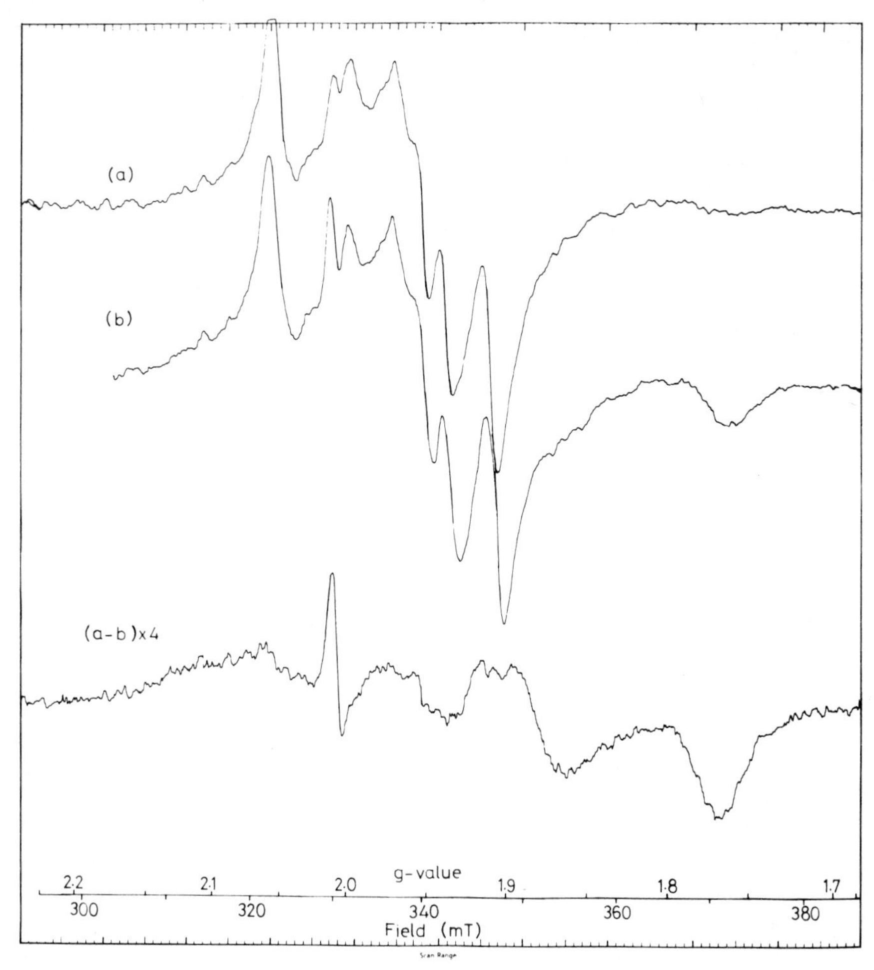

Figure 5. The EPR spectra of the electron-accepting center X in spinach photosystem I particles. Photosystem I particles were reduced with sodium dithionite at pH 10.0 and frozen in the dark. *(a)* EPR spectrum at 10K in the dark. *(b)* EPR spectrum at 10K during illumination; *(c)* Spectrum *(b)* – *(a)*, the light minus dark spectrum, showing the spectrum due to X (g = 2.08, 1.88, and 1.78) and the electron donor P700 (g = 2.00). The P700 spectrum is distorted by overmodulation and power saturation effects.

mary electron acceptor complex of photosystem I, providing stabilization of the initial charge separation. Transfer of electrons from P700 to this complex involves at least one other component, thought to be a chlorophyll molecule. The kinetics of the forward reaction cannot yet be measured satisfactorily, but seem likely to be in the picosecond time range (25).

3.1 Chemical Identity of the Components A, B, and X

Centers A and B were initially identified as Fe-S centers on the basis of their EPR spectra, but because of the relatively simple spectrum of center A were thought to be [2Fe-2S] centers. However, EPR spectra of photosystem I particles treated with dimethylsulfoxide, a treatment that appears to neutralize the effect of the protein environment on spectral shape, clearly indicated that all the Fe-S centers present in photosystem I were [4Fe-4S] centers (48). This conclusion was supported by Fe and labile sulfide analysis, which indicated the presence of 10 or 11 Fe atoms and acid labile S atoms per P700 in highly purified photosystem I particles (49), and by EPR studies of an Fe^{57}-enriched preparation (50).

The identity of centers A and B has now been fully confirmed by Mössbauer studies of Fe^{57}-enriched photosystem I particles from the blue-green alga *Chlorogloea fritschii* (Figs. 6 and 7) (47, 50). Figure 6 shows the spectra obtained from oxidized and reduced samples of photosystem I particles at different temperatures in zero applied magnetic field. In the oxidized centers spin coupling results in a nonmagnetic ground state. The only method of distinguishing [2Fe-2S] and [4Fe-4S] centers in the oxidized state is the chemical shift. In [2Fe-2S] centers the Fe is formally Fe^{3+} and the chemical shift ($\delta \sim$ 0.26 mm/sec at 77K and below) is typical of Fe^{3+} in a tetrahedral S environment. In the oxidized [4Fe-4S] centers the Fe has formal valence of $2Fe^{3+}$ and $2Fe^{2+}$ with $\delta \simeq 0.42$ mm/sec at 77K and below. The oxidized photosystem I preparations have a chemical shift very similar to that of the [4Fe-4S] center, as does the uncharged nonmagnetic component in the reduced samples. The reduced samples had A and B reduced; X was oxidized as monitored by parallel EPR measurements. The 77K spectra are consistent with the proposal that the observed changes result from the reduction of ferredoxinlike centers. The 195K spectra clearly show that the changes are not compatible with these centers being [2Fe-2S] centers. The [2Fe-2S] ferredoxins show well-separated quadrupole doublets at this temperature, arising from distinct Fe^{2+} and Fe^{3+} atoms within the center. The 4.2K spectra are more complex, the reduced spectrum showing a superposition of the quadrupole doublet seen in the oxidized spectrum, and a broad region of absorption over the range -2.0 to $+4.0$ mm/sec. This spectrum can be resolved by the application of an external magnetic field which decouples the electronic and

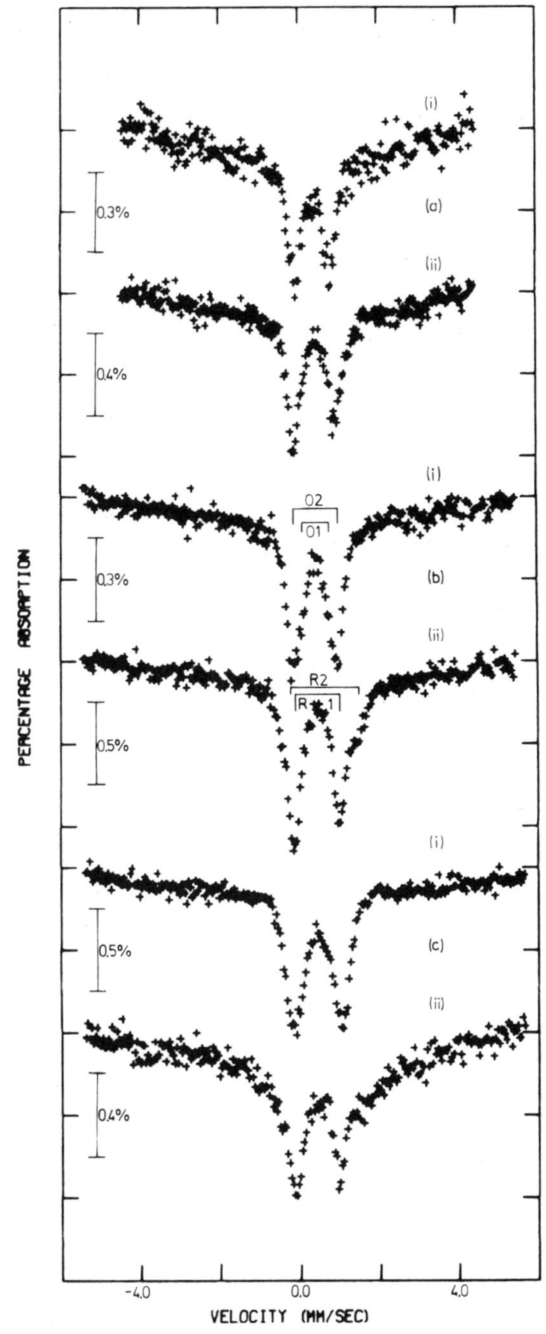

Figure 6. Mossbauer spectra of Fe^{57}-enriched photosystem I particles from *Chlorogloea fritschii.* (i) Oxidized samples; (ii) samples with centers A and B reduced. The spectra were taken in zero applied magnetic field at 195K *(a)*, 77K *(b)*, and 4.2K *(c)*.

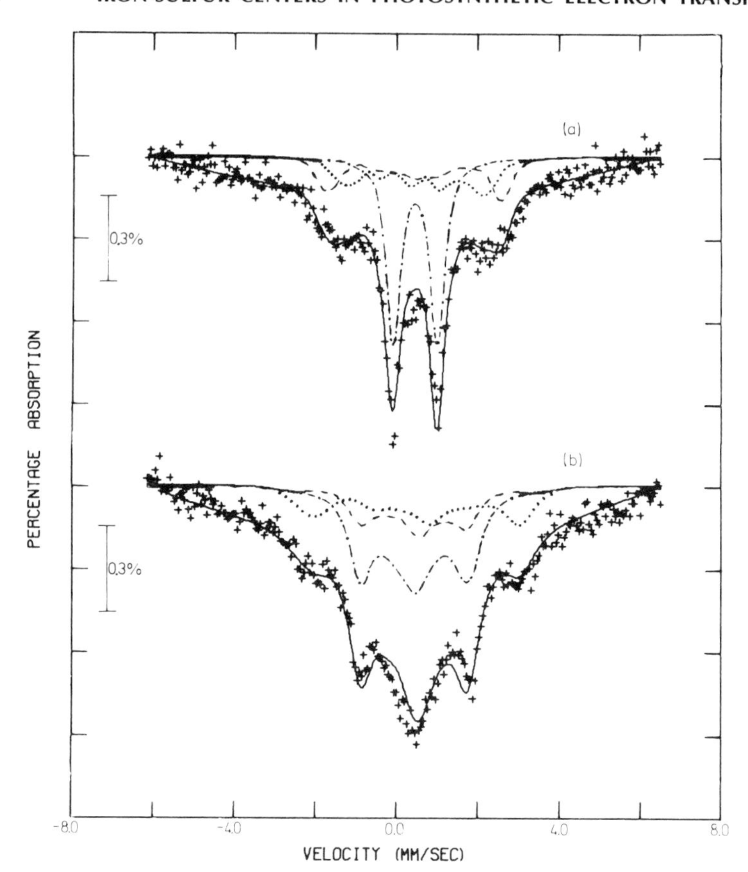

Figure 7. Mossbauer spectra of Fe^{57}-enriched photosystem I particles from *Chlorogloea fritschii* with centers A and B reduced at 4.2K in magnetic fields of 0.5 tesla *(a)* and 6.0 teslas *(b)* applied parallel to the γ-ray direction. Computer simulations have been fitted to the spectra. -·-·-) Nonmagnetic Fe-S component (oxidized); . . . . .), -----) subspectra from the reduced Fe-S center components based on the parameters of *Bacillus stearothermophilus*; ———) total simulation based on an oxidized to reduced ratio of 1 : 1 with the addition of a parabolic background.

nuclear spins, sharpening the magnetic hyperfine spectrum. The oxidized spectrum is essentially unchanged by low magnetic field; comparison of Figure 6 (cl) and Figure 7*a* most clearly shows the difference between oxidized and reduced spectra. Computer simulations of the reduced spectra were carried out assuming the spectra consisted of contributions from unreduced centers with the properties of the oxidized spectrum and from reduced centers with the properties of reduced [4Fe-4S] centers similar to those in *Bacillus steorothermophilus* ferredoxin. For both the low and high magnetic field spectra the major features of the spectrum are well reproduced by this procedure, providing strong evidence that the changes observed on reduction of centers A and B result from the reduction of [4Fe-4S] centers. These conclusions have recently been confirmed by work with *Plectonema boreanum* preparations by Isaakidou and colleagues (51).

In subsequent work (47) Evans and co-workers have extended their experiments to samples with (1) only center A reduced, (2) with A and B reduced, or (3) with A, B, and X reduced. They have found that in (1) $\approx 24\%$ of the Fe is affected, in (2) $\approx 48\%$ is affected, and in (3) $\approx 61\%$ is affected. These results show that all the Fe in the sample can be accounted for as [4Fe-4S] centers, excluding the possibility that X is a different type of Fe compound. It could not, for example, be an Fe-quinone of the type observed in the bacterial reaction center, where the Mössbauer data clearly identify the Fe as Fe^{2+} (52). The increased reduction of the [4Fe-4S] centers in parallel with the reduction of X is very strong evidence that X is indeed a center of this type. The quantitative relationship observed in the extent of reduction of the Fe as each center is reduced suggests the presence of equivalent amounts of each center. However, even in the most reduced preparations 35% of the Fe centers remain in the oxidized form. These may be centers that are in some way dissociated from the photosynthetic preparation by the detergent treatments, or they may indicate that X has two centers, one of which is not reduced by the procedure used in these experiments.

These experiments suggest that there are 3 or 4 [4Fe-4S] centers associated with each reaction center. Quantitative EPR estimations indicate equivalent amounts of A, B, and X, with 1 A per reaction center. The values obtained for X were not, however, as accurate as those for A (46). The Fe determinations of Golbeck and co-workers (49) of 10–11 Fe atoms per P700 would suggest that there are not more than 3 centers. However, in all these experiments it is possible that damaged reaction centers with functional P700 but no Fe-S centers, or Fe-S centers associated with inactivated P700, may produce errors. So it is not possible to decide conclusively how many centers are present, although it is clear that at least three are involved in the operation of photosystem I.

3.2 Optical Detection of Photosystem I Fe-S Centers

At about the same time that the bound Fe-S proteins were first detected, an optically detected absorption change in the 430 nm region was detected by Hiyama and Ke (53). The component giving rise to this signal (designated P430) was suggested to be the primary electron acceptor of photosystem I. The flash-induced signal showed parallel decay kinetics to $P700^{+}$. The spectrum has some resemblance to that of soluble ferredoxin. Ke and Beinert (54) suggested that P430 and center A were the same component, but only because neither was photoinduced when P700 was oxidized. More recently, Hiyama and Fork (55) have proposed that P430 and X are the same on the basis of the decay kinetics at room temperature in blue-green algal particles. However, both Goldbeck and colleagues (56) and Shuvalov and co-workers (57) obtained spectra of a component observed under conditions similar to those in which X is observed, that is, in samples with A and B chemically reduced; these spectra were different from the spectrum of P430. The absorption change of P430 is very small, and is observed only after signal averaging of repetitive flash experiments. No low-temperature spectra of it have been obtained. While it seems probable that kinetic changes of P430 do represent changes in the redox state of the Fe-S centers of the electron-acceptor complex of photosystem I, it is by no means certain that this is correct. It is certainly unclear that P430 can be ascribed to one specific center.

3.3 Properties of the Low-Potential Fe-S Centers of Photosystem I

In terms of their proposed function as electron acceptors in photosystem I, one of the most important properties of these centers is their oxidation reduction potential. If they are to function as carriers in an electron-transport chain that reduces NADP, they must have very low potentials. This has been shown to be the case for centers A and B by redox potential titrations, which show that in spinach photosystem I preparations made either with detergents or mechanical disruption of the chloroplasts, center A has $Em_{10} \approx -540$ mV and center B has $E_{m10} \approx -590$ mV (58, 59). The two groups that made these measurements have obtained results differing by only 20 mV, which is within the error of the technique. Figure 8 shows the EPR spectra of spinach photosystem I particles obtained from such a redox titration. They show that while initially center A is reduced, giving the spectrum seen following low-temperature photoreduction, the spectrum slowly changes as center B is reduced and the $g = 1.86$ component of the center A spectrum is lost, apparently shifting to $g = 1.89$ to give the distorted spectrum seen in fully reduced samples. The $g = 2.05$ component of the spectra also clearly reflects reduction of both center A and B. This effect has been interpreted as being the result of magnetic interaction between the two centers, suggesting that the centers might be analogous to the two centers of bacterial ferredoxin such as that of *Clostridium pasteurianum.* However, Mössbauer spectroscopy does not support this interpretation, the centers behaving like those of ferredoxins with only one center. Most work on these centers has been done with spinach or the blue-green alga *Chlorogloea fritschii* (60), with some comparative work on other species (61), all of which showed similar behavior, with essentially only center A reduced by low-temperature illumination and

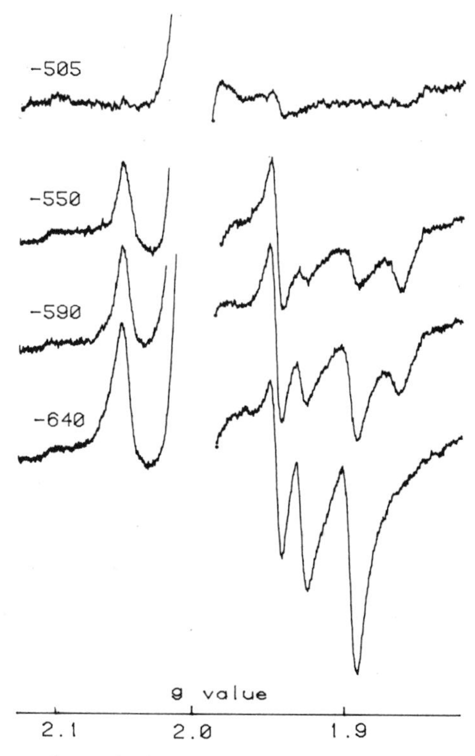

Figure 8. EPR spectra of spinach photosystem I particles poised at different oxidation reduction potentials, showing the sequential reduction of centers A and B and the loss of the g = 1.86 signal of center A as B is reduced.

similar redox properties for A and B. However, recently Cammack and coworkers (62) reported that in preparations of *Phormidium laminosum* low-temperature illumination results in reduction of a mixture of centers A and B. Similar results have also been reported in *Dunalliela parva* (1), barley (63), and spinach preparations treated with glycerol (64). In *P. laminosum* the relative potentials of A and B were apparently affected, and center B was more readily reduced by dithionite; the relative potentials were also altered by glycerol treatment of spinach preparations. These experiments made it possible to obtain a spectrum of center B unaffected by center A, either by chemical reduction or by the more rapid low-temperature decay of photoreduced center A in barley. Figure 9 shows this spectrum with $g_x = 1.886$, $g_y = 1.935$, and $g_z = 2.065$. It is clear that when photosystem I, prepared in the dark under conditions where P700 is reduced and A and B are oxidized, is illuminated at low temperature, an electron is transferred to either a center A

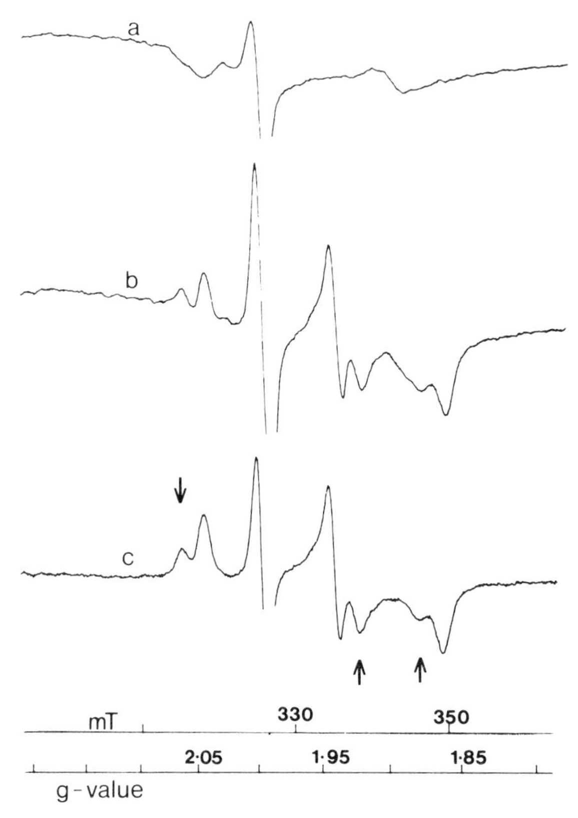

Figure 9. Low-temperature photoreduction of centers A and B in barley chloroplasts. *(a)* EPR spectrum at 15K of barley chloroplasts reduced with sodium ascorbate and frozen in the dark. *(b)* Spectrum *(a)* after illumination for 30 sec at 15K. *(c)* Spectra *(b)* – *(a)* showing the EPR spectrum of centers A and B (arrowed). The signal at $g = 2.00$ is caused by photooxidized P700.

or center B in different reaction centers, as there is no evidence of magnetic interaction between centers in such samples. It is not clear what controls which center is reduced. Although the redox potentials may have some effect, the extent of reduction of B is less in spinach than might be expected, and the reduction of A more than would be expected in *P. laminosum*. Other factors must therefore be involved; if, for example, the two centers are on different polypeptides, physical organization of the reaction center may be an important factor. At room temperature it seems likely that the electron would move between the centers with an average distribution reflecting the redox potentials.

The redox potential of X has not been directly determined. It may be inferred that its potential is lower than that of A or B, as it remains oxidized under conditions where these centers are reduced. In redox titrations of spinach particles using sodium dithionite as reducing agent, little or no reduction of X is observed. The redox states of A, B, and X can be monitored indirectly by observing the response of P700 to illumination using either EPR or optical detection. Using EPR detection and parallel measurements of the redox states of A and B, we found that the normally irreversible low-temperature photooxidation of P700 became reversible as center B was reduced. Demeter and Ke (65), using electrochemical reduction and optical detection, found the reversible photooxidation of P700 titrated in ~530 mV and then decreased again ~730 mV. They proposed that P700 photooxidation became reversible as center A was reduced; however, the EPR data show that it actually occurs as center B is reduced, the more oxidized potentials observed by Demeter and Ke probably resulting from the effects of glycerol on the potential of A and B (64). They suggested that the loss of P700 photooxidation below −700 mV was the result of reduction of X. It seems likely that this interpretation is correct, but confirmation by a parallel EPR experiment is required. The low-temperature photoreduction of center A in some reaction centers and center B in others raises the possibility that they are in separate reaction centers. However, it is very unlikely that this is the case, as in spinach preparations very little reduction of center B is seen when both centers are oxidized initially. If center A is partially reduced before freezing, low-temperature illumination results in extensive reduction of center B (Fig. 4) (43). Because of the overlap of the potentials and possible variations in the magnetic interactions between the centers, it is difficult to obtain complete reduction of both centers in these experiments. This, together with the fact that the saturation characteristics of P700 are altered by reducing agents and detergents, led to considerable controversy about the quantitative importances of the photoreduction of B and X (1). However, quantitative EPR determinations of the relative amounts of these centers and of electron transfer from P700 to X (40, 39), together with the Mössbauer data (47, 50), clearly demonstrate the importance of these centers. In preparations in which center B is most easily reduced chemically, illumination of samples with B already partially reduced results in greater reduction of center A (Fig. 10) (63).

In recent years it has become apparent that many membrane-bound components show specific orientation relative to the membrane surface. In chloroplasts oriented either by exposure to high magnetic field or by drying in thin films, it has been found that the EPR spectra of centers A and B and the Rieske center show specific orientation of the g tensors (Fig. 11) (66, 67). Dismukes and Sauer (66), using magnetic orientation, found orientation of B and X but not of A. Prince and co-workers (67), using thin films, found orientation of A, B, and the Rieske center but did not investigate X. The dif-

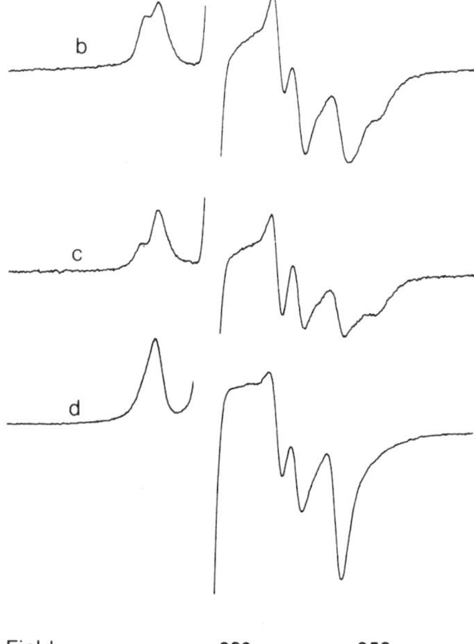

Figure 10. *(a)* EPR spectrum at 15K of barley chloroplasts partially reduced with sodium dithionite, showing center B and the Rieske center, *(b)* Spectrum *(a)* after illumination for 30 sec at 15K. *(c)* Spectrum *(b)* – *(a)*, showing the photoreduction of A and some B and the distortion of the spectrum by the interaction between the centers when both A and B are reduced. *(d)* A similar sample illuminated at room temperature and frozen under illumination, showing spectra when A and B are both fully reduced.

ferent results for center A may arise because Dismukes and Sauer only presented data for the EPR magnetic field parallel or perpendicular to the plane of the membrane. The more complete data of Prince and his group shows that no intensity differences would be seen at these angles. Table 1 shows the orientations relative to the plane of the membrane for A, B, and the Rieske center. Dismukes and Sauer found that X is oriented in the membrane with $g_y = 1.90$ and $g_z = 2.09$, essentially in the plane of the membrane, and $g_x = 1.78$ perpendicular to the plane of the membrane. They concluded that these orientations were consistent with X^- being the paramagnetic partner of $P700^+$ in the early stages of electron transfer, leading to the observation of spin-polarized EPR signals.

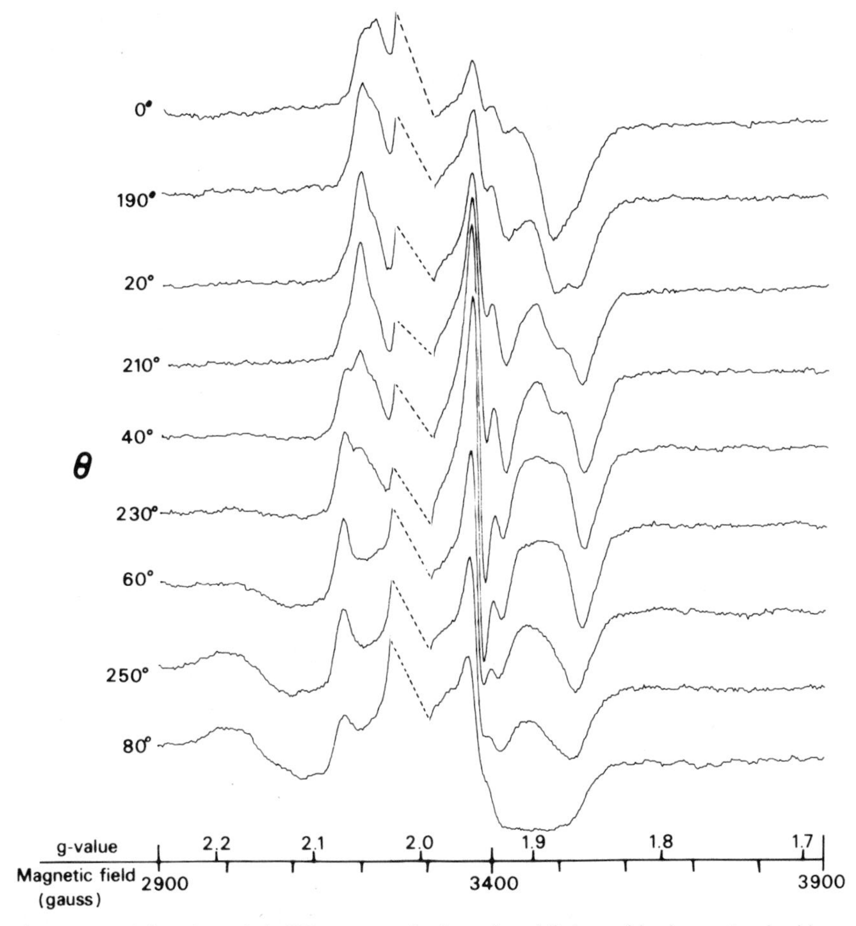

Figure 11. Light minus dark EPR spectra of oriented multilayers of broken spinach chloroplasts. The spectra were recorded at 10° intervals in the dark, and chloroplasts were then removed from the EPR cavity and warmed to room temperature, then illuminated with white light during freezing. Another series of spectra were recorded at the same angles. Light minus dark spectra were obtained by subtraction of spectra at each angle shown. (Reproduced from reference 67 with permission.)

These results suggest that the Fe-S centers, like other components of the reaction center, have an essentially rigid organization, which is presumably important for the efficient operation of the photochemical process. However, the data are less helpful for model building than those for porphyrin compounds, since the Fe-S centers of the photosystem I electron-acceptor complex are [4Fe-4S] centers, which are essentially cubic. (Figure 9 in ref. 67 is misleading in implying that they are flat boxes that can be oriented to the

Table 1 Orientation Relative to the membrane of EPR Signals of Fe-S Centers in Oriented Spinach Chloroplasts[a]

g Value	Orientation Relative to Membrane Plane	Orientation Relative to Membrane Normal
Rieske Center		
2.03	0	90
1.90	90	0
Center A		
2.05	25	65
1.94	40	50
1.86	40	50
Center B		
2.07	50	40
1.92	90	0
1.89	0	90

[a]Adapted from reference 67.

membrane surface.) There is also no satisfactory theoretical model allowing the attribution of magnetic axes to structural axes for this symmetrical complex. The Rieske center is probably a [2Fe-2S] center; with this the data are more helpful, as they suggest that the [Fe–Fe] axis is parallel to the membrane surface (67).

4 THE SOLUBLE [2Fe-2S] FERREDOXIN

The soluble chloroplast ferredoxin is the classical [2Fe-2S] ferredoxin. It was first isolated in 1952 as the met-hemoglobin reducing factor (68). It has been purified and characterized from many sources, with amino acid sequences (69) and, recently, X-ray structure readily available (70). It has characteristic EPR and optical spectra; however, it does not seem to be possible to detect it as an electron-transport component in vivo by flash absorption spectroscopy (71). Its function appears to be biochemically well defined as an electron carrier between the chloroplast membrane-bound acceptors of photosystem I and the flavoprotein-ferredoxin NADP reductase and also in cyclic electron transport (28, 29). It forms a complex with the flavoprotein in solution with altered spectroscopic characteristics (72). In this complex, reduction with sodium dithionite no longer induced the normal reduced spectra. Its specific binding to a membrane site involving the reductases may explain the problems of detection in vivo, or it may possibly be seen as P430. Shuvalov (73) obtained a P430 spectrum with peaks at 420, 445, and 717, which are characteristic of the soluble [2Fe-2S] protein, rather than [4Fe-4S] centers, suggesting that P430 may also include this protein as well as centers A, B, or X.

5 THE FUNCTION OF THE LOW-POTENTIAL Fe-S CENTERS AND THE STRUCTURE OF THE PHOTOSYSTEM I REACTION CENTER

The results described in this chapter clearly show that the photosynthetic membrane system of oxygenic photosynthetic organisms contains 3 very-low-potential [4Fe-4S] Fe-S centers, and that these centers undergo photochemical reactions at low temperatures. They were in fact discovered as a result of the search for the primary electron acceptor of photosystem I. It seemed reasonable to presume that the first electron acceptor was a component that was photoreduced as the reaction center chlorophyll was photooxidized at temperatures as low as 4.2K by a fast electron tunneling mechanism (faster than the time resolution of any technique so far applied to it). The discovery of center A under such conditions led to the proposal that it was the primary electron acceptor. However, as described above, the situation is much more complex with multiple Fe-S centers. It has also been found, first in bacteria and more recently in photosystems I and II, that these centers or the analogous quinone complexes are secondary acceptors that serve to stabilize the initial photochemical charge separation into the time domain of the subsequent electron-transport reactions, that is, to the microseconds to milliseconds time range. To avoid conflict with the earlier literature it has become widespread practice to refer to those acceptors as the *stable primary electron-acceptor complex* and to the chlorophyll or pheophytin acceptors as *transient intermediates.* It should be stressed that *stable* in this context is a relative term, meaning stable for perhaps milliseconds at room temperature, and milliseconds to hours at low temperature, as opposed to the picosecond lifetimes of the chlorophyll intermediate.

The observation that electron-transfer components function at low temperature is clearly strong evidence that they are reaction center components. However, it is essential to show that the components are associated with the correct reaction center and that the reactions observed at low temperature are the same as those that occur at room temperature.

There are a number of clear lines of evidence that associate centers A, B, and X with the photosystem I reaction center. Isolation of photosystem I enriched fractions, as determined by P700 to total chlorophyll, either by relatively simple mechanical procedures (6) or detergent fractionation with Triton X-100 (40), Digitonin, or Lauryl diethylamine oxide (73), results in enrichment of the Fe-S centers. Treatment with sodium dodecylsulfate (SDS) results in purification of a P700 protein complex. However, such preparations, or preparations made with other detergents and treated with SDS, lack the EPR signal of the Fe-S centers and the photochemical activity associated with them (74–76). They do, however, retain the P700 to the chlorophyll-acceptor photochemical reaction. Treatment of photosystem I preparations with ferricyanide and urea also inactivates the Fe-S centers with parallel loss of photosynthetic activity.

The operation of photosystems I and II can be separated by the ability of photosystem I to use light of wavelengths of 700–730 nm. The low-temperature photoreduction of center A is observed following illumination of preparations with 730 nm light, confirming its role as a photosystem I acceptor (1). The very low potentials of the centers indicate a role in photosystem I, as the equivalent acceptors in photosystem II can be shown by indirect means to have potentials more oxidized than -400 mV (16). Finally, the quantitative relationship between P700 oxidation and reduction of A or X is very strong evidence for their role in photosystem I.

There is some disagreement about the redox potential of the reaction center chlorophyll of photosystem I, which might cast a slight doubt on the interpretation of the results. The redox potential of P700 has been measured by a number of workers to be $\approx +480$ to $\approx +520$ mV (77, 78). On the other hand we found a potential of $+375$ mV for the midpoint potential of the low-temperature electron donor to center A (79). We obtained variable results using different optical techniques at room temperature: a value of $\approx +500$ mV when we used conventional dual wavelength measurements, but $\sim +360$ mV if we actually scanned the P700 spectrum. We concluded that the more oxidized value was erroneous because of artifacts arising from the oxidation of bulk chlorophyll. However, Setif and Mathis (80) have recently obtained a value of $+500$ mV using measurements of flash-induced P700 photooxidation, but did obtain a value of about $+400$ mV for samples measured after freezing and in SDS-treated material. It seems likely that the different values obtained are the result of technical differences, perhaps leading to modification of the P700 environment and redox potential, but the possibility that the low-temperature experiment is not measuring P700 cannot be entirely excluded, or equally, that many room temperature experiments are not measuring photosystem I as defined by the low-temperature work.

Although much of the work on Fe-S centers has involved their low-temperature photoreactions, they can be shown to undergo photoreduction at room temperature not only in the presence of strong reducing agents (6), but also in unfractionated chloroplasts with water as the electron donor (66, 81).

In the absence of any satisfactory room-temperature procedure for distinguishing the three centers and for providing kinetic measurements of electron transfer through them, they can only be placed in the electron-transport chain by a deductive procedure based on their properties. This has resulted in the proposal of three possible configurations (Fig. 12). Two of these would be functionally indistinguishable (Fig. 12*A*,*B*).

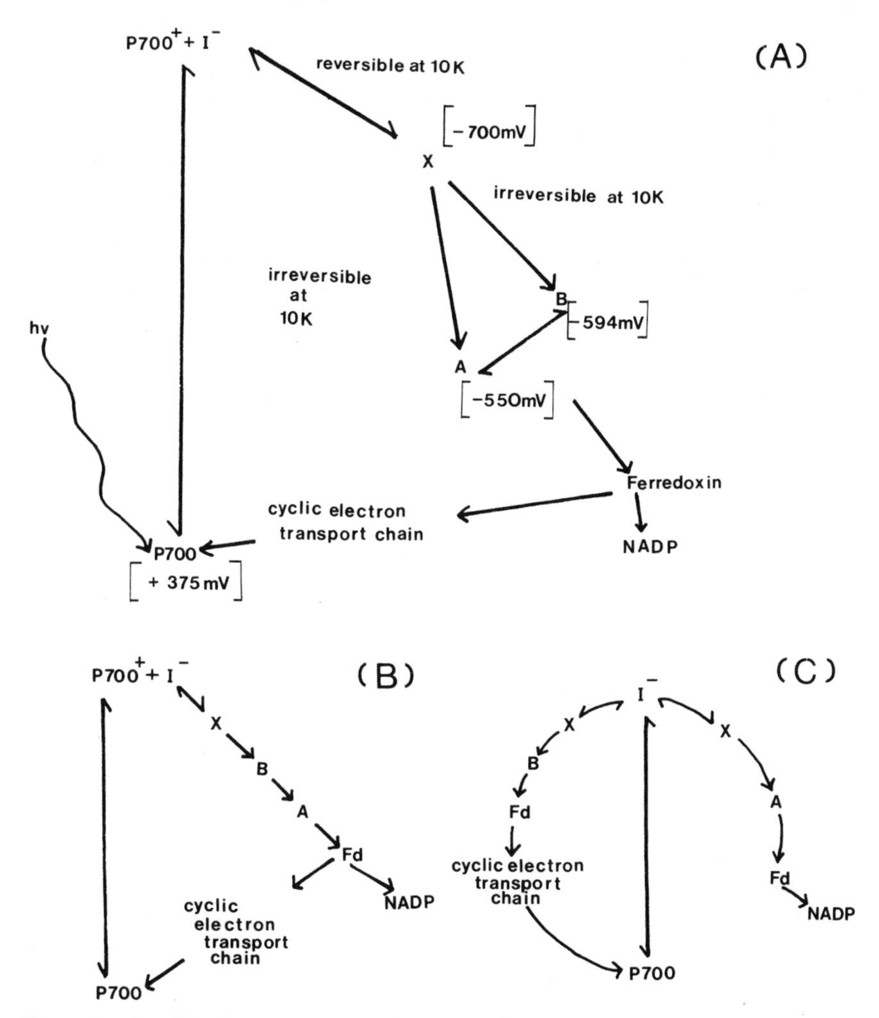

Figure 12. Possible electron-transport schemes in photosystem I. *(A)* Parallel operation of centers A and B; *(B)* Linear electron flow through the bound Fe-S centers. *(C)* Separate electron-acceptor complexes for cyclic and noncyclic electron transport.

Two observations suggest that P700, the chlorophyll intermediate, and X are very close together, and that X is the first in the chain of Fe-S centers to accept electrons. The initial charge separation between P700 and the intermediary chlorophyll electron carrier, and subsequently between P700 and X, gives rise to chemically induced dynamic electron polarization effects (82). Also, the reduction of the intermediary chlorophyll electron carrier produces changes in the spectrum of X which can be interpreted as arising from magnetic interactions between the two reduced components (24). Both effects require close proximity of the centers. The redox properties of the system in which X has a lower potential than A or B also suggest that X is the first acceptor. In reaction centers where B has a much lower potential than A and is only photoreduced at low temperature after A is reduced, a linear sequence such as that shown in Figure 12*b* would seem to be logical. However, this does not fit so well with those reaction centers, such as in barley, where either A or B may be reduced, or with the behavior of A and B, which shows that they interact rather in the way that the two centers of Clostridial ferredoxin behave. A model in which A and B act as parallel acceptors from X may therefore be more appropriate (Fig. 12*a*). Another possibility that has been suggested is that A and B function in separate electron-transport chains, for example, A in noncyclic and B in cyclic ones (26) (Fig. 12*c*). The close physical relationship of the two centers and their very close redox potentials suggest that electrons would be readily transferred between them, making it unlikely that they would be a branching point, particularly as the soluble ferredoxin appears to be involved in both chains. An extreme version of this model, in which two X centers are present (as suggested by the Mössbauer data), one donating to A and one to B, seems to be excluded by the fact that reversible electron transfer between P700 and X is only seen as both A and B are reduced. Arnon and co-workers (81) presented experiments on the effect of acceptors on A and B reduction, which they interpreted as showing that only B was involved in NADP reduction. This interpretation is based on a failure to understand that the $g = 1.86$ signal of A disappears as B is reduced because of the spin-spin interaction, not because A is oxidized. A model of the type shown in Figure 12*a* is perhaps most compatible with the EPR data. It also fits quite well with models of the photosystem I reaction center inferred from more indirect procedures. Sauer and colleagues (83), using kinetic optical measurements of the rereduction of P700 following flash-induced photo-oxidation, found that the rate of return of the electron from the acceptors to P700 depends on the redox state of the acceptors. When all the acceptors were oxidized, they found that two electrons could be transferred to acceptors with a relatively slow decay rate back to P700. These correspond to centers A and B. If these are reduced, a faster decay rate is seen, corresponding to electron transfer from X. If X was reduced, they found a very fast charge recombination from an acceptor that has now been identified as the chlorophyll intermediary acceptor.

5.1 The Structure of the Photosystem I Reaction Center

Structural requirements for the photosystem I reaction center have been largely based on SDS gel analysis of chloroplasts (84). These indicate that photosystem I includes a chlorophyll protein CP1 with a molecular weight of ≈ 100 kilodaltons, two peptides of 15–20 kilodaltons, and probably other small peptides. CP1 is thought to contain P700 and the immediate antenna chlorophyll. If chlorophyll is completely solubilized before electrophoresis, CP1 is replaced on the gel by a polypeptide of $\simeq 60$K daltons (85). The two small peptides that are enriched in photosystem I particles are suggested to be the bound ferredoxins, and other peptides are involved in plastocyanin binding and oxidation (86). These proposals are supported by work with mutants defective in photosystem I (85). Models based on this work, such as that of Bengis and Nelson (88), suggest that CP1 or an oligomeric array of CP1 forms a transmembrane bridge, with P700 near the inside of the membrane, accessible to electron donation from plastocyanin, and the Fe-S protein peptides on the outer edge to reduce the soluble ferredoxin. While such a transmembrane arrangement is clearly essential to allow the well-established transmembrane charge separation and electron-transport pathway, evidence from EPR measurements of magnetic interaction between components and the tunneling time for back reactions suggests that the reaction center is considerably more compact than such models allow. The observation of CIDEP effects following flash-induced charge separation (82), interpreted as arising from the radical pairs $P700^+$ and the reduced chlorophyll intermediary electron carrier and $P700^+ X^-$, requires that these components should be within 10–20 Å of each other, while the magnetic interaction between the chlorophyll intermediate and X^- observed in steady-state experiments (24) also requires that they should be within 10 Å. The short back reaction time from X^- to $P700^+$ also suggests a short tunneling distance. There is no apparent interaction between A or B and X, so these centers are further removed, a probability supported by the long back reaction time from A^- to $P700^+$. However, centers A and B interact strongly, as do the centers of 2 ([4Fe-4S]) ferredoxins, and it seems likely that these two centers are associated with a much more compact reaction center, perhaps only 40 Å across (Fig. 13). Such a model is not compatible with the polypeptide arrangements suggested previously. A sufficiently compact arrangement might be achieved if CP1 were in fact the complete reaction center, or by rearrangement of the more complex peptide grouping to bring the active sites into close contact. It does seem unlikely that the reaction center spans the membrane, as shown in con-

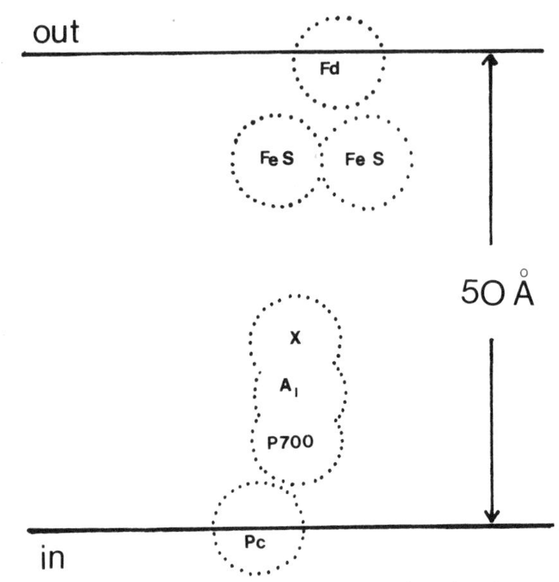

Figure 13. A model of the photosystem I reaction center based on interactions between the components observed by EPR. The dotted circles represent 10 Å spheres within which magnetic interactions would be expected to be observed. The thickness of the membrane is still a matter of some controversy, with values between 40 and 80 Å reported in the recent literature. There is no evidence for magnetic interaction between plastocyanin and P700 or soluble ferredoxin and the bound reaction center components, indicating that their active centers are at least 20–30 Å from the reaction center. This may suggest a greater membrane thickness than is shown here. Fd, soluble ferredoxin; Fe-S, bound iron-sulfur centers A and B; X, bound Fe-S center X; A_1 chlorophyll intermediary electron carrier; P700, the reaction center chlorophyll; Pc, plastocyanin.

ventional diagrams. Recent results also suggest that a more compact model must be drawn for the bacterial reaction center, in which all four electron-transfer components, B890, I, Q_1, and Q_2, are within 10 Å of a single Fe atom (90).

6 THE RIESKE Fe-S CENTER IN OXYGENIC PHOTOSYNTHETIC ORGANISMS

The only well-characterized Fe-S center in oxygenic photosynthetic organisms outside the photosystem I reaction center is the Rieske Fe-S center, which gives rise to an EPR signal at $q = 1.90$ and probably at $g = 2.03$ (2, 67, 88) (Fig. 14). This type of center was first identified as a component of the mito-

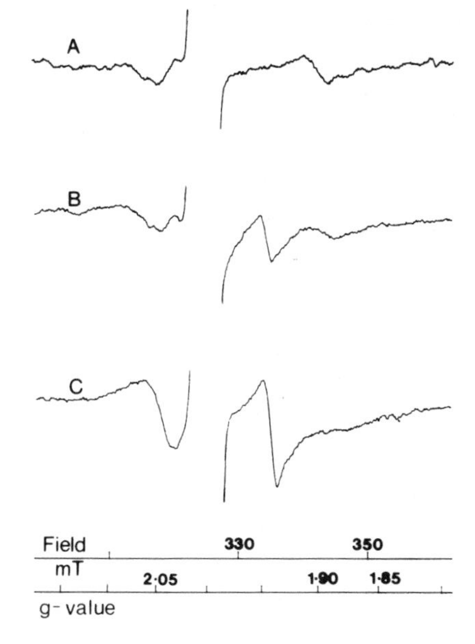

Figure 14. EPR spectra of the Rieske center in barley chloroplasts. *(a)* EPR spectrum at 16K of chloroplasts frozen in the dark. *(b)* Chloroplasts frozen in the dark after addition of 33 μM DBMIB. *(c)* Chloroplasts frozen under illumination after addition of 33 μM DBMIB. The spectra show the normal g = 1.90 and 2.03 signals of the Rieske center and the modified signal at g = 1.95 seen in the presence of DBMIB.

chondrial electron-transport chain functioning in the cytochrome *bc* complex (3, 4). It is thought to be a [2Fe-2S] center (89). A component with similar properties was subsequently identified in photosynthetic bacteria (8, 10, 11) and in chloroplasts (2). It is characterized by a very oxidized midpoint redox potential (Table 2), which is pH dependent at least at alkaline pH (11, 90).

The Rieske center appears to function in chloroplast electron transport as a carrier between plastoquinone and cytochrome *f*. A mutant of Lemna (91) deficient in electron transport between plastoquinone and cytochrome *f* lacks the EPR signal of this component. The center is reduced by electron donors donating after the DCMU inhibition site. Its steady-state redox changes in il-

Table 2 Oxidation-Reduction Potentials of the Rieske Center

Organism	E_m (mV)	pK	Ref.
Spinach chloroplasts	+290	—	2
Chlorobium Sp.	+160 (pH 7.0)	>6.5	78
Chromatium	+285 (pH 8.0)	7.4	8,1
Rdp. spheroides	+285 (pH 7.0)	8.0	90
Rdp. capsulata	+310 (pH 7.0)	—	9
Rhodospirillum rubrum	+160 (pH 7.0)	—	100

luminated chloroplasts show the same responses as do cytochrome *f*, plastocyanin, and P700 (92). That is, with photosystem II functional it is reduced, while inhibition of electron transport from photosystem II, for example, by DCMU, results in oxidation by photosystem I. It is also reduced in the dark by reduced soluble ferredoxin in an Antimycin A sensitive reaction, indicating involvement in the cyclic electron-transport chain as well as in the noncyclic system (93).

It is probably a component of the cytochrome *bf* complex; although reported to be absent from one preparation of this complex (1), it is present if the complex is prepared with digitonin as the only detergent (22). The EPR spectrum of this center is very unusual, with only the g_y component at $g = 1.89$ normally being detected. A new signal at $g = 2.03$ has recently been reported, which may be the g_z component (67, 88). It has recently been reported that the EPR spectrum of the Rieske center is altered in chloroplasts by exposure to the quinone antagonist DBMIB (94). In the presence of DMBIB under conditions where the Rieske protein is normally reduced, no $g = 1.89$ signal was observed, and a new large signal at $g = 1.94$ was seen instead. A small effect on the signal was also induced by another quinone antagonist, UHDBT (93). These results suggest that the Rieske signal may arise as the result of modification of a normal Fe-S $g = 1.94$ signal by interaction with a quinone. However, although we have reproduced these effects with chloroplasts, we did not observe any effect of DBMIB on the Rieske signal in the isolated cytochrome *bf* complex (22). The origin of the 1.94 signal in the presence of DBMIB must therefore remain an open question: though it may arise from modification of the Rieske signal, it may equally well arise from a previously undetected component in the chloroplast electron-transport chain.

7 OTHER Fe-S CENTERS IN OXYGENIC ORGANISMS

There are a small number of reports of EPR signals characteristic of Fe-S proteins found in preparations of photosynthetic organisms that have not yet been shown to have any function in photosynthesis. A component with a large signal at $g = 1.92$ at liquid helium temperatures has been found in preparations of some blue-green algae (Fig. 15) (95, 96). It has a midpoint redox potential of -270 mV. The signal is not present in purified photosystem I preparations, but was found in photosystem II particles of similar enrichment (96). The component is reduced by steady-state illumination of these particles at room temperature. It does not, however, show any low-temperature light-induced redox changes. These properties would be compatible with those of a carrier in the intermediary electron-transport chain close to photosystem II or in the cyclic system. However, the thylakoid system

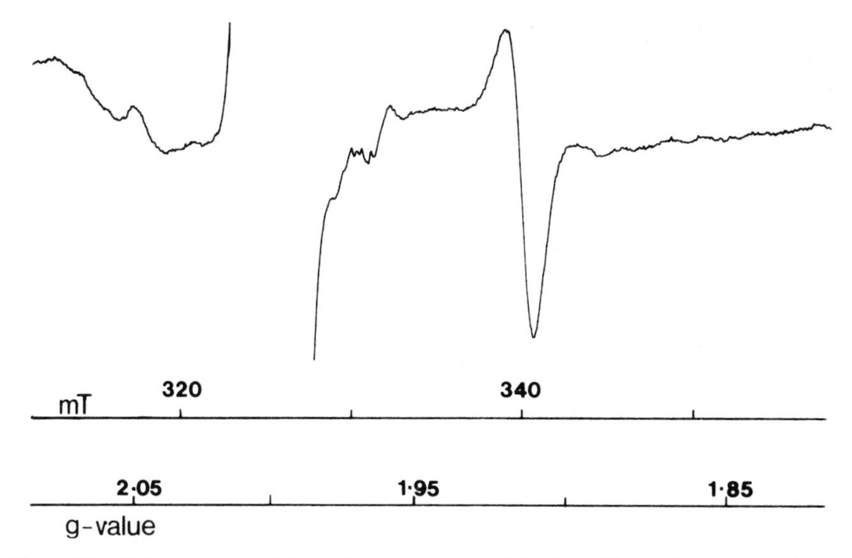

Figure 15. EPR spectrum at 15K of photosystem II particles from *Phormidium luridum* frozen under illumination, showing the signals of an Fe-S center at g = 2.05 and 1.92.

of blue-green algae also contains the respiratory chain. The signal may arise from a component of this respiratory chain, the photoreduction merely reflecting changes in the overall redox state of the sample, a possibility which, of course, affects all steady-state illumination experiments.

8 Fe-S CENTERS IN PHOTOSYNTHETIC BACTERIA

The photosynthetic membranes of both green and purple photosynthetic bacteria contain a number of bound Fe-S centers. At the present time it seems that most of the centers identified are only peripherally involved with photosynthesis, being components of either succinic dehydrogenase or possibly NADH dehydrogenase (97). In purple bacteria only the Rieske Fe-S center has been shown to be directly involved in photosynthetic electron flow (10). In the green bacteria two membrane-bound centers are probably involved, the Rieske center and a low-potential center in the electron-acceptor complex (11, 13).

Other Fe-S proteins normally considered as soluble components may in fact be membrane associated. The high-potential Fe protein of *Chromatium,* which is readily isolated as a soluble protein, is present in chromatophore membranes (8, 98); the yield of many of the soluble 4-Fe or 8-Fe ferredoxins from purple bacteria is also increased by detergent or organic solvent extraction of the membranes (99). One of the ferredoxins from *Chlorobium* is involved in photosynthetic electron transport in the same way as the analogous chloroplast ferredoxin (33), and illumination of whole cells of this organism results in reduction of this ferredoxin in vivo (12).

8.1 The Rieske Center

The Rieske center has been identified in all groups of photosynthetic bacteria that have been investigated, that is, in purple sulfur bacteria (97), purple non-sulfur bacteria (8), and green bacteria (11). In all the purple bacteria it has been found to have a high redox potential of ~ +280 mV (Table 2), similar to that of the analogous component in mitochondria and chloroplasts. However, in the green bacterium *Chlorobium* the center was found to have a more reduced potential +160 mV at pH 7.0, and this potential showed a pH dependence of 60 mV/pH unit, indicating the involvement of a proton in reduction of the center (11). Subsequently it was found that the center also has a pH-dependent midpoint potential at alkaline pH in mitochondria and purple bacteria (90). This led to proposals that the center functions as a H_2 carrier in the electron-transport chain. However, it has now been shown that a specific quinone Qz is the electron donor to the Rieske protein in purple bacteria. It seems possible that the pH dependence reflects either an interaction of the Fe-S center with this quinone, or problems of equilibration in the redox titrations, with electron donation to the Rieske depending on the membrane-bound quinone.

Early experiments showed that room-temperature illumination of chromatophores from purple bacteria oxidized the Rieske center unless an electron donor such as ascorbate was present, indicating a role in photosynthetic electron transport (8). Recently it has been proposed that the center is the electron donor to cytochrome *c*. This has been supported by a study involving correlation of the kinetics of cytochrome *b* and *c* redox changes and the effect of the inhibitors antimycin and UHDBT with redox changes of the Rieske center as determined by EPR measurements (10). These experiments show that antimycin blocks the oxidation of cytochrome *b* and reduction of the Rieske center, while UHDBT, which as in chloroplasts induced small changes in the EPR spectrum of the Rieske center, blocks the oxidation of the center by cytochrome *c*. This, together with the effects of these inhibitors on the cytochrome reactions, supports the view that the Rieske center is the immediate donor to the membrane-bound cytochrome *c*.

8.2 Centers Giving Rise to $g = 1.94$ EPR Signals

Chromatophore preparations contain a number of Fe-S centers giving rise to EPR signals at $g = 1.94$. Redox titrations have allowed the identification of two or three centers in most cases. The midpoint potentials of these centers (1), together with the presence of a center with a $g = 2.01$ signal in the oxidized state, suggest that they arise from succinic dehydrogenase. Solubilization of this enzyme produces a fraction containing these centers which is similar to the mitochondrial enzyme (97). The depleted membrane fractions retain some $g = 1.94$ component, which may represent the NADH dehydrogenase, although no evidence to support this suggestion is available.

9 THE GREEN PHOTOSYNTHETIC BACTERIA

The green photosynthetic bacterium *Chlorobium thiosulfatophilum* differs from the purple bacteria in that vesicle preparations from it have the ability to photoreduce NAD or soluble ferredoxin by a noncyclic electron-transport system (34). These vesicle preparations contain two Fe-S centers with EPR signals at $g = 1.94$ and potentials of 0 and -170 mV, which are probably succinic dehydrogenase, and two centers at $g = 1.94$ and $g = 1.90$ with potentials of -550 mV (11, 13). The $g = 1.94$ center spectrum is similar to that of the soluble ferredoxin with $E_m = -575$ mV isolated from this organism, and may be a bound fraction of this protein. The $g = 1.90$ component appears to be the electron acceptor for a photosystem in the vesicles (13). It is irreversibly reduced on illumination of vesicles at liquid helium temperature (Fig. 16), with the parallel oxidation of a component assumed to be the reaction center chlorophyll giving rise to a radical signal at $g = 2.00$. The reaction center chlorophyll associated with this reaction is oxidized with $E_m = +220$ mV, and the ability to photooxidize it is lost with $E_m \simeq -550$ mV as the acceptor is reduced. Knaff and co-workers (36) observed reaction center photooxidation to low potentials in *C. thiosulfatophilum* vesicle preparations, but were unable to detect the reduction of the $g = 1.90$ component. *C. thiosulfatophilum* preparations are very difficult to work with, and the $g = 1.90$ component is easily lost as a result of freezing and thawing the preparation or exposure to oxygen. Isolated reaction centers that have been prepared from a closely related organism *Prosthecochloris estuaris* seem to be essentially the same as those from purple bacteria (37). In vesicle preparations it is possible to see changes in the $g = 2.00$ reaction center chlorophyll radical, corresponding to both the irreversible reaction center oxidation associated with the reduction of the $g = 1.90$ center and a reversible photoreaction similar to that seen in the isolated reaction center. It is not yet clear whether there are in fact two different reaction centers, one rather similar to photosystem I of plants and the other like the purple bacterial reaction center, or a single center that functions differently in vesicles and in purified reaction center preparations.

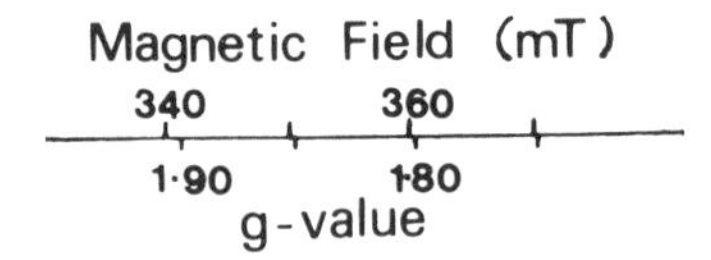

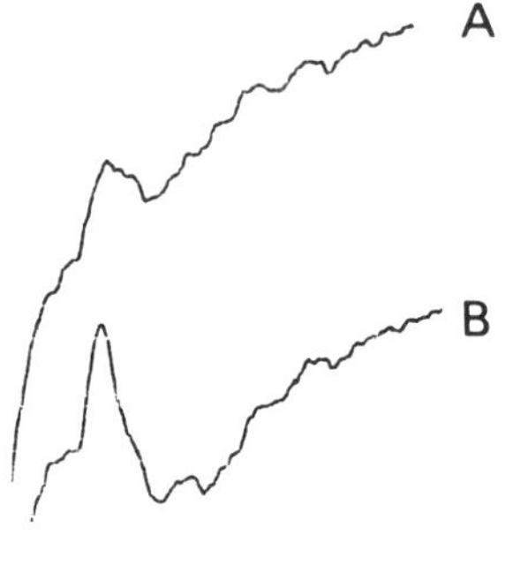

Figure 16. EPR spectrum of *Chlorobium thiosulfatophilum* membrane particles, showing reduction of the $g = 1.90$ component. *(A)* Spectrum at 15K of particles poised at +75 mV in the dark. *(B)* The same sample after illumination at 15K. *(C)* Light minus dark difference spectrum (×2).

10 CONCLUDING REMARKS

The past decade has seen the initial detection and subsequent characterization of the bound Fe-S centers in the photosynthetic apparatus. In the oxygenic photosynthetic organisms the rather small number of clearly identifiable centers has allowed definition of their function in photosynthesis by physical techniques. However, the absence of good techniques for measuring redox changes at room temperature means that details of their function are still open to investigation. Further understanding of their integration into the membrane system will come initially from identification of specific peptides carrying the centers and eventually from structural analysis by X-ray or NMR techniques.

In photosynthetic bacteria only the role of the Rieske center in cyclic electron transport in purple bacteria is well established. In green bacteria a role in noncyclic electron transport for reduction of soluble ferredoxin and NAD also seems likely. However, the mechanism of ferredoxin reduction in purple bacteria and of cyclic electron transport for ATP synthesis in green bacteria are as yet undefined and may well involve as yet unidentified Fe-S centers.

ACKNOWLEDGMENTS

I am grateful to my colleagues, J. H. A. Nugent and P. Heathcote, for critical reading of the manuscript and assistance with preparation of the figures, to A. W. Rutherford, for the scheme of electron transport in purple bacteria, and to Professor C. E. Johnson, for the Mössbauer spectra.

REFERENCES

1. R. Malkin and A. J. Bearden, *Biochim. Biophys. Acta,* **505,** 147–181 (1978).
2. R. Malkin and P. J. Aparicio, *Biochem. Biophys. Res. Commun.,* **63,** 1157–1160 (1975).
3. J. S. Rieske, R. E. Hansen, and W. Z. Zaugg, *J. Biol. Chem.,* **239,** 3017–3022 (1964).
4. J. S. Rieske, W. S. Zaugg, and R. E. Hansen, *J. Biol. Chem.,* **239,** 3023–3030 (1964).
5. R. Malkin and A. J. Bearden, *Proc. Natl. Acad. Sci. USA,* **68,** 16–19 (1971).
6. M. C. W. Evans, A. Telfer, and A. V. Lord, *Biochim. Biophys. Acta,* **267,** 530–537 (1972).
7. M. C. W. Evans, C. K. Sihra, J. R. Bolton, and R. Cammack, *Nature,* **256,** 668–670 (1975).
8. M. C. W. Evans, A. V. Lord, and S. G. Reeves, *Biochem. J.,* **138,** 177–183 (1974).
9. R. C. Prince, J. S. Leigh, and P. L. Dutton, *Biochem. Soc. Trans.,* **2,** 950–953 (1974).
10. J. R. Bowyer, P. L. Dutton, R. C. Prince, and A. R. Crofts, *Biochim. Biophys. Acta,* **592,** 445–460 (1980).
11. D. B. Knaff and R. Malkin, *Biochim. Biophys. Acta,* **430,** 244–252 (1976).
12. J. V. Jennings, Ph.D. Thesis, "The Iron-Sulphur Proteins of *Chlorobium thiosulfatophilum,*" University of London, 1977.
13. J. V. Jennings and M. C. W. Evans, *FEBS Lett.,* **75,** 33–36 (1977).
14. J. R. Norris and J. J. Katz, in *The Photosynthetic Bacteria,* R. K. Clayton and W. R. Sistrom, eds., Plenum, New York, 1978, Chapter 21.
15. J. Amesz and L. N. M. Duysens, in *Primary Processes of Photosynthesis,* J. Barber, ed., Elsevier, Amsterdam, 1977, Chapter 4.
16. B. R. Velthuys, *Ann. Rev. Plant Physiol.,* **30,** 545–568 (1980).
17. V. V. Klimov, S. I. Allakhverdieu, S. Demeter, and A. A. Krasnovsky, *Dokl. Akad. Nauk,* **249,** 227–230 (1980).
18. V. V. Klimov, E. Dolan, and B. Ke, *FEBS Lett.,* **112,** 97–100 (1980).
19. H. J. Van Gorkum, *Biochim. Biophys. Acta,* **347,** 439–442 (1974).
20. D. B. Knaff and D. I. Arnon, *Proc. Natl. Acad. Sci. USA,* **63,** 963–969 (1969).
21. N. Nelson and J. Neumann, *J. Biol. Chem.,* **247,** 1817–1824 (1972).
22. P. R. Rich, P. Heathcote, M. C. W. Evans, and D. S. Bendall, *FEBS Lett.,* **116,** 51–56 (1980).
23. B. Bouges-Bocquet, *Biochim. Biophys. Acta,* **314,** 250–256 (1973).
24. P. Heathcote, K. N. Timofeev, and M. C. W. Evans, *FEBS Lett.,* **101,** 105–109 (1979).
25. V. A. Shuvalov, A. V. Klevanik, A. V. Sharkov, P. G. Kryukov, and B. Ke, *FEBS Lett.,* **107,** 313–316 (1979).
26. J. R. Bolton, in *Primary Processes of Photosynthesis,* J. Barber, ed., Elsevier, Amsterdam, 1977, Chapter 5.

27. J. H. Golbeck, S. Lien, and A. San Pietro, in *Encyclopedia of Plant Physiology New Series Vol. 5,* A. Trebst and M. Avron, eds., Springer-Verlag, Berlin, 1977, Chapter 1b.
28. M. Shin and D. I. Arnon, *J. Biol. Chem.*, **240,** 1405-1411 (1965).
29. G. Hind, J. D. Mills, and R. E. Slovacek, *Proceedings 4th International Photosynthesis Congress,* D. O. Hall, J. Coombs, and T. W. Goodwin, eds., Biochemical Society, London, 1978, pp. 591-599.
30. R. K. Clayton, in *The Photosynthetic Bacteria,* R. K. Clayton and W. R. Sistrom, eds., Plenum, New York, 1978, Chapter 20.
31. W. W. Parson and V. A. Shuvalov, *Abstracts 5th International Congress of Photosynthesis,* 1980, p. 439.
32. P. L. Dutton and R. C. Prince, in *The Photosynthetic Bacteria,* R. K. Clayton and W. R. Sistrom, eds., Plenum, New York, 1978, Chapter 28.
33. D. L. Keister and N. J. Yike, *Biochemistry,* **6,** 3847-3857 (1967).
34. B. B. Buchanan and M. C. W. Evans, *Biochim. Biophys. Acta,* **180,** 123-129 (1969).
35. K. Kusai and T. Yamanaka, *Biochim. Biophys. Acta,* **325,** 304-314 (1973).
36. D. B. Knaff, J. M. Olson, and R. C. Prince, *FEBS Lett.*, **98,** 285-289 (1979).
37. T. Swarthoff and J. Amesz, *Biochim. Biophys. Acta,* **548,** 427-432 (1979).
38. B. Commoner, J. J. Heise, and J. Townsend, *Proc. Natl. Acad. Sci. USA,* **42,** 710-718 (1956).
39. A. J. Bearden and R. Malkin, *Biochim. Biophys. Acta,* **283,** 456-468 (1972).
40. D. L. Williams Smith, P. Heathcote, C. K. Sihra, and M. C. W. Evans, *Biochem. J.*, **170,** 365-371 (1978).
41. J. T. Warden and J. R. Bolton, *J. Am. Chem. Soc.*, **95,** 6435-6436 (1973).
42. L. L. Shipman, T. M. Cotton, J. R. Norris, and J. J. Katz, *Proc. Natl. Acad. Sci. USA,* **73,** 1791-1794 (1976).
43. P. Heathcote, D. L. Williams-Smith, C. K. Sihra, and M. C. W. Evans, *Biochim. Biophys. Acta,* **503,** 333-342 (1978).
44. A. R. McIntosh, M. Chu, and J. R. Bolton, *Biochim. Biophys. Acta,* **376,** 308-314 (1975).
45. M. C. W. Evans, C. K. Sihra, and R. Cammack, *Biochem. J.*, **158,** 71-77 (1976).
46. P. Heathcote, D. L. Williams-Smith, and M. C. W. Evans, *Biochem. J.*, **170,** 373-378 (1978).
47. E. H. Evans, J. D. Rush, C. E. Johnson, M. C. W. Evans, and D. P. E. Dickson, *Eur. J. Biochem.*, **118,** 81-84 (1981).
48. R. Cammack and M. C. W. Evans, *Biochem. Biophys. Res. Commun.*, **67,** 544-549 (1975).
49. J. H. Golbeck, S. Lien, and A. San Pietro, *Arch. Biochem. Biophys.*, **178,** 140-150 (1977).
50. E. H. Evans, J. D. Rush, C. E. Johnson, and M. C. W. Evans, *Biochem. J.*, **182,** 861-865 (1979).
51. J. Isaakidou, G. Papageorgiou, V. Petrouleas, A. Simonopoulous, A. Kostikas, and C. Dismukes, *Abstracts of the 5th International Photosynthesis Congress,* 1980, p. 272.
52. G. Feher and M. Okamura, in *The Photosynthetic Bacteria,* R. K. Clayton and W. R. Sistrom, eds., Plenum, New York, 1978, Chapter 19.
53. T. Hiyama and B. Ke, *Proc. Natl. Acad. Sci. USA,* **68,** 1010-1013 (1971).
54. B. Ke and H. Beinert, *Biochim. Biophys. Acta,* **305,** 689-693 (1973).
55. T. Hiyama and D. C. Fork, *Arch. Biochem. Biophys.*, **199,** 488-496 (1980).

56. J. H. Goldbeck, B. R. Velthuys, and B. Kok, *Biochim. Biophys. Acta,* **504,** 226–230 (1978).
57. V. A. Shuvalov, E. Dolan, and B. Ke, *Proc. Natl. Acad. Sci. USA,* **76,** 770–773 (1979).
58. M. C. W. Evans, S. G. Reeves, and R. Cammack, *FEBS Lett.,* **49,** 111–114 (1974).
59. B. Ke, R. E. Hansen, and H. Beinert, *Proc. Natl. Acad. Sci. USA,* **70,** 2941–2945 (1973).
60. E. H. Evans, R. Cammack, and M. C. W. Evans, *Biochem. Biophys. Res. Commun.,* **68,** 1212–1218 (1976).
61. M. C. W. Evans, S. G. Reeves, and A. Telfer, *Biochem. Biophys. Res. Commun.,* **51,** 593–596 (1973).
62. R. Cammack, M. D. Ryan, and A. C. Stewart, *FEBS Lett.,* **107,** 422–426 (1979).
63. J. H. A. Nugent, B. L. Möller, and M. C. W. Evans, *Biochim. Biophys. Acta,* **634,** 249–255 (1981).
64. M. C. W. Evans and P. Heathcote, *Biochim. Biophys. Acta,* **590,** 89–96 (1980).
65. S. Demeter and B. Ke, *Biochim. Biophys. Acta,* **462,** 770–774 (1977).
66. G. C. Dismukes and K. Sauer, *Biochim. Biophys. Acta,* **504,** 431 (1978).
67. R. C. Prince, M. S. Crowder, and A. J. Bearden, *Biochim. Biophys. Acta,* **592,** 323–337 (1980).
68. H. E. Davenport, R. Hill, and F. R. Whatley, *Proc. R. Soc. B,* **139,** 346–349 (1951).
69. D. O. Hall, K. K. Rao, and R. Cammack, *Sci. Prog.,* **62,** 285–317 (1975).
70. K. Fukuyama, T. Hase, S. Matsumoto, T. Tsukihara, Y. Katsube, N. Tanaka, M. Kakudo, K. Wada, and H. Matsubara, *Nature,* **286,** 522–524 (1980).
71. B. Bouges-Bocquet, *Biochim. Biophys. Acta,* **590,** 223–233 (1980).
72. N. Nelson and J. Neumann, *Biochem. Biophys. Res. Commun.,* **30,** 142–147 (1968).
73. R. Malkin, *Arch. Biochem. Biophys.,* **169,** 77–83 (1975).
74. N. Nelson, C. Bengis, B. L. Silver, D. Getz, and M. C. W. Evans, *FEBS Lett.,* **58,** 363–365 (1975).
75. R. Malkin, A. J. Bearden, F. A. Hunter, R. S. Alberte, and J. P. Thornber, *Biochim. Biophys. Acta,* **430,** 389–394 (1976).
76. P. Mathis, K. Sauer, and R. Remy, *FEBS Lett.,* **88,** 275–278 (1978).
77. B. Ke, K. Sugahara, and E. R. Shaw, *Biochim. Biophys. Acta,* **488,** 12–25 (1975).
78. D. B. Knaff and R. Malkin, *Arch. Biochem. Biophys.,* **159,** 555–562 (1973).
79. M. C. W. Evans, C. K. Sihra, and A. R. Slabas, *Biochem. J.,* **162,** 75–85 (1977).
80. P. Setif and P. Mathis, *Arch. Biochem. Biophys.,* **204,** 477–485 (1980).
81. D. I. Arnon, H. Y. Tsujimoto, and T. Hiyama, *Proc. Natl. Acad. Sci. USA,* **74,** 3826–3830 (1977).
82. C. Dismukes, A. McGuire, R. Friesner, and K. Sauer, *Rev. Chem. Intermed.,* **3,** 59–88 (1979).
83. K. Sauer, P. Mathis, S. Acker, and J. Van Best, *Biochim. Biophys. Acta,* **503,** 120–134 (1978).
84. J. P. Thornber, *Ann. Rev. Plant Physiol.,* **26,** 127–158 (1975).
85. D. von Wettstein, B. L. Moller, G. Hoyer-Hansen, and D. Simpson, in *Origin of Chloroplasts,* J. A. Schiff and R. Y. Stanier, eds., Elsevier-North Holland, Amsterdam, 1980.
86. C. Bengis and N. Nelson, *J. Biol. Chem.,* **252,** 4564–4569 (1977).
87. M. K. Bowman, C. Wraight, and J. R. Norris, *Abstracts 5th International Congress on Photosynthesis,* 1980, p. 83.

88. J. H. A. Nugent and M. C. W. Evans, in preparation.
89. J. S. Rieske, D. H. MacLennan, and R. Coleman, *Biochem. Biophys. Res. Commun.*, **15,** 338-344 (1965).
90. R. C. Prince and P. L. Dutton, *FEBS Lett.*, **65,** 117-119 (1976).
91. R. Malkin and H. B. Posner, *Biochim. Biophys. Acta,* **501,** 552-554 (1978).
92. J. Whitmarsh and W. A. Cramer, *Proc. Natl. Acad. Sci. USA,* **76,** 4417-4420 (1979).
93. R. Malkin, *Abstracts 5th International Congress on Photosynthesis,* 1980, p. 360.
94. R. K. Chain and R. Malkin, *Arch. Biochem. Biophys.*, **197,** 52-56 (1979).
95. R. Cammack, L. J. Luijk, J. J. Maguire, I. V. Fry, and L. Packer, *Biochim. Biophys. Acta,* **548,** 267-275 (1979).
96. J. H. A. Nugent, A. C. Stewart, and M. C. W. Evans, *Biochim. Biophys. Acta,* **635,** 488-497 (1981).
97. W. J. Ingledew and R. C. Prince, *Arch. Biochem. Biophys.*, **178,** 303-307 (1977).
98. P. L. Dutton and J. S. Leigh, *Biochim. Biophys. Acta,* **314,** 178-190 (1973).
99. B. B. Buchanan and D. I. Arnon, *Meth. Enzymol.*, **23,** 413-440 (1971).
100. R. Carrithers, D. C. Yoch, and D. I. Arnon, *J. Biol. Chem.*, **252,** 7461-7467 (1977).

CHAPTER 7

Iron-Sulfur Clusters in the Mitochondrial Electron-Transport Chain

TOMOKO OHNISHI

Department of Biochemistry and Biophysics
University of Pennsylvania
Philadelphia, Pennsylvania

JOHN C. SALERNO

Department of Biology
Rensselaer Polytechnic Institute
Troy, New York

CONTENTS

1 INTRODUCTION

The multiple Fe-S clusters of mitochondria are primarily involved in electron transfer in the dehydrogenase segments of the respiratory chain. Major exceptions include the Rieske Fe-S cluster (1) in the cytochrome bc_1 region and the Fe-S cluster of aconitase (2). Scheme I depicts the redox components of the mitochondrial respiratory chain on the electron-transfer pathways from three major substrates to O_2. This scheme is essentially the same as one presented in 1979 (3), except that distinct species of bound ubiquinone (4–9) have now been resolved from bulk ubiquinone.

The initial phase of investigation, in which individual Fe-S centers were identified and their basic physicochemical parameters obtained, is nearing completion, although some interesting questions remain unanswered. In addition, the central dogma of the chemiosmotic hypothesis (10, 11) has become generally accepted; it seems certain that a proton electrochemical gradient across the inner mitochondrial membrane is an obligatory intermediate between electron transfer and ATP synthesis.

The mechanism of redox-linked proton translocation is still unknown. Although the loop theory proposed by Mitchell has greatly stimulated research in this area, only ubiquinone has emerged as a plausible mobile H_2 carrier in mitochondria (12, 13). No such H_2 carrier is available in site 0 (transhydrogenase) (14, 15), site I (3), or in cytochrome oxidase (16). Thus ion pump mechanisms rather than loops appear to be the most likely candidates for energy transducing mechanisms at these sites.

Many recent reviews have stressed the identification and characterization of dehydrogenase electron carriers (3, 17–19). We have therefore focused our attention on the spatial organization of Fe-S cluster containing regions of the respiratory chain. This includes the orientation and distances of components relative to each other and to the inner and outer surfaces of the membrane.

2 CHARACTERISTICS OF Fe-S CLUSTERS

2.1 Succinate Dehydrogenase (SDH)

The first soluble SDH purified in 1955 contained one covalently bound FAD and 2–4 atoms each of nonheme Fe and acid labile S per flavin: it only catalyzed electron transfer from succinate to nonphysiological electron acceptors (20, 21). Subsequently, a more intact SDH preparation was isolated (22), which retained reconstitutive activity (i.e., can bind to SDH-depleted particles and restore electron-transfer activity from succinate to O_2) in addition to the nonphysiological electron transfer. This enzyme was found to con-

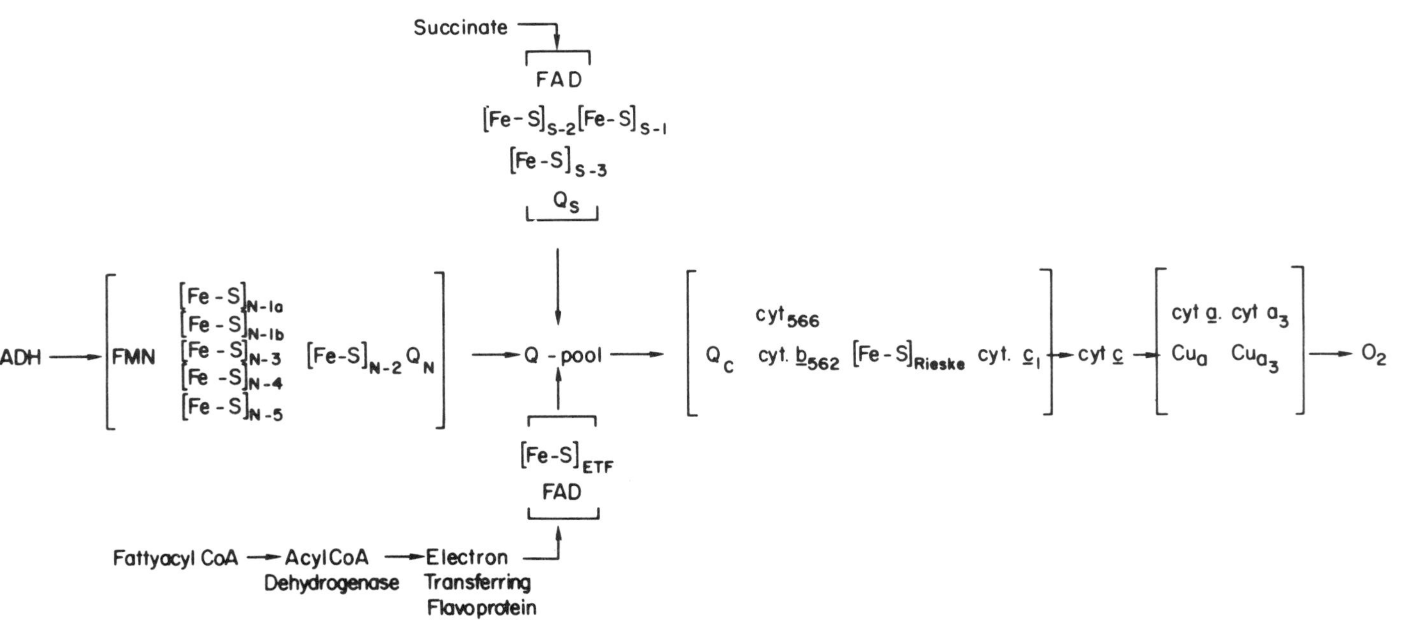

Scheme I. Respiratory chain redox components present in the inner mitochondrial membrane. Fe-S clusters associated with NADH-UQ and succinate-UQ reductase segments are designated wth suffixes N-x and S-x, respectively. Q_S, Q_N, and Q_C are protein-associated pools of ubiquinone in succinate-UQ NADH-UQ, and ubiquinol-cytochrome *c* reductase segments, respectively, which can be distinguished from the bulk ubiquinone pool.

tain 8 nonheme Fe atoms and 8 acid labile sulfides per flavin, but it was only 30–40% pure. In 1971 a pure SDH preparation was isolated from succinate-ubiquinone (UQ) reductase (complex II),* using chaotropic agents (23, 24). The molecular weight of this enzyme was ≈97,000 and it contained 7–8 g.-atoms each of nonheme Fe and acid labile sulfide and one molecule of covalently bound flavin (25). It was also shown that the SDH molecule is composed of 2 nonidentical subunits: namely, the flavo Fe-S protein (FP) and the Fe-S protein (IP), with molecular weights of 70,000 and 27,000, respectively. The FP contains one FAD, 4 nonheme Fe atoms, and 4 acid labile sulfides; IP contains approximately 3–4 nonheme Fe atoms and 3–4 acid labile sulfides. This enzyme was only ≈20% active in reconstituting respiratory chain electron transport.

In 1977 an essentially pure and reconstitutively fully active SDH was isolated from complex II by an alkaline extraction together with butanol treatment under anaerobic conditions (28). Subsequently a similar high quality SDH preparation was isolated from succinate-cytochrome *c* reductase (SCR), with higher SDH yield (29). A fully active and pure SDH preparation has also been obtained using chaotropic reagents under strictly anaerobic conditions (27). These pure and reconstitutively fully active bovine heart SDH preparations all contain, on the average, 1 covalently bound FAD, 8 nonheme Fe atoms, and 8 acid labile sulfides per molecule.

From the EPR characterization of paramagnetic redox centers in various SDH preparations and in complex II, the following composition of SDH redox components was proposed. The FP subunit contains two spin-coupled binuclear $[2Fe\text{-}2S]^{+1(+1,+2)}$ Fe-S clusters, namely S-1 (30) and S-2 (31), in addition to a covalently bound FAD. The IP subunit contains 1 tetranuclear Fe-S cluster $[4Fe\text{-}4S]^{+3(+2,+3)}$, S-3 (32); this signal had been attributed to a complex II iron-sulfur cluster (33). Fe-S core extrusion and interprotein core transfer experiments (34) independently demonstrated that 2 binuclear and 1 tetranuclear Fe-S clusters are present in the SDH molecule as a whole, supporting the above-proposed 3 Fe-S cluster composition of SDH.

Cluster S-1 is a binuclear Fe-S component (35, 36), which gives an EPR spectrum of rhombic symmetry with g values of $g_z = 2.025$, $g_y = 1.93$, and $g_x = 1.905$ in the reduced state (37). Its spin relaxation is relatively slow, so EPR signals are readily saturated at temperatures below 30°K (31, 32). The midpoint redox potential of this cluster is ≈0 mV in the soluble SDH (31); thus it is quantitatively reduced with a high concentration of succinate. Its spin concentration is approximately equivalent to that of flavin.

The presence of the second ferredoxin (Fd) type Fe-S cluster (S-2) was

*Complex II consists of SDH and two additional polypeptides of smaller molecular weights (26, 27).

recognized when EPR spectra of older-type reconstitutively active SDH preparations reduced with succinate and with dithionite, respectively, were compared at temperatures below 20°K (3, 37). The potentiometric resolution of the $g = 1.94$ signal recorded below 15°K indicated the presence of an additional Fd-type Fe-S cluster with a measured midpoint potential of ~ -400 mV in all soluble dehydrogenase preparations and -260 mV in particulate preparations, with an n value equal to one (3). The succinate-reduced and dithionite-reduced older-type reconstitutively active SDH showed a similar spin intensity above 40°K, but in the lower temperature range the dithionite-reduced SDH gave 1.5–1.75 spins/flavin, compared to 0.8–1.0 for the succinate-reduced enzyme (3, 19). Beinert and co-workers (38) also obtained a spin concentration of 1.2–1.75 equivalents for the dithionite-reduced enzyme relative to that of cluster S-1. The spin concentration of the $g = 1.94$ type signal of dithionite-reduced enzyme (both S-1 and S-2 are paramagnetic) appeared to vary considerably depending on the SDH preparations used. As discussed later, clusters S-1 and S-2 are spin coupled; thus the spin relaxation behavior of the dithionite-reduced enzyme also varies considerably depending on the kind of enzyme preparation. In the reconstitutively inactive SDH the dithionite-reduced enzyme showed slower relaxation, similar to the succinate-reduced enzyme in the higher temperature range. The line shape of cluster S-2 was therefore estimated from the difference between the spectra of the dithionite-reduced and succinate-reduced reconstitutively inactive enzyme at temperatures above 25°K, with minimal effects from the spin coupling between S-1 and S-2. The line shape and the principal g values of S-1 and S-2 were found to be almost identical (37). It was also found that at extremely low temperatures (<6.5°K) reconstitutively inactive SDH exhibits a splitting of the central resonance of 23 G and broadening of both the g_z and g_x peaks. In contrast, in the older-type reconstitutively active enzymes only broadening of the central signal (splitting was not resolved) was observed in the same temperature range. The reversible conversion of the enzyme from the reconstitutively active to the inactive form and vice versa accompanied the change in the low-temperature spectral pattern described above (31, 39). In all SDH preparations a dynamic spin-spin interaction was observed between clusters S-1 and S-2, namely, an enhancement of S-1 spin relaxation on S-2 reduction. This observation was supported by other investigators (38).

More recently, Albracht (40, 41) reported that the EPR absorption responsible for the second binuclear Fe-S cluster, S-2, is not detectable in the SDH molecule in the most intact preparations; this finding was confirmed by other investigators. Based on this evidence, together with some additional data from his own laboratory (40), Albracht concluded that the apparent existence of two binuclear clusters in earlier SDH preparations is an isolation artifact and that the intact SDH molecule contains only one of each of the binuclear and tetranuclear clusters. He proposed that the spin relaxation enhancement is caused by a protein conformational change induced by the reduction of the flavin to the fully reduced state. The enhancement of the S-1 spin relaxation, however, is caused by an $n = 1$ redox component with a midpoint potential of ≈ -400 mV in the soluble SDH preparations (31). Direct redox titration of the SDH flavin free radical indicates that the fully reduced form of the flavin titrates with a midpoint potential of -81 mV at pH 7 and an n value close to 2 (42). Thus neither midpoint potential nor n value support the hypothesis proposed by Albracht. It is important to emphasize the fact that, as presented in Figure 1, a dramatic enhancement of S-1 relaxation is also seen in the most intact SDH preparation [2 orders of magnitude enhancement in the half-saturation parameter* ($P_{1/2}$): from 0.07 mW to 6.0 mW at 12°K (T. Ohnishi, H. Blum, C. Y. Yu, and L. Yu, unpublished)]. We believe that the apparent nondetectability of the S-2 signal in the intact state of SDH is most likely caused by the spin coupling between S-1 and S-2, which renders the total EPR detectable spin concentration of S-1 plus S-2 not higher than the flavin concentration in this system. More details of the spin

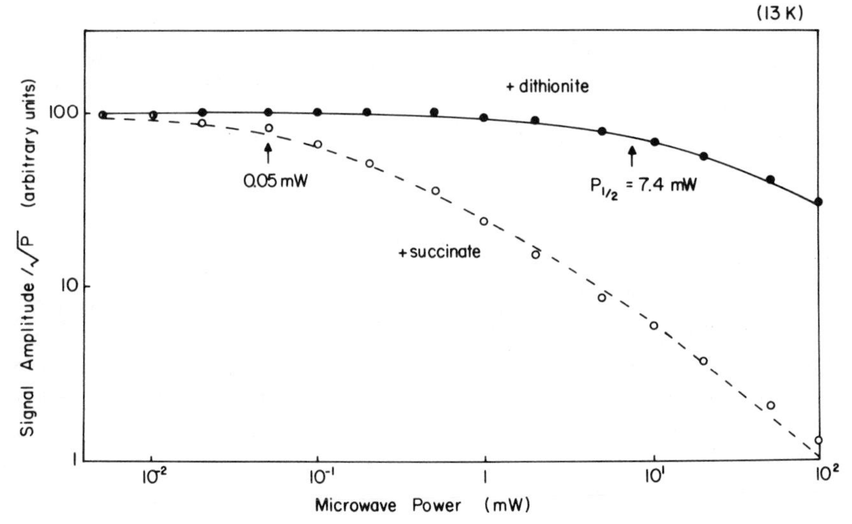

Figure 1. Power saturation behavior of the $g = 1.93$ EPR signals of a fully active and pure SDH preparation [prepared by Yu and Yu (29)]. The enzyme (final flavin concentration 41.5 μM) in 50 mM HEPES buffer (50 mM) was reduced with 23 mM succinate or 10 mM dithionite, respectively. Sample temperature was 13°K.

*$P_{1/2}$ is a half-saturation parameter; a quantitative definition was given in reference 43. The faster the spin relaxation the greater is the microwave power $P_{1/2}$ needed for saturation.

interactions are discussed in a separate section. Coles and co-workers also proposed an EPR-silent cluster S-2, based on the existence of two binuclear and one tetranuclear clusters in the intact SDH preparation, using an independent experimental approach (34).

Cluster S-3 is paramagnetic in the oxidized state (HIPIP-type cluster), and shows relatively isotropic highly temperature-sensitive EPR signals with g values of $g_z = 2.015$, $g_y = 2.014$, and $g_x = 1.990$ (4). The midpoint redox potential of cluster S-3 is +65 mV in complex II and is greater than 120 mV in the intact mitochondrial system (3). It has been generally accepted that this cluster is a tetranuclear Fe-S species, because 4 nonheme Fe atoms and 4 acid labile sulfides are otherwise unaccounted for and because its EPR characteristics resemble other HIPIP-type clusters. Recently, Fe-S clusters having a novel trinuclear [3Fe-3S] structure have been identified in various Fe-S proteins (44–46) that exhibit EPR characteristics similar to those of center S-3. Extrusion of this cluster in tetranuclear form (approximately one per SDH molecule), however, indicates that it is tetranuclear rather than trinuclear in structure (34). Cluster S-3 is stable in the membrane-bound state, and the spin concentration was found to be equal to that of the flavin in complex II preparations (32, 38). Cluster S-3 becomes extremely labile toward oxidants once the enzyme is solubilized, and its EPR signals are detectable only in the reconstitutively active form of SDH (33). It is generally accepted that cluster S-3 is an integral component of the SDH molecule, based on various lines of indirect evidence (3).

2.2 NADH-UQ Oxidoreductase Segment

Based on two-dimensional SDS/polyacrylamide-gel electrophoresis and specific immunoprecipitation studies, Ragan and his co-workers (47) resolved the NADH-UQ oxidoreductase complex (complex I) (48) into 26 polypeptides with varying molecular weights. NADH-UQ reductase is much more complicated than succinate-UQ reductase, because in addition to the soluble NADH dehydrogenase segment, which is analogous to the succinate dehydrogenase, it contains components that are required for energy coupling at site I.

Resolution of this complex enzyme using chaotropic reagents in combination with ammonium sulfate fractionation has greatly facilitated the structural and topographical analysis of Fe-S clusters in the resolved system (49, 50), because this method allows the separation of subfractions that retain equivalent nonheme Fe and acid labile sulfides and display some distinct EPR signals (50). This method resolved a flavo-Fe-S protein (FP) and an Fe-S protein (IP) in soluble form and a hydrophobic fraction in an insoluble form. The FP fraction corresponds to soluble NADH dehydrogenase and contains three polypeptides (subunits I–III), with molecular weights of $\approx$51,000, $\approx$24,000, and $\approx$9000, respectively (51–54). By a stronger chaotropic treatment combined with freeze-thawing, the FP fraction was resolved into two subfractions, namely, subunit I and II + III. Subunit I apparently contains 4 nonheme Fe atoms and equivalent acid labile sulfides, and subunit II + III contains about 2 each of nonheme Fe atoms and acid labile sulfides (50). Amino acid composition of these subunits (52) revealed that subunit I contains about 6 cysteine units. One essential cysteine is required for substrate binding; the remaining five cysteinyl residues are sufficient for one tetranuclear cluster, but not for two binuclear clusters. Subunit III does not contain enough cysteines to accommodate an Fe-S cluster. Subunit II contains cysteines sufficient for one binuclear Fe-S cluster. The hydrophilic Fe-S subunit (IP) was also further resolved into two subfractions, IP-I and IP-II. The former is mostly composed of a 75,000 dalton polypeptide, and the latter contains 49,000, 51,000 and 29,000 dalton polypeptides. Both IP subfractions contain nonheme Fe and acid labile sulfide at equivalent concentrations (50). EPR analysis of these subfractions is described in the following section. About 80% of the total protein and 50% of the nonheme Fe and acid labile sulfide of complex I are located in the hydrophobic polypeptide subunits. Further resolution of the hydrophobic fractions remains for investigation.

In the NADH-UQ reductase segment of the respiratory chain, 6 distinct EPR-detectable Fe-S clusters have been reported to date (see scheme I). The EPR spectrum of cluster N-1 was detected as early as 1960 (30). An EPR spectrum of cluster N-1 with rhombic symmetry and g values of 2.022, 1.938, and 1.923 was reported in complex I using EPR measurements at temperatures below 77°K (55). It has a resonance absorption approximately equal to the flavin concentration. This cluster is completely reducible with NADH in complex I. The $g = 1.94$ species in intact mitochondrial membrane systems was potentiometrically resolved into two $n = 1$ components with midpoint potentials of -240 ± 20 mV and -380 ± 20 mV at pH 7.2; these two components were designated N-1b and N-1a, respectively (3). These two components differ in their response to phosphate potential (56) and in the pH dependence of their midpoint potentials (57). Both clusters exhibit EPR spectra with rhombic symmetry in beef heart mitochondria, while N-1a in the pigeon heart system exhibits a spectrum of axial symmetry. More recently, Albracht and colleagues proposed the presence of two N-1 type clusters of axial symmetry, based on computer simulation of their EPR lineshapes using complex I (58) and beef heart SMP (59). They reported that the two clusters were each present at 0.25 spin equivalents per FMN, and were fully reducible with NADH. The two species were designated clusters 1a and 1b. More recently, Ohnishi and co-workers (60) have studied the N-1 type Fe-S species by the combined application of potentiometric analysis and computer lineshape simulations using complex I preparations isolated in Hatefi's laboratory (48). One component was titrated with a midpoint redox potential of −335 mV at pH 8.0 with an n value equal to one. The EPR spectrum of this component was obtained as a difference spectrum (Fig. 2*C*) of complex I poised at appropriate potentials, for example, −407 mV (Fig. 2*A*) and −286 mV (Fig. 2*B*). This component shows a typical binuclear $g = 1.94$

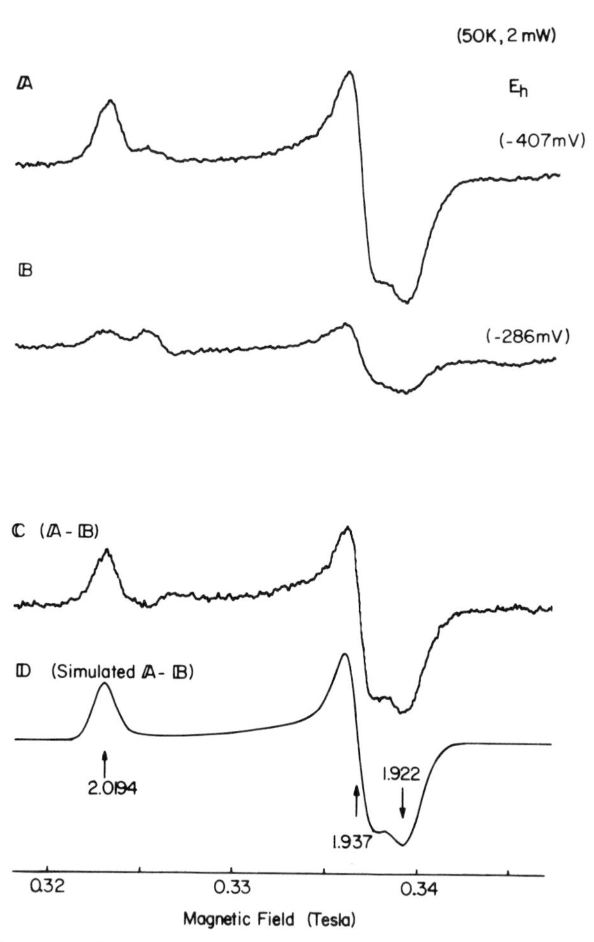

Figure 2. EPR spectra of Fe-S cluster N-Ib obtained from potentiometric resolution in complex I, and its computer simulation. Simulation parameters used were g values shown in the figure and linewidths of $L_z = 7.5 \times 10^{-4}$ tesla, $L_y = 7.6 \times 10^{-4}$ tesla, and $L_x = 1.05 \times 10^{-3}$ tesla (60).

type EPR spectrum, which can be simulated as a single rhombic component with g values of $g_z = 2.019$, $g_y = 1.937$, and $g_x = 1.922$ and with linewidths of $L\hat{z} = 7.5 \times 10^{-4}$, $L\hat{y} = 7.6 \times 10^{-4}$, $L\hat{x} = 1.05 \times 10^{-3}$ tesla (Fig. 2*D*). This component is reducible with NADH and is tentatively assigned to cluster N-1b. These g values and relative spin concentrations are almost identical to those reported earlier by Orme-Johnson and co-workers (55) in complex I reduced with NADH.

The EPR spectrum of an additional N-1 type cluster with apparent g values of 2.03, 1.95, and 1.91 was observed at redox potentials below −450 mV. The quantitation of spectra poised at lowest E_h (−518 mV) obtained at pH 8 has revealed that 1.5 times more spins are seen relative to that of N-1b. We tentatively identified this very low E_m (< -500 mV) component as cluster N-1a. The detailed spectral lineshape of this component, however, is difficult to simulate as a single component; it appears that either cluster N-1a is partially modified during the isolation of complex I or that there is a spin-spin interaction between cluster N-1a and N-1b (or with another component) similar to the interaction of clusters S-1 and S-2 in succinate dehydrogenase. The existence of cluster N-1a has been questioned by Albracht and colleagues (58, 59) because no additional resonance absorbance was detected on addition of dithionite together with methyl viologen to the NADH-reduced complex I. However, to obtain this very low potential (E_h) range, an extremely reducing environment is required with mediator dyes such as triquat (1,1-trimethylene-2,2-pyridilium dibromide) and ICI (1,1′-dimethyl-2,2′-bipyridilium) (60). In the earlier studies using complex I prepared in King's laboratory, a high-potential component ($E_{m8.0} = -150$ mV) was assigned to a modified cluster N-1b, and the −390 mV component to N-1a. The high-potential component is now considered to be an extraneous impurity (not N-1b), because this component has been almost completely removed from complex I. The −390 mV component seems to be N-1b (with an E_m value ≈ -60 mV lower than that in the complex I isolated in Hatefi's laboratory), and the N-1a component was apparently not seen because of its extremely low E_m value.

These results identify cluster N-1b as an intrinsic component of complex I present in a concentration equivalent to that of FMN, and displaying a rhombic spectrum. Clusters 1a and 1b reported by Albracht (58) appear to correspond to cluster N-1b of Ohnishi and co-workers (60).*

*Hearshen and Dunham's analysis of the data of Albracht and Ohnishi indicates that a single rhombic component with g strain is sufficient to account for the observed spectra at present signal to noise ratios. D. O. Hearshen, W. R. Dunham, S. P. J. Albracht, T. Ohnishi, and H. Beinert (1981) FEBS Letts, *133*, 287-290.

Cluster N-2 has the highest midpoint redox potential among Fe-S clusters in this segment of the respiratory chain (3). The cluster is present in approximately equal concentration to FMN, and displays an EPR spectrum of axial symmetry with derivative peaks at 2.054 and 1.922 (55). This cluster shows a phosphate potential- and pH-dependent midpoint potential, indicating its involvement in energy coupling at site I (56, 57).

The final identification of the EPR spectra of clusters N-3, N-4, and N-5 was achieved by computer simulations conducted by Albracht and his group (58). Orme-Johnson and colleagues distinguished EPR signals from 2 distinct Fe-S species, and assigned g values of (2.100, 1.886, and 1.862) for N-3 and (2.103, —, and 1.864) for N-4. Subsequent potentiometric analysis of these clusters in pigeon heart submitochondrial particles (SMP) (61) indicated that the $g = 1.86$ and 1.89 signals correspond to the g_x peak of N-3 and N-4, respectively, based on their different midpoint potentials. Ohnishi attributed the slight shift of the g_z peak in the titration to the overlapping of two components with slightly different g values in the same fashion as Orme-Johnson and co-workers (55). More recently, Albracht and colleagues (58) conducted computer simulations of the spectral lineshape of individual clusters, assuming that each cluster has a spin concentration equal to that of the flavin. He came to the conclusion that the g_z peak of N-3 is 2.037 rather than 2.10. The $g = 2.037$ signal was initially missed by other workers (55, 61)*; its identity as the g_z peak of N-3 was confirmed by the demonstration of concurrent power saturation of the 2.037 and 1.86 signals of complex I poised at an appropriate redox potential (3).

The revised assignment of the N-3 and N-4 spectra necessitated a reinterpretation of the spectra of clusters N-5 and N-6. Signals observed at very low temperatures could be attributed to partially saturated N-4 signals and an additional cluster N-5 with an extremely fast spin relaxation, discarding cluster N-6 (58). The spin concentration of N-5 is less than 0.25 per FMN, while all other clusters seem to be present at a concentration equivalent to FMN in complex I. All clusters except N-1a and N-1b appear to be of tetranuclear structure (36), although the recently identified trinuclear structure (44–46) has not been completely excluded.

Recently Singer and colleagues (62) reported that binuclear and tetranuclear Fe-S clusters in the complex I segment are present in a ratio of 2:1, based on core extrusion experiments conducted on a large molecular weight NADH dehydrogenase (63).

*Albracht and co-workers (58) named the component with g values of (2.037, 1.92–1.93, 1.863) cluster N-4, and the component with g values of (2.103, 1.93–1.94, 1.884) cluster N-3. We have, however, retained our original nomenclature for N-3 and N-4, with the revised g_z value.

2.3 Electron-Transferring Flavo Oxidoreductase

The Fd-type EPR signal with $g_z = 2.08$ and $g_x = 1.89$ was first recognized by Ohnishi and co-workers at temperatures below 40°K in pigeon heart mitochondria or SMP, and its midpoint potential was determined to be $\sim +40$ mV (64).

Subsequently, Ruzicka and Beinert purified an Fe-S flavoprotein from beef heart mitochondria, following the EPR signals through the purification steps (65). The molecular weight is ~70,000; it contains 1 FAD, and 4–5 atoms each of nonheme Fe and acid labile sulfide. Acid treatment of the enzyme releases FAD, Fe, and S* in the ratio of 1:4:4. The enzyme gives an EPR spectrum of rhombic symmetry at $g_z = 2.086$, $g_y = 1.939$, and $g_x = 1.886$, and a flavin free radical signal at $g = 2$. Spin quantitation of the Fe-S signal showed approximately one spin per flavin. Thus the enzyme was proposed to contain a single tetranuclear $[4Fe\text{-}4S]^{+1(+1,+2)}$ Fe-S cluster (65). This Fe-S cluster functions as a redox component transferring electrons for the β-oxidation pathway of fatty acids.

2.4 The Rieske Fe-S Cluster

Recently, the Rieske Fe-S protein (66) has been isolated in a reconstitutively active form (67), and resolution and reconstitution studies using succinate-cytochrome *c* reductase complex showed that this Fe-S cluster is essential for ubiquinol-cytochrome *c* reductase activity (68). Removal of the Fe-S protein eliminates rapid reduction of cytochrome c_1 by succinate in the presence and absence of antimycin; it eliminates the rapid reduction of cytochrome *b* only in the presence of antimycin, while the deletion does not affect the reduction of cytochrome *b* in the absence of antimycin (69). The Fe-S protein is also required to demonstrate the oxidant-induced reduction of cytochrome *b* in the presence of antimycin (70).

A ubiquinone analog, UHDBT, appears to inhibit electron transfer from the Rieske Fe-S cluster to cytochrome c_1 (71, 72) and oxidant-induced reduction of cytochrome *b* (70). This inhibitor shifts the g_y value of the Rieske Fe-S cluster from 1.90 to 1.89, and causes a positive shift of its midpoint redox potential by 70 mV, suggesting the Rieske Fe-S protein as a possible candidate for a ubiquinone binding site. This Fe-S protein shows interesting spectral lineshape changes on lowering of the redox potential; it was suggested that this is associated with redox change of either cytochrome *b* or UQ (73, 74). The redox potential range for the lineshape change was found to be pH dependent (60 mV/pH) (75), favoring UQ as the component involved in the phenomena. This is discussed later.

3 INTERACTIONS AMONG INTRINSIC REDOX COMPONENTS

The interpretation of much of the data on electron carriers in the complexes of the respiratory chain hinges on the assumption that the properties of each electron carrier are more or less independent of the redox state of the other electron carriers. This assumption allows great simplification in the analysis of data, and can sometimes be justified after the fact. However, it can be shown to be unjustified in other instances, and can in principle lead to important conceptual errors in the study of the biochemical function of the electron carriers. When interactions do occur, ambiguity in the interpretation of results can be introduced. On the other hand, interactions can provide important clues to the structure and function of complex electron-carrying proteins.

Several kinds of interaction can be postulated. Redox interactions have been proposed to account for the apparent E_m's of heme groups in cytochrome oxidase (76, 77), *b* cytochromes (16, 78), and Fe-S clusters in succinate dehydrogenase (37). Briefly, if two groups interact in this way, reduction of one group causes the midpoint potential of the other group to change by an additive factor Δ. This might be caused by direct electrostatic interaction between the reducing electrons, or it might be mediated by redox-linked protein conformational changes. Clearly, analogous interactions could induce changes in the pK's of associated ionizable groups.

The reduction of one group might also trigger spectral changes in another group. This has been postulated in cytochrome oxidase (77), complex III (73–75), and complex I (3, 79). A 5 nm shift in a heme absorption band near 600 nm corresponds to a change in energy level spacing of $\approx 140\ \text{cm}^{-1}$. This sort of change in the spacing of the ferrous Fe *d* orbitals of a binuclear Fe-S cluster would lead to a shift of ≈ 5 G in the position of the central ESR line in the $g \sim 1.94$ region. Spectral shifts could conceivably be caused by either Coulomb interactions or conformational changes.

Magnetic interactions between adjacent paramagnetic groups are also of importance. Although compared to electrostatic interactions they are far too weak to play a role in the chemistry of the respiratory complexes, the possibility of magnetic interactions often must be considered to adequately interpret EPR data from multielectron carrier complexes. When magnetic interactions can be detected, they can provide important structural clues.

Two main types of magnetic interactions can be observed (80). The dipole-dipole interaction, $\mathbf{g}_1\mathbf{g}_2\,\beta^2\,\mathbf{S}_1\mathbf{S}_2/\mathrm{r}^3 - \mathbf{g}_1\mathbf{g}_2\,\beta^2\,(\mathbf{S}_1\cdot\mathbf{r})\,(\mathbf{S}_2\cdot\mathbf{r})/\mathrm{r}^5$, is a measure of the strengths of the interacting dipoles and the average of the cube of the separation of the interacting unpaired electrons (r^3). The exchange interaction, $-2\mathbf{S}_1\cdot\mathbf{J}\cdot\mathbf{S}_2$, is a measure of the orbital overlap. Unfortunately, because interpretation of the spectra of spin-coupled species can be difficult, all the theoretically available information is not always obtained.

3.1 Succinate-UQ Oxidoreductase Segment

The spin relaxation of the covalently bound FAD semiquinone is faster in SDH preparations, in which the $g = 1.94$ species can be observed, than the relaxation of flavin semiquinones in metalfree flavoproteins such as flavodoxin. This is true even in preparations such as reconstitutively inactive SDH in which center S-3 is modified to an EPR-inactive form. If the binuclear clusters giving rise to the $g = 1.94$ signals are destroyed by incubation at low pH (pH 4), however, the relaxation of the SDH flavin reverts to the slower rate characteristic of metalfree flavoproteins. This suggests that the relatively rapid relaxation of SDH flavin is caused by magnetic interactions with the binuclear Fe-S clusters. Since no effects of spin coupling can be observed on the lineshape of the radical, the FAD–Fe/S distance is at least 12 Å. The enhanced relaxation of the radical is consistent with an FAD–Fe/S distance of 12–18 Å (42).

As mentioned previously, the EPR spectra of fully reduced SDH preparations exhibit a number of interesting features not observed after reduction with succinate. In the former both clusters S-1 and S-2 are paramagnetic, while in the latter only S-1 is paramagnetic. The type of change seen after complete reduction depends on the SDH preparation used. In addition to the enhanced relaxation observed at low temperatures ($T < 12°$K), lineshape changes are observed, particularly in the $g = 1.94$ region, in the reconstitutively inactive preparations. These range from broadening of the central resonance at 10–12°K to well-resolved splitting of ≈ 23 G as the temperature is lowered further (37).

In the earlier reconstitutively active SDH prepared according to King (22), the extra spin intensity was not observable at higher temperatures ($T >$ 30°K). A greater relaxation enhancement was observed after addition of dithionite than that seen in reconstitutively inactive enzymes. Broadening of the central peak was observed at low temperature, but splitting could not be resolved.

In intact systems, such as succinate-cytochrome *c* reductase, mitochondria, and submitochondrial particles, no additional spin intensity can be observed after dithionite reduction, but pronounced relaxation enhancement is seen. Recent soluble preparations of high purity and full reconstitutive activity are apparently similar to the intact membrane-bound enzyme in this respect, as shown earlier (Fig. 1) (40, 41).

The effects of complete reduction of the enzyme on the spectral properties of center S-1 demonstrate that these properties depend on the redox state of some other group in the enzyme. The appearance of the lineshape changes at $g = 1.94$ and the onset of relaxation enhancement during reductive titration coincide with the appearance of additional spin intensity associated with $g = 1.94$ species in preparations where this is observed. The potential at which this occurs is far below the midpoint potentials of the other prosthetic groups (center S-3 and flavin) in the enzyme. We have therefore postulated that the spectral effects (lineshape changes and relaxation enhancement) arise from magnetic interactions between two binuclear Fe-S clusters, S-1 and S-2. The splitting and broadening of the central peak at low temperatures can be simulated by assuming dipolar coupling between two similar clusters (37); small changes in distance or angular parameters are sufficient to reduce resolution from splitting to only broadening. Modest amounts of isotropic exchange coupling do not improve or degrade the fit substantially (if the g tensors are parallel, exchange terms have no effect on the spectra), but do introduce some additional uncertainty into the distance used for the dipolar coupling. When a point approximation is used, this distance is found to be 9–12 Å.

Enhanced relaxation is observed in the 8 Fe ferredoxins when both identical clusters are reduced compared to the half-reduced state (81). For this reason it is not clear whether the relaxation enhancement is purely a function of the magnetic interaction (possibly by producing a four level system) or, alternatively, whether S-2 has a fast relaxation mechanism independent of S-1, which can then transmit energy from S-1 to the lattice through a cross-relaxation mechanism. The fact that enhanced relaxation is observed in all SDH preparations and in mitochondria suggests that S-1 and S-2 are always present and are always spin coupled. It is not clear why additional spin concentration is not always observed, however. The Leigh effect (82), in which the EPR intensity of coupled spins is diminished, does not by itself explain the full magnitude of the spin concentration reduction in this system.

In addition to the magnetic interactions between the components of SDH, there may be important electrostatic interactions that contribute to the function of the enzyme. The additional observable spin concentration and/or enhanced relaxation associated with the reduction of center S-2 titrates with an apparent E_m of -400 mV in the soluble enzymes and -260 mV in membrane-bound systems. Since this is far below the midpoint potential of the succinate/fumarate couple, doubt has been expressed as to the function of center S-2 in electron transfer.

The observation of a low-potential component in potentiometric titration experiments can have more than one explanation, however. If S-1 and S-2 were two spectroscopically indistinguishable Fe-S clusters with the same midpoint potential, and reduction of one cluster led to a 224 mV drop in the midpoint of the other by anticooperativity, potentiometric titration of the system would reveal two components separated by 260 mV (36 mV of the separation arising from a statistical term 120 $\log_{10}2$), as shown in Figure 3*A*. In this case half of S-1 would titrate at the higher potential and half at the lower potential; the same would be true of S-2. If the fast relaxation of the S-1–S-2 system arises purely from spin-spin interaction, this may be the case in SDH. If S-2 has intrinsically rapid relaxation, S-1 and S-2 must have different E_m's, since no rapidly relaxing component is present after succinate reduction. S-2 might still have a midpoint as high as −90 mV in the presence of oxidized S-1, however. In that case only a few percent of the S-2 would titrate with the high-potential component. An interaction of ~ −170 mV would then provide the −260 mV separation of the components (in this case the statistical term is negligible) (Fig. 3*B*).

S-2 might thus have a function in electron transfer in either case. The constraint is changed from the statement that S-2 is nonreducible by substrate to the statement that S-1 and S-2 are not simultaneously reducible by substrate. Electron transfer between S-1 and S-2 is clearly not ruled out by the latter; it suggests, however, that if S-2 functions in electron transfer it must be in series rather than in parallel with S-1 (83).

As previously mentioned, electrons from succinate are transferred by SDH to ubiquinone in the native system. Soluble SDH preparations are incapable of transferring electrons to ubiquinone, possibly in part because electron transfer between the $n = 1$ electron-carrying Fe-S groups and $n = 2$ ubiquinone is impeded by the instability of the ubisemiquinone. This makes the transfer of the first electron into UQ thermodynamically extremely unfavorable.

Beinert and co-workers (4) observed EPR signals near $g = 2.04$, 1.98, and 1.96 at low temperatures in submitochondrial particles at intermediate stages of reduction. These could also be observed as kinetic transients in complex II. Sands' group simulated the spectra, assuming dipolar coupling between two radical species or between a radical and oxidized Fe-S cluster S-3. The signals were abolished by ubiquinone extraction and restored by readdition of UQ, implying that one of the interacting partners is a UQ radical.

Ingledew and co-workers eliminated center S-3 and SDH flavin as the other interacting species on the basis of potentiometric titrations (84, 85). More advanced experiments utilizing computer simulation of the ESR spectra have demonstrated that the two interacting species are electrochemically very similar and that any redox interactions between them are relatively unimportant (6).

The results of these investigations suggest that the signals arise from a pair of ubisemiquinone molecules, which we designate as Qs, associated with succinate dehydrogenase. These UQ molecules are probably bound to a membrane protein that binds the semiquinone form much more tightly than the quinol or quinone, since the apparent stability constant of the radicals is near

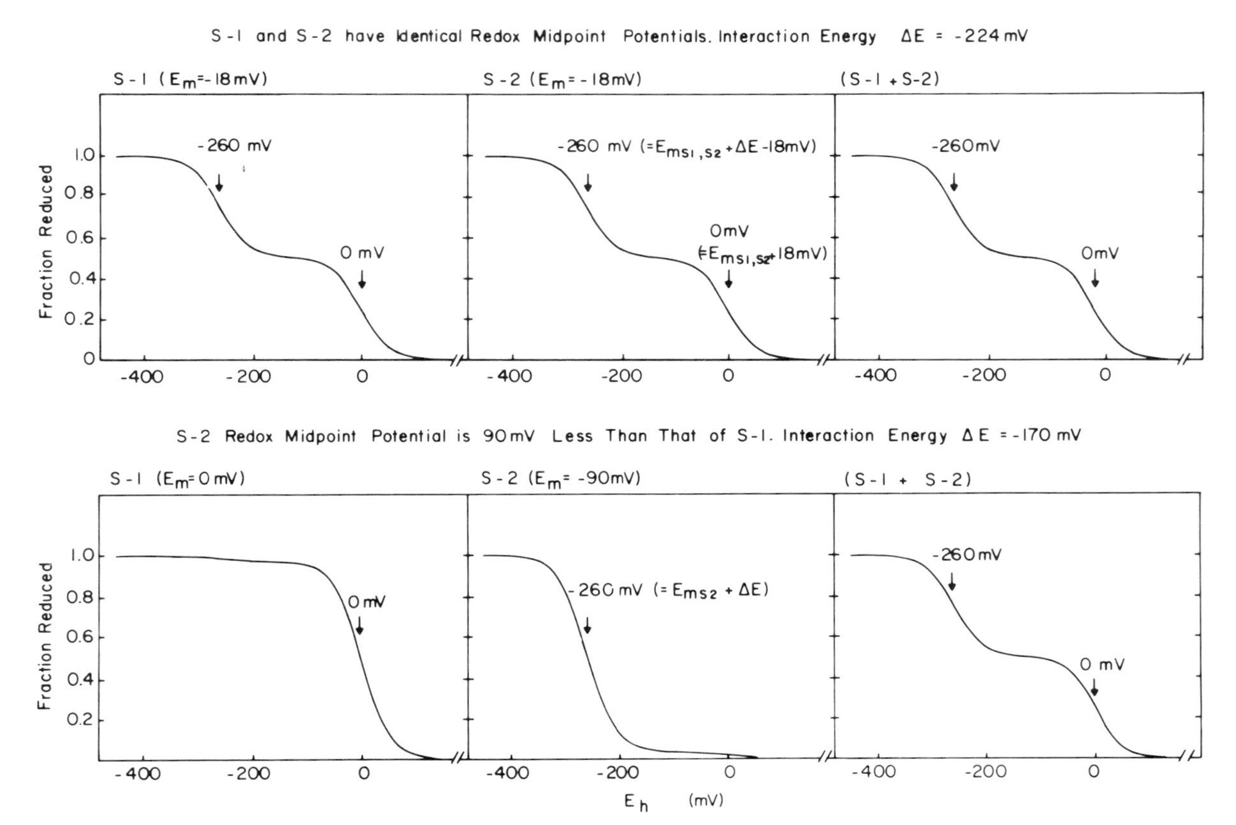

Figure 3. Possible redox interactions between clusters S-1 and S-2 in succinate dehydrogenase.

1 at pH 7.0. The interacting semiquinones are separated by ≈8 Å [a distance of 7.7 Å was used by Ruzicka et al. (4) in the best-fit case].

3.2 NADH-UQ Oxidoreductase Segment

NADH-UQ reductase is much more complex than succinate-UQ reductase, because in addition to the low molecular weight NADH dehydrogenase segment, which is analogous to succinate dehydrogenase in function, it contains the molecular machinery for site I energy conservation. The complexity of this system makes it more difficult to sort out the interactions between the components and to arrive at unambiguous interpretations of the sometimes complex behavior of the enzyme. Several observations suggest interactions between the individual electron carriers. The earliest reported phenomenon of this sort was the enhancement of the spin relaxation of the flavin $g = 2.00$ signal by a nearby transition metal, reported by Beinert in 1965 (86). This is analogous to the enhancement of the relaxation of the covalently bound SDH flavin by the binuclear Fe-S clusters S-1 and/or S-2 (42).

More recently, a shift in the position of the peak near $g = 2.10$ in the ESR spectrum of complex I was observed as the E_h was lowered (55, 61). Under conditions where only a small fraction of the low-potential Fe-S clusters are reduced, the peak position is near $g = 2.100$; on further reduction the peak position shifts to $g = 2.103$. This was first attributed to overlapping of two components (clusters N-3 and N-4) with slightly different EPR spectra and midpoint potentials. Later, however, Albracht and his group (58) showed that the g_z position corresponding to the g_x feature near 1.86, attributed to N-3, was at 2.037. Furthermore, all the intensity near 2.10 is necessary to account for the intensity at $g = 1.88$, attributed to N-4, and the spin concentration associated with a small shift of the $g = 2.10$ signal is much less than one per FMN. It is therefore unlikely that two such clusters with slightly different g_z values exist in complex I. Two components could conceivably result from inhomogeneity in the preparation, which could in turn result from either isolation artifact or isomerism.

An alternative explanation for the shift in the $g = 2.10$ peak involves interaction between the Fe-S clusters. The shift of the peak here might arise from a single Fe-S cluster, identical in all enzyme molecules. On reduction of a second redox component elsewhere in the enzyme, the g_z peak position of center N-4 would shift slightly, possibly because of coulomb and/or magnetic interactions with the additional unpaired electron. A coulomb interaction, as we have pointed out, might give rise to small spectral shifts by changing the spacings of orbitals in the Fe-S group by a few hundred reciprocal centimeters; a similar effect could be mediated by protein conformational changes.

Detailed potentiometric titration of complex I has revealed another case of interaction between electron carriers. While the other Fe-S clusters titrate as nearly ideal $n = 1$ components, the $g = 1.86$ feature of the EPR spectrum of cluster N-3 behaves in an anomolous fashion (3, 79). Below its midpoint potential the signal increases with reduction, as expected for an $n = 1$ component that is paramagnetic in the reduced state. As the potential is lowered further, however, the signal begins to decrease, reaches a local minimum, and eventually increases again to a maximum. This dip in the titration curve was attributed to spin-spin interaction between center N-3 and the FMN semiquinone. When the FMN is in the fully oxidized or fully reduced state, no interaction would be observed, since the flavin would be diamagnetic, and the signal at $g = 1.86$ would be seen. When the FMN is in the semiquinone state, the presence of a strong interaction between the flavin radical and center N-3 could cause a loss of both signals. In either case the dip would result, mirroring the typical bell-shaped semiquinone titration curve. This interpretation was strengthened by the observation of weak "half-field" signals near $g = 3.9$, which could correspond to the so-called $\Delta m_s = 2$ transition in a system of two coupled $S = 1/2$ species, at appropriate values of E_h. Direct potentiometric titration of the flavin $g = 2.00$ signal recently showed that the dip potential of the $N-3$ ($g = 1.86$) titration corresponds to the peak of the bell-shaped flavin free radical titration curve (Ohnishi et al., unpublished).

Another possibility can also be considered to explain the dip. If the flavin semiquinone stability constant is much less than one, $n = 2$ behavior is expected in titrating the FMN from oxidized to fully reduced forms. A redox interaction between FMN and center N-3 such that reduction of one leads to a drop in the midpoint potential of the other could then explain the dip in the titration of N-3. If N-3 has a midpoint potential E_1 in the presence of oxidized FMN, it then has a potential of $E_1 + \Delta$ in the presence of reduced FMN. FMN has a midpoint E_2 for the $n = 2$ FMN/$FMNH_2$ couple when N-3 is oxidized, and $E_2 + \Delta/2$ when N-3 is reduced. For $E_1 > E_2$ reduction first leads to an $n = 1$ reduction of N-3. As FMN is reduced, the Δ term leads to oxidation of N-3.

Fortunately, these two possibilities can be distinguished by titration of the flavin semiquinone. The two models predict a different relationship between the flavin midpoint potential and the dip midpoint. Also, one depends on the flavin semiquinone being present in significant quantities, while the other effect is enhanced by semiquinone instability. Investigations along these lines are currently being conducted.

Spin coupling was also suggested between cluster N-1a and N-1b (or another Fe-S cluster), based on the spectral lineshape alteration at extremely low redox potentials (60).

The possibility of interactions of several kinds makes it important to carefully evaluate the stoichiometry of observed EPR signals. Some of these signals, accounting for much less than 1 spin/mol, may arise from Fe-S clusters in molecules in a different state. This is discussed more fully in the section on proton translocation mechanisms.

3.3 UQ-Cytochrome c Oxidoreductase Segment

The Rieske Fe-S cluster in complex III has been reported to undergo an interesting change in EPR lineshape as the potential is lowered through a range considerably below its midpoint potential (73–75). On the basis of the potential at which the change occurs, ubiquinone and/or the *b* cytochromes have been considered as possibilities for the "control" of the lineshape change. The recent discovery of the bound quinones (Q_s and Q_c) makes Q_c or possibly an as-yet-undiscovered bound quinone in this region attractive candidates for the role. The possible physiological relevance of interactions in this region and in site I are discussed in the section on redox-linked proton translocation.

4 ORIENTATION OF Fe-S CLUSTERS AND ASSOCIATED REDOX COMPONENTS IN THE MITOCHONDRIAL MEMBRANE

Recently, considerable effort has been expended in the study of the orientation of intrinsic chromophores in the mitochondrial inner membrane. Initial efforts focused on the heme groups of cytochrome oxidase and the *b* and *c* cytochromes, using oriented multilamellar preparations prepared by centrifugation of mitochondria or submitochondrial particles onto thin plastic sheets and partial dehydration under conditions of controlled humidity (87–89). This produces a sample consisting of a large number of approximately parallel sheets of membrane. While several other methods have been successfully used to prepare oriented samples in other systems, this has proved the method of choice with mitochondria.

The angular dependence of the low-temperature EPR spectra of these samples has provided information on the orientation of Fe-S and bound quinone groups (90). The orientation of coordinate systems related to magnetic properties such as the principal axes of the *g* or *A* tensors or the vector between two coupled species can be extracted with the aid of computer simulation (91). The simulation must take into account the orientations of the magnetic tensors and the degree of disorder in the system. Disorder is introduced as a gaussian distribution of allowed orientations, containing contributions from disorder in the alignment of the individual membrane sheets, of the proteins in the membranes, and of chromophores in proteins. Typical values of the disorder parameter are 15°–35° in "good" samples. If information is available relating the structure of the group to the magnetic parameters, the orientation of the group can in turn be deduced.

The Rieske Fe-S cluster and cluster S-1 in succinate dehydrogenase have their *g* tensors oriented so that the direction corresponding to the lowest numerical *g* value corresponds to the normal to the membrane. As this implies, the other two principal axes lie in the membrane. Although it is difficult to get the entire multilayer to equilibrate with added oxidant or reductant without destroying order, results with NADH-reduced samples suggest that of the other binuclear Fe-S centers at least N-1b has the same orientation.

The model of Gibson and co-workers (92) has been generally successful in accounting for the magnetic properties of binuclear Fe-S clusters. The symmetry of the EPR spectra and the structure of the binuclear cluster suggest that the *z* direction (corresponding to the *g* value above 2.00) in the *g* tensors lies near the Fe–Fe axis of the cluster. We thus expect the Fe–Fe axes of the clusters to lie in the plane of the membrane.

The spectra of the tetranuclear Fe-S clusters in these oriented samples are also indicative of a high degree of order, with apparent dichroic ratios ($A_\perp - A_{/\!/}/A_{/\!/} + A_\perp$) often in excess of 0.9. In contrast to optical spectra, however, the position as well as the transition probability of an EPR line are orientation dependent; furthermore, the spectra are first derivative with respect to field. Thus the orientation dependence and dichroism observed cannot be directly related to linear dichroism in optical experiments. A computer simulation is useful in minimizing errors in interpretation (91).

Unfortunately, the cubane structure of the tetranuclear clusters provides no clue to the relationship of the *g* tensors to the molecular coordinate system. We know, however, that the orientation of each such cluster is highly ordered with respect to the membrane. Surprisingly, *g* tensors of all the tetranuclear clusters of NADH dehydrogenase are oriented so that one axis is normal to the membrane and two lie in the membrane plane (see Fig. 4). This is not always the case for other groups [for example, the *a* heme of cytochrome oxidase (89, 91)].

The orientations of the *g* tensors of clusters N-2, N-3, and N-4 can be thought of as being connected by a series of 90° rotations about one of the *g* tensor's principal axes lying in the membrane plane. The cubane structures themselves, having at most nearly T_d symmetry, are not invariant to such rotations, but if the cylindrical symmetry of the sample is considered, we see that the spectra will always be invariant to rotations about the membrane

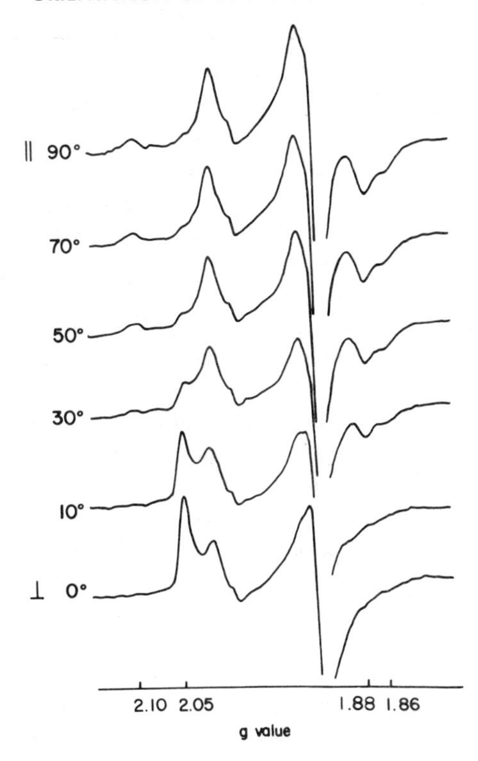

Figure 4. EPR spectra of multilayers of beef heart mitochondria reduced with dithionite, recorded at various orientations relative to the Zeeman field. EPR conditions: microwave power, 5 mW; modulation amplitude, 1×10^{-3} tesla, at 100 KHz; sample temperature, 13°K.

normal. Thus the orientations of these clusters with respect to the membrane are probably structurally nearly equivalent.

The orientations of both the tetranuclear and binuclear Fe-S clusters must be determined by the secondary and tertiary structure of the membrane proteins of which they are the prosthetic groups. For example, if the clusters were carried between two or more sections of helix or random coil running normal to the membrane, the binuclear clusters would be expected to have the Fe–Fe axis in the membrane plane, while the tetranuclear clusters would be restricted to a few possible orientations consistent with the observed allowed orientations of their g tensors. The structures of integral membrane proteins such as bacteriorhodopsin and cytochrome oxidase are believed to involve sections of polypeptide chain containing mainly hydrophobic residues running through the normal to the membrane and connected by loops of polypeptides exposed to the aqueous phase (93–95). It is interesting that the heme groups of integral membrane heme proteins such as cytochrome oxidase, cytochrome c_1, and the b cytochromes (88) are oriented so that the heme normal lies in the membrane plane. This is the orientation in which a heme could be most easily suspended between two polypeptide segments normal to the membrane. The orientation of the other two principal axes is not fixed.

Exceptions to these rules are of two general kinds. EPR spectra of loosely associated proteins such as cytochrome *c* or ETF Fe-S cluster show little or no orientation dependence (89, 90). The spectra of cytochrome b_5, which is attached to the membranes of microsomes only by a hydrophobic tail, also show little orientation dependence (96). This might be caused by dissociation of peripheral proteins from the membrane during sample preparation or freezing. Alternatively, since the constraints imposed by the membrane on the polypeptide chains are not the same as for integral proteins, the membrane normal might lie near the so-called magic angle with respect to the *g* tensors of these centers. Species with this special orientation should have spectra with much less angular dependence than species with other orientations. The spectra still should not be completely orientation independent. Although the effects of sample orientation are more subtle than for other cases, careful study should be able to distinguish between the possibilities outlined above.

The other important case is the mammalian cytochromes P-450, which as a class, are integral membrane proteins having their heme groups parallel to the membranes in which they are imbedded (96, 97). These enzymes transfer electrons from hydrophilic donors (NADPH) to hydrophobic substrates in the course of hydroxylation reactions with either a soluble Fe-S protein or a membrane-bound flavoprotein functioning as their immediate reductant. Experiments using paramagnetic probes have shown that the heme group lies at or near the surface of the membrane (98). It is thus possible that in this case the axial ligand is attached to a loop of protein in the aqueous phase, resulting in a very different orientation from those found for the deeply buried hemes of the *a* and *b* cytochromes and cytochrome c_1.

When magnetic interactions of sufficient strength to resolve line splitting are present, oriented multilayers can be used to provide information on the orientation of the vector connecting two paramagnetic species with respect to the membrane providing that it is possible to resolve the dipolar component of the interaction, which contains the angular information. The angular dependence of the dipolar interaction is $(1 - 3\cos^2\theta)$, where θ is the angle between the Zeeman (applied) magnetic field and the vector connecting the two species. In solution all values of θ are present. The EPR spectra of a pair of identical isotropic dipole-coupled species consist of four lines at $\pm$D and $\pm$2D from the position of the species in the absence of coupling; the outer pair of lines arises from molecules oriented so that 0° and thus $(1 - 3\cos^2$

$\theta) = \pm 2$; the more intense inner lines arise from the many possible orientations for which $(1 - 3\cos^2\theta) = \pm 1$.

The dipole-coupled quinone pair (Q_s), probably functioning as the immediate electron acceptor from SDH, is closely approximated by this model when both quinones are in the semiquinone redox state. The small anisotropy ($g_x = 2.0041$, $g_y = 2.0066$, $g_z = 2.0066$,) in the g tensors of the bound semiquinones, arising from the angular momentum contribution of the oxygen p orbitals, is completely dominated by the dipolar coupling. Although the best spectral fits are obtained when the rings are edge to edge, the possibility of inhomogeneity (populations of Q_s with slightly different values of D) makes it impossible to draw definitive conclusions from present data. While exchange interactions between such nearly identical species cannot contribute to the gross features of the spectra (such as splitting), if the two quinone rings have somewhat different orientations with respect to the membrane, exchange coupling could make this difficult to detect.

Although there is uncertainty in the orientation of the individual quinone rings, the orientation of the quinone–quinone vector is well established (90). When an oriented multilayer sample is used, the intensity of the outer pair of lines reaches a sharp maximum when the magnetic field is aligned along the axis of the stack (normal to the plane) (see Fig. 5). The inner lines are invisible at this orientation; they reach a maximum when the field has an in-plane orientation. Clearly, $\theta = 0$ when the field is along the membrane normal; thus the quinones are oriented in a transmembrane fashion. Using computer simulations, we can state that the vector connecting the two quinones is

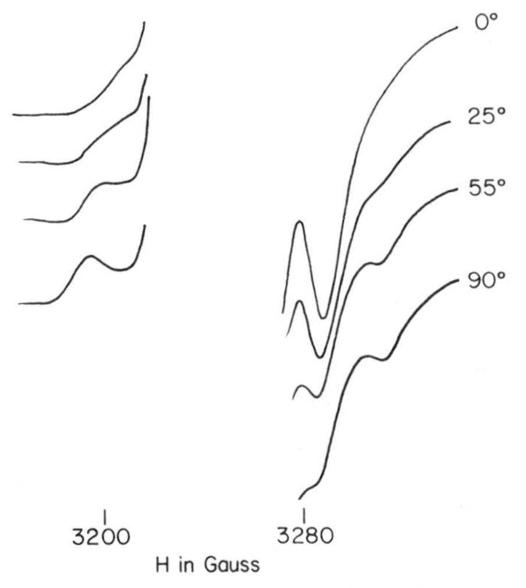

Figure 5. EPR spectra recorded at 12°K of oriented multilayers prepared from beef heart mitochondria. The multilayers were prepared in the presence of 100 μM antimycin; after drying they were dipped in a mixture of 10 mM fumarate and 20 mM ascorbate, inserted in a quartz sample tube, and frozen in liquid N_2. The signals in the wings are caused by the spin-coupled ubiquinone pair. The large central peak arises from Fe-S cluster S-3.

within 10° of the normal to the plane. With much less certainty we can deduce from the computer simulations that the most likely ring orientation is with the ring normal lying in the membrane plane.

5 TOPOGRAPHICAL DISTRIBUTION OF Fe-S CLUSTERS AND ASSOCIATED REDOX COMPONENTS IN THE MITOCHONDRIAL MEMBRANE

Various biophysical and biochemical approaches are useful in the determination of the spatial arrangement of redox components in the mitochondrial inner membrane to obtain a coherent picture of electron- and proton-transfer mechanisms. Approaches such as surface labeling (99) or cross-linking (100), combined with gel electrophoresis, give information on the arrangement of protein subunits of dehydrogenases and other components relative to the neighboring subunits and to the mitochondrial membrane. Determination of the distance from extrinsic probes attached to the surface of the protein or membrane to intrinsic redox components using fluorescence energy transfer has proved useful (101, 102). Case and Leigh devised a novel procedure to directly measure the distance of redox centers from the cytoplasmic or matrix surface of the mitochondrial membrane by utilizing the alteration of EPR parameters of intrinsic redox components by membrane-impermeable extrinsic paramagnetic probes, such as Gd and Ni complexes (103). Unfortunately, these have had only limited applicability because of their relatively small magnetic moment. The technique has been greatly extended by the development of more quantitative analysis and, more importantly, by the use of stronger paramagnetic probes, such as Dy complexes (104). Dy has a total angular momentum quantum number (J) of $^{15}\!/_2$ and has an extremely short spin-lattice relaxation time [$T_1 = 7.1 \times 10^{-11} \exp(10.4/T)$ sec] at low temperatures (8–40°K) (105).

The interaction of Dy with intrinsic redox components takes the form of enhanced relaxation (increased $P_{1/2}$) and/or line broadening (increased ΔH) because of spin-spin interaction between intrinsic redox components and the dysprosium complex. The saturation parameter ($P_{1/2}$) is often taken as the microwave power needed for half-saturation of the EPR signal. More rigorously, $P_{1/2}$ is the incident microwave power at which the saturation condition $\gamma^2 H^2{}_1 T_1 T_2 = 1$ is satisfied, where γ is the magnetogyric ratio and the value of H_1 is $H_{1/2}$, the microwave magnetic field corresponding to $P_{1/2}$ (43). When defined in this way, the value of $P_{1/2}$ cannot be directly determined from saturation curves, but must be estimated using simulations. To convert EPR parameters to distances, we have analyzed the effects of various dysprosium complexes on the $[4Fe\text{-}4S]^{+3(+2,+3)}$ cluster in the isolated *Pseudomonas gelatinosa* Fe-S protein (105) and mammalian cytochrome *c* (106, 107), where we can estimate the distance between the redox center and dysprosium ions based on X-ray-crystallographic information (108, 109). In these systems effective paramagnetic probes appeared to bind to the surface of the protein. Thus we used a model where $1/T_1$ is proportional to r^{-6} (r is the distance from the active Fe to the Dy ion). In the temperature range of these EPR measurements T_2 of the intrinsic redox components is constant; thus $\Delta P_{1/2}$ is proportional to $1/T_1$. We have obtained a parameter for distance measurements, $\Delta P_{1/2} = 4.12 \times 10^8 \times r^{-6} \exp(-12.5/T)$, from the studies on *R. gelatinosa* Fe-S protein (105). Here $\Delta P_{1/2}$ is the enhancement of $P_{1/2}$ per millimole of the paramagnetic probe, and T is the sample temperature. To estimate effective distances from the protein surface to the paramagnetic cluster, we reduce r by 5 Å to account for the size of the Dy-EDTA complex, and by 3 Å for $Dy(NO_3)_3$.

5.1 Succinate-UQ Reductase Segment

The arrangement of complex II subunits in the mitochondrial inner membrane was examined by surface labeling of mitochondria or SMP using [^{35}S] diazobenzene sulfonate combined with immunoprecipitation (99). Complex II was shown to span the mitochondrial inner membrane, with both the FP and IP subunits of SDH on the matrix side of the membrane, but with IP partly shielded from water. One of the two small peptides (27, 28) of complex II [13,500 dalton component, which is likely to be the Q binding protein (29)] was shown to be exposed to the cytosolic surface but not to the matrix surface of the inner membrane. More recently, the same group examined the interaction of complex II subunits with phospholipids having reactive nitrene in the head group region of the bilayer and on the methyl terminus of one of the fatty acid chains (110). The FP subunit in complex II was found to be held above the bilayer, and thus was not labeled by either probe. The IP subunit and two small molecular weight subunits were shown to be inserted into the interior of the lipid bilayer, since they were labeled by both probes. Based on these results, a schematic picture of the subunit arrangement of complex II has been presented by Capaldi's group; this information provides the framework for Figure 6.

The distance of SDH redox components from the barrier surface of complex II, SMP, and mitochondria was estimated using paramagnetic probes. As presented in Table 1 (111), cluster S-3 is located ≈ 12 Å away from the complex II surface and ≈ 16 Å from the matrix surface of the mitochondrial inner membrane, but greater than 33 Å from the cytosolic surface. These results suggest a partly buried location of the cluster S-3 on the matrix side of the inner membrane, as illustrated in Figure 3. Covalently bound SDH flavin

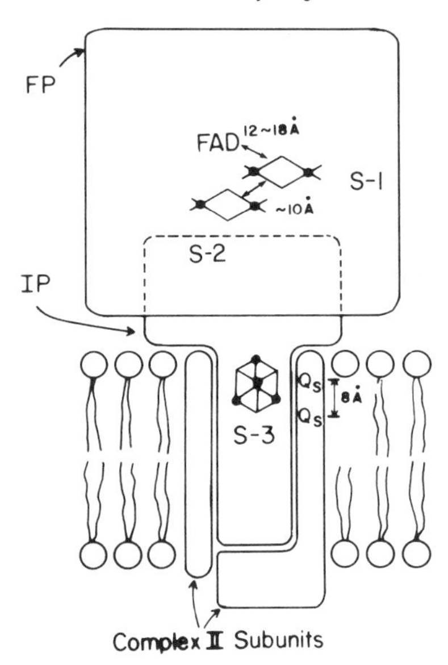

Figure 6. Schematic presentation of the topographical distribution of redox components of succinate dehydrogenase, covalently bound FAD, and clusters S-1, S-2, and S-3, as well as a pair of bound UQ species associated with the succinate-UQ reductase segment of the respiratory chain.

Table 1 Distance Calculation of Fe-S Cluster S-3 From the Dy Barrier Surface in Various Systems

	$P_{1/2}$ (mW)	$P_{1/2}$ (mW/mM)	Effective Distance (Å)[a]
(A) Succinate-ubiquinone reductase (complex II)			
Control (+10 mM La-EDTA)	8.2		
+5 mM Dy-EDTA	24.0	3.2	12
+10 mM Dy-EDTA	37.0	2.9	12
+10 mM $Dy(NO_3)_3$	86.0	7.8	11
(B) Beef heart submitochondrial particles			
Control (+5 mM La-EDTA)	0.8		
+2 mM $Dy(NO_3)_3$	3.0	1.1	17
+5 mM $Dy(NO_3)_3$	8.0	1.4	16
(C) Beef heart mitochondria			
Control	2.6		
+5 mM $Dy(NO_3)_3$	2.6	(<0.05)	>33

[a] Effective distances were calculated using the expression $\Delta P_{1/2} = 4.12 \times 10^8 \times r^{-6} \exp(-12.5/T)$ (mW/mM), at 7°K and subtracting 5 or 3 Å for the EDTA or $(NO_3)_3$ complexes, respectively. Reproduced with permission from reference 106.

and cluster S-1 in the FP subunit have been found to be ~25 Å and ≈22 Å, respectively, away from the protein surface in both soluble SDH and in complex II (Ohnishi et al., unpublished). These results, combined with information on the spatial relationship between S-1 and S-2 (37) and between flavin and S-1 (40), suggest that the prosthetic groups in the FP subunit are arranged as illustrated in Figure 6. The most probable orientation of the Fe-S clusters relative to the mitochondrial membrane is also shown in Figure 6. More detailed studies are now in progress.

5.2 NADH-UQ Reductase Segment

A considerable amount of information is also available on the topographical arrangement of subunit peptides and redox components in the NADH-UQ reductase segment. However, because of the complexity of the system a great deal more work is required at different levels of structural organization.

The distribution of constituent polypeptides in the mitochondrial inner membrane has been studied by the surface labeling of complex I, SMP, or whole mitochondria with the impermeable probes, diazobenzene [^{35}S] sulfonate or lactoperoxidase-catalyzed [^{125}I] iodination, followed by SDS/polyacrylamide gel electrophoresis (112). None of the three FP subunits is exposed to the surface, even in isolated complex I, although the substrate (NADH) binding site resides on the 51,000 dalton subunit (113). The large molecular weight hydrophilic subunits (75,000, 49,000, and 29,000 daltons) of the IP fraction are transmembranous. Several hydrophobic protein subunits are not exposed to either side of the mitochondrial inner membrane surface, but are in contact with the hydrophobic interior of the membrane (112).

The topographical distribution of Fe-S clusters of this segment of the respiratory chain is presently being investigated using various preparations such as complex I, SMP, and mitochondria. Most of the Fe-S clusters appear to be deeply buried, even in isolated complex I, and are thus not exposed to bulk solvent in vesicular preparations of either sidedness. Preliminary results indicate that clusters N-2 and N-3, for example, are more than 30 Å away from the cytosolic and matrix side surface of the mitochondrial inner membrane (Ohnishi et al., unpublished).

5.3 Rieske Fe-S Cluster

Topographical assignment of the Rieske Fe-S cluster is also important because of its suggested role for Q binding (71, 72). EPR signals of the Rieske Fe-S cluster exhibit both spectral broadening and relaxation enhancement in the presence of 10 *mM* $Dy(NO_3)_3$, when added on either the matrix or cytosolic surface of the mitochondrial inner membrane (Fig. 7), indicating

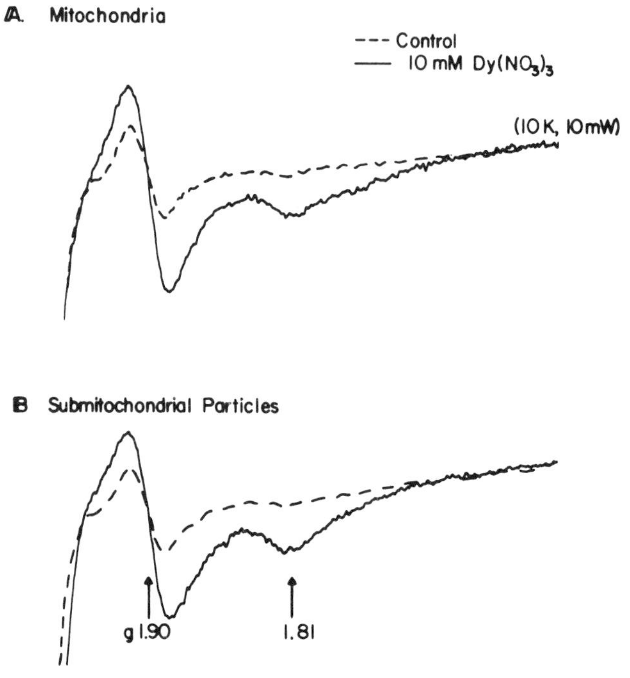

Figure 7. EPR spectra of the Rieske Fe-S cluster examined with and without 10 *mM* $Dy(NO_3)_3$ in uncoupled pigeon heart mitochondria poised potentiometrically at E_h = 150 mV and in uncoupled pigeon heart SMP reduced with 5 *mM* ascorbate.

its location in the middle of the inner membrane. Distance measurements conducted with the isolated Rieske Fe-S protein (111) showed that the cluster is 19 Å from the barrier surface of the protein; it is also deeply buried within the molecule.

6 ENERGY TRANSDUCTION

It is now well established that the high-energy intermediate that links electron transfer to ATP synthesis is a proton electrochemical gradient (10, 11). Even before the nature of this intermediate became known, it was predicted that the apparent oxidation reduction potentials of electron carriers associated with a coupling site should depend on the ratio of [ATP] to [ADP] [Pi]. This was at first accounted for by postulating a chemical intermediate. Of the many models incorporating a high-energy compound of this type, the most sophisticated were perhaps the two potential transductase models of DeVault (114). These models are of more than historical interest, since they form a useful basis from which various proton translocation schemes can be analyzed.

Mitchell (10) proposed that loops consisting of alternate transmembrane H_2- and electron-carrying arms were responsible for proton translocation. A small mobile hydrogen carrier would be reduced on one side of the membrane, the N side (*n*egatively charged side = matrix side), acquiring an equal number of protons and electrons. Moving to the P side (*p*ositively charged side = cytoplasmic side) of the membrane, the H_2 carrier would be oxidized and return to the N side in the oxidized unprotonated state. The protons would be disgorged into the P side medium and the electrons would return to the N side via a series of electron carriers.

Since in the classical loop model transmembrane movement of the proton is simultaneous with transmembrane electron movement, the action of the H_2-carrying arm is not electrogenic, but does directly contribute to the proton concentration gradient. The electron-carrying arm transfers electrons alone from the P side to the N side in the electrogenic step. The loop model implies that the intermediate causes of the ATP dependence of the apparent midpoint potentials of electron carriers are transmembrane electric potential ($\Delta\psi$) and pH gradients (ΔpH). Several authors (10, 114–116) have shown that the apparent E_0's of loop components are affected by transmembrane movement of charge against $\Delta\psi$ or proton translocation against ΔpH. For example, if E_1 is the E_h imposed by a substrate couple at A in Figure 8, the effective E_h at C is $E_1 + R'T/NF\,(\Delta\text{pH}) - \Delta\psi$ [ΔpH = (N side pH) − (P side pH); $\Delta\psi$ = (N side potential) − (P side potential)]. Similarly, the effective E_h at A if a potential E_2 is imposed at C is $E_2 - R'T/NF\,(\Delta\text{pH}) + \Delta\psi$. In this classical loop the effects of $\Delta\psi$ are linear.

Mitchell has described proton pumps that are foreshortened loops (117) in which access to one or both sides is provided by channels which may or may not be "proton wells" (see Fig. 9). A proton well is a channel that admits protons but excludes other charged species. Therefore such a channel could partially convert the local $\Delta\psi$ into an internal pH gradient. Clearly, if the ionizable group of a pH-dependent component lies in a proton well, nonlinear dependence of E_m on $\Delta\psi$ will be observed if the pK of either the oxidized or reduced forms lies within the range of local pH obtainable by varying $\Delta\psi$. This form of nonlinearity could be easily detected by varying the bulk pH.

The loop model predicts a site ratio of one net proton translocated per electron transferred, driven by a fall in the potential energy of the electrons. On this basis the site I loop hypothesis was supported and extended by Garland

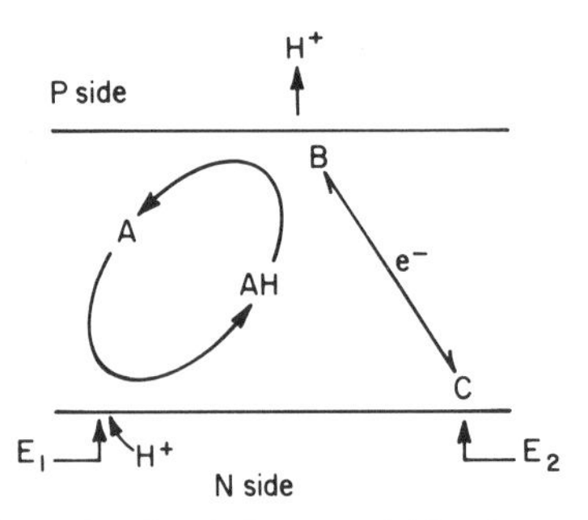

Figure 8. Simple loop model for the coupling of proton translocation to electron transport. A is a mobile H_2 carrier, B and C are electron carriers, with an H^+/e^- ratio of 1.

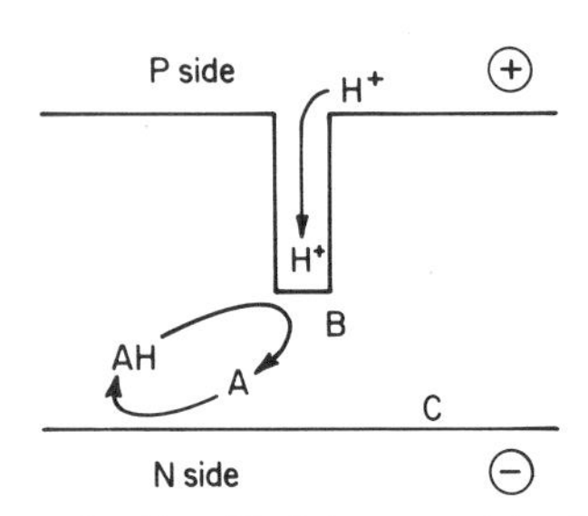

Figure 9. Simplified loop model with a proton well.

and co-workers (116), who directly measured one proton translocated per electron transferred from NADH to UQ.

Proton translocation need not be carried out by loops, however. Bacteriorhodopsin contains a light-driven proton pump that operates without H_2 or electron carriers (118, 119). Numerous proposals have been made in which proton translocation is coupled to electron transfer by nebulous "ion pump" mechanisms; Mitchell has argued that such pumps are topologically identical to loops (117). Although a class of looplike model pumps can be constructed, other types of pumps related to DeVault's transductase models appear equally feasible (114). An example of this is the site I pump proposed by Skulachev (120).

NADH-UQ reductase contains a group of at least three nearly isopotential Fe-S clusters (N-1b, N-3, N-4) with apparent E_m's of ≈ -250 mV. A fourth Fe-S cluster, N-1a, has an apparent E_m of -380 mV. Cluster N-2, with an apparent E_m of 20 mV, is the obvious terminal electron carrier. The gap in potential between the isopotential group and N-2 suggests that energy transduction takes place in this site. In addition, the E_m's of centers N-2 and N-1a are pH dependent between pH 6 and 9 (57), signifying that reduction is coupled to the uptake of a proton in this pH range. Experimental data relevant to site I energy conservation were reviewed extensively in reference 3.

Gutman and co-workers (121) observed ATP-driven reoxidation of center N-2 in submitochondrial particles that were pretreated with a minimal amount of NADH in the presence of piericidin A. This is often referred to as a lowering of the apparent E_m of cluster N-2, and was confirmed by Ohnishi and colleagues (56). Ohnishi showed that in mitochondria or SMP poised at relatively high potential ($< +50$ mV) with the succinate/fumarate couple or with redox mediators, ATP addition causes reduction of cluster N-2. As pointed out be DeVault (114), the opposite effects of ATP on cluster N-2 poised from the high-potential side or low-potential side imply that N-2 is part of the site I energy conservation mechanism.

Early observations of ATP-dependent E_m's were interpreted in terms of chemical intermediate models. Mitchell (115) and Garland and co-workers (116) proposed chemiosmotic models for site I energy conservation based on the loop principle. In these schemes NADH reduces FMN on the N side of the membrane. FMN functions as the H_2 carrier, reducing cluster N-1 on the P side with the release of two protons; the electrons are transferred back to the N side to cluster N-2. This was consistent both with the data of Gutman and colleagues (121) and the experimentally determined site ratio of 1 H^+/e^- found by Garland's group (116).

DeVault (114) pointed out that, while chemiosmotic theory can account for the so-called *ATP-dependent midpoint potential shifts* of cluster N-2 as well as the older chemical intermediate theory, cluster N-2 cannot be on the N side of the loop and still account for the fact that the redox poise of N-2 is ATP sensitive when poised from the NADH/NAD or succinate/fumarate sides.

Other ATP effects have been reported in this region. Ohnishi reported oxidation of cluster N-1a in mitochondria poised with low-potential redox mediators (56). Singer and Gutman reported reduction of center N-1 in mitochondria poised with the NADH/NAD couple (122); they attributed this behavior to the involvement of the Fe-S cluster in a site I loop. Careful analysis reveals, however, that the observations of Gutman and co-workers are inconsistent with the function of an electron carrier in a site I loop; Ohnishi's observations are consistent with most site I energy conservation models. Ingledew and Ohnishi (57) subsequently reported ATP-dependent oxidation of cluster N-4 in beef heart submitochondrial particles when the redox potential of the system was poised with the $NADH/NAD^+$ couple. These observations suggest possible roles for cluster N-1a and/or N-4 in energy conservation, but until more data are available only the involvement of cluster N-2 can be regarded as completely established.

Although it seems virtually certain that site I functions by a chemiosmotic mechanism in the general sense (coupling of electron transfer to ATP synthesis via a proton electrochemical gradient), evidence is accumulating against the original loop mechanism. Experiments designed to determine the sidedness of electron-transfer components by measuring the reduction of ferricyanide by NADH dehydrogenase indicate that no electron carriers below the rotenone inhibition site can reduce ferricyanide on the cytosolic side (P side) of the mitochondrial membrane (123). Experiments with extrinsic paramagnetic probes indicate that none of the Fe-S clusters in this region is near the cytosolic surface of the membrane (124). [For example, both clusters N-2 and N-3 appear to be located further than 30 Å from both surfaces of the mitochondrial inner membrane (Ohnishi et al., unpublished).] Mitchell has proposed a proton pump mechanism that is essentially a foreshortened loop in which access to the cytosol would be provided by a deep pore or proton well (117).

Recently, a stoichiometry of two protons translocated per electron was reported for site I (125–127). This stoichiometry is inconsistent with any loop-type formulation of site I (unless two loops with two different H_2 carriers were postulated!). The correct stoichiometry is still a matter of controversy. A potentially more serious defect in site I loop formulations (including looplike pumps) is the lack of a good candidate for the role of H_2 carrier. The FMN cast in this role by Mitchell and Garland is almost certainly the site of interaction with NADH. Resolution of complex I into subunits has provided evidence that the FMN resides on a small molecular weight NADH dehydrogenase (49–53) that is almost completely surrounded by other subunits (112), as described in a preceding section. It is thus extremely unlikely that the FMN functions as a mobile H_2 carrier. Free ubiquinone, while an excellent candidate for a site II H_2 carrier, has a midpoint potential far too positive (by ≈ 300 mV) to be seriously considered for site I. If FMN can be ruled out, an unknown H_2 carrier must therefore be postulated if the loop formulation is to be retained.

To tightly couple electron transfer to proton translocation, it is necessary to couple the accessibility of at least one electron carrier (to electron carriers on the high- and low-potential sides of the coupling site) to the accessibility of at least one proton carrier to the in and out sides of the membrane. To accomplish this, at least one electron-carrying group must exist in (at least) two states. In one state the carrier must equilibrate with other electron carriers on the low-potential side of the coupling site; in the other state the carrier must equilibrate with carriers on the high-potential side. In the former state a proton-carrying group must equilibrate with the N side aqueous phase; in the latter state the proton-carrying group must equilibrate with the P side aqueous phase. (We ignore the reverse case, since reduction generally favors protonation.) In no case can either electron or proton carriers equilibrate with both pools simultaneously. In addition, transitions between the states can be allowed only in certain redox/protonation states, and the electron and proton carriers can never change their accessibility independently.

In a loop this is accomplished by placing both the electron- and proton-carrying functions together on a small mobile molecule (the H_2 carrier). The site of reduction of the H_2 carrier is then placed on the N side of the membrane and the site of oxidation is placed on the P side: in one "state," the H_2 carrier is bound at its reduction site on the N side; the other state corresponds to H_2 carrier binding at the oxidation site on the P side. The H_2 carrier moves between the sites only in the protonated reduced and unprotonated oxidized states. Looplike pumps differ from this scheme only in that access to one or both sides is provided by a pore or channel.

Protons can be translocated without the aid of a H_2 carrier per se if the necessary accessibility coupling can be provided by a specific series of allowed conformations in the membrane protein to which the proton- and electron-carrying groups are bound. The minimal model for a proton pump of this type (Fig. 10) would consist of a single electron carrier and associated ionizable group (114, 128). The transducer could exist in two states, A and B. Since protonation is, in general, favored by reduction, we postulate that in state A the electron carrier equilibrates on the low-potential side of the coupling site and the ionizable group on the N side of the membrane, while in state B the electron carrier equilibrates on the high-potential side and the ionizable group on the P side. Transitions between the A and B states are allowed only when reduced and protonated or when oxidized and unprotonated. The other transitions must be forbidden, since they would result in uncoupling. The pK's of the ionizable group and the E_m's of the electron carrier may be different in state A and state B; the most efficient design would match the pK's and E_m's to the prevailing pH's and E_h's under normal operating conditions. As pointed out independently in the reviews of Ohnishi (3) and of Wikstrom (16, 77), the minimal model for a proton pump of this type (a *translocase*) is essentially equivalent to the transductase models of DeVault, except that the formation of a chemical intermediate is replaced by the translocation of a proton.

The response of the translocase electron carrier to $\Delta\psi$ and ΔpH can be predicted. The dependence of the apparent midpoint potential of the carrier

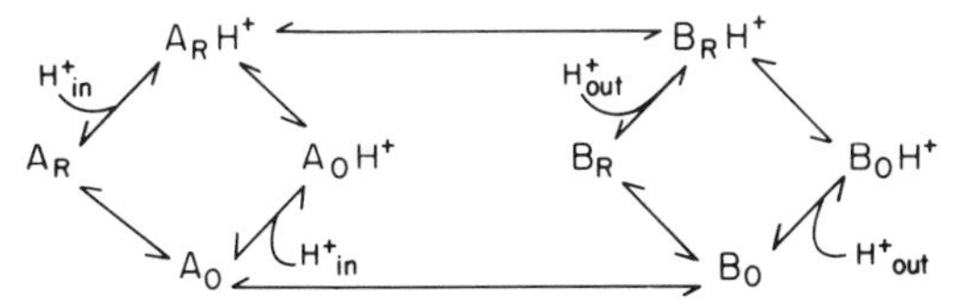

Figure 10. Minimal model for translocase type pump. In state A the pump equilibrates on the N side and with the low-potential pool at E_1; in state B pump equilibrates on the P side and with the high-potential pool at E_2.

on $\Delta\psi$ and ΔpH is extremely dependent on the choice of parameters, and in general quite different from the dependencies expected for loop components. In the minimal model the parameters E'_{0A} and E'_{0B} specify the values of E'_0 for the unprotonated couples in the A and B states, respectively. Similarly, pK_{AR}, pK_{AO}, pK_{BR}, and pK_{BO} specify the PK's of the four possible conformational redox states, where K is the equilibrium constant for the $A_RH^+ \rightleftharpoons B_RH^+$ transition when $\Delta\psi = 0$; α is a parameter that describes the electrogenic character of the $A_RH^+ \rightleftharpoons B_RH^+$ equilibrium, and thus its dependence on $\Delta\psi$. When $\alpha = 0$, the transition $A_RH^+ \rightleftharpoons B_RH^+$ is completely nonelectrogenic, hence $\Delta\psi$ insensitive. When $\alpha = -1/60$, one positive charge moves across $\Delta\psi$ (from the N to the P side in the $A_RH^+ \rightleftharpoons B_RH^+$ transition). If $A_RH^+ \rightleftharpoons B_RH^+$ is not electrogenic, $A_O \rightleftharpoons B_O$ must be electrogenic.

Generally, the response of the apparent midpoint potential of a pump component poised from either side to $\Delta\psi$ is nonlinear for most values of K and α. Table 2 summarizes the response of proton pumps with various values of α and K to $\Delta\psi$. Knowledge of the response of a transducer component to $\Delta\psi$ (especially at more than one pH) could be used to distinguish between loop and pump components, to test for minimal model behavior in a pump, and to determine which steps in proton pump operation are electrogenic.

The response of cluster N-2 to ATP and its pH-dependent E_m make it a logical candidate for a translocase component. Obviously, it would be useful to know the complete dependence of the redox poise of N-2 on $\Delta\psi$ and ΔpH

Table 2 Effect of Membrane Potential on the Apparent E_m of a Minimal Model Redox-Linked Proton Pump

	Poised from High-Potential Side; E_m Increases with $-\Delta\psi$	Poised from Low-Potential Side; E_m Decreases with $-\Delta\psi$
$\alpha \geq 0$	E_m linear with $\Delta\psi$ for $\Delta\psi \gg 0$	E_m independent of $\Delta\psi$ for $\Delta\psi \gg 0$
	E_m and $\Delta\psi$ independent for $\Delta\psi \ll 0$	E_m linear with $\Delta\psi$ for $\Delta\psi \ll 0$
$\alpha \sim -1/120$	E_m linear with $\Delta\psi$ for $\lvert\Delta\psi\rvert \gg 0$ when poised from high and low sides, with a central region of steeper ($k > 1$) or shallower ($k < 1$) dependence; linear throughout for $\alpha = -1/120$, $k = 1$.	
$\alpha \leq -1/60$	E_m linear with $\Delta\psi$ for $\Delta\psi \ll 0$	E_m independent of $\Delta\psi$ for $\Delta\psi \ll 0$
	E_m $\Delta\psi$ independent for $\Delta\psi \gg 0$	E_m linear with $\Delta\psi$ for $\Delta\psi \gg 0$

when poised from the high- and low-potential sides. While this level of information is not yet available, current data provide some interesting implications. Both Gutman (122) and Ohnishi (56) agreed that N-2 is oxidized after ATP addition when poised from the low-potential sides. The apparent shift in midpoint potential appears greater when poised from the low-potential side ($\approx$100 mV vs. 30–50 mV). Referring to Table 2, we see that for $\alpha = 0$ we expect a greater effect for negative $\Delta\psi$ (the physiological direction) when the system is poised from the low side, as was observed. When $\alpha = -1/60$, the effect would be greater from the high-potential side. Thus, if N-2 functions as a minimal model translocase at site I, we expect the $A_O \rightleftharpoons B_O$ transition to be the primary electrogenic step, unless the protons equilibrate through channels with special characteristics ("proton well" or "proton knoll"). In these special cases $\Delta\psi$-driven local pH gradients would complicate the analysis of experimental results. Additional information (such as the pH dependence of ATP-induced E_m shifts) would therefore be needed.

A proton well converts a large part of $\Delta\psi$ into the local ΔpH. This would concentrate most of $\Delta\psi$ across the transducer, resulting in extremely nonlinear dependence of the apparent E_m on $\Delta\psi$. Although this behavior could be complex, we should still expect the $A_RH^+ \rightleftharpoons B_RH^+$ transition specified above to be electrogenic.

If the channel into which protons are released has a low affinity for protons (proton knoll), the portion of $\Delta\psi$ across the transducer could be small. The major effect of $\Delta\psi$ on the apparent E_m would be through a local proton activity gradient, and neither of the A $\rightarrow$ B transitions would need to be very electrogenic. In this case knowledge of the response of E_m to ΔpH would be especially important.

The actual mechanism of energy transduction need not conform to a minimal model, however. Many possible series of ordered accessibility changes could provide coupling between electron transfer and proton translocation. The involvement of multiple electron carriers is a possibility strengthened by the reported response of clusters N-1a and N-4 to ATP. These clusters might be involved as stacked minimal models or in a more complex mechanism in which accessibility changes are controlled by the redox states of several components ("switching unit"). These possibilities would be favored if the higher site ratios now being debated for site I prove correct.

Similar proton pumps may function elsewhere in the respiratory chain and in other energy-conserving electron-transfer chains. In particular, Wikstrom (129) has proposed that cytochrome oxidase and the *b* cytochromes are examples of conformation-linked or translocase-type proton pumps, and provided detailed hypotheses describing these pumps in terms of a minimal model system. Site II in particular is likely to be an area of increasing controversy, since it has the only plausible mobile H_2 carrier in mitochondria, ubiquinone; and various Q cycle schemes have recently received widespread attention.

It is not necessarily the case that a site II proton pump would conform to the minimal translocase model. Interactions such as the shift in the Rieske center EPR spectrum (73–75) when another component is reduced and the multiple effects of inhibitors such as antimycin suggest that a site II proton pump might instead be controlled cooperatively be several electron carriers. The requirements for the coupling of proton and electron accessibility could be met by a molecular switching unit involving the *b* cytochromes, bound quinone, the Rieske center, and/or cytochrome c_1 (130). If the bc_1 complex exists in several discrete states such that an electron carrier (e.g., Rieske center) is accessible to its oxidant (cytochrome c_1) or reductant (probably a bound quinone) but not both simultaneously, the requirements for a tightly coupled proton pump can be met, provided that the accessibility of an associated ionizable group to the P or N side media depends on the state of the switching unit. Many such models can be constructed (128).

The multiple conformational states that a proton pump (which lacks a mobile H_2 carrier) must be able to assume raise questions about minority species and the spectral shifts discussed earlier under interactions. It is not unlikely that the spectra of electron carriers in a proton pump would be conformation dependent; it is thus possible that spectral species present in small relative concentrations arise not from contaminating proteins, but from previously reported electron carriers in minority (but physiologically important) conformational states. If an electron carrier is part of a more complex switching unit (130) controlling accessibility, either as part of a proton-translocating apparatus or to control the flow of electrons at a branch point in the respiratory chain (e.g., center 'o' in Mitchell's Q cycle scheme for proton translocation (13), the redox state of another carrier could control transitions between conformational states affecting both carriers. Thus, for example, the shift in the EPR spectrum of the Rieske Fe-S center as the potential is lowered (73–75) might reflect a shift in the equilibrium between conformational states as another component(s) is/are reduced (bound quinone and/or cytochromes *b*); this sort of conformational change could be of critical importance in energy conservation and the regulation of electron transfer at branch points.

ACKNOWLEDGMENTS

The authors wish to express their thanks to Dr. John R. Bowyer and Dr. Haywood Blum for critically reading the manuscript and for stimulating discussions. They also thank Dr. J. R. Bowyer for simulating redox titrations of two interacting redox components and for providing the data presented in Figure 3. Authors would like to thank Dr. H. Beinert for his comments on the manuscript of this article.

The unpublished work cited in this review, which has been conducted in Ohnishi's laboratory, is supported by NSF grant PCM 78-16779 and NIH grant GM 12202.

REFERENCES

1. J. S. Rieske, *Biochim. Biophys. Acta,* **456,** 195–247 (1976).
2. F. J. Ruzicka and H. Beinert, *J. Biol. Chem.,* **253,** 2514–2517 (1978).
3. T. Ohnishi, in *Membrane Proteins in Energy Transduction,* R. A. Capaldi, ed., Marcel Dekker, New York (1979), pp. 1–87.
4. F. J. Ruzicka, H. Beinert, K. L. Schepler, W. K. Dunham, and R. H. Sands, *Proc. Natl. Acad. Sci. USA,* **72,** 2886–2890 (1975).
5. A. A. Konstantinov and E. K. Ruuge, *FEBS Lett.,* **81,** 137–141 (1977).
6. J. C. Salerno and T. Ohnishi, *Biochem. J.,* **192,** 769–781 (1980).
7. T. Ohnishi and B. L. Trumpower, *J. Biol. Chem.,* **255,** 3278–3284 (1980).
8. S. Nagaoka, C. A. Yu, L. Yu, and T. E. King, *Arch Biochem. Biophys.,* **204,** 59–70 (1980).
9. S. DeVries, J. E. Berden, and E. C. Slater, *FEBS Lett.,* **122,** 143–148 (1979).
10. P. Mitchell, *Chemiosmotic Coupling in Oxidative and Photosynthetic Phosphorylation,* Glynn Research Ltd., Bodmin, United Kingdom, 1966, p. 192.
11. P. Mitchell, *Les Prix Novel en 1978,* The Nobel Foundation, Stockholm, 1979, pp. 137–172.
12. D. E. Green, *Comp. Biochem. Physiol.,* **4,** 81–122 (1962).
13. P. Mitchell, *J. Theor. Biol.,* **62,** 327–367 (1976).
14. J. Rydstrom, *Biochim. Biophys. Acta,* **463,** 155–184 (1977).
15. W. M. Anderson, W. T. Fowler, R. M. Pennington, and R. R. Fisher, *J. Biol. Chem.,* **256,** 1888–1895 (1981).
16. M. F. D. Wikström, and K. Krab, *Biochim. Biophys. Acta,* **549,** 177–222 (1979).
17. H. Beinert in *Iron Sulfur Proteins,* Vol. III, W. Lovenberg, ed., Academic, New York, pp. 61–100.
18. H. Beinert, *Meth. Enzymol.,* **54,** 133–150 (1978).
19. T. Ohnishi, *Mitochondria and Microsomes,* C. P. Lee, G. Schatz, and G. Dallner, eds., Addison-Wesley, Reading, Massachusetts, 1981, pp. 191–216.
20. E. G. Kearney, and T. P. Singer, *J. Biol. Chem.,* **219,** 963–975 (1956).
21. T. Y. Wang, C. L. Tsou, and Y. L. Wang, *Sic. Sinica Peking,* **5,** 73–90 (1956).
22. T. E. King, *J. Biol. Chem.,* **238,** 4036–4051 (1963).
23. K. A. Davis and Y. Hatefi, *Biochemistry,* **10,** 2509–2516 (1971).
24. W. G. Hanstein, K. A. Davis, M. A. Ghalamber, and Y. Hatefi, *Biochemistry,* **10,** 2517–2524 (1971).

25. P. Hemmerich, A. Ehrenberg, W. H. Walker, L. E. G. Erikson, J. Salach, P. Bader, and T. P. Singer, *FEBS Lett.*, **3,** 37–42 (1969).
26. B. A. Ackrell, M. B. Ball, and E. B. Kearney, *J. Biol. Chem.*, **255,** 2761–2769 (1980).
27. Y. Hatefi and Y. M. Galante, *J. Biol. Chem.*, **255,** 5530–5537 (1980).
28. B. A. C. Ackrell, E. B. Kearney, and C. J. Coles, *J. Biol. Chem.*, **252,** 6963–6965 (1977).
29. C. A. Yu, and L. Yu, *Biochim. Biophys. Acta,* **591,** 409–420 (1980).
30. H. Beinert, and R. H. Sands, *Biochem. Biophys. Res. Commun.*, **3,** 41–46 (1960).
31. T. Ohnishi, J. C. Salerno, D. B. Winter, J. Lim, C. A. Yu, L. Yu, and T. E. King, *J. Biol. Chem.*, **251,** 2094–2104 (1976).
32. T. Ohnishi, J. Lim, D. B. Winter, and T. E. King, *J. Biol. Chem.*, **251,** 2105–2109 (1976).
33. H. Beinert, B. A. C. Ackrell, E. B. Kearney, and T. P. Singer, *Eur. J. Biochem.*, **54,** 185–194 (1975).
34. C. J. Coles, R. H. Holm, D. M. Kurz, W. H. Orme-Johnson, J. Rawlings, T. P. Singer, and G. B. Wong, *Proc. Natl. Acad. Sci. USA,* **76,** 3805–3808 (1979).
35. J. C. Salerno, T. Ohnishi, H. Blum, and J. S. Leigh, *Biochim. Biophys. Acta,* **494,** 191–197 (1977).
36. S. P. J. Albracht and J. Subramanian, *Biochim. Biophys. Acta,* **462,** 36–48 (1977).
37. J. C. Salerno, J. Lim, T. E. King, H. Blum, and T. Ohnishi, *J. Biol. Chem.*, **254,** 4828–4835 (1979).
38. H. Beinert, B. A. C. Ackrell, A. D. Vinogradov, E. Kearney, and T. P. Singer, *Arch. Biochem. Biophys.*, **182,** 95–106 (1977).
39. T. Ohnishi, J. S. Leigh, D. B. Winter, J. Lim, and T. E. King, *Biochem. Biophys Res. Commun.*, **61,** 1026–1035 (1974).
40. S. P. J. Albracht, *Biochim. Biophys. Acta,* **612,** 11–28 (1980).
41. S. P. J. Albracht, *Proc. Eur. Bioenerg. Conf.*, First, Patron Editore, Bologna, Italy, 1980, pp. 39–40.
42. T. Ohnishi, T. E. King, J. C. Salerno, H. Blum, J. R. Bowyer, and T. Maida, *J. Biol. Chem.*, **256,** 5577–5582 (1981).
43. H. Blum and T. Ohnishi, *Biochim. Biophys. Acta,* **621,** 9–18 (1980).
44. M. H. Emptage, J. A. Kent, B. H. Huyuh, J. Rawlings, W. H. Orme-Johnson, and E. Münck, *J. Biol. Chem.*, **255,** 795–1796 (1980).
45. D. Stout, W. Ghosh, V. Pattahhi, and A. H. Robbins, *J. Biol. Chem.*, **255,** 1797–1800 (1980).
46. I. Moura, B. H. Huyuh, R. P. Hansinger, J. LeGall, A. V. Xavier, and E. Münck, *J. Biol. Chem.*, **255,** 2493–2498 (1980).
47. C. Heron, S. Smith, and C. I. Ragan, *Biochem. J.*, **181,** 435–443 (1979).
48. Y. Hatefi, A. G. Haavik, and D. E. Griffiths, *J. Biol. Chem.*, **237,** 1676–1680 (1962).
49. Y. Hatefi and K. E. Stempel, *J. Biol. Chem.*, **244,** 2350–2357 (1969).
50. C. I. Ragan, Y. M. Galante, Y. Hatefi, and T. Ohnishi, *in press* for Biochemistry.
51. G. Dooijewaard, E. C. Slater, P. J. van Dijk, and G. J. M. DeBruin, *Biochim. Biophys. Acta,* **503,** 405–424 (1978).
52. Y. M. Galante and Y. Hatefi, *Arch. Biochem. Biophys.*, **192,** 559–568 (1979).
53. C. Heron, S. Smith, and C. I. Ragan, *Biochem. J.*, **181,** 435–443 (1979).
54. W. Widger, Ph.D. Thesis, Department of Chemistry, State University of New York at Albany, 1979.

55. N. R. Orme-Johnson, R. E. Hansen, and H. Beinert, *J. Biol. Chem.*, **249**, 1922-1927 (1974).
56. T. Ohnishi, *Eur. J. Biochem.*, **64**, 91-103 (1976).
57. W. J. Ingledew and T. Ohnishi, *Biochem. J.*, **186**, 111-117 (1980).
58. S. P. J. Albracht, G. Dooijewaard, F. J. Leeuwerik, and B. van Swol, *Biochim. Biophys. Acta*, **459**, 300-317 (1977).
59. S. P. J. Albracht, F. J. Leeuwerik, and B. van Swol, *FEBS Lett.*, **104**, 197-200 (1979).
60. T. Ohnishi, H. Blum, Y. Galante, and Y. Hatefi, *J. Biol. Chem.*, **256**, 9216-9220 (1981).
61. T. Ohnishi, *Biochim. Biophys. Acta*, **387**, 475-490 (1975).
62. C. Paech, J. G. Reynolds, T. P. Singer, and R. H. Holm, *J. Biol. Chem.*, **256**, 3167-3170 (1981).
63. R. L. Ringler, S. Minakami, and T. P. Singer, *J. Biol. Chem.*, **238**, 801-811 (1963).
64. T. Ohnishi, D. F. Wilson, T. Asakura, and B. Chance, *Biochem. Biophys. Res. Commun.*, **46**, 1631-1638 (1972).
65. H. Beinert, in *Structure and Function of Energy Transducing Membranes*, D. Van Dam and B. F. Van Gelder, eds., Elsevier, Amsterdam, pp. 11-22.
66. J. S. Rieske, R. F. Hansen, and W. S. Zaugg, *J. Biol. Chem.*, **239**, 3017-3022 (1964).
67. B. L. Trumpower and C. A. Edwards, *J. Biol. Chem.*, **254**, 8697-8706 (1979).
68. B. L. Trumpower, C. A. Edwards, and T. Ohnishi, *J. Biol. Chem.*, **255**, 7489-7493 (1980).
69. B. L. Trumpower, *J. Bioenerg. Biomembr.*, **13**, 1-24 (1981).
70. J. R. Bowyer and B. L. Trumpower, *FEBS Lett.*, **115**, 171-174 (1980).
71. J. R. Bowyer, P. L. Dutton, R. C. Prince, and A. R. Croft, *Biochim. Biophys. Acta*, **592**, 445-460 (1980).
72. J. R. Bowyer, in *Function of Quinones in Energy Conserving Systems*, B. L. Trumpower, ed., Academic, New York, in press, 1981.
73. N. R. Orme-Johnson, R. E. Hansen, and H. Beinert, *J. Biol. Chem.*, **249**, 1928-1939 (1974).
74. J. N. Siedow, S. Power, F. F. DeLa Rosa, and G. Palmer, *J. Biol. Chem.*, **253**, 2392-2399 (1978).
75. S. De Vries, S. P. J. Albracht, and F. J. Leeuwik, *Biochim. Biophys. Acta*, **546**, 316-333 (1979).
76. B. G. Malmström, *Q. Rev. Biophys.*, **6**, 389-431 (1974).
77. M. K. F. Wikström, H. J. Harmon, W. J. Ingledew, and B. Chance, *FEBS Lett.*, **65**, 259-277 (1976).
78. G. Von Jagow and W. D. Engel, *FEBS Lett.*, **111**, 1-5 (1980).
79. J. C. Salerno, T. Ohnishi, J. Lim, W. R. Wedger, and T. E. King, *Biochem. Biophys. Res. Commun.*, **75**, 618-624 (1977).
80. A. Abragam and B. Bleaney, *Electron Paramagnetic Resonance of Transition Ions*, Clardeon Press, Oxford, 1970, p. 560.
81. H. Rupp, K. K. Rao, D. O. Hall, and R. Cammack, *Biochim. Biophys. Acta*, **537**, 255-269 (1978).
82. J. S. Leigh, *J. Chem. Phys.*, **52**, 2608-2612 (1970).
83. J. C. Salerno, Ph.D. Thesis, University of Pennsylvania, Philadelphia, 1977.
84. W. J. Ingledew, J. C. Salerno, and T. Ohnishi, *Arch. Biochem. Biophys.*, **177**, 176-184 (1976).

85. W. J. Ingledew and T. Ohnishi, *Biochem. J.*, **164,** 617-620 (1977).
86. H. Beinert, in *Oxidases and Related Redox Systems,* Vol. I, T. E. King, H. S. Mason, and M. Morrison, eds., Plenum, Elmwood, New York, 1965, pp. 198.
87. J. K. Blasie, M. Erecinska, S. Samuels, and J. S. Leigh, *Biochim. Biophys. Acta,* **501,** 33-52 (1978).
88. M. Erecinska, D. F. Wilson, and J. K. Blasie, *Biochim. Biophys. Acta,* **501,** 63-71 (1978).
89. H. Blum, H. J. Harmon, J. S. Leigh, J. C. Salerno, and B. Chance, *Biochim. Biophys. Acta,* **502,** 1-10 (1978).
90. J. C. Salerno, H. Blum, and T. Ohnishi, *Biochim. Biophys. Acta,* **547,** 270-281 (1979).
91. H. Blum, J. C. Salerno, and J. S. Leigh, *J. Magn. Resonance,* **30,** 385-391 (1978).
92. J. F. Gibson, D. O. Hall, J. F. Thornley, and F. Whatley, *Proc. Natl. Acad. Sci. USA,* **56,** 987-990 (1966).
93. R. Henderson, *J. Mol. Biol.*, **93,** 123-138 (1975).
94. Yu. A. Ovchinnikov, N. G. Abdulaev, M. Yu. Feigina, A. V. Kislev, and N. A. Lobanov, *FEBS Lett.*, **100,** 219-224 (1979).
95. R. Henderson, R. A. Capaldi, and J. S. Leigh, *J. Mol. Biol.*, **112,** 631-648 (1977).
96. D. M. Tiede, P. R. Rich, and W. D. Bonner, *Biochim. Biophys. Acta,* **506,** 307-315 (1979).
97. H. Blum, J. S. Leigh, J. C. Salerno, and T. Ohnishi, *Arch. Biochem. Biophys.*, **187,** 153-157 (1978).
98. J. C. Salerno, J. R. Lancaster, J. D. Lambeth and H. Kamin, *Fed. Proc.*, **39,** 1140 (1980).
99. R. L. Bell, J. Sweetland, B. Ludwid, and R. A. Capaldi, *Proc. Natl. Acad. Sci. USA,* **76,** 741-745 (1979).
100. R. J. Smith and R. A. Capaldi, *Biochemistry,* **16,** 2629-2633 (1977).
101. J. M. Vanderkooi, R. Landesberg, G. W. Hayden, and C. S. Owen, *Eur. J. Biochem.*, **81,** 339-347 (1977).
102. M. E. Docter, A. Steinemann, and G. Schatz, *J. Biol. Chem.*, **253,** 311-317 (1978).
103. G. D. Case and J. S. Leigh, *Biochem. J.*, **160,** 769-673 (1976).
104. T. Sarna, J. S. Hyde, and H. M. Swartz, *Science,* **192,** 1132-1134 (1976).
105. H. Blum, M. A. Cusanovich, W. V. Sweeney, and T. Ohnishi, *J. Biol. Chem.*, **256,** 2199-2206 (1980).
106. T. Ohnishi, H. Blum, J. S. Leigh, and J. C. Salerno, in *Membrane Bioenergetics,* C. P. Lee, G. Schatz, and L. Ernster, eds., Addison-Wesley, Reading, Massachusetts, 1979, pp. 21-30.
107. H. Blum, J. S. Leigh, and T. Ohnishi, *Biochim. Biophys. Acta,* **626,** 31-40 (1980).
108. E. T. Adman, *Biochim. Biophys. Acta,* **549,** 107-144 (1979).
109. R. E. Dickerson, T. Takano, D. Eisenberg, O. B. Kallai, L. Samson, A. Cooper, and E. Margoliash, *J. Biol. Chem.*, **246,** 1511-1535 (1971).
110. J. Girdlestone, R. Bisson, and R. A. Capaldi, *Biochemistry,* **20,** 152-156 (1981).
111. T. Ohnishi, H. Blum, H. J. Harmon, and T. Hompo, in *Interaction Between Iron and Proteins in Oxygen and Electron Transport,* C. Ho and W. Eaton, eds., Elsevier-North Holland, New York, in press, 1982.
112. S. Smith and C. I. Ragan, *Biochem. J.*, **185,** 315-326 (1980).
113. S. Chen and R. J. Guillory, *Fed. Proc.*, **39,** 2057 (1980).
114. D. DeVault, *J. Theor. Biol.*, **62,** 115-139 (1976).

115. P. Mitchell, in *Mitochondria/Biomembranes,* Vol. 28, S. G. van der Bergh et al., eds., North Holland, Amsterdam, 1972, pp. 358–370.

116. P. Garland, R. A. Clegg, J. A. Downie, T. A. Gray, H. G. Lawford, and J. Skyrme, in *Mitochondria/Biomembranes,* Vol. 28, S. G. van der Bergh et al., eds., North Holland, Amsterdam, 1972, pp. 105–117.

117. P. Mitchell, in *Oxidases and Related Redox Systems,* T. E. King, H. S. Mason, and M. Morrison, eds., Plenum, Elmwood, New York, in press, 1982.

118. E. Racker and W. Stoeckenius, *J. Biol. Chem.,* **249,** 662–663 (1974).

119. N. Sone, Y. Takeuchi, M. Yoshida, and K. Ohno, *J. Biochem. (Toyko),* **82,** 1751–1758 (1977).

120. V. P. Skulachev, in *Proceedings of the 10th FEBS Meeting,* Paris, Y. Raoul, ed., Associated Scientific, Amsterdam, 1976, pp. 225–238.

121. M. Gutman, T. P. Singer, and H. Beinert, *Biochemistry,* **11,** 556–562 (1972).

122. T. P. Singer and M. Gutman, in *Horizone in Biochemistry and Biophysics,* Vol. I, E. Quagliariello, F. Palmieri, and T. P. Singer, eds., Addison-Wesley, Reading, Massachusetts, 1974, pp. 261–302.

123. M. Klingenberg and M. Buchholz, *Eur. J. Biochem.,* **13,** 247–252 (1970).

124. G. D. Case, T. Ohnishi, and J. S. Leigh, *Biochem. J.,* **160,** 785–795 (1976).

125. H. Rottenberg and M. Gutman, *Biochemistry,* **16,** 3220–3226 (1977).

126. T. Pozzan, V. Miconi, F. DiVirgilio, and G. F. Azzone, *J. Biol. Chem.,* **254,** 10200–10205 (1979).

127. A. L. Lehninger, B. Reynafarje, and A. Alexandre, in *Cation Flux Across Biomembranes,* Y. Mukohata and L. Packer, eds., Academic, New York, 1979, pp. 343–354.

128. J. C. Salerno, submitted for publication, (1981).

129. M. Wikstrom, *Current Topics in Membrane Transport,* in press, 1981.

130. T. Ohnishi, H. Blum, and J. C. Salerno, in *Function of Quinones in Energy Conserving Systems,* B. L. Trumpower, ed., Academic, New York, 1982, *in press.*

CHAPTER 8

Iron-Sulfur Complexes of Ferredoxin as a Storage Form of Iron in *Clostridium pasteurianum*

RUDOLF K. THAUER
PETER SCHÖNHEIT

Laboratory for Microbiology, Fachbereich Biologie, Philipps-University Marburg, Marburg

CONTENTS

1 INTRODUCTION

Most organisms, with the exception of maybe a few lactic acid bacteria, are dependent on Fe for growth. The element is required for the synthesis of a variety of different Fe proteins such as cytochromes, Fe-S proteins, and aconitase, most of which are indispensible to a functioning metabolism (1, 2).

Iron is one of the most abundant elements in the earth's crust, being present to the extent of $\approx 4\%$ in typical soil (3). In spite of this fact, the transition metal is not always available for living cells. A major problem is posed by the relative insolubility of ferric hydroxide and other Fe compounds, such as iron sulfide, from which Fe must be extracted. Consequently, most organisms have developed solubilization and transport mechanisms for Fe, as well as the ability to store the transition metal within the cell under conditions of Fe sufficiency (1, 4).

Storage of Fe in eucaryotes has been extensively investigated. In animals, green plants, and fungi Fe is stored in the form of ferritin, a red-brown water-soluble protein (5). Ferritin contains 17–23% Fe as a dense core of hydrated ferric hydroxide, 7 nm in diameter, surrounded by a protein coat consisting of 24 subunits of molecular weight 18,500 in a cubic array. The core is readily visible in the electron microscope. Ferritin does not contain acid labile S. Another storage form of Fe in eucaryotes, hemosiderin, is water insoluble and seems to originate from degraded ferritin (6). Release of Fe from ferritin is probably caused by its reduction to the ferrous state (4, 5). The existence of Fe storage material in procaryotes is less established. However, many bacteria can grow for a considerable period of time on media completely devoid of Fe, provided that they had previously been cultured under conditions of Fe sufficiency (7, 8). This clearly indicates that these bacteria have the ability to store Fe. Whether they have specific Fe storage proteins is less evident.

In several procaryotes, including *Escherichia coli* (9, 10), *Proteus mirabilis* (11), *Mycoplasma capricolum* (12), and *Azotobacter vinelandii* (13), Fe-rich proteins, which have some similarities to ferritin of higher organisms, have recently been identified. The protein of *A. vinelandii* is a *b*-type cytochrome containing ≈ 100 mol nonheme Fe/mol heme Fe and no labile S (13, 14). Fe constitutes 13–20% of the weight of this protein, and forms an electron dense core of 5.5 nm in diameter. SDS gel electrophoresis reveals a single subunit of molecular weight 17,000, slightly smaller than that of mammalian ferritin. The function of the ferritinlike species in procaryotes has not yet been elucidated. The high Fe content makes these proteins, however, prime candidates for an Fe storage protein, although other functions, such as an electron storage protein, have also been envisaged (13).

In *Clostridium pasteurianum* the electron carrier ferredoxin has been shown to be used as an intracellular source of Fe during periods of Fe deprivation (8). Ferredoxin from *C. pasteurianum* is an Fe-S protein with a molecular weight of 6000, containing 8 nonheme Fe and 8 acid-labile S atoms organized in 2 [4Fe-4S] clusters (15). The Fe content of 7.5% is not as high as that of ferritins, but is considerably higher than that of any other known Fe proteins. Ferredoxin is present in *C. pasteurianum* at a concentration of 300 nmol/g cells (dry weight) (8, 16).

In the following sections the evidence for an Fe storage function of ferredoxin in *C. pasteurianum* is discussed, following a summary of its well-established function as an electron carrier. For details the reader is referred to the review of Yoch and Carithers (17).

2. FUNCTION OF FERREDOXIN AS AN ELECTRON CARRIER

As mentioned above, the ferredoxin of *C. pasteurianum* is an Fe-S protein with two [4Fe-4S] clusters. The clusters are assumed to have a structure similar to the [4Fe-4S] clusters of HIPIP from *Chromatium vinosum* and of the [8Fe-8S] ferredoxin from *Peptococcus aerogenes*, the structures of which have been determined by X-ray crystallography. The 2 [4Fe-4S] clusters can be reduced by one electron each, and function at about the same negative redox potential ($E^{0\prime}$) of -400 mV (15, 17).

Ferredoxin plays a central role in almost every aspect of the redox-related metabolism of *C. pasteurianum*. It is involved as electron carrier in catabolism, and links catabolic reactions of fermentation to numerous biosynthetic reactions. *C. pasteurianum* is a strict anaerobic Gram-positive endospore-forming bacterium that can grow on glucose as a sole energy source, and it has the ability to use protons as electron acceptors in fermentation. Fermentation products are acetate, butyrate, carbon dioxide, and large amounts of H_2. The specific function of ferredoxin in the catabolism is to transfer electrons generated in catabolic oxidation processes to protons. Enzymes involved are NADH-ferredoxin-oxidoreductase, pyruvate-ferredoxin-oxidoreductase, and ferredoxin-hydrogenase (Fig. 1) (18, 19a).

The anabolism of *C. pasteurianum* is characterized by its ability to use N_2 as N source, sulfate as S source, and carbon dioxide as precursor of reduced C-1 compounds, such as C-2 and C-8 of purines and the *S*-methyl group of methionine. The assimilation of N_2, sulfate, and carbon dioxide are reductive processes in which reduced ferredoxin serves as electron donor. The nitrogenase, bisulfite-reductase, and carbon dioxide-reductase of *C. pasteurianum* have been shown to be specific for ferredoxin as coenzyme. Last

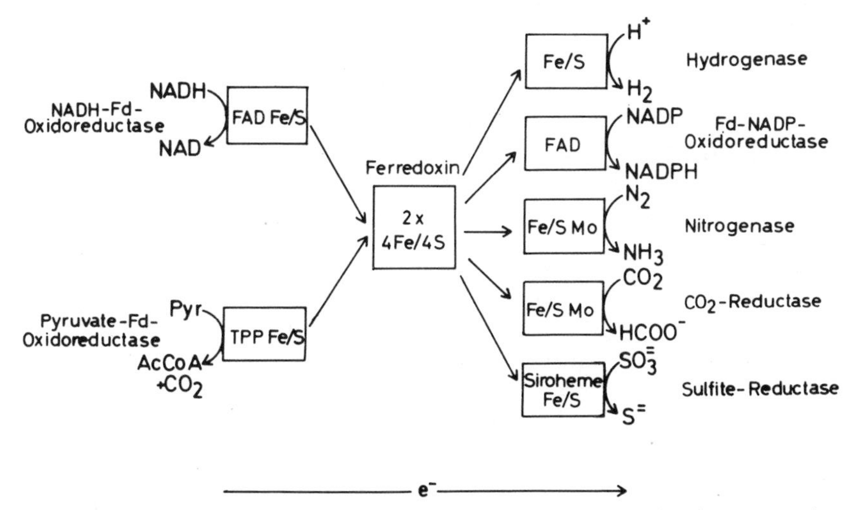

Figure 1. Ferredoxin-mediated redox processes in *C. pasteurianum.* Flavodoxin can substitute for ferredoxin in its function as electron carrier. See also Table 1. Fd, ferredoxin; TPP, thiamine pyrophosphate; Fe, nonheme Fe; S, acid-labile sulfur; Mo, molybdenum; FAD, flavin adenine dinucleotide.

but not least, NADP in *C. pasteurianum* is reduced via ferredoxin (Fig. 1) (17, 19b).

In spite of the central role of ferredoxin in catabolism and anabolism, the presence of ferredoxin in *C. pasteurianum* is not essential for growth of the organism. After growth on media that are low in Fe, cells contain flavodoxin rather than ferredoxin. The flavodoxin of *C. pasteurianum* is a flavoprotein with a molecular weight of 14,600. It does not contain Fe, and can substitute as an electron carrier for ferredoxin in all redox reactions (20, 21).

3 FUNCTION OF FERREDOXIN AS AN Fe STORAGE PROTEIN

3.1 Fe Proteins and Fe Storage in *C. pasteurianum*

Clostridium pasteurianum is dependent on Fe for growth, mainly for the synthesis of Fe-S proteins. Cells whose growth is limited by the Fe concentration in the medium contain almost equal amounts of Fe and acid-labile S (22), indicating that most of the Fe is bound in Fe-S clusters. In addition to ferredoxin (23, 24) *C. pasteurianum* is known to contain the following Fe-S proteins (Table 1): two pyruvate-ferredoxin-oxidoreductases (49); two hydro-

Table 1. Fe-S Proteins Known to Be Present in *C. pasteurianum*[a]

Protein	Fe/S^{2-} Content	M_γ	Function	Comments	Refs.[b]
Hydrogenase	12 Fe/12 S^{2-}	60,000	catabolic, H_2 formation		25–28
Pyruvate-ferredoxin-oxidoreductase	6 Fe/3 S^{2-}[c]	240,000	catabolic, acetyl-CoA formation	contains 1 TPP/6 Fe	29
NADH-ferredoxin-oxidoreductase			catabolic, NADH reoxidation	contains FAD	30, 31
Component I	?	250,000			
Component II	Fe/S^{2-}	160,000			
Nitrogenase			anabolic, N_2 fixation	contains 1 Mo/24 Fe	32–35
Mo-Fe Protein	24 Fe/24 S^{2-}	220,000			
Fe Protein	4 Fe/4 S^{2-}	50,000–60,000			
CO_2-reductase	24 Fe/24 S^{2-}	118,000	anabolic, formation of C-1 compounds	contains 1 Mo/24 Fe	36
Sulfite reductase (assimilatory)	16 Fe/14 S^{2-}[d]	680,000	anabolic, SO_4^{2-} assimilation	contains 1 siroheme/4 Fe, 4 FAD and 4 FMN/mol	37–39
Glutamate synthase (GOGAT)	8 Fe/8 S^{2-}[e]	200,000	anabolic, NH_3 assimilation	contains 1 FAD and 1 FMN	40–42
"Paramagnetic protein"	2 Fe/2 S^{2-}	25,000	catabolic, pyruvate oxidation (?)		43, 44
Rubredoxin	1 Fe/ no S^{2-}	6,000	electron carrier		45, 46
Ferredoxin	8 Fe/8 S^{2-}	6,000	electron carrier,[f] Fe storage		28, 47, 48

[a]For a recent review see Yoch and Carithers (17).
[b]Fe/S^{2-} content and M_r values.
[c]Assumed to be the same as in the enzyme of *Clostridium acidi-urici* (29).
[d]Assumed to be the same as in the enzyme of *Escherichia coli* (37, 38).
[e]Assumed to be the same as in the enzymes of *Escherichia coli* and *Klebsiella aerogenes* (40, 41).
[f]Can be substituted by flavodoxin in its function as electron carrier (20, 21).

genases (27, 50); NADH-ferredoxin-oxidoreductase (18); carbon dioxide-reductase (19b, 51); bisulfite-reductase (52), which also contains siroheme; glutamate synthase (53); rubredoxin (45, 54), which does not contain acid-labile S; and, when grown in the absence of NH_3, nitrogenase (55). A paramagnetic protein containing 2 Fe/2 S^{2-} of unknown function is found (43). Probably glutamine-phosphoribosyl pyrophosphate amidotransferase from *C. pasteurianum* is also an Fe-S protein (56). Except for ferredoxin none of these proteins can be replaced by non-Fe proteins. *C. pasteurianum* can thus grow normally only if enough Fe is available for the synthesis of these proteins.

The bacteria can, however, grow for a considerable period of time in the absence of exogenous Fe, provided that the organism had previously been cultured in its presence (8). An example is given in Figure 2; the *Clostridium* was grown in a batch culture on a glucose medium containing 10μM Fe. As can be seen, all the Fe present in the medium was taken up by the cells before 50% of the final cell density was reached. The bacterium then continued to normally grow for one doubling time on the Fe depleted medium.

During growth in the presence of Fe the cells contained up to 12–14 μmol

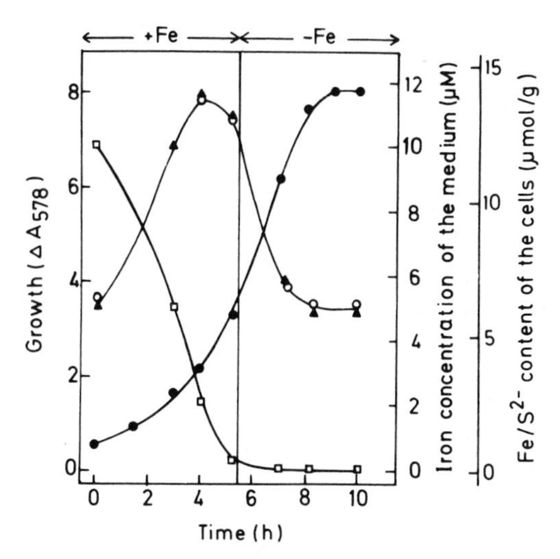

Figure 2. Growth of *C. pasteurianum* after Fe depletion of the medium. The medium contained 10 μM Fe initially and was inoculated with 10% of a late exponential culture grown under the same conditions. At the time indicated the Fe and the acid-labile S contents of the cells were determined. ●) Growth; □) Fe concentration in the medium; ▲) Fe content of the cells in micromoles per gram dried weight; ○) acid-labile S in micromoles per gram dried weight; ← + Fe→) medium contains Fe; ← – Fe→) medium is depleted of in Fe. For experimental details see Schönheit, Brandis, and Thauer (8).

Fe and 11–13 μmol acid-labile S/g dried cells. During growth in the absence of exogenous Fe the Fe-S content decreased to 6 μmol/g, and then growth ceased as a result of Fe insufficiency (Fig. 2). The decrease in the Fe-S content by 6–8 μmol/g indicates that during growth in the presence of Fe an excess of 6–8 μmol/g of Fe is taken up by the cells and incorporated into Fe-S complexes.

The ability of *C. pasteurianum* to grow in the absence of exogenous Fe can have one of the following explanations:

1. *C. pasteurianum* contains Fe stored in the form of Fe-S complexes, which, under conditions of Fe deprivation, release Fe for the synthesis of Fe-S enzymes, whose activity limits the growth rate.
2. *C. pasteurianum* does not contain Fe storage material. Then synthesis of all Fe proteins should stop as soon as Fe is depleted from the medium. This is only possible if none of the Fe enzymes is growth-rate limiting.

In the first case synthesis of essential Fe-S proteins during growth should continue in the absence of Fe. In the second case synthesis should not be observed.

To decide between the two possibilities, the synthesis of the Fe-S protein pyruvate-ferredoxin-oxidoreductase and of NADH-ferredoxin-oxidoreductase was studied because the activities of these catabolic enzymes, rather than of other Fe-S enzymes, are probably growth-rate limiting (57). It was found that, during growth of *C. pasteurianum* in the absence of exogenous Fe, pyruvate-ferredoxin-oxidoreductase continued to be synthesized (8). The activity of this enzyme increased parallel to growth in the absence of Fe (Fig. 3). The same holds true for NADH-ferredoxin-oxidoreductase (unpublished results). This is only possible if concomitantly an endogenous Fe storage is degraded. The evidence thus points to the presence of Fe storage material in *C. pasteurianum.*

3.2 Ferredoxin Degradation During Growth Under Conditions of Fe Deprivation

During growth of *C. pasteurianum* in the presence of Fe the cells contain ≈0.3 μmol ferredoxin/g cells (dry weight), in which 2.4 μmol Fe is bound (8). The Fe-S protein is synthesized in the growing cells as long as Fe can be detected in the medium (Fig. 3). Only when the exogenous Fe concentration is lower than 1 *μM* does ferredoxin synthesis stop. During growth in the absence of Fe ferredoxin is completely degraded without apparent excretion of Fe into the medium. [The term *ferredoxin degradation* is used here both for the formation of apoferredoxin (protein free of Fe and sulfide) and for the

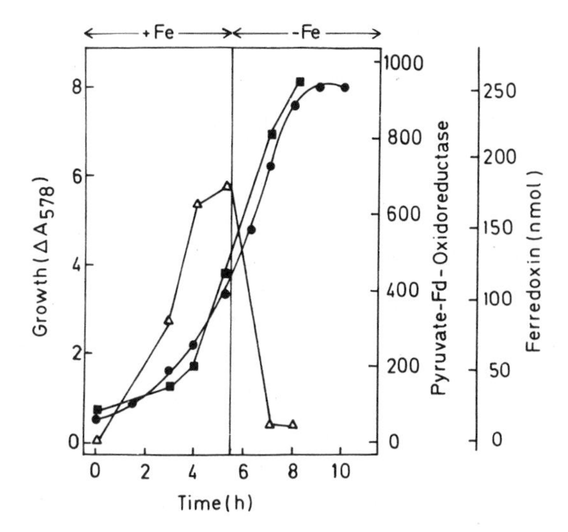

Figure 3. Synthesis of pyruvate-ferredoxin-oxidoreductase and degradation of ferredoxin in growing cells of *C. pasteurianum* under conditions of Fe deprivation. At the times indicated the pyruvate-ferredoxin-oxidoreductase activity and the ferredoxin content in the cells of 1 l culture were determined. ●) Growth; ■) pyruvate-ferredoxin-oxidoreductase; Δ) ferredoxin; ←+Fe→) medium contains Fe; ←−Fe→) medium is depleted of Fe. For experimental details see Schönheit, Brandis, and Thauer (8) and Figure 2.

formation of amino acids from ferredoxin via proteolysis.] Flavodoxin is synthesized in this period to substitute for ferredoxin as an electron carrier.

At the end of growth in the absence of Fe the cells, which are now ferredoxinfree, still contain 6 μmol Fe/g. This is evidently the minimal amount of Fe required to sustain growth. In ferredoxin-containing cells 2.4 μmol Fe/g is bound in ferredoxin. This amount of Fe is thus sufficient for the synthesis of 0.4 g of ferredoxinfree cells. The findings that ferredoxin is degraded during periods of Fe deprivation and that the Fe released during degradation is sufficient to enable significant growth in the absence of exogenous Fe clearly indicate that in *C. pasteurianum* ferredoxin has, in addition to its function as an electron carrier, the function of an Fe storage protein.

Cells with 0.3 μmol ferredoxin/g contain approximately twice the amount of Fe (12–14 μmol/g) as ferredoxinfree cells (6 μmol/g) (Fig. 2). The higher amount cannot solely be explained by the presence of ferredoxin, which accounts for only 2.4 μmol Fe/g. This finding indicates either that Fe storage proteins other than ferredoxin exist, or that many Fe proteins are present in the ferredoxin-containing cells in excess amounts. In the latter case these Fe-S proteins should not be synthesized during growth under conditions of Fe deprivation. Indeed, synthesis of hydrogenase and carbon dioxide-reductase, for example, stops as soon as Fe is depleted from the medium. Thus additional Fe storage proteins need not necessarily be assumed (22).

3.3 Possible Mechanisms of Fe Release from Ferredoxin

The mechanism of Fe release from ferredoxin is uncertain. It is not known how ferredoxin is degraded. Two different mechanisms can be envisaged: (1) Ferredoxin is degraded only to the level of apoprotein, which is reused for the synthesis of ferredoxin when exogenous Fe becomes available again (58). This mechanism would imply the operation of a "dechalatase," which specifically removes the Fe, or of an enzyme, which removes the whole Fe-S cluster; (2) ferredoxin is completely degraded to the level of amino acids. Either the holoprotein or the apoprotein could be attacked by proteases. Proteolysis of holoferredoxin would have to involve a specific protease, but that of the apoprotein could be achieved by an unspecific enzyme (59–62).

Experiments were conducted to determine whether ferredoxin resynthesis in cells in which ferredoxin had previously been degraded starts from the apoprotein level or from the amino acid level. Both Fe and U^{14} C-glucose were added to the growing culture when ferredoxin degradation was completed. Both the synthesis of ferredoxin and the incorporation of ^{14}C were followed. Immediately after the addition of Fe ferredoxin was rapidly synthesized (Fig. 4). The ferredoxin formed had the same specific radioactivity per mole of C as the C atoms of U^{14} C-glucose (63). The latter finding indicates that resynthesis of the carbon skeleton of ferredoxin originated from glucose, and excludes a resynthesis of ferredoxin from previously accumulated apoferredoxin; an apoprotein pool formed before the addition of U^{14} C-glucose should not become labeled. Since it is not very likely that ferredoxin is resynthesized from the glucose level if the apoferredoxin is still available, it must be concluded that ferredoxin degradation proceeds beyond the level of apoferredoxin when Fe is released from the Fe-S clusters.

Probably Fe release from ferritin does not involve degradation of the apoferritin, and Fe storage in ferritin does not necessarily involve protein synthesis (4, 5). Thus Fe storage in ferredoxin proceeds via a totally different mechanism.

ACKNOWLEDGMENTS

This work was supported by a grant from the Deutsche Forschungsgemeinschaft and by the Fonds der Chemischen Industrie.

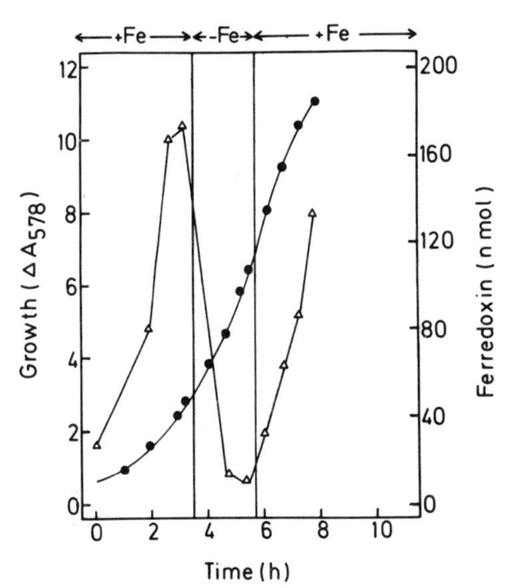

Figure 4. Ferredoxin resynthesis in growing *C. pasteurianum* after addition of Fe (100 μM) to the Fe-depleted medium. The growth medium contained 8.5 μM Fe initially. The medium was inoculated with 10% of a late exponential culture grown on a medium with 30 μM Fe initially. After 5.8 hr of growth 100 μmol $FeCl_2$ was added per liter. At the time indicated the ferredoxin content in the cells of a 1 l culture was determined. ●) Growth; Δ) ferredoxin; ←+Fe→) medium contains Fe; ←−Fe→) medium is depleted of Fe. For experimental details see Schönheit, Brandis, and Thauer (8) and Figure 2.

REFERENCES

1. A. Jacobs and M. Worwood, eds., *Iron in Biochemistry and Medicine,* Academic, New York-London, 1974.
2. J. B. Neilands, ed., *Microbial Iron Metabolism,* Academic, New York-London, 1974.
3. H. J. M. Bowen, *Trace Elements in Biochemistry,* Academic, New York-London, 1966.
4. T. Emery, in *Metal Ions in Biological Systems,* Vol. 7, H. Sigel, ed., Marcel Dekker, New York-Basel, 1978, pp. 77-126.
5. P. Aisen and I. Listowsky, *Ann. Rev. Biochem.,* **49,** 357 (1980).
6. G. W. Richter, *Am. J. Pathol.,* **91,** 363 (1978).
7. P. A. Light and R. A. Clegg, in *Microbial Iron Metabolism,* J. B. Neilands, ed., Academic, New York-London, 1974, pp. 35-64.
8. P. Schönheit, A. Brandis, and R. K. Thauer, *Arch. Microbiol.,* **120,** 73 (1979).
9. E. R. Bauminger, S. G. Cohen, D. P. E. Dickson, A. Levy, S. Ofer, and J. Yariv, *J. Phys. (Paris),* **40,** 523 (1979).
10. E. R. Bauminger, S. G. Cohen, D. P. E. Dickson, A. Levy, S. Ofer, and J. Yariv, *Biochim. Biophys. Acta,* **623,** 237 (1980).

11. D. P. E. Dickson and S. Rottem, *Eur. J. Biochem.*, **101,** 291 (1979).
12. E. R. Bauminger, S. G. Cohen, F. Labenski de Kanter, A. Levy, S. Ofer, M. Kessel, and S. Rottem, *J. Bacteriol.*, **141,** 378 (1980).
13. E. I. Stiefel and G. D. Watt, *Nature,* **279,** 81 (1979).
14. W. A. Bulen, J. R. Le Comte, and S. Lough, *Biochem. Biophys. Res. Commun.*, **54,** 1274 (1973).
15. W. V. Sweeney and J. C. Rabinowitz, *Ann. Rev. Biochem.*, **49**, 139 (1980).
16. P. Schönheit, C. Wäscher, and R. K. Thauer, *FEBS Lett.*, **89,** 219 (1978).
17. D. C. Yoch and R. P. Carithers, *Microbiol. Rev.*, **43,** 384 (1979).
18. K. Jungermann, R. K. Thauer, G. Leimenstoll, and K. Decker, *Biochim. Biophys. Acta,* **305,** 268 (1973).
19a. R. K. Thauer, K. Jungermann, and K. Decker, *Bacteriol. Rev.*, **41,** 100 (1977).
19b. R. K. Thauer, G. Fuchs, and K. Jungermann, in *Iron-Sulfur Proteins,* Vol. 3, W. Lovenberg, Academic, New York-San Francisco-London, 1977, pp. 121-156.
20. E. Knight, Jr., and R. W. F. Hardy, *J. Biol. Chem.*, **241,** 2752 (1966).
21. E. Knight, Jr., and R. W. F. Hardy, *J. Biol. Chem.*, **242,** 1370 (1967).
22. A. Brandis, Thesis, Philipps-Universität, Marburg, 1978.
23. L. E. Mortenson, in *The Bacteria,* Vol. 3, I. C. Gunsalus and R. Stanier, eds., Academic, New York, 1962, pp. 119-166.
24. L. E. Mortenson, *Ann. Rev. Microbiol.*, **17,** 115 (1963).
25. J. S. Chen and L. E. Mortenson, *Biochim. Biophys. Acta,* **371,** 283 (1974).
26. L. E. Mortenson and J. S. Chen, in *Microbial Iron Metabolism,* J. B. Neilands, ed., Academic, New York, 1974, pp. 231-282.
27. J. S. Chen, in *Hydrogenases: Their Catalytic Activity, Structure and Function,* H. G. Schlegel and K. Schneider, eds., Erich Goltze KG, Göttingen, 1978, pp. 57-81.
28. L. E. Mortenson and G. Nakos, in *Iron-Sulfur Proteins,* Vol. 1, W. Lovenberg, ed., Academic, New York, 1973, pp. 37-64.
29. K. Uyeda and J. S. Rabinowitz, *J. Biol. Chem.*, **246,** 3111 (1971).
30. K. Jungermann, M. Kern, N. Katz, and R. K. Thauer, unpublished.
31. K. Jungermann, M. Kern, V. Riebeling, and R. K. Thauer, in *Microbial Production and Utilization of Gases,* H. G. Schlegel, G. Gottschalk, and N. Pfennig, eds., E. Goltze KG, Göttingen, 1976, pp. 85-96.
32. T. C. Huang, W. G. Zumft, and L. E. Mortenson, *J. Bacteriol.*, **113,** 884 (1973).
33. L. E. Mortenson, W. G. Zumft, T. C. Huang, and G. Palmer, *Biochem. Soc. Trans.*, **1,** 35 (1973).
34. W. H. Orme-Johnson and L. C. Davis, in *Iron-Sulfur Proteins,* Vol. 3, W. Lovenberg, ed., Academic, New York, 1977, pp. 15-60.
35. L. E. Mortenson and N. F. Thorneley, *Ann. Rev. Biochem.*, **48,** 387 (1979).
36. P. A. Scherer and R. K. Thauer, *Eur. J. Biochem.*, **85,** 125 (1978).
37. L. M. Siegel and P. S. Davis, *J. Biol. Chem.*, **249,** 1587 (1974).
38. E. J. Faeder, S. Davis, and L. M. Siegel, *J. Biol. Chem.*, **249,** 1599 (1974).
39. L. M. Siegel, in *Metabolic Pathways,* Vol. 7, D. M. Greenberg, ed., *Metabolism of Sulfur Compounds,* Academic, New York, 1975, pp. 217-286.
40. R. E. Miller and E. R. Stadtman, *J. Biol. Chem.*, **247,** 7407 (1972).

41. P. P. Trotta, K. E. B. Platzer, R. H. Haschemeyer, and A. Meister, *Proc. Natl. Acad. Sci. USA,* **71,** 4607 (1974).
42. R. E. Miller, in *Microbial Iron Metabolism,* J. B. Neilands, ed., Academic, New York-London, 1974, pp. 283-302.
43. R. W. F. Hardy, F. Knight, Jr., C. C. McDonald, and A. J. D'Eustachio, in *Non-heme Iron Proteins: Role of Energy Conversion,* A. San Pietro, ed., Antioch, Yellow Springs, Ohio, 1965, pp. 275-282.
44. J. Cardenas, L. E. Mortenson, and D. C. Yoch, *Biochim. Biophys. Acta,* **434,** 244 (1976).
45. W. Lovenberg and B. E. Sobel, *Proc. Natl. Acad. Sci. USA,* **54,** 193 (1965).
46. W. A. Eaton and W. Lovenberg, in *Iron-Sulfur Proteins,* Vol. 2, W. Lovenberg, ed., Academic, New York-London, 1973, pp. 131-162.
47. J. S. Hong and J. C. Rabinowitz, *J. Biol. Chem.,* **245,** 4982 (1970).
48. R. Malkin, in *Iron-Sulfur Proteins,* Vol. 2, W. Lovenberg, ed., Academic, New York-London, 1973, pp. 1-26.
49. F. D. Sauer, R. S. Bush, and L. L. Stevenson, *Biochim. Biophys. Acta,* **445,** 518 (1976).
50. J. S. Chen and D. K. Blanchard, *Biochem. Biophys. Res. Commun.,* **84,** 1144 (1978).
51. K. Jungermann, H. Kirchniawy, and R. K. Thauer, *Biochem. Biophys. Res. Commun.,* **41,** 682 (1970).
52. E. J. Laishley, P. -M. Lin, and H. D. Peck, Jr., *Can. J. Microbiol.,* **17,** 889 (1971).
53. R. H. Dainty, *Biochem. J.,* **126,** 1055 (1972).
54. W. Lovenberg, in *Microbial Iron Metabolism,* J. B. Neilands, ed., Academic, New York-London, 1974, pp. 161-185.
55. R. W. F. Hardy and R. C. Burns, in *Iron-Sulfur Proteins,* Vol. 1, W. Lovenberg, ed., Academic, New York-London, 1973, pp. 65-110.
56. J. Y. Wong, E. Meyer, and R. L. Switzer, *J. Biol. Chem.,* **252,** 7424 (1977).
57. F. J. Tewes and R. K. Thauer, in *Anaerobes and Anaerobic Infections,* G. Gottschalk, N. Pfennig, and H. Werner, eds., Gustav Fischer Verlag, Stuttgart-New York, 1980, pp. 97-104.
58. J. W. Brodrick and J. C. Rabinowitz, in *Iron-Sulfur Proteins,* Vol. 3, W. Lovenberg, ed., Academic, New York-San Francisco-London, 1977, pp. 101-119.
59. A. L. Goldberg and A. C. St. John, *Ann. Rev. Biochem.,* **45,** 747 (1976).
60. R. L. Switzer, *Ann. Rev. Microbiol.,* **31,** 135 (1977).
61. D. Wolf, in *Advances in Microbial Physiology,* Vol. 21, A. H. Rose and J. G. Morris, eds., Academic, London-New York, 1980, pp. 267-338.
62. M. R. Maurizi and R. L. Switzer, in *Current Topics in Cellular Regulation,* Vol. 16, B. L. Horecker and E. R. Stadtman, eds., Academic, New York-London, 1980, pp. 163-224.
63. M. Brenneke, Thesis, Philipps-Universität, Marburg, 1979.

CHAPTER 9

X-Ray Absorption Studies of Fe-S Proteins and Related Compounds

BOON-KENG TEO

Bell Laboratories
Murray Hill, New Jersey

R. G. SHULMAN

Department of Molecular Biophysics and Biochemistry
Yale University
New Haven, Connecticut

CONTENTS

1 INTRODUCTION

The recent availability of high fluxes of X-rays from synchrotron radiation has allowed X-ray absorption measurements to be made of dilute metal ions in metalloenzymes (1, 2). Because synchrotron radiation generates X-rays smoothly over the wavelength range, monochromators can use this source to provide a tunable, even source of X-rays. The intensity of monochromatic X-rays available from synchrotron radiation is $\approx 10^4$–10^5 times greater than that previously available from the broad Bremsstrahlung radiation of X-ray tubes. The counting rates for detecting absorption are proportional to the X-ray flux, so the great improvements in signal-to-noise ratios have made qualitative differences in the kinds of experiments that can be performed. Since spectroscopic studies advance very rapidly, even when small improvements of counting rates are made, it is really quite appropriate to describe X-ray absorption studies since 1974 as being in a revolutionary period.

The basis of the X-ray absorption experiments has been known for over 50 years (3), although our understanding and experimental techniques are improving continually (4). When a sample is placed in the X-ray beam, X-rays of the proper energy are absorbed by inducing a transition of the core electrons to the outer empty energy levels. In the cases discussed here the transitions are from the $1s$ innermost electron of Fe to empty bound states. The spectra are naturally divided into two regions. Going to higher energy, a rapid increase in the absorption is first observed in the so-called *edge region,* where the transitions are from the $1s$ state to empty localized states. These transitions have been described lately in terms of the bound states of the central atom interacting via bonds with the ligands (5). As the X-ray energy is increased, the excited electrons acquire enough energy to be ionized and to move out into the surrounding medium. In this, the second spectral region, transitions are to delocalized states and the probability of absorption is modulated by backscattering from the neighbors. This modulation, the so-called *extended X-ray absorption fine structure* (EXAFS), can be used to determine distances from the absorbing atom to its neighbors, as discussed below (2–4).

The Fe-S proteins have been very suitable subjects for X-ray absorption studies because the backscattering is dominated by the S ligands and the nearby Fe atoms. In practice it has been possible to fit data while completely neglecting scattering from other, farther neighbors. With this geometry the problem of determining bond lengths simplifies to the determination of one unknown, an Fe–S distance, in rubredoxin, and two unknowns, the Fe–S and Fe–Fe distances, in the other Fe-S proteins. This simplification occurs because the backscattering by the nearest S and Fe neighbors is so much greater than that by the lighter atoms of the protein (i.e., C, N, O, and H) that these contributions to the backscattering can be neglected.

Another consideration that intensifies our interest in determining the bond lengths in the prosthetic groups of these proteins is the availability of model Fe-S compounds whose structures have been determined by single-crystal X-ray crystallography and whose physical properties were known to be extremely similar to those of the Fe-S proteins. These were used as model systems in which the methods of analyzing the EXAFS spectra were developed. When the structures determined by the EXAFS studies were compared with the accurate crystallographic determinations, the agreement was excellent. With the confidence in the methodology generated by this agreement the EXAFS method was then applied to determining the structures of the Fe-S prosthetic groups in the enzymes. The results are discussed in the following sections.

2 ABSORPTION EDGE

Transitions from the core electrons to the empty localized valence orbitals give rise to absorptions at or below the so-called *absorption edge.* It is a good first approximation to regard these transitions as having the character of the central, absorbing ion, and to consider them as one electron transitions between free ion states (5). The orbitals involved in these transitions and the applicable selection rules have been derived in this free ion approximation and applied to X-ray absorption edge spectra of the weakly covalent Fe group fluorides (5a). The data were in agreement with the interpretation, and the salient features are as follows. The transitions are from the innermost $1s$ electron of the Fe group to the lowest empty orbitals, which in order of increasing energies are $3d$, $4s$, and $4p$. In the ionic limit the selection rules for the one electron transition are that $\Delta l = \pm 1$, so $1s \rightarrow 4p$ transitions are allowed, while $1s \rightarrow 3d$ and $1s \rightarrow 4s$ are forbidden. The forbiddance is removed by mixing of the allowed $4p$ final state into $3d$ and $4s$, with the result that $1s \rightarrow 3d$ is a weak transition observed at the low-energy side of the absorption edge, $1s \rightarrow 4s$ is detected as a shoulder on the edge, and the edge itself is associated with the allowed transition $1s \rightarrow 4p$. In crystals where the Fe group ion is at a site of cubic symmetry it has been possible to resolve and assign all three of these transitions at the absorption edge. It addition a discrete absorption was observed at higher energies which was tentatively assigned to the $1s \rightarrow 5p$ transition. With these well-resolved transitions it was possible to test the assignments by comparing the energy intervals with those known for the free ions, with excellent agreement. The comparison depended on an interesting property of the final state, in which one electron has been promoted from the $1s$ to the erstwhile empty $3d$, $4s$, or $4p$ state; namely, that these final states have outer electron configurations that are identical to those of the same ionic state of the next higher atomic number. For example, while the outermost configuration of Fe^{3+} was $3d^5$ (beyond the Ar core), when a $1s$ electron was excited to the valence orbitals there were six outermost electrons to be accommodated, while the net charge was still 3. Obviously, this corresponds to Co^{+3}, in which the apparently larger nuclear charge has been created not by one extra positive charge in the nucleus but rather by one less negative charge in the innermost $1s$ shell from which an electron has been excited. When the site became less symmetric in the Fe group fluorides, the spectral resolution was lowered with the result that, while these transitions could be assigned, they no longer could be completely resolved.

In addition to the selection rule that $\Delta l = \pm 1$, it was also possible to select the multiplet of the final state and the total spin of that state; these selection rules are discussed in the original report (5a).

For the present purpose of studying Fe-S proteins the dependence of the energy of the edge on the formal charge is relevant. It was observed that the energy of the absorption edge, defined as the steeply rising part of the absorption heading for its maximum, is 5 eV higher in ferric fluoride than in ferrous fluoride (5a). The $1s \rightarrow 3d$ transition was ≈ 1 V higher in the ferric fluoride complexes, showing less effect of charge, as expected. In an ionic compound the reason for this difference is that the larger nuclear charge attracts the electron more strongly; the difference in ionization potential is 13.6 V. In the free ion, in the coulombic approximation the difference between energies from the $1s$ to the $4p$ state should be, and in fact is, ~ 5 V. Hence in the very ionic fluorides the edge energies are characteristic of the charge of the central ion, with the more positive ion having the higher energy. Beyond the very ionic fluoride complexes this correlation is not observed. Although differences in the edge spectra are observed among different compounds of the same metal ion, the spectral features cannot be simply explained by the free ion properties. The molecular orbital (MO) explanation of the breakdown of the free ion picture is shown in Figure 1, which generalizes the free ion energy levels involved in the one electron transition to include bonding with the ligands. The schematic MOs are shown only for the $3d$ state; similar bonding and antibonding orbitals exist for the $4s$ and $4p$ states. In addition the $3d$ and $4p$ molecular orbitals are split into σ and π bonds, which are split further by lower symmetry environments. All this helps to explain, qualitatively, why the edge absorption spectra become less well resolved as the symmetry is lowered.

When we turn to quite strongly covalent complexes of Fe such as the Fe-S prosthetic groups, the theoretical situation becomes even more complicated, and the edge spectra are less well resolved. The simplest case, oxidized and reduced rubredoxin and its model compound, give the edge spectra shown in Figure 2. The $1s \rightarrow 3d$ transitions are well resolved, but the rest of the edge

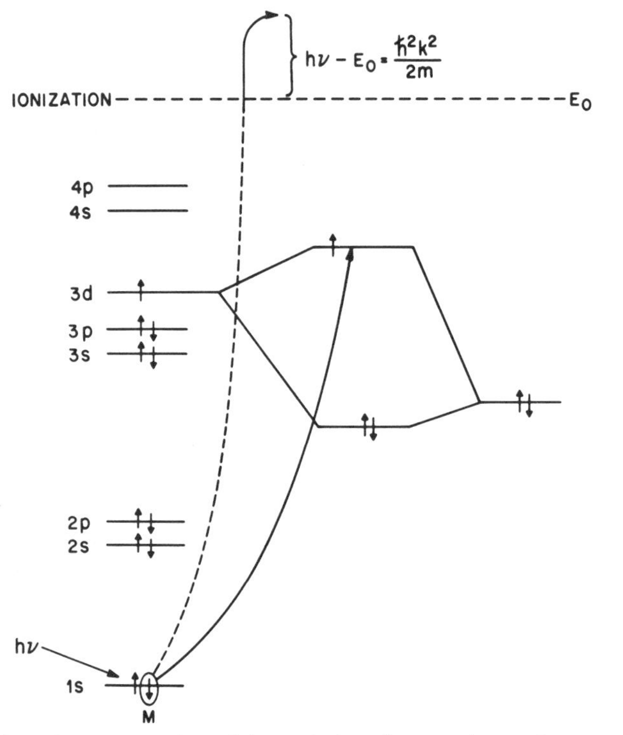

Figure 1. Schematic representations of the excitation of a core electron from a 1*s* orbital (*K* shell) to a valence 3*d* orbital (solid curve) via absorption of a photon ($h\nu$). At sufficiently high photon energy, ionization (dashed curve) occurs, and the outgoing electron possesses kinetic energy ($h\nu - E_0$).

region is not. There is a small shift of ≈ 1 eV to higher energies of the $1s \rightarrow 3d$ absorption in the oxidized (ferric) form, similar to that observed in the iron fluorides but slightly smaller. Note that this shift of ≈ 1 eV is close to the error limits in these experiments, which were performed several years ago at the Stanford Synchrotron Radiation Laboratory and which consisted of measuring the absolute energies of absorption under parasitic running conditions.

The broader absorption of the edge itself is, as mentioned above, not well resolved. It is quite possible to see that both oxidized rubredoxin and the oxidized form of the model compound $[Fe(S_2\text{-}O\text{-}xyl)_2]^-$ give very similar spectra, whereas a different, but again almost identical, pair of spectra are obtained from the reduced forms of both compounds. Hence these edge spectra can be regarded as phenomenological indications of the oxidation state, although definite assignments of peaks are only possible for the $1s \rightarrow 3d$ absorption.

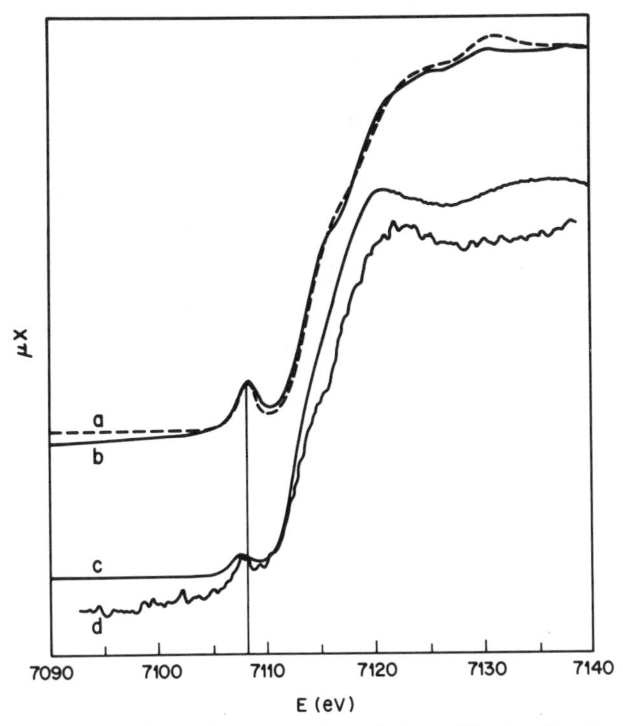

Figure 2. Comparisons of the absorption edges of the oxidized and reduced forms of rubredoxin from *P. aerogenes* and of $Fe(S_2\text{-O-xyl})_2$. The curves are (*a*) $[Fe(S_2\text{-O-xyl})_2]^{-1}$; (*b*) Rub_{ox}; (*c*) $[Fe(S_2\text{-O-xyl})_2]^{2-}$; (*d*) Rub_{red}. The faint vertical line through the first peak in the spectra of the oxidized forms (curves *a* and *b*) shows that in the analogous reduced forms (curves *c* and *d*) the 1*s*-3*d* transitions are shifted ~0.7 V to lower energies.

At this point we indicate the additional theoretical complexities that presently prevent peak assignments. The transition matrix element depends on the amount of metal ion *p*-electron character in the final bound state, since only the $s \rightarrow p$ transition is allowed. Furthermore, it has previously been shown that the $1s \rightarrow 3d$ and $1s \rightarrow 4s$ transitions become allowed to the extent that 4*p* metal ion character has been mixed into them. The molecular orbitals describing the bonding and antibonding states take the form

$$\psi_b = (1 - \gamma^2)^{1/2}\phi_L + \gamma\phi_M$$

$$\psi_a = \lambda\phi_L + (1 - \lambda^2)^{1/2}\phi_M$$

where $\lambda \simeq \gamma$ is the degree of covalent mixing. In the ionic cases $\lambda \ll 1$, and the X-ray transition is mainly from 1*s* to the antibonding state ψ_a. However, when covalency is appreciable, as in the present Fe-S complexes, these transitions are to bonding orbitals that are appreciably metal in character and consequently these transitions are intense and cannot be neglected. Since the bonding-antibonding energy splitting is several volts, which is just slightly larger than the natural linewidth, at these energies ≈ 1 V, these additional transitions can fill in the holes in the spectrum and produce the kind of broad spectrum shown in Figure 3.

Having discarded, for the present, any possibility of assigning peaks in the broad intense absorption band, we now try to see what information can be obtained from the observed absorption edges. Figure 3 shows the previously unpublished edge spectra of the oxidized and reduced states of typical 2-Fe, 4-Fe, and 8-Fe proteins of the Fe-S class. The first surprising feature is that

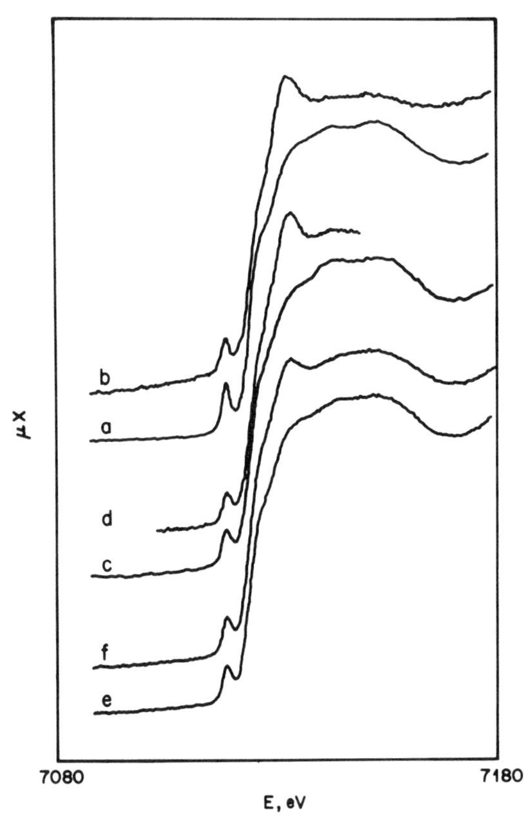

Figure 3. X-Ray absorption edge spectra of Rhubarb Fd., oxidized (*a*); Rhubarb Fd., reduced (*b*); HIPIP, oxidized (*c*); HIPIP, reduced (*d*); Fd. P. Aerogenes, oxidized (*e*); Fd. P. Aerogenes, reduced (*f*).

all the air-stable, oxidized forms have almost identical spectra, which closely resemble the oxidized form of rubredoxin, while all the dithionite-treated, reduced states are also similar to each other and to the corresponding reduced form of rubredoxin. This surprise is accentuated when we realize that the formal charges of these clusters do not depend on dithionite reduction but are as given in Table 1. Since X-ray absorption is an extremely fast reaction, it should respond to the charges on the Fe atoms, influenced slightly by the distribution of final states. We would expect from this that the Fe^{+3} and Fe^{+2} states of Fe, surrounded by a tetrahedron of S atoms, would have X-ray absorption edges similar to oxidized and reduced rubredoxin, respectively. Since the absorption measurement is instantaneous, we would expect that the spectra of these two oxidation states would be like vectors, which could be added together in proportions reflecting the formal charge states of the Fe ions to produce the observed spectra. For example, the oxidized state of the [2Fe-2S] ferrodoxin (rhubarb) should resemble a superposition of two spectra similar to those in oxidized rubredoxin, which it does, while the spectra of the one electron reduced form should look like a superposition of equal parts of the oxidized and reduced rubredoxin spectrum. It does not; in fact it strongly resembles the reduced form of rubredoxin. In all of the other multi-Fe proteins and models (see Figure 3a–3f), similar results are observed. The absorption edges of the air-stable, oxidized form resemble that of oxidized rubredoxin, and the edges of the dithionite-reduced forms resemble that of dithionite-reduced rubredoxin. The formal charges of the oxidized forms range from +3.0 to +2.5 (see Table 1). On the basis of formal charge we would expect that the reduced forms of HIPIP and spinach ferrodoxin should resemble the oxidized form of bacterial Fd, which is clearly not the case, as can be seen in Figure 3. Instead, the two reduced forms resemble each other and the other two reduced forms, while the oxidized form is different and

Table 1 Formal Charges of Fe-S Clusters

No. of Fe Atoms	Formal Charge				
	+3.00	+2.75	+2.50	+2.25	+2.00
1	Rub. (ox)				Rub (red)
2	Fd, plant (ox)		Fd, plant (red)		
4		HIPIP (ox)	HIPIP (red)		
8			Fd, bacterial (ox)	Fd, bacterial (red)	

resembles all the other oxidized forms. These results suggest that the formal charge is a poor measure of the "actual" charge on the Fe atoms, and that for these highly covalently bonded systems the charges may be delocalized over the entire cluster in which the sulfurs absorb the electrons.

A second surprising result is seen in the $1s \rightarrow 3d$ transition of the [2Fe-2S] ferrodoxin. We have examined this resonance very carefully looking for two lines, separated by ~ 1 eV, coming from the different types of Fe atoms. Although we did see a difference between the positions of these two peaks in the oxidized and reduced forms of rubredoxin and its models, the reduced 2-Fe ferrodoxin did not show two distinct lines in the $1s \rightarrow 3d$ region. This result is consistent with the description given above, in which the rest of the edge spectrum did not seem to be a sum of oxidized and reduced forms. However, it is not consistent with published analyses of the NMR and Mössbauer studies of 2-Fe ferrodoxins that were interpreted in terms of 1 of the 2 Fe ions being Fe^{2+} and the other Fe^{3+}. We are not able to resolve these differences at this time.

Finally, what description of the electronic distribution is consistent with these results? Qualitatively it would be first that the Fe atoms are all nearly identical in each oxidation state (oxidized vs. reduced) of the protein; second, that they respond similarly in their charge state to the electronic free energy of the outside world (i.e., either an O_2 or dithionite environment); and third, that the different numbers of electrons accounting for the different formal charges must be found elsewhere, presumably delocalized over (and perhaps beyond) the cluster core. This novel description for these highly covalent Fe-S clusters can be tested in the future.

3 EXAFS STUDIES

The extended X-ray absorption fine structure (EXAFS) refers to the oscillatory variation of the X-ray absorption as a function of photon energy beyond the absorption edge. The absorption, normally expressed in terms of the absorption coefficient μ, can be determined from a measurement of the attenuation of X-rays on their passage through a material. As mentioned above, when the X-ray photon energy E is increased through the binding energy of some core level of an atom in the material, an abrupt increase, known as the *absorption edge,* occurs in the absorption coefficient. For isolated atoms the absorption coefficient decreases monotonically as a function of energy beyond the edge. For atoms either in a molecule or embedded in a condensed phase, the variation of absorption coefficient at energies above the absorption edge displays a complex fine structure called *EXAFS.*

Figure 4 shows the X-ray absorption spectrum of rubredoxin, which shows

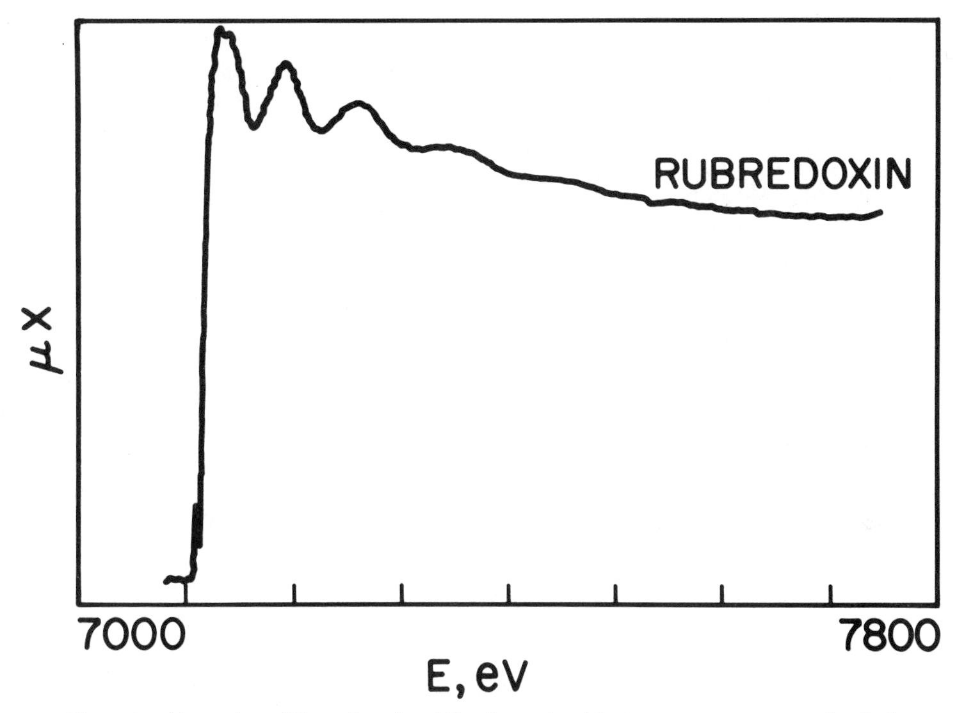

Figure 4. Absorption of X-rays by a lyophilized sample of Peptococcus aerogenes rubredoxin measured in transmission experiment. The sample was packed into a holder 0.75 × 2.0 × 40 mm in which the beam traversed the narrowest dimension. About 25 mg of protein was used.

the $1s \rightarrow 3d$ transition before the edge and the EXAFS modulations above the edge. EXAFS spectra generally refer to the region 40–1000 eV above the absorption edge.

Transmission is just one of several modes of EXAFS measurements. The fluorescence technique involves the measurement of the fluorescence radiation (over some solid angle) at right angle to the incident beam (6). For dilute biological systems this method removes the background absorption due to other constituents, thereby improving the sensitivity by orders of magnitude. Both methods require no vacuum technique.

From a qualitative viewpoint the probability that an X-ray photon will be absorbed by a core electron depends on both the initial and the final states of the electron. The initial state is the localized core level. The final state is that of the ejected photoelectron, which can be represented as an outgoing spherical wave originating from the X-ray absorbing atom. If the absorbing atom is surrounded by a neighboring atom, the outgoing photoelectron wave is back-scattered by the neighboring atom, thereby producing an incoming electron wave. The final state is then the sum of the outgoing and all the incoming waves, one per each neighboring atom. It is the interference between the outgoing and the incoming waves that gives rise to the sinusoidal variation of μ vs. E known as EXAFS (Fig. 4).

The frequency of each EXAFS wave depends on the distance between the absorbing atom and the neighboring atom, since the photoelectron wave must travel from the absorber to the scatterer and back. During the trip the photoelectron actually experiences a phase shift (coulombic interaction) of the absorber twice (i.e., once going out and once coming back) and a phase shift of the scatterer once (scattering). On the other hand the amplitude of each EXAFS wave depends on the number and the backscattering power of the neighboring atom, as well as on its bonding to and distance from the absorber (*vide infra*). From an analysis of the scattering profiles, we can semi-quantitatively assess the types and numbers of atoms surrounding the absorber.

To analyze EXAFS spectra, we first convert the absorption coefficient $\mu(E)$ into EXAFS $\chi(E)$ by the equation

$$\chi(E) = \frac{\mu(E) - \mu_0(E)}{\mu_0(E)} \tag{1}$$

where $\mu_0(E)$ is the background absorption, which can be approximated by fitting a smooth curve (polynomial or spline) through the observed $\mu(E)$. This procedure is often called *background removal.* To relate $\chi(E)$ to structural parameters, it is necessary to convert the energy E into the photoelectron wavevector k via the equation

$$k = \sqrt{\frac{2m}{h^2}(E - E_0)} \tag{2}$$

This transformation of $\chi(E)$ in E space gives rise to $\chi(k)$ in k space, as shown in Figure 5*a*, where (4)

$$\chi(k) = \sum_j N_j S_j(k) F_j(k) e^{-2\sigma_j^2 k^2} e^{-2r_j/\lambda(k)} \frac{\sin[2kr_j + \phi_j(k)]}{kr_j^2} \tag{4}$$

Here $F_j(k)$ is the backscattering amplitude from each of the N_j neighboring atoms of the jth type with a Debye-Waller factor of σ_j [to account for thermal vibration (assuming harmonic vibration) and static disorder (assuming gaussian pair distribution)] and at a distance r_j away; $\phi_j(k)$ is the total phase shift experienced by the photoelectron. The term $e^{-2r_j/\lambda}$ arises from inelastic losses in the scattering process (caused by neighboring atoms and the medium in between), with λ being the electron mean free path. The term $S_j(k)$ is the amplitude reduction factor due to many-body effects such as shake up/off processes at the central atom. It is clear that each EXAFS wave

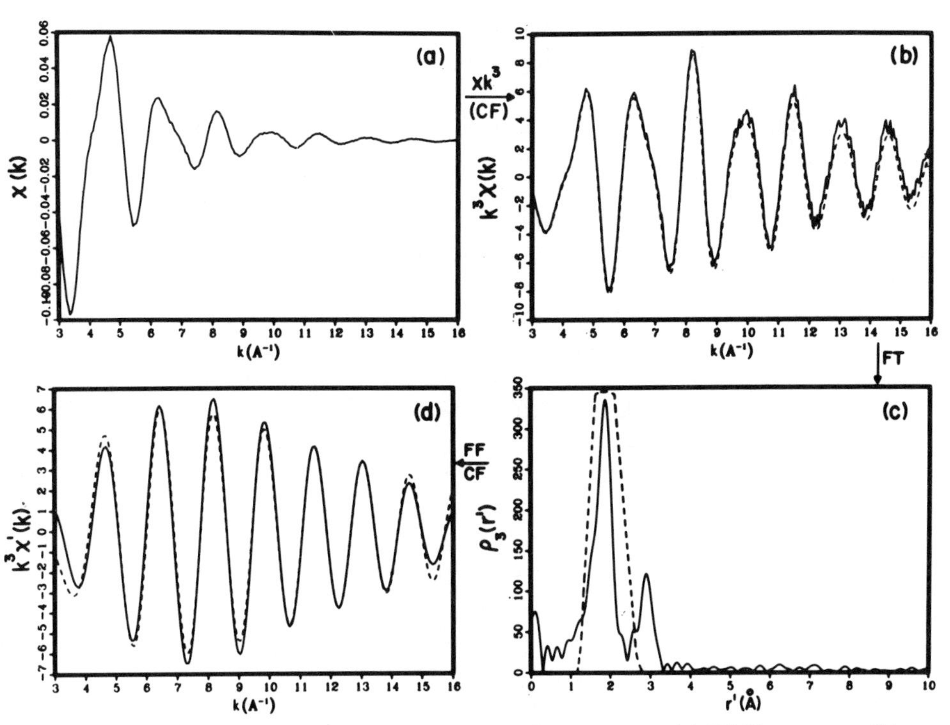

Figure 5. Data reduction and data analysis in EXAFS spectroscopy: (*a*) EXAFS spectrum $\chi(k)$ versus k after background removal; (*b*) the solid curve is the weighted EXAFS spectrum $k^3\chi(k)$ versus k [after multiplying $\chi(k)$ by k^3]. The dashed curve represents an attempt to fit the data with a two-distance model by the curve-fitting (CF) technique; (*c*) Fourier transformation (FT) of the weighted EXAFS spectrum in momentum (k) space into the radial distribution function $\rho_3(r')$ versus r' in distance space. r' is related to the true distance r by a "phase shift" $\alpha = r - r'$. The dashed curve is the window function used to filter the major peak in Fouirer-filtering (FF); (*d*) Fourier-filtered EXAFS spectrum $k^3\chi'(k)$ versus k (solid curve) of the major peak in (*c*) after backtransforming into k space. The dashed curve attempts to fit the filtered data with a single-distance model.

is determined by the backscattering amplitude $N_jF_j(k)$, modified by the reduction factors $S_j(k)$, e, and $e^{-2r_j/\lambda}$, and the $1/kr_j^2$ distance dependence, and the sinusoidal oscillation, which is a function of interatomic distances $2kr_j$ and the phase shift $\phi_j(k)$. This is the generally accepted short-range single electron single scattering theory.

It should be emphasized that, while the amplitude function $F_b(k)$ depends only on the type of backscatter, the phase function contains contributions from both the absorber and the backscatter:

$$\phi_{ab}^l(k) = \phi_a^l(k) + \phi_b(k) - l\pi \tag{5}$$

where $l = 1$ for the K and L_I edges and $l = 2$ or 0 for the $L_{II,III}$ edges. Here $\phi_a = 2\delta_l'$ is the l phase shift of the absorber and $\phi_b = \theta$ is the phase of the backscattering amplitude (7).

Structural determinations via EXAFS depend on the feasibility of resolving the data into individual waves corresponding to the different types of neighbors of the absorbing atom. This can be accomplished by either curve-fitting or Fourier transform techniques. Curve-fitting involves a best fitting of the data with a sum of individual waves modeled by some empirical equations, each of which contains appropriate structural parameters for each type of neighbor (Fig. 5*b*). On the other hand the Fourier transform technique provides a photoelectron scattering profile as a function of the radial distance from the absorber (Fig. 5*c*). In such a radial distribution function the positions of the peaks are related to the distance between the absorber and the neighboring atoms, and the sizes of the peaks are related to the numbers and types of the neighboring atoms.

It is obvious that the Fourier transform technique has the advantage of providing a simple physical picture—radial distribution function—of the local structure around the absorber, whereas curve-fitting methods can provide higher resolution and more accurate results, especially for systems with closely spaced interatomic distances.

A compromise of these two approaches is Fourier filtering following by the curve-fitting technique. It involves Fourier transforming the $k^n\chi(k)$ data into the distance space, selecting the distance range of interest with some smooth window (dashed curve in Fig. 5*c*), and backtransforming the data to k space (Fig. 5*d*). The resulting "filtered" EXAFS spectrum $k^n\chi'(k)$ can then be fitted with simpler models (dashed curve in Fig. 5*d*). This procedure has the additional advantage of simultaneous removal of the high-frequency noise and the residual background as well as providing equally-spaced data points in k space.

EXAFS has been used to probe the prosthetic group of Fe-S proteins, and studies have been made of prototypes of nonheme Fe-S proteins containing 1-4 Fe atoms (8). The minimal prosthetic groups are $Fe(SR)_4$ in rubredoxin, $Fe_2S_2(SR)_4$ in plant ferredoxins (Fd), $Fe_3S_s(SR)_6$ in the Fe-S protein III from *Azotobacter vinelandii* (actually this protein contains a 3-Fe and a 4-Fe cluster), and $Fe_4S_4(SR)_4$ in high potential Fe proteins (HIPIP) and bacterial ferredoxins (actually this protein has two 4-Fe clusters), as tabulated in Table 2. The Fe EXAFS of these proteins are dominated by neighboring S and Fe atoms of the active sites (Fig. 6). The amplitude envelope of the monomeric $Fe(SR)_4$ species varies smoothly with k (Fig. 6*g*), indicative of a single-shell system with one type of distance (9). In contrast the amplitude envelope of the 2-Fe and 4-Fe oligomers exhibits a "beat" node at $k \approx 7$ Å^{-1}, which is characteristic of two-shell systems with two types of distances (Fig. 6) (10). The frequency at lower k region, which reflects the shorter Fe–S

Table 2 Four Distinct Prototypes of Cluster Structures Found in Nonheme Fe-S proteins[a]

Simplified Formula	Idealized Symmetry	Fe	S^b	S^t	SR	Fe-Fe/Fe-S
$Fe(SR)_4$	T_d	1	0	0	4	0/4
$Fe_2S_2(SR)_4$	D_{2h}	2	2	0	4	1/4
$Fe_3S_3(SR)_6$	D_{3h}	3	3	0	6	2/4
$Fe_4S_4(SR)_4$	T_d	4	0	4	4	3/4

[a]The numbers of Fe, doubly bridging S, triply bridging S, and terminal S atoms are listed under Fe, S^b, S^t and SR, respectively. The ratio of the numbers of Fe-Fe vs. Fe-S bonds is given under Fe-Fe/Fe-S.

bonds, is lower than the frequency at larger k region, which is indicative of the longer Fe-Fe bonds. Fourier transforms of the $\chi(k)k^3$ data for both solid and solution states reveal only one peak for the monomer (Fig. 6) but two peaks for the two oligomers (Fig. 6). The major peak can be assigned to the Fe-S bonds, and the minor peak at a larger distance in each of the oligomers can be assigned to the Fe-Fe bonds.

EXAFS has provided an excellent opportunity for correlating structural parameters with redox states of the proteins, both in the solid state and in solution, as well as comparison between the proteins and the model compounds. This is demonstrated in Table 3 where we compare 1-Fe, 2-Fe, 4-Fe, and 8-Fe proteins in different oxidation states with their respective model compounds (10).

Generally speaking, the EXAFS results are in good agreement with single-crystal structural data. For the model compounds the average Fe-S and Fe-Fe bond lengths agree to better than 0.03 Å. For protein systems where crystallographical data are available, the EXAFS results generally fall within the range of the scattered bond lengths derived from protein crystallography. A detailed comparison of the average molecular parameters of the models with those of the proteins provides strong direct structural evidence that the model compounds are excellent representations of the active sites of the proteins.

In most applications EXAFS provides only the average distances. Nevertheless, the Debye-Waller factor σ can indicate the spread of the distances, since in general σ has two components σ_{stat} and σ_{vib}, arising from static disorder and thermal vibrations, respectively (assuming small disorders with a symmetric pair distribution function for static disorder and harmonic vibration for thermal disorder):

$$\sigma^2 = \sigma_{stat}^2 + \sigma_{vib}^2 \tag{6}$$

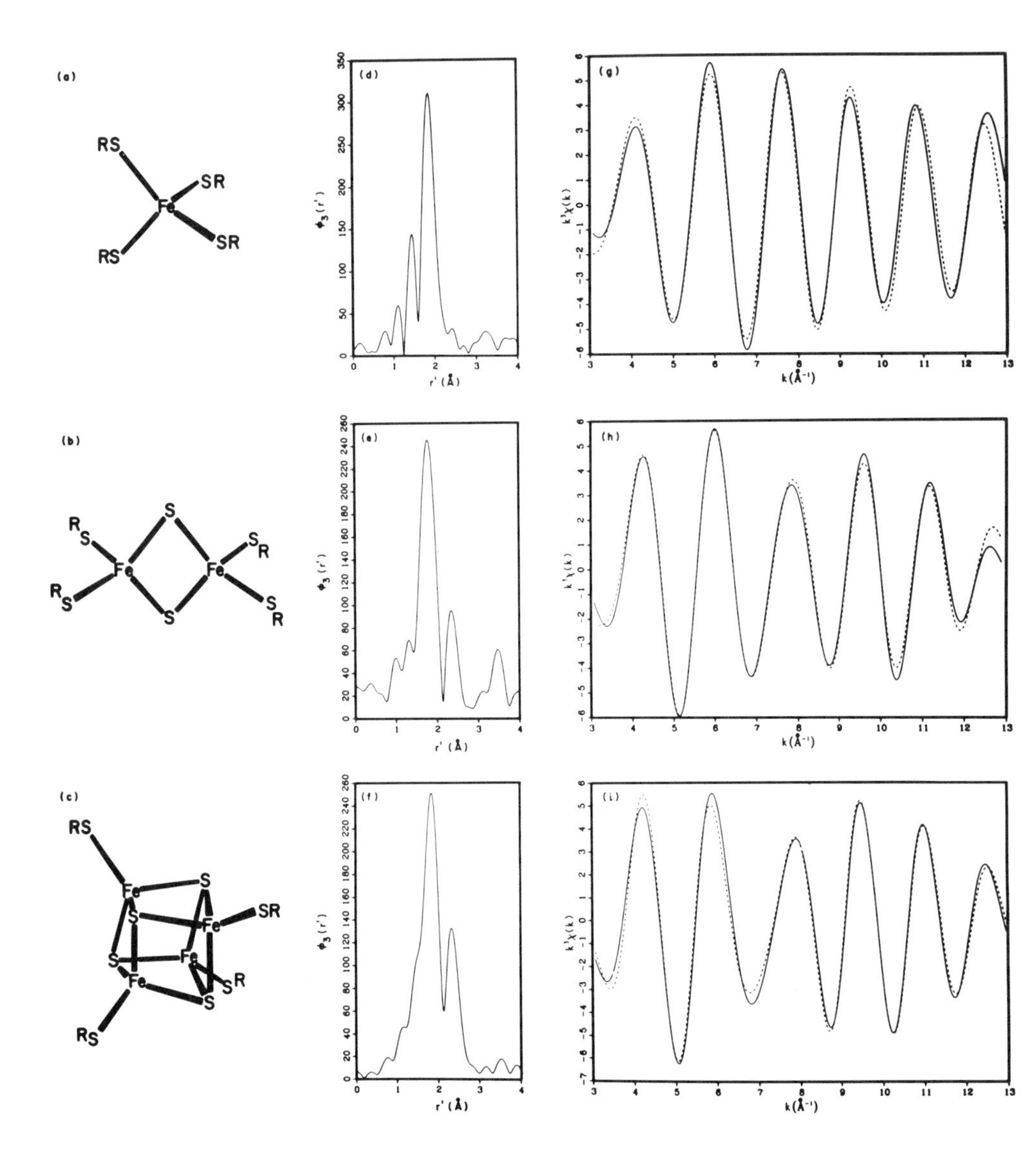

Figure 6. *Left*: the three prototypes of iron-sulfur protein active sites: (*a*) monomer, (*b*) dimer, and (*c*) tetramer. *Middle*: the Fourier transforms of the corresponding EXAFS spectra ($k = 3 \sim 14$ Å^{-1}). *Right*: the corresponding Fourier filtered (window: $r' = 0.9 \sim 3.5$ Å). EXAFS spectra (solid curves) and the theoretical fits (dashed curves). The major peak in each case is assigned to the Fe-S distances. The minor peak at a larger distance for the oligomers are Fe-Fe distances. The minor peaks to the left of the Fe-S peak are due to residual background and/or Fourier truncation.

Table 3. Least-Squares Refined Interatomic Distances (Å) and Debye-Waller Factors (Å) with Fitting Errors for Fe-S Models (1–4) and Proteins (5–14)[a–c] and Their Comparisons with Crystallographic Results (Bond Lengths Only)

no.	compd		EXAFS Fe–S	EXAFS Fe–Fe	diffraction Fe–S	diffraction Fe–Fe
			Model Compounds			
1	$[Fe(S_2\text{-}o\text{-xyl})_2]^-$	r	2.279 (13)		2.267 (2)	
		σ	0.043 (15)			
2	$[Fe(S_2\text{-}o\text{-xyl})_2]^{2-}$	r	2.340 (14)		2.356 (5)	
		σ	0.053 (17)			
3	$[Fe_2S_2(S_2\text{-}o\text{-xyl})_2]^{2-}$	r	2.234 (15)	2.704 (23)	2.257 (2)	2.698 (1)
		σ	0.070 (11)	0.070 (10)		
4	$[Fe_4S_4(S\text{-benzyl})_4]^{2-}$	r	2.270 (13)	2.717 (24)	2.286 (2)	2.747 (2)
		σ	0.064 (10)	0.093 (9)		
			Proteins			
5	Rub_{ox}(solid)	r	2.265 (13)		2.24	
		σ	0.049 (15)			
6	Rub_{ox}(soln)	r	2.256 (16)			
		σ	0.047 (18)			
7	Rub_{red}(soln)	r	2.32 (2)			
		σ	0.057 (25)			
8	plant Fd_{ox}(solid)	r	2.227 (15)	2.696 (47)		
		σ	0.063 (14)	0.078 (16)		
9	plant Fd_{ox}(soln)[d]	r	2.233 (22)	2.726 (40)		
		σ	0.063 (19)	0.057 (18)		
10	plant Fd_{red}(soln)[d]	r	2.241 (28)	2.762 (48)		
		σ	0.059 (22)	0.076 (31)		
11	$HIPIP_{ox}$(soln)	r	2.262 (13)	2.705 (26)	2.24 (5)	2.73 (4)
		σ	0.060 (11)	0.088 (9)		
12	$HIPIP_{red}$(soln)	r	2.251 (13)	2.659 (50)	2.30 (7)	2.81 (5)
		σ	0.001 (27)	0.088 (17)		
13	bact Fd_{ox}(soln)	r	2.249 (16)	2.727 (35)	2.27 (20)	2.85 (10)
		σ	0.063 (15)	0.092 (13)		
14	bact Fd_{red}(soln)	r	2.262 (14)	2.744 (32)		
		σ	0.062 (11)	0.098 (13)		

[a] Abbreviations: Rub, rubredoxin; Fd, ferredoxin; bact, bacterial; HIPIP, high-potential iron protein; ox, oxidized; red, reduced; soln, solution. [b] The fitting errors for each parameter (in parentheses) were obtained by changing that particular parameter, while least-squares refining the others within the same term, until the χ^2 contribution from that particular term is doubled. All parameters associated with the other term (except the overall scale factor) were held constant. [c] Other systematic errors including background removal and Fourier filtering may give rise to uncertainties of 1, 2, and 10% in r(Fe–S), r(Fe–Fe), and σ, respectively. [d] The data for **4** and **5** are poorer so that the errors should probably be doubled.

For a two distance system with m bonds at a distance r_m and n bonds at a distance r_n,

$$\sigma_{\text{stat}} \approx \frac{\sqrt{mn}}{m+n} \Delta r = \frac{\sqrt{mn}}{m+n} |r_m - r_n| \tag{7}$$

For example, in the monomeric models $[Fe(S_2\text{-O-xyl})_2]^{-,2-}$ we found the spread Δr of the Fe–S bond lengths to be 0.00(4) and 0.06(4) Å, respectively, if we assume an $m:n = 2:2$ model and a σ_{vib} of 0.045 Å (based on the Fe–S vibrational frequency of 314 cm^{-1}). These values are in good agreement with the X-ray crystallographic results of 0.00(1) and 0.04(1) Å, respectively (9).

The average Fe-S distances in rubredoxin, as determined by EXAFS, are 2.27(1) and 2.32(2) Å for the oxidized and reduced states, respectively (9). Protein crystallography for the oxidized form of rubredoxin initially revealed two kinds of Fe–S bonds with three distances of normal bond lengths at r_3 = 2.30(av) Å and one unusually short distance at r_1 = 2.05(3) Å (11). Based on the Debye-Waller factor of 0.049(15) Å from a single distance fit and a reasonable σ_{vib} of 0.045 Å calculated from an Fe–S stretching frequency of 314 cm^{-1}, we concluded that the 4 Fe–S bonds [av 2.27(1) Å] in rubredoxin are chemically equivalent to within 0.04 Å (with a limit of +0.06 or −0.04 Å) for either the 3:1 or the 2:2 ($m:n$) model [see eqs. (6, 9)]. On the other hand, if we fit the data with a two distance model with three Fe–S bonds at r_3 and one Fe–S bond at r_1, a broad least squares residuals minimum (curve A), shown in Figure 7, is obtained when we allow all parameters to vary. The reason is that strong correlation exists between σ and the distance spread Δr. To make a better estimate of the difference $\Delta r = r_1 - r_3$, we must fix σ at some reasonable value such as σ_{vib} (curves B–D correspond to fixing σ at 0.030, 0.045 (σ_{vib}), and 0.049 Å, respectively) (9). In fact comparisons with model structures suggest that such estimates represent an upper limit of the true disparity in distances. If the distance spread is great enough to produce a beat node in the EXAFS amplitude, the correlation between σ and Δr diminishes such that it is possible to resolve the individual distances. Consistent with the EXAFS results, further crystallographic refinements have revised the distances such that $r_3 - r_1 = 0.10$ Å. It is now clear that the 4 Fe–S bonds in rubredoxin in both solid and solution states are equivalent to within a range of better than 0.10 Å. Similar conclusions were also reached by Stern and co-workers (12).

For the oxidized plant ferredoxin in powder form we found average Fe–S and Fe–Fe bond distances at 2.23(2) and 2.70(4) Å, respectively (10), very similar to the values for the model compound. Assuming an $m:n = 2:2$ model and a reasonable $\sigma_{\text{vib}} = 0.045$ Å for the Fe–S bonds, we estimate the nonequivalency of the Fe–S bonds in both oxidized and reduced plant ferrodoxin ($\sigma = 0.06$ Å) to be 0.08 Å. This latter value is even smaller than that of

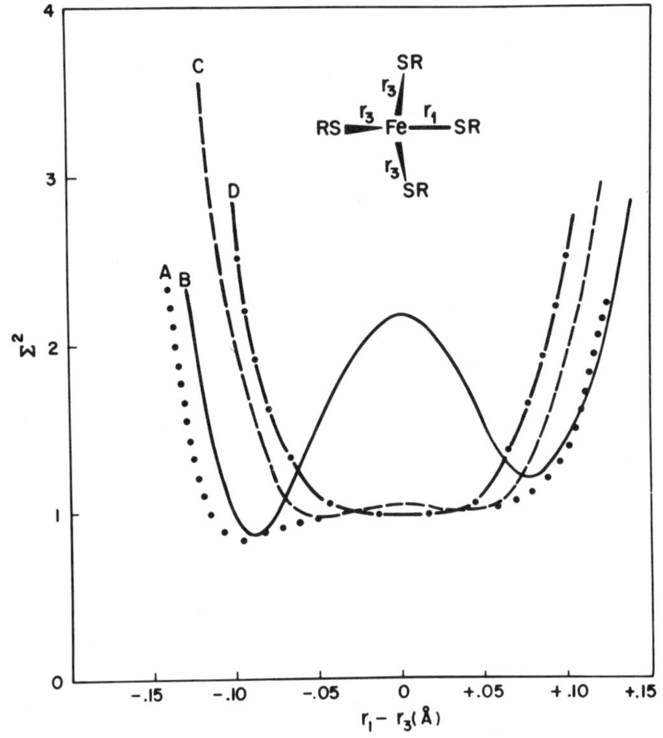

Figure 7. The chi-squares minimization of the curve fitting of Fourier filtered EXAFS data of rubredoxin with a single-shell two-distance model. Σ^2, the sum of squares of the least-squares residuals, is plotted as a function of $(r_1 - r_3)$ where r_1 and r_3 are the distances of one and three Fe-S bonds, respectively. The outer curve $A(\cdots)$ was calculated by fixing values of $r_1 - r_3$ and varying the four parameters $\frac{1}{4}(r_1 + 3r_3)$, the Debye-Waller factor σ, the scale factor, and the energy threshold E_0. The other three curves (B, C, and D) show the effect of fixing σ at 0.030 (-), 0.045 (- - -), and 0.049 (-·-·-·-) Å while allowing the remaining three parameters to vary.

0.12 Å, which was found for the model compound, suggesting that the 2 Fe atoms in both states are chemically equivalent (at least at room temperature).

For ferredoxin containing [4Fe-4S*] clusters, the Fe–S and Fe–Fe distances range from 2.25(2) to 2.26(2) Å and from 2.66(4) to 2.74(4) Å, respectively (10). In fact the Fe–S and Fe–Fe distances in reduced HIPIP (2.25 and 2.66 Å) and oxidized bacteria ferrodoxin (2.25 and 2.73 Å) are very similar to model tetramer (2.27 and 2.72 Å), which is in accord with the three stage hypothesis (13).

We find no significant changes (within 0.01 Å) in the structural parameters on dissolution of the proteins in solution. The Debye-Waller factors are also in good agreement in the two phases. These findings indicate that the active sites of the proteins undergo little, if any, structural changes in going from the powder to the solution phase. Any solvation effects on the redox properties must therefore affect only the proteins and not the active sites.

Since these Fe-S proteins function as electron carriers, it is important to correlate structure changes, if any, with the redox behavior. For each structural type there are small but significant increases in the average Fe–S and Fe–Fe distances on redox. The magnitude (per electron) of such variation, however, decreases in going from monomeric to dimeric to tetrameric Fe-S clusters because of the increasing number of Fe-S and Fe-Fe bonds involved.

For rubredoxin one electron reduction causes a lengthening of the four Fe–S bonds by 0.06 Å. For ferrodoxins containing [2Fe-2S*] and [4Fe-4S*] clusters, however, the changes in the average Fe–S and Fe–Fe distances are much smaller. For example, on reduction the Fe–S and Fe–Fe distances in bacterial ferrodoxin change by only 0.01 and 0.01 Å (2.26 vs. 2.25 Å, 2.74 vs. 2.73 Å), respectively. The small changes upon reduction disagree with protein crystallography at 2 Å resolution, where changes of +0.06 [2.30(6) vs. 2.24(5) Å] and +0.07 [2.81(5) vs. 2.73(4) Å] Å in Fe–S and Fe–Fe distances have been found. The EXAFS findings, however, are in accord with the corresponding changes of +0.03 and +0.01 Å on reduction of $[Fe_4S_4(SPh)_4]^{2-}$ to its trianion. We emphasize that these changes are within the limits of our present EXAFS accuracy (0.02 Å in Fe–S and 0.04 Å in Fe–Fe distances). Furthermore, though EXAFS does not allow the determination of the sense of distortion of the Fe_4S_4 cube, the small spreads of Fe–S and Fe–Fe distances indicated by the Debye-Waller factors rule out any drastic structural distortion. The implication is that any strain energy, if present, probably lies within the polypeptide region (either through a localized or a delocalized mechanism) rather than being stored in the redox centers, since the prosthetic group, in the protein, behaves exactly like the model compounds, upon reduction.

A recent X-ray crystallographic study (14) of the ferredoxinlike protein (Fe-S protein III) from *Azotobacter vinelandii* at 2.5 Å revealed two distinctly different clusters separated by 12 Å. The larger cluster consists of a tetranuclear Fe_4S_4 core ligated to the protein at each Fe (via a cysteine S), while the smaller cluster is a planar Fe_3S_3 core which makes six linkages (5 cysteines and 1 glutamic acid residue) to the protein, as shown in Figure 8. The 4-Fe center is structurally analogous to other $Fe_4S_4(SR)_4$ clusters with *bonding* Fe–Fe distances of 2.7 Å. The 3-Fe center, on the other hand, has *nonbonding* Fe ··· Fe distances of 4.1, 4.2, and 4.4 Å (14a) [in an earlier communication (14b) the Fe ··· Fe distances were reported to be 3.17, 3.65, and 3.72 Å]. The conformation of the 3-Fe cluster is a twist-boat with a unique Fe atom that has a terminal cysteinyl and glutamic acid linkages (rather than two cysteinyl linkages) to the protein. An alternative view of the 3-Fe cluster

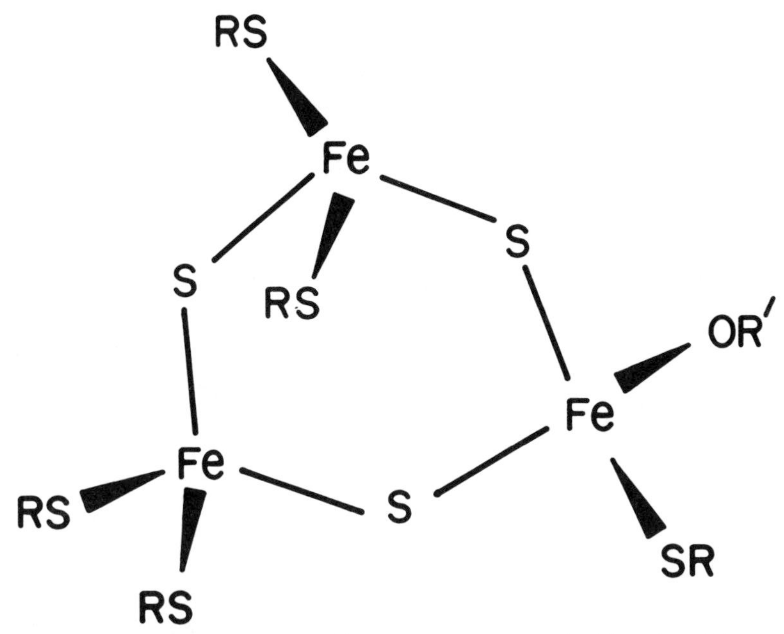

Figure 8. A 3Fe-3S cluster with five sulfur and one oxygen terminal ligands and three bridging sulfur ligands.

is a $Fe_2S_2(SR)_4$ dimer with one of the doubly bridging S atoms (the acid-labile S) replaced by a $FeS_2(SR)(OR')$ moiety where Sr and OR′ represent cysteine and glutamic acid ligands, respectively. The most notable result of such a formal insertion is the lengthening of the Fe–Fe bond distance of 2.7 Å in the dimer to a nonbonding Fe ··· Fe distance of 4.4 Å in the 3-Fe cluster. The two new nonbonding Fe ··· Fe distances are 4.1 and 4.2 Å.

From a recent EXAFS study (15) of the Mo site (the Mo-Fe protein as well as the Fe-Mo cofactor) in the N_2 fixation enzyme nitrogenase, Hodgson and co-workers concluded that the Mo atom is bonded to 3.8 S atoms at 2.35 Å, 3.0 Fe atoms at 2.72 Å, and 1-2 S atoms at ~2.55 Å (15a). Two structural models (see Fig. 9) were proposed for the active site of nitrogenase. Synthetic approaches taken by various groups (16) have produced [Mo-Fe-S] clusters resembling these models. They include structures consisting of two $MoFe_3S_4$ cubes (see model I in Fig. 9) linked through the Mo atoms via one sulfide and two mercaptide (16a), three mercaptide (16b–e), or $Fe(SR)_6$ (16f) bridges on one hand, and trinuclear [Mo-Fe-S] structures (see model II in Fig. 9) such as $[Cl_2FeS_2MoS_2FeCl_2]^{2-}$ (16g) on the other. A different interpretation (17), however, led to a yet unknown cluster model shown in Figure 10*a*. This model, in which the Mo bridges two Fe_4S_4 cubes via 4 S ligands, is consistent

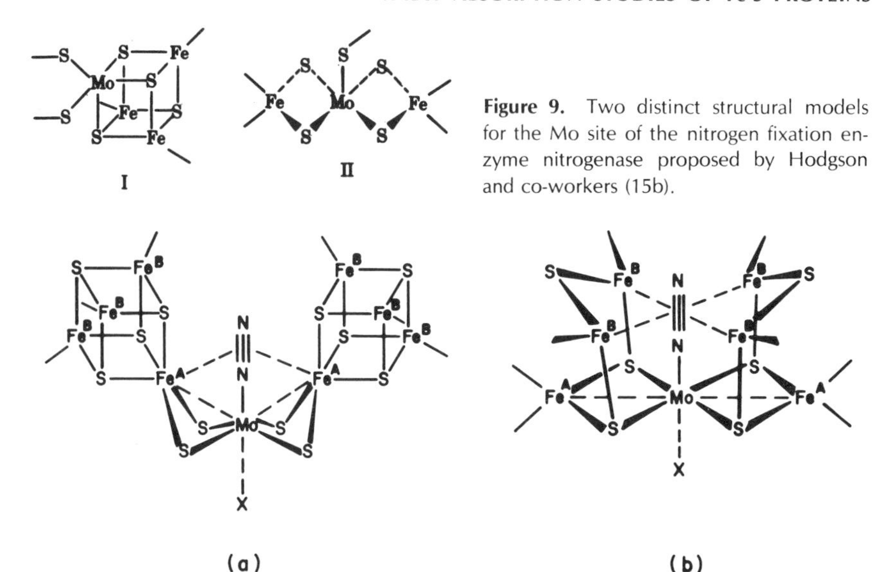

Figure 9. Two distinct structural models for the Mo site of the nitrogen fixation enzyme nitrogenase proposed by Hodgson and co-workers (15b).

Figure 10. Two alternative cluster models for the Mo site of nitrogenase: (*a*) from Ref. 17a and (*b*) Ref. 2k. The cysteine sulfurs (or the SR groups) on the irons are omitted for clarity. *X* represents either a sulfur or a lighter atom from a ligand whose nature or existence remains to be determined.

with the EXAFS data as well as other spectroscopic evidence. The Mo-S distances of 2.35 Å observed in nitrogenase are more consistent with a pseudo-octahedral rather than a tetrahedral coordination, where Mo-S bonds of 2.20–2.25 Å are commonly observed (16g, 17b). Since we now know that the trinuclear Fe-S cluster $Fe_sS_3(SR)_6$ with three bridging sulfido ligands is present in some Fe-S proteins (14), it is natural to consider the models in Figure 10*b* as the active site of nitrogenase. This new structural model, in which the Mo bridges two Fe_3S_3 units, possesses most of the characteristics of model *a*. Strong interactions between the Mo and the two Fe^A atoms result in the presumably diamagnetic $Fe^AS_2MoS_2Fe^A$ unit, which can give rise to simple quadrupole doublets in the Mössbauer spectra. The spins (total $S = 3/2$) reside primarily on the Fe^B atoms. Model *b* has the distinct feature of a variable-size cage through which N_2 can σ bond to the Mo and π bond to the 4 high-spin Fe^B atoms. The net result would be a significant weakening and activation of the N-N bond. Injection of electrons in a stepwise manner via the two Fe_3S_3 units and successive protonation of the terminal N atom would give rise to the intermediates Mo-N≡N → Mo-N=NH → Mo-N-NH_2, and eventually produce ammonia. We note that other similar but distinct types of model structures have also been suggested by Lu and co-workers in their

Fuzhou model II (17c), and by Tsai and co-workers in the Xiamen model (17d). Although these proposed models are structurally quite distinct, they are all consistent with the EXAFS findings. This points to the danger of overstretching EXAFS information or overinterpreting EXAFS data of unknown systems, particularly in light of the large uncertainties often encountered in the determination of coordination numbers. This also emphasizes however, that the Fe-S proteins with 1, 2, 4, or 8 Fe atoms have been ideal systems for EXAFS study because the nature, as well as the numbers, of the ligands were known and the EXAFS studies were used to refine their distances.

REFERENCES

1. (a) B. M. Kincaid and P. Eisenberger, *Phys. Rev. Lett.*, **34,** 1361 (1975); (b) H. Winick and A. Bienenstock, *Ann. Rev. Nucl. Part. Sci.*, **28,** 33 (1978); (c) I. Lindau and H. Winick, *J. Vac. Sci. Technol.*, **15,** 977 (1978); (d) R. E. Watson and M. L. Perlman, *Science,* **199,** 1295 (1978); (e) B. W. Batterman and N. W. Ashcroft, *Science,* **206,** 157 (1979).
2. (a) E. A. Stern, *Contemp. Phys.*, **19,** 289 (1978); (b) P. Eisenberger and B. M. Kincaid, *Science,* **200,** 1441 (1978); (c) R. G. Shulman, P. Eisenberger, and B. M. Kincaid, *Ann. Rev. Biophys. Bioeng.*, **7,** 559 (1978); (d) D. R. Sandstrom and F. W. Lytle, *Ann. Rev. Phys. Chem.*, **30,** 215 (1979); (e) S. P. Cramer and K. O. Hodgson, *Prog. Inorg. Chem.*, **25,** 1 (1979); (f) T. M. Hayes, *J. Non-Cryst. Solids,* **31,** 57 (1978); (g) J. Wong, in *Metallic Glasses,* H. J. Guntherodt, ed., Springer-Verlag, Berlin (1980); (h) *Synchrotron Radiation Research,* H. Winick and S. Doniach, eds., Plenum, New York, 1980; (i) P. A. Lee, P. H. Citrin, P. Eisenberger, and P. M. Kincaid, *Rev. Mod. Phys.*, in press; (j) B. K. Teo, *Acc. Chem. Res.*, **13,** 412 (1980); (k) *EXAFS Spectroscopy: Techniques and Applications,* B. K. Teo and D. C. Joy, eds., Plenum, New York, 1981.
3. R. de L. Kronig, *Z. Physik.*, **70,** 317 (1931); **75,** 191, 468 (1932).
4. (a) E. A. Stern, *Phys. Rev. B.*, **10,** 3027 (1974); (b) E. A. Stern, D. E. Sayers, and F. W. Lytle, *Phys. Rev., B,* **11,** 4836 (1975), and references cited therein; (c) C. A. Ashley and S Doniach, *Phys. Rev., B,* **11,** 1279 (1975); (d) P. A. Lee and G. Beni, *Phys. Rev., B,* **15,** 2862 (1977); (e) P. A. Lee and J. B. Pendry, *Phys. Rev., B,* **11,** 2795 (1975).
5. (a) R. G. Shulman, Y. Yafet, P. Eisenberger, and W. E. Blumberg, *Proc. Natl. Acad. Sci., USA,* **73,** 1384 (1976); (b) F. W. Lytle, P. S. P. Wei, R. B. Greegor, G. H. Via, and J. H. Sinfelt, *J. Chem. Phys.*, **70,** 4849 (1979); (c) L. Powers, W. E. Blumberg, B. Chance, C. H. Barlow, J. S. Leigh, Jr., J. Smith, T. Yonetani, S. Vik, and J. Peisach, *Biochim. Biophys. Acta,* **546,** 520 (1979); (d) F. W. Kutzler, C. R. Natoli, D. K. Misemer, S. Doniach, and K. O. Hodgson, *J. Chem. Phys.*, **73,** 3274 (1980).
6. (a) J. Jaklevic, J. A. Kirby, M. P. Klein, A. S. Robertson, G. S. Brown, and P. Eisenberger, *Solid State Commun.*, **23,** 679 (1977); (b) F. S. Goulding, J. M. Jaklevic, and A. C. Thompson, SSRL Report No. 78/04, May 1978; (c) E. A. Stern and S. M. Heald, *Rev. Sci. Instrum.*, **50,** 1579 (1979).
7. B. K. Teo and P. A. Lee, *J. Am. Chem. Soc.*, **101,** 2815 (1979)
8. For an excellent review see R. H. Holm, *Acc. Chem. Res.*, **10,** 427 (1977), and references cited therein.

9. R. G. Shulman, P. Eisenberger, B. K. Teo, B. M. Kincaid, and G. S. Brown, *J. Mol. Biol.*, **124,** 305 (1978), R. G. Shulman, P. Eisenberger, W. E. Blumberg and N. A. Stombaugh, *Proc. Natl. Acad. Sci. (USA)* **72,** 4003 (1975).

10. B. K. Teo, R. G. Shulman, G. S. Brown, and A. E. Meixner, *J. Am. Chem. Soc.*, **101,** 5624 (1979).

11. K. D. Watenbaugh, L. C. Sieker, J. R. Herriot, and L. H. Jensen, *Acta Crystallogr., B,* **29,** 943 (1973).

12. (a) B. Bunker and E. A. Stern, *Biophys. J.*, **19,** 253 (1977); (b) D. E. Sayers, E. A. Stern, and J. R. Herriott, *J. Chem. Phys.*, **64,** 427 (1976).

13. (a) C. W. Carter et al., *Proc. Natl. Acad. Sci., USA,* **69,** 3526 (1972); (b) C. W. Carter et al., *J. Biol. Chem.*, **249,** 6339 (1974).

14. (a) C. D. Stout, private communication; (b) C. D. Stout, D. Ghosh, V. Pattabhi, and A. Robbins, *J. Biol. Chem.*, **255,** 1797 (1980).

15. (a) T. E. Wolff, J. M. Berg, C. Warrick, K. O. Hodgson, R. H. Holm, and R. B. Frankel, *J. Am. Chem Soc.*, **100,** 4630 (1978); (b) S. P. Cramer, K. O. Hodgson, W. O. Gillum, and L. E. Mortenson, *J. Am. Chem. Soc.*, **100,** 3398 (1978); (c) S. P. Cramer, W. O. Gillum, K. O. Hodgson, L. E. Mortenson, E. I. Stiefel, J. R. Chisnell, W. J. Brill, and V. K. Shah, *J. Am. Chem. Soc.*, **100,** 3814 (1978).

16. (a) T. E. Wolff, J. M. Berg, C. Warrick, K. O. Hodgson, and R. H. Holm, *J. Am. Chem. Soc.*, **100,** 4630 (1978); (b) T. E. Wolff, J. M. Berg, K. O. Hodgson, R. B. Frankel, and R. H. Holm, *J. Am. Chem. Soc.*, **101,** 4140 (1979); (c) G. Christou, C. D. Garner, F. E. Mabbs, and T. J. King, *J. Chem. Soc., Chem. Commun.*, 740 (1978); (d) G. Christou, C. D. Garner, F. E. Mabbs, and M. G. B. Drew, *J. Chem. Soc., Chem. Commun.*, **91,** (1979); (e) S. R. Acott, G. Christou, C. D. Garner, T. J. King, F. E. Mabbs, and R. M. Miller, *Inorg. Chim. Acta,* **35,** L337 (1979); (f) T. E. Wolff, J. M. Berg, P. P. Power, K. O. Hodgson, R. H. Holm, and R. B. Frankel, *Inorg. Chim. Acta,* **101,** 5454 (1979); (g) D. Coucouvanis, N. C. Baenziger, E. D. Simhon, P. Stremple, D. Swenson, A. Simopoulos, A. Kostikas, V. Petrouleas, and V. Papaefthymiou, *J. Am. Chim. Soc.*, **102,** 1732 (1980).

17. (a) B. K. Teo and B. A. Averill, *Biochem. Biophys. Res. Commun.*, **88,** 1454 (1979); (b) R. H. Tieckelmann, H. C. Silvis, T. A. Kent, B. H. Huynh, J. V. Waszczak, B. K. Teo, and B. A. Averill, *J. Am. Chem. Soc.*, **102,** 5550 (1980); (c) J. X. Lu, Fujian Institute of Research on the Structure of Matter (PRC), private communication; (d) K. R. Tsai, Xiamen University (PRC), private communication.

CHAPTER 10

Low-Temperature Magnetic Circular Dichroism Studies of Iron-Sulfur Proteins

MICHAEL K. JOHNSON

Department of Chemistry, Princeton University
Princeton, New Jersey

A. EDWARD ROBINSON
ANDREW J. THOMSON

School of Chemical Sciences, University of East Anglia,
Norwich, United Kingdom

CONTENTS

1 INTRODUCTION

During the last five years we in Norwich have been developing the use of magnetic circular dichroism (MCD) spectroscopy as an optical probe of the electronic ground state of metalloproteins, especially those containing heme and Fe-S centers. To furnish information about the nature of the ground state, it is necessary to make measurements on samples cooled to temperatures of 1.5 K and in magnetic fields of up to at least 5 Tesla. Apart from early reports of experiments on rubredoxin (2) and on some [2Fe-2S] proteins (24) carried out by one of us in Paris, no other laboratories have undertaken this type of work in spite of the potential value of making measurements in this way. In this chapter we describe the results of our own experiments and illustrate the type of information we can obtain.

This novel use of MCD is possible because paramagnetic chromophores invariably give temperature-dependent MCD spectra, whereas diamagnetic species yield MCD signals that are independent of temperature and linearly dependent on magnetic field intensity. For paramagnetic species the MCD intensity can be increased by a factor of up to 70 on going from room temperature to 4.2 K. Hence low-temperature MCD spectroscopy is specifically a sensitive optical probe for paramagnets, and can be used to observe paramagnets in the presence of overlapping absorption bands arising from other chromophores. By careful monitoring of the magnetic field and temperature dependence of the MCD signal, information concerning the nature of the ground state is obtained, and in certain circumstances ground-state g values can be estimated (1). Therefore MCD spectroscopy can supply information similar to that obtained from EPR, Mössbauer, and magnetic susceptibility studies, although the theoretical basis is as yet not so well developed. Furthermore it has certain advantages over each of these techniques. It is less limited than EPR, in that systems with an even number of unpaired electrons can be investigated, or Mössbauer, where ^{57}Fe enrichment is normally required. Magnetic susceptibility studies can be difficult to interpret for multicentered proteins, whereas with MCD spectroscopy the magnetic properties of individual centers can be investigated separately, provided that each has at least one isolated MCD band. However, as with all these techniques, MCD is most powerful when used in conjunction with one of more of the others.

The richness of detail exhibited by low-temperature MCD spectra of Fe-S proteins, particularly compared to the corresponding absorption spectra, illustrates the complexity of their electronic spectra. However, since each type of cluster gives characteristic spectra, this complexity is very useful in providing a good means of identifying Fe-S clusters, but indicates the difficulty of the problem of assigning electronic transitions in all but the simplest case, such as rubredoxin (2).

We have used the technique of low-temperature MCD spectroscopy to study a range of 1-Fe (2), 2-Fe (3, 24), and 4-Fe and 8-Fe (3) proteins as well as several novel cluster types, notably the 3-Fe cluster in Fd II from *Desulfovibrio gigas* (4) and the P and cofactor centers in nitrogenase from *Klebsiella pneumoniae* (5). It has been shown that MCD spectra at ultralow temperatures can provide an excellent spectroscopic fingerprint of cluster type, revealing unique clusters in the nitrogenase protein. The electronic properties of the paramagnetic ground states can be studied. Thus the spin of the reduced form of the 3-Fe center in Fd II is determined to be $S = 2$. We have been able to characterize the MCD spectra and magnetization curves of the three accessible oxidation states of the [4Fe-4S] cluster in several proteins. The MCD spectra enable a clear distinction to be made between [3Fe-3S] and [4Fe-4S] clusters. This has allowed us to establish that the so-called *superoxidized* derivative of *Clostridium pasteurianum* Fd is a [3Fe-3S] cluster (6). This result has since been confirmed by resonance Raman spectroscopy (see Chapter 11). The observation that [3Fe-3S] clusters can be generated by oxidative damage to [4Fe-4S] clusters is of great significance and raises many interesting questions about the physiological role of [3Fe-3S] clusters.

2 ANALYSIS OF LOW-TEMPERATURE MCD DATA

The information content of low-temperature MCD data can be divided into two areas, both of which are useful in characterizing Fe-S clusters. First, there is the form of the spectrum (i.e. the number, sign, and intensity of the peaks), which is indicative of cluster-type. Second, there is the form of the temperature T and magnetic field B dependence of the spectrum. By plotting signal intensity as a function of B/T, we can obtain magnetization curves for individual paramagnetic centers. We have discussed the theoretical basis for analyzing such curves for hemoproteins (1), and more recently applied the technique to the investigation of the magnetic properties of the EPR-undetectable heme a_3 in oxidized cytochrome c oxidase (7). The problems in analyzing such data for Fe-S proteins are more acute due to the relative weakness of their low-temperature MCD signals and the absence of single crystal polarization data for bands in the UV/visible spectrum. The results presented in this chapter are the first reported MCD magnetization curves for Fe-S proteins. The origin and analysis of such curves are discussed below.

An MCD experiment measures the difference in the absorption of left and right circularly polarized light, $\Delta\epsilon(=\epsilon_L - \epsilon_R)$, as a function of wavelength for a sample in a uniform longitudinal magnetic field. The strong temperature dependence of MCD signals from paramagnetic species is a consequence of the Boltzmann population distribution among the Zeeman-split sublevels of the electronic ground state. A differential absorption of left-minus-right circularly polarized light arises because electronic transitions from the Zeeman sublevels are circularly polarized (8, 9). When the temperature is high or $g\beta B/kT \ll 1$, where g is the ground state g factor, β is the Bohr magneton, B is the magnetic flux density, k is Boltzmann's constant, and T is the absolute temperature, the MCD signal intensity is directly proportional to B/T. In this case the MCD signal is obeying Curie's law, and its magnitude is expressed by the so-called C term (8). However, when $g\beta B/kT \geq 1$, $\Delta\epsilon$ becomes nonlinear as a function of B/T. Eventually, at ultralow temperature and high field, there is population of only the lowest Zeeman sublevel. In this situation $\Delta\epsilon$ is then independent of B/T; the MCD signal is said to be saturated and the paramagnetic center is fully magnetized. Very few MCD experiments involving paramagnetic species have been conducted at sufficiently low temperature and high magnetic field to achieve signal saturation, although the information content of such magnetization curves is high.

The theoretical interpretation of MCD magnetization curves for an isolated Kramers doublet, $S = 1/2$ ground state, is now developed. This is largely the result of an important paper by Schatz and colleagues (10). For a detailed discussion of both the application and experimental justification of such theoretical analysis for biological systems, the reader is referred elsewhere (1, 3).

By studying the temperature dependence of an MCD spectrum in the Curie law region, it is in principle possible to obtain the ground state g value by measuring the ratio of the slope of the Curie law plot to the dipole strength determined from the absorption spectrum. This is the procedure that has been widely adopted by inorganic spectroscopists studying well-defined transition metal complexes. However, this procedure is not useful for the study of metalloproteins. For example, the absorption band of the component being investigated may be obscured by those of other metal centers or chromophores. In addition, it is necessary to have a secure assignment of the excited state and to know certain parameters of the excited state, especially the magnitude of the spin-orbit coupling. This is rarely possibly for the rather complex metal chromophores present in proteins, and will certainly not be possible for Fe-S proteins other than rubredoxin. However, estimates of ground state g factors can be made rather reliably and accurately from the

MCD spectrum if it is measured down to the saturation limit. In general the form of the MCD magnetization curve for a rhombic chromophore depends on the three ground state g factors and the polarization of the electronic transition being observed. Thus in effect the requirement to know excited state spin-orbit parameters is replaced by another parameter, the anisotropy of the polarization (denoted m_z/m_+ for an axial chromophore, where m_z^2 and m_+^2 are the transition dipole moments in the molecular z and xy directions, respectively), which can be fitted to an experimental spectrum. Moreover, if the ground state g factor anisotropy is small, the fit is almost independent of the polarization ratio. Similarly, if the g factor anisotropy is very large, approximating to $g_z \neq 0$, $g_x = g_y = 0$, the MCD magnetization curve depends only on the x, y polarized electric dipole transition intensity and no parameter is required in the fitting. If the polarization of the electronic transition is known from independent evidence, as, for example, in the case of hemes, where the visible and Soret region bands are x, y polarized but the g factors are rhombic, a good estimate of the g factors and their anisotropy can be obtained (1).

Apart from fitting theoretical curves to the experimental magnetization curves, we find that there is a simple procedure that gives an estimate of g factors from magnetization curves in certain limiting cases (1). The ratio of the asymptotic limit of the curve to the initial slope gives a quantity that we term the *intercept* I on the x axis. If the ground state is isotropic, $I = 1/g$ and there is no dependence of the curve on the polarization of the electronic transition. If the ground state is completely axial, such that $g_\parallel = 4S$ and $g_\perp = 0$, then $I = 3/2g_\parallel$, and the magnetization curve has contributions only from x, y polarized transitions. Thus the abscissa represents a scale of $1/g_{\text{isotropic}}$ or $3/2g_\parallel$ for a completely axial paramagnet.

Although rigorous theoretical expressions have yet to be worked out, it is possible to recognize spin states with $S > 1/2$ from MCD magnetization curves in several ways. First, a g factor measured by the intercept value I is significantly different from 2.0. In certain instances the S value can be estimated from the intercept value, provided that both ground state rhombic distortions and field-induced mixing of zero-field components are assumed to be small. Second, plots of $\Delta\epsilon$ as a function of $1/T$ only become linear when kT is much greater than the spread of the zero-field components. This enables estimates to be made of the zero-field splitting parameters. Third, plots of $\Delta\epsilon$ as a function of B/T can lead to a "nested" set of curves rather than to the smooth curve obtained for an isolated doublet ground state (1). By "nested" we mean that the data points for each temperature studied fall on separate curves. This situation arises if zero-field components are populated over the temperature range of the experiment.

3 EXPERIMENTAL CONSIDERATIONS

The technique of low-temperature MCD spectroscopy is new in the study of Fe-S proteins, so a few remarks are required on experimental aspects. The approximate condition that must be fulfilled to completely saturate the MCD signal from a doublet ground state is $g\beta B \geq 4kT$. If $g = 2$, it is clear that magnetic fields of up to 5 Tesla and temperatures as low as 1.5 K are required. Therefore superconducting magnets are used with the facility to have the sample at the center of the magnet, immersed in a bath of pumped liquid helium (see Fig. 1). The use of high magnetic fields can affect the performance of certain dichrograph components, and great care needs to be taken to shield such components, particularly for the high sensitivities that are needed to obtain MCD spectra of Fe-S proteins. The quality of the magnetization data depends critically on the accuracy of both magnetic field and temperature measurement. Since shielding can affect the magnetic field experienced by the sample, it is essential to calibrate accurately the field at the sample by using a Hall probe. In this work sample temperatures were measured by means of carbon glass resistors and cryogenic linear temperature sensors placed both above and below the sample. Such an arrangement enables measurement to an accuracy of $<0.5\%$ over the temperature range 1.5–300 K.

The frozen samples need to be in the form of a glass to allow transmittance of light, and for this purpose protein solutions are made up to be 50% in ethylene glycol or are saturated with sucrose. Before commencing a study of a particular protein it is necessary to check that no deleterious effects result from such dilutions by monitoring natural CD, absorption, and EPR spectra with and without the glassing agent. The optical quality of the glass produced on rapid freezing by immersion into liquid helium is assessed in two ways. First, the MCD spectra are measured at positive, negative, and zero field. The MCD spectra at positive and negative field must form mirror images of one another reflected in the zero-field baseline. Second, any depolarization of the light beam by the glass is assessed by measuring the natural CD of a sample of D-tris (ethylenediamine) cobaltIII chloride placed after the magnet, with and without the sample in position, in the absence of an applied field. Techniques for both aerobic and anaerobic sample manipulation, suitable for low-temperature MCD experiments, have been described elsewhere (3, 5). Particularly when samples are handled anaerobically, it is necessary to be sure of the redox state of the protein under investigation. Therefore it is advisable routinely to perform parallel EPR experiments as well as to check both the low-temperature absorption and natural CD spectra of the frozen sample.

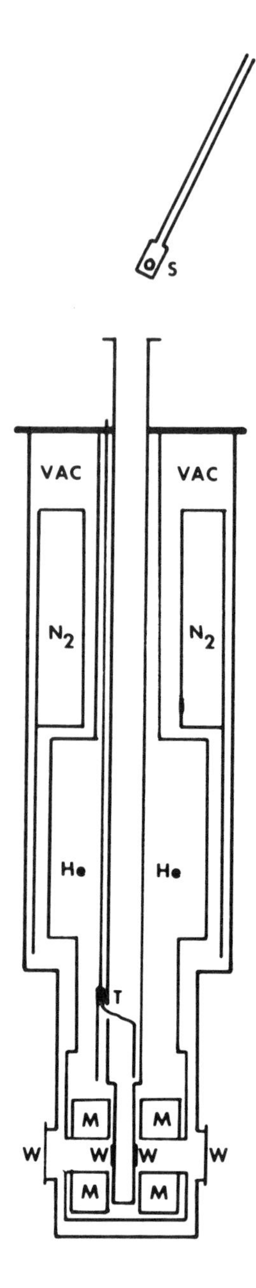

Figure 1. Schematic cross-section of Oxford Instruments split-coil superconducting magnet, type SM4. (S) Sample holder, lowered down into the sample space; (W) optical windows; (M) coils of superconducting magnet; (T) tube for transferring liquid H from main reservoir to sample chamber; the latter can be pumped down to $\approx$ 1.5 K, or the He can be removed and it can be warmed to 300 K.

4 APPLICATION OF LOW-TEMPERATURE MCD TO Fe-S PROTEINS

4.1 Rubredoxins

The crystal structure of rubredoxin from *Clostridium pasteurianum* shows it to be a single Fe atom liganded by four cysteine residues in an approximately tetrahedral geometry (11). Originally it was suggested, on the basis of X-ray crystallography, that the coordination geometry about the Fe atom was rather severely distorted, with an unusually short Fe–S distance of 2.05 Å compared with the other three distances of 2.24, 2.32, and 2.34 Å. Subsequently, EXAFS measurements (12, 13) have cast doubt on the accuracy of these distances, and it is now clear that all the Fe–S bond lengths fall within the range of 2.30 ± 0.04 Å. Thus the metal center is a single Fe atom that can adopt either the high-spin ferrous or ferric states. The relative simplicity of this structure has meant that its electronic properties are rather well understood. A detailed analysis of the EPR spectrum of the rubredoxin from *Pseudomonas oleovorans* (14) concluded that, in the oxidized state, the ferric ion is in the high-spin state with an environment that imposes a virtually completely rhombic crystal field on the ground state. Thus the state, with total spin $S = 5/2$, consists of three pairs of Kramers doublets, each separated by 6.2 cm^{-1}, giving a total spread of 12.4 cm^{-1}. The lowest Kramers doublet has effective g values of 9.42, 1.25, and 0.9. The EPR spectrum of the Rd from *Clostridium pasteurianum* is very similar to that from *Pseudomonas oleovorans* (15). In the reduced state no EPR signals are detectable, but the ferrous ion has been shown by Mössbauer spectroscopy (16, 17) to be in the high-spin state with $S = 2$.

The low-temperature MCD of the oxidized form of Rd from *Clostridium pasteurianum,* in the wavelength range 350–650 nm at temperatures of 300, 17.2, 10.5, and 6.1 K and at a magnetic field of 0.83 Tesla, was studied by Rivoal and co-workers (2). It was possible to make a reliable assignment of the visible spectrum, showing that the bands arise from charge-transfer transitions from RS^- to Fe^{3+}. The region from 425 to 625 nm is dominated by the one electron transition from an S p_π orbital to a metal d orbital derived from the e subset in tetrahedral symmetry. This assignment is unambiguous, because three MCD bands, two of positive sign and one of negative sign, are expected to arise from this one electron transition. The three bands are a consequence of the combined action of a low-symmetry field and spin-orbit coupling separating the excited state into three components. The next three MCD bands in order of higher energy probably result from the next higher energy one electron transition. Thus this assignment shows the essential simplicity of the electronic spectrum of a single ferric ion in an environment of one type of S atom, namely an aliphatic thiol.

Furthermore, it was possible to estimate, from the moment of the MCD spectrum, the relative contributions of the spin-orbit perturbation and the low-symmetry crystal field components to the separations of the three peaks of the lowest-energy one electron transition. It was concluded that the structure arose mainly as a result of a predominantly axial distortion to the excited state. This confirmed a view proposed by Eaton and Lovenberg (18) on the basis of the polarized single-crystal absorption spectrum of Rd from *Clostridium pasteurianum*.

The temperature dependence of the MCD spectrum, although limited in range because of the nature of the MCD cryostat then available, was analyzed. It was pointed out that it was consistent with the analysis of the EPR spectrum. Extraction of accurate zero-field splitting parameters was not possible owing to the limited set of data points. However, it was clear from the analysis of the data that saturation experiments on samples down to pumped-helium temperatures would provide invaluable data, especially a link between EPR and optical spectra (2). Thus these experiments were the inspiration for all the later series of experiments carried out at Norwich and reported in this section.

As examples of Rd MCD spectra we choose to show the low-temperature spectra of the oxidized and reduced forms of two types of Rd extracted from *Desulfovibrio gigas*. These data have not been published before, and were obtained on samples prepared by I. Moura, J. J. G. Moura, A. V. Xavier, and J. LeGall in Marseilles and Lisbon (19). One of these proteins from *D. gigas*, which contains a single Fe atom per 4 cysteine residues bound to the metal center, is virtually identical in its EPR and Mössbauer characteristics to that of the Rd from *Clostridium pasteurianum* (20). The other protein, called *desulforedoxin*, since it lacks acid-labile S, is isolated as a dimer of two identical subunits, with 1 Fe atom and 4 cysteines per monomer (19). Of particular note is the spacing of the 4 cysteines; residues 9 and 12 occur in the Rd sequence pattern, namely cys-X-X-cys, whereas the other 2 cysteines are adjacent, in positions 28 and 29 of the sequence. The EPR and Mössbauer spectra of oxidized desulforedoxin have been analyzed in terms of a high-spin ferric ion with zero-field splitting of the $S = 5/2$ ground state that is close to axial, leaving the $M_s = \pm 1/2$ level with effective g factors of 4.09, 7.70, and 1.80 lowest in energy (20). This contrasts with the situation found in Rd, in which the ground state is almost completely rhombically distorted. The dithionite-reduced state of desulforedoxin contains the high-spin ferrous ion with $S = 2$. Mössbauer analysis of both Rd and desulforedoxin (21, 20) shows that the ground state in the former case is a pure d_{z^2} orbital and in the latter a $d_{x^2-y^2}$ orbital. Also, the parameters describing the zero-field splitting have opposite signs for the two proteins. Both of these facts suggest an essentially similar coordination number and type for the Fe atoms in the two proteins but differences in the low-symmetry distortions.

The low-temperature MCD spectra of the oxidized forms of the *D. gigas* Rd and desulforedoxin bear out these points in broad terms (see Fig. 2). There is an overall similarity between the two sets of spectra, leaving little doubt that the ferric ion is coordinated by four cysteine ligands in both cases.

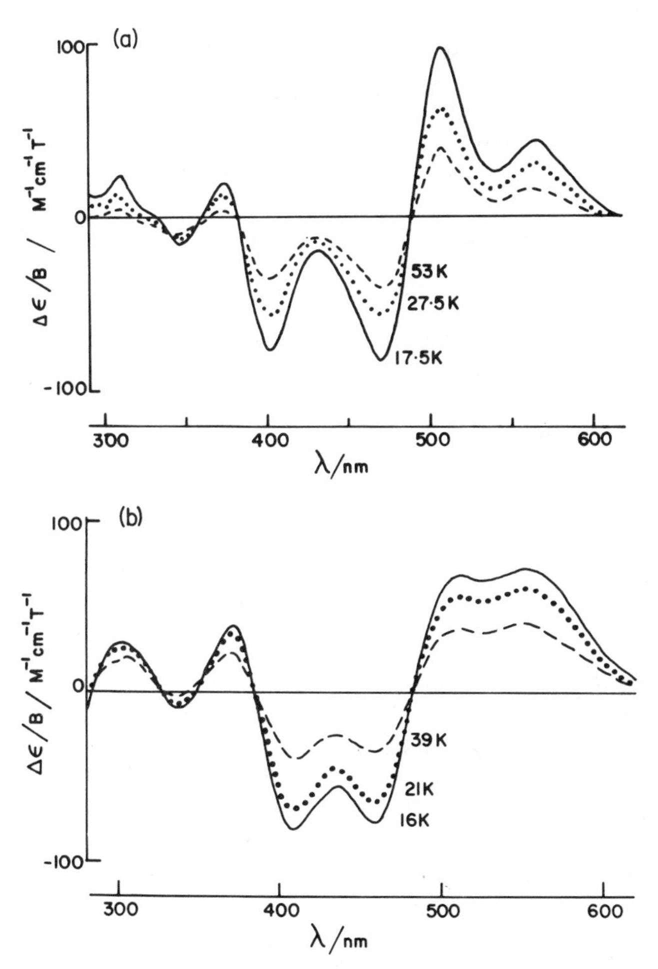

Figure 2. MCD spectra of oxidized *D. gigas* Rd *(a)* and oxidized *D. gigas* desulforedoxin *(b)* measured in 0.01 *M* Tris buffer at pH 7.6 saturated with sucrose. Magnetic field 5 Tesla; $\Delta\epsilon$ expressed per unit Tesla. (I. Moura, J. J. G. Moura, A. V. Xavier, J. LeGall, J. P. Springall, and A. J. Thomson, unpublished data).

The spectra of Rd from *D. gigas* and from *Clostridium pasteurianum* (2) are identical. However, we have shown that the separations of the first three major bands in the MCD spectrum of oxidized Rd arise from a one electron charge-transfer transition from cys-S^- to Fe(III), and further that the positive peak at 565 nm is separated from the crossing point at 487 nm by a predominantly axial distortion of the excited state, rather than by spin-orbit coupling (2). In the MCD spectrum of oxidized desulforedoxin the equivalent peaks are less well separated, suggesting that the axial distortion is less in this case. The Mössbauer spectra are unable to distinguish the environments of the 2 Fe atoms in desulforedoxin. Therefore we take it that both will give identical optical spectra.

The low-temperature MCD spectra of the dithionite-reduced forms of the two proteins are given in Figure 3. No assignment of these has been attempted, although the intense nature of the bands suggests they correspond to Fe^{2+} to ligand, cys-S^-, charge-transfer transitions. Once again, distinct differences between the forms of the two spectra are apparent. Magnetization curves of this oxidation state should be especially informative concerning the zero-field splitting parameters of the ground state, providing a useful check on the Mössbauer analysis. We are in the process of gathering such data.

4.2 [2Fe-2S] Proteins

The structure of [2Fe-2S] sites in proteins has long been assumed by the similarity of their spectroscopic and magnetic properties to structurally well-characterized $Fe_2S_2(SR)_4^{2-}$ model complexes (ref. 22 and references therein). The recent X-ray crystallographic study of *Spirulina platensis* ferredoxin (23) has confirmed the structure as consisting of 2 Fe atoms bridged by 2 sulfides, with cysteinyl S completing the approximately tetrahedral coordination about the Fe atoms.

Variable-temperature MCD studies on oxidized ferredoxins from spinach and *Spirulina maxima* show an MCD signal that is temperature independent and weak compared to the natural CD (24). A diamagnetic $S = 0$ ground state is clearly indicated in good agreement with magnetic susceptibility data (25), and this also accounts for the absence of an EPR signal. Mössbauer spectroscopy demonstrates that oxidized [2Fe-2S] proteins contain two high-spin $S = 5/2$ Fe(III) ions (ref. 26 and references therein), which has led to the proposal that the Fe atoms are antiferromagnetically coupled (25). The MCD spectra of oxidized spinach and *Spirulina maxima* Fd at room temperature and at 18°K are identical within experimental error (24). Therefore it appears that the small population of the next highest zero-field component of

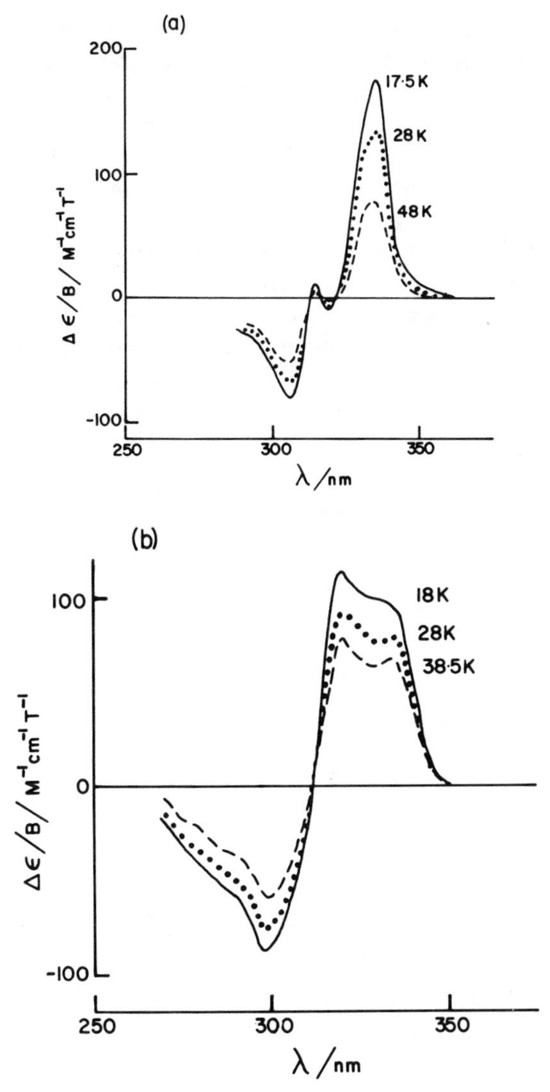

Figure 3. MCD spectra of dithionite-reduced *D. gigas* Rd *(a)* and dithionite-reduced *D. gigas* desulforedoxin *(b)*. Conditions of measurement as in Fig. 2, except for the anaerobic addition of sodium dithionite. (I. Moura, J. J. G. Moura, A. V. Xavier, J. LeGall, J. P. Springall, and A. J. Thomson, unpublished data).

the ground state, namely that with $S = 1$, expected on the basis of the exchange-coupling magnitude estimated by susceptibility, does not contribute sufficiently to the MCD spectrum to be detectable. Hence only lower limits on the magnitude of the antiferromagnetic coupling can be obtained from MCD studies.

Figure 4 shows the low-temperature MCD spectrum of reduced *Spirulina maxima* ferredoxin. The spectrum is rich in detail and exhibits marked temperature dependence. MCD magnetization curves constructed at 330 and 670 nm are presented in Figure 5. In both instances the experimental points fall on or close to the theoretical curve computed for $g_{\parallel} = 2.05$, $g_{\perp} = 1.93$, and $m_z/m_+ = -1.0$. The intercept values I at both wavelengths investigated are the same and yield an isotropic g factor of 1.96, which represents the average of the EPR-determined g factors. These results are in good agreement with EPR, magnetic susceptibility, and Mössbauer data for reduced [2Fe-2S] proteins, which show an $S = ½$ ground state from an antiferromagnetically coupled pair of high-spin Fe(II) and Fe(III), with no electronic states below 200 cm^{-1}. Of the three reduced [2Fe-2S] proteins that have

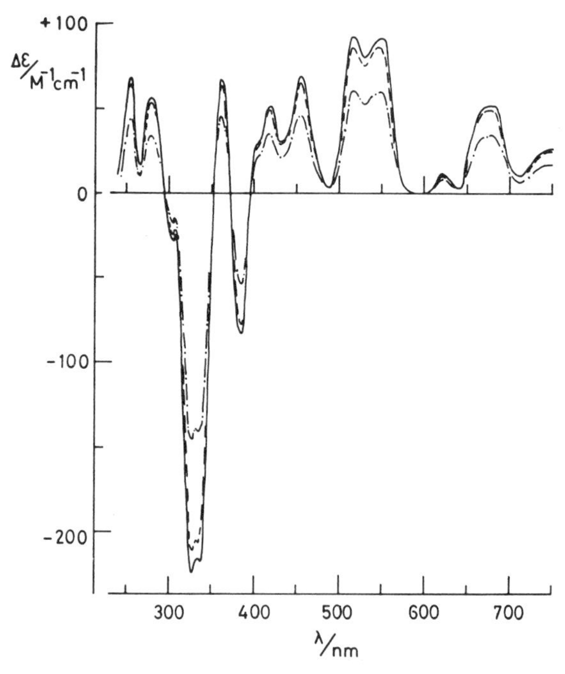

Figure 4. MCD of reduced *Spirulina maxima* Fd in the presence of excess sodium dithionite. Conditions: 1.39 mm pathlength, 50% ethylene glycol, aqueous Tris/HCl buffer, magnetic field 5 Tesla. —— 1.54 K; --- 2.08 K; —·— 4.22 K. Protein concentration 350 μM.

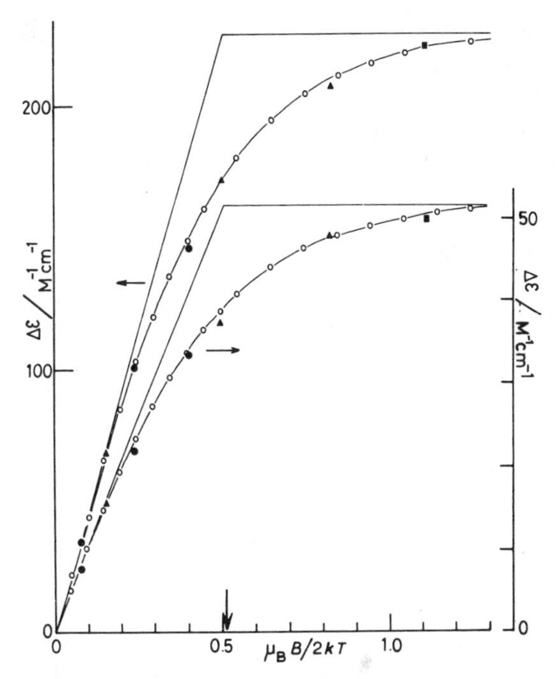

Figure 5. MCD magnetization curves of reduced *Spirulina maxima* Fd. Upper curve λ = 330 nm; lower curve λ = 670 nm. ■) 1.54 K, B = 5 Tesla; ▲) 2.08 K; B = 5, 3, and 1 Tesla; ●) 4.22 K, B = 5, 3, and 1 Tesla. Open circles indicate the theoretical curve computed with parameters $g_{\parallel} = 2.05$, $g_{\perp} = 1.93$, $m_z/m_+ = -1.0$. The arrow indicates intercept value $I = 0.51$.

been subjected to low-temperature MCD investigation, the spectra of spinach and *Spirulina maxima* ferredoxins are remarkably similar. However, there are marked differences in the spectra of adrenodoxin (24). This suggests the existence of at least two distinct classes of [2Fe-2S] proteins, typified by spinach ferredoxin and adrenodoxin, with differences presumably related to the structural parameters of the active centers. This conclusion is substantiated by several pieces of spectroscopic and magnetic evidence:

1. The magnetic coupling is at least 50% stronger in adrenodoxin ($J_{ox} \geq$ 350 cm^{-1}, $J_{red} \geq$ 250 cm^{-1}) than in spinach ferredoxin (25, 27).
2. The EPR spectra for the reduced proteins is significantly different, with rhombic and axial spectra for spinach ferredoxin and adrenodoxin respectively. Since EPR spectra of reduced adrenodoxin can be observed at higher temperatures than those of reduced spinach ferredoxin, the electron spin relaxation is slower for adrenodoxin (ref. 28 and references therein).
3. The difference in electron spin relaxation time shows up in the Möss-

bauer spectra of reduced proteins, since magnetic hyperfine structure for spinach ferredoxin is only observed at low temperatures, whereas it is apparent up to 244 K for adrenodoxin (26).

4. As indicated in Chapter 11, there are significant differences in the resonance Raman of the oxidized proteins, indicating structural differences in the chromophores.

4.3 [3Fe-3S] Proteins

The work of Münck and his co-workers in establishing the presence of the novel [3Fe-3S] cluster types in Fe-S proteins has been one of the major achievements of Mössbauer spectroscopy as applied to biological systems. The reader is referred to Chapter 4 for a detailed discussion. The center was initially identified in a ferredoxin from *Azotobacter vinelandii* as the low-potential cluster, with the high-potential center being a conventional HIPIP-type [4Fe-4S] cluster (29). X-Ray crystallographic investigation of the same ferredoxin supports this interpretation, suggesting that one of the Fe atoms in the [3Fe-3S] center is coordinated by both a cysteine residue and an O atom supplied by either an exogenous ligand, such as hydroxyl, or a glutamate residue. If substantiated, this is the first reported case of an Fe-S cluster that is not exclusively bound by cysteine residues. The structural aspects of the [3Fe-3S] center in *Azotobacter vinelandii* ferredoxin are discussed at greater length in Chapter 3.

The list of Fe-S proteins now believed to contain a [3Fe-3S] center is growing rapidly. Spectroscopic evidence suggests that such clusters occur in Fd I and II from *Desulfovibrio gigas* (30), aconitase from beef heart (29), glutamate synthase from *Azotobacter vinelandii* (29), a ferredoxin from *Thermus thermophilus* (see Chapter 4) and *Methanosarcina barkeri* Fd (see Chapter 5). The EPR characteristics of [3Fe-3S] centers are an almost isotropic $g = 2.01$ signal for the oxidized cluster with no observable EPR signal on $1e^-$ reduction. However, characterization of [3Fe-3S] clusters by EPR has up to now been impeded because an analogous $g = 2.01$ signal is observed for superoxidized [4Fe-4S] centers in ferredoxins (31). This has led to numerous literature assignments of isotropic $g = 2.01$ EPR signals to $[4Fe\text{-}4S]^{3+}$ HIPIP-type centers (for examples see ref. 32 and 33), despite the fact that conventional HIPIP-type $[4Fe\text{-}4S]^{3+}$ clusters exhibit essentially axial EPR signals with $g_{\parallel} = 2.12$ and $g_{\perp} = 2.04$ (28). The important contribution of low-temperature MCD studies has clarified the situation by establishing that [4Fe-4S] Fd-type centers can undergo breakdown to [3Fe-3S] clusters on ferricyanide oxidation. Therefore it would appear that an almost isotropic $g = 2.01$ EPR signal is characteristic of an oxidized [3Fe-3S] cluster. Thus the presence of [3Fe-3S] clusters must be suspected in bacterial ferredoxins from

Rhodospirillum rubrum (34), *Corynebacterium autotrophicum* (35), *Mycobacterium flavum* (33), and *Spirillum lipoferum* (33), in both mammalian and bacterial succinate dehydrogenases (36–38) and a nitrate reductase from *Micrococcus denitrificans* (39).

Up to the present time low-temperature MCD studies have largely been confined to the tetrameric Fd II from *D. gigas,* which consists solely of [3Fe-3S] clusters (30). Figure 6 shows both the 4.22 K absorption and low-temperature MCD spectra for oxidized Fd II from *D. gigas*. In common with most other Fe-S proteins, the low-temperature absorption has only minor additional features compared to the room-temperature spectrum (40). In contrast the low-temperature MCD is very structured, with several positive and negative peaks for the wavelength range investigated. On the basis of absorption intensity, it is rational to assign the weak bands in the 550–750 nm region to intra-*d* shell transitions and the more intense bands in the 250–550 nm region to charge-transfer transitions from S to the ferric core. The form of the spectrum is quite distinct from that of any other Fe-S cluster type. Most importantly, the spectrum is different from that of oxidized HIPIP from *Chromatium* (3). The temperature dependence exhibited by the MCD signal confirms the paramagnetic nature of the ground state, in agreement with Mössbauer and EPR studies (30, 41). The MCD magnetization

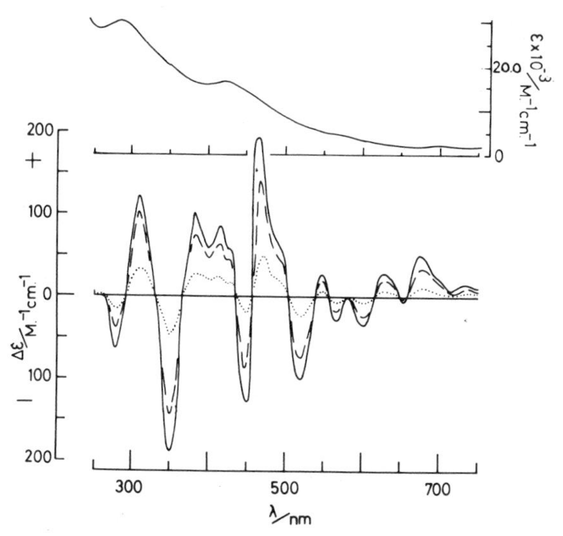

Figure 6. Upper panel: absorption spectrum at 4.2 K of oxidized *D. gigas* Fd II. Lower panel: MCD spectra of oxidized *D. gigas* Fd II. —— 1.53 K; --- 4.22 K; ······ 20.0 K. Magnetic field = 5.1 Tesla, pathlength = 1.053 mm. The sample was dissolved in pH 7.6 buffer diluted to 50% by volume with ethylene glycol.

curve, shown in Figure 8*a* for the temperature range 1.52–22.5 K, indicates that the signal intensity is a smooth function of B/T. This behavior is characteristic of a doublet ground state with no low-lying zero-field components thermally accessible over the temperature range of the experiment. The intercept value I is close to 0.5, giving an isotropic g factor of 2.00. Clearly the MCD signal originates from transitions out of the same $S = \frac{1}{2}$ ground state that gives rise to the isotropic $g = 2.01$ EPR signal. Preliminary investigation of the detailed temperature dependence of the MCD signal over the range 10–165 K suggests the first presence of an excited state at $\sim 80\ cm^{-1}$ above the ground state (4).

On dithionite reduction the form of the low-temperature MCD spectrum is completely changed, as shown in Figure 7. The number of peaks decreases, pointing to a lower density of excited electronic states in the reduced form. The pattern of bands is similar in some respects to that of reduced [4Fe-4S] Fd-type centers. However, these cluster types are easily differentiated by the form of their MCD magnetization curves. The magnetization curve, shown in Figure 8*b*, is of particular interest, since no EPR signals are observed from this redox state. Mössbauer data does reveal that the ground state is paramagnetic with electron spin $S \geq 1$ and integer (30). Comparison of Figure 8*a* and *b* shows that the magnetization curves are quite different for the oxidized

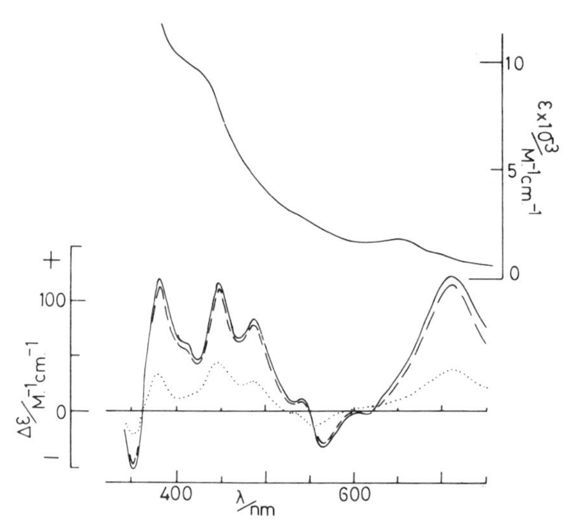

Figure 7. Upper panel: absorption spectrum at 4.2 K of reduced *D. gigas* Fd II. Lower panel: MCD spectra of reduced *D. gigas* Fd II. —— 1.50 K; --- 4.22 K; ······ 20.0 K. Magnetic field = 5.1 Tesla, pathlength = 1.103 mm. The sample was dissolved in pH 7.6 buffer, made anaerobic, and reduced for 35 min with solid sodium dithionite. The sample was diluted to 50% by volume with ethylene glycol.

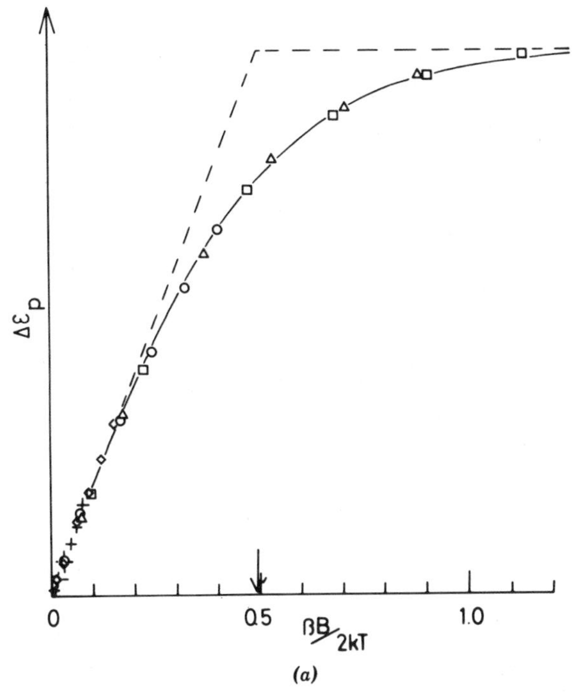

(a)

Figure 8. *(a)* MCD magnetization curve of oxidized *D. gigas* Fd II. $\Delta\epsilon_p$ is the $\Delta\epsilon$ value measured from the peak at 448 nm to the trough at 466 nm, expressed in arbitrary units. □) 1.5 K; △) 1.945 K; ○) 4.22 K; ◇) 11.1 K; +) 22.5 K. Magnetic fields were between 0 and 5.1 Tesla for each of these temperatures. The solid line is the curve computed for an isotropic system with $g = 2.01$ according to reference 10. The arrow indicates intercept value $I = 0.495$. *(b)* MCD magnetization curves of reduced *D. gigas* Fd II. $\Delta\epsilon_p$ is the $\Delta\epsilon$ value measured from the peak at 380 nm to the trough at 351 nm, expressed in arbitrary units. □) 1.53 K; △) 1.995 K; ○) 4.22 K; ◇) 9.5 K; +) 21.9 K. Magnetic fields were between 0 and 5.1 Tesla for each of these temperatures. The solid line is the curve computed for $g_{\parallel} = 8.00$, $g_{\perp} = 0.20$, $m_z/m_+ = 0.20$ according to reference 10. The arrow indicates intercept value for 1.53 K and 1.995 K data, $I = 0.185$. The inset shows the zero-field splitting of a state with spin $S = 2$ in the presence of an axial distortion and zero rhombic distortion. The energy levels are given by a spin Hamiltonian $\hat{H}$ of the following form: $\hat{H} = D\hat{S}_z^2 + g_0 \cdot \beta \cdot (B_z \cdot \hat{S}_z + B_x \cdot \hat{S}_x + B_y \cdot \hat{S}_y)$. Here D is taken to be negative in sign to place the $M_s = \pm 2$ levels lowest, and g_0 is given the free spin value of 2.00. For applied fields such that $g_0 \beta B \ll 3D$ and in the absence of a rhombic distortion, it is possible to describe the magnetic properties of the two doublets separately using a spin Hamiltonian, with an effective spin $\tilde{S} = 1/2$: $\hat{H}_e = \beta \tilde{S} \cdot \tilde{g} \cdot B$. The resulting effective g values, $\tilde{g}$, are given in the diagram.

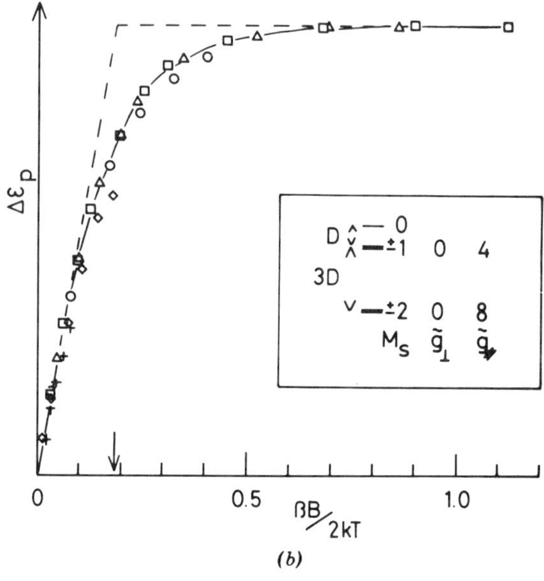

(b)

Figure 8. *(Continued)*

and reduced forms. For the reduced case the data points for temperatures in the 1.530–21.9 K range form a nested set of curves. Such nesting implies that low-lying excited states become thermally populated as the temperature is raised, and that these states give weaker MCD signals than the lowest state. However, the magnetization curves measured at 1.530 and 1.995 K are completely superposable, which implies that below 2 K only the lowest level is populated. Furthermore, because these curves approach an asymptotic limit, this ground state must be an electronic doublet. The steepness of the magnetization is indicated by the intercept value $I = 0.185$. Assuming a completely axial ground state, such an intercept value yields anisotropic g values, $g_{\|} = 8.1$ and $g_{\perp} = 0$. A least-squares fitting procedure gives $g_{\|} = 8.0$, $g_{\perp} = 0.20$, and $m_z/m_+ = 0.20$, and Figure 8*b* shows the excellence of fit. Hence we conclude that the electronic ground state of reduced Fd II has $S = 2$ with a predominantly axial zero-field distortion, leaving $M_s = \pm 2$ as the lowest-lying level. Such an energy level scheme is shown as an inset in Figure 8*b*. Preliminary plots of the MCD intensity over the range 10–121 K as a function of $1/T$ indicate an upper limit of ≈ 15 cm^{-1} for the spread of the zero-field components of the ground state and a lower limit of 85 cm^{-1} for the closest electronic excited state (4). It should be recognized that this analysis assumes that any rhombic splitting of the ground state is small, and that is also neglects any second-order Zeeman effects involving field-induced mixing of zero-field components. The fact that the magnetization curves at 1.530 and 1.995 K are superposable, with both undergoing complete saturation, strongly suggests that the latter assumption is valid.

We have undertaken low-temperature MCD studies on the trimeric Fd I from *D. gigas,* although magnetization curves are still to be obtained. The form of spectrum for the oxidized protein is essentially analogous to that for oxidized Fd II except for differences in the relative intensity of some bands. In the reduced state the spectrum appears to be a superposition of reduced Fd II and of a reduced Fd-type [4Fe-4S] center (42). The latter cluster would be diamagnetic in the oxidized state, hence would have a negligible low-temperature MCD compared to a paramagnetic center. The implication that Fd I from *D. gigas* contains both [3Fe-3S] and [4Fe-4S] Fd-type clusters is in accord with Mössbauer (30), EPR (41), and resonance Raman studies (see Chapter 8). Since both the trimeric Fd I and tetrameric Fd II contain only one type of peptide chain, this result shows that the same polypeptide chain can accommodate either [3Fe-3S] or [4Fe-4S] clusters.

The limitations of room-temperature MCD and CD spectroscopy in Fe-S cluster identification are illustrated by a recent study on aconitase (43). However, it was concluded that aconitase does not contain a [4Fe-4S] center. Low-temperature MCD studies on aconitase and other [3Fe-3S] proteins will be of great interest.

4.4 [4Fe-4S] Proteins

The presence of cubane-type [4Fe-4S] clusters has been established by X-ray crystallography for HIPIPs in *Chromatium* (44–46) and *Azotobacter vinelandii* Fd (47) as well as for the ferredoxin from *Peptococcus aerogenes* (48, 49). The latter contains two low-potential clusters separated by ~12 Å. EPR (26, 28) and Mössbauer (26) spectroscopy have been extensively used both to characterize their electronic properties and to provide spectroscopic criteria to enable their recognition in more complex proteins. The electronic and magnetic properties of these clusters have been rationalized by the three oxidation state hypothesis of Carter and co-workers (50). This suggests that [4Fe-4S] centers can adopt one of three possible redox states, each differing by one electron. These are denoted by C^-, C^{2-}, and C^{3-}, where C represents $Fe_4S_4(S\text{-cys})_4$ with S-cys as the anion as cysteine.* HIPIP-type

*The IUPAC nomenclature on clusters recommends that the cluster type and charge be indicated by the core only with the charge on the core, after subtraction of the cysteine ligands. Thus the C^-, C^{2-}, and C^{3-} states of the 4-Fe clusters become $[4Fe\text{-}4S]^{3+}$ $[4Fe\text{-}4S]^{2+}$, and $[4Fe\text{-}4S]^{1+}$, respectively. See *Eur. J. Biochem.* **93,** 427–430 (1979).

centers cycle between C^- and C^{2-} with a redox potential $E° \approx +350$ mV, and the low-potential Fd-type centers cycle between C^{2-} and C^{3-} with redox potentials between −350 and −450 mV. These two types of $[4Fe\text{-}4S]^{n+}$ center are discussed separately below.

4.4.1 The $[4Fe\text{-}4S]^{3+,2+}$ Centers in HIPIP. Oxidized HIPIP, corresponding to the C^- state, is paramagnetic with spin $S = 1/2$ and gives rise to an essentially axial EPR signal with $g_{\parallel} = 2.12$ and $g_{\perp} = 2.04$. Additional structure in the spectrum suggests a very weak, rhombic EPR component with $g_1 = 2.086$, $g_2 = 2.055$, and $g_3 = 2.040$ (52). The paramagnetic nature of this state is further demonstrated by the temperature dependence of the MCD spectrum of oxidized *Chromatium* HIPIP, shown in Figure 9*a*. The spectrum is extremely complex compared to the room temperature MCD (54), showing many peaks in the 300–800 nm region. MCD magnetization curves (see Figure 2 of ref. 3) for temperatures of 1.50–47.0 K are smooth functions of B/T, indicating that no excited electronic states are significantly populated over this range. The experimental points lie very close to the theoretical curve computed for $g_{\parallel} = 2.12$, $g_{\perp} = 2.04$, and $m_z/m_+ = -1.0$. The intercept value I (0.49) gives a good estimate of the average ground state g value.

One electron reduction of oxidized *Chromatium* HIPIP leads to reduced HIPIP, denoted C^{2-}; the low-temperature MCD spectrum is shown in Figure 9*b*. The virtual coincidence of the spectra at 4.2 and 20 K proves the diamagnetism of this state, $S = 0$, and shows the complete absence of any paramagnetic impurity absorbing in this spectral region. The low-temperature spectra show the same features, though better resolved, as the room-temperature spectrum, with good agreement in terms of $\Delta\epsilon$ values. This result explains the lack of EPR signal for this redox state and is in good accord with conclusions based on Mössbauer (26) and magnetic susceptibility studies (52). An investigation of the temperature dependence of MCD signals from both oxidized and reduced HIPIP to higher values than those described here may lead to estimates for the energies of low-lying excited states, if appreciable thermal population of paramagnetic excited states can be obtained below about 200 cm^{-1}. Studies of the temperature dependence of intensities of the EPR signals indicate the presence of an excited state at 160 ± 10 cm^{-1} above the ground state for oxidized HIPIP from *Chromatium* (55). Magnetic susceptibility measurements reveal a state ~200 cm^{-1} above the ground state for the oxidized protein. In contrast, no electronic states are observed for the reduced protein for at least 400 cm^{-1} above the ground state.

Cammack has reported EPR evidence which shows that a further one electron reduction of HIPIP is possible, by the addition of 80% dimethylsulfoxide

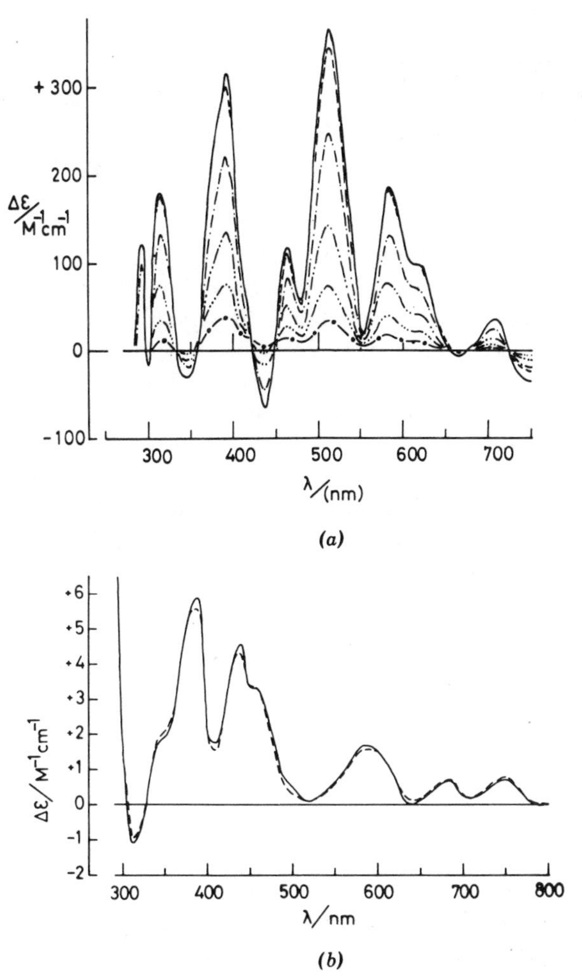

Figure 9. *(a)* MCD spectrum of oxidized *Chromatium* HIPIP. Pathlength = 1.13 mm, 50% *v/v* ethylene glycol, aqueous Tris/HCl, pH 7.4, buffer, magnetic field = 5 Tesla. —— 1.48 K; --- 2.07 K; —·—· 4.22 K; —··— 10 K; —···— 20 K; —o—o— 50 K. Protein concentration = 192 μM. *(b)* MCD spectrum of reduced *Chromatium* HIPIP. Pathlength = 1.08 mm, 50% *v/v* ethylene glycol, aqueous Tris/HCl, pH 7.4, buffer, magnetic field = 5 Tesla. —— 4.22 K; --- 20 K. Protein concentration = 422 μM.

(51), to give the so-called *superreduced* or C^{3-} state. This result has been interpreted in terms of unfolding of the protein in 80% dimethylsulfoxide, and suggests that the polypeptide chain is very effective in stabilizing the structure of the [4Fe-4S] cluster in the C^{-} state. This conclusion is substantiated by the fact that it has not been possible to prepare stable, crystalline models of [4Fe-4S] clusters in the C^{-} state (60).

4.4.2 The $[4Fe\text{-}4S]^{2+,1+}$ Centers in Ferredoxins. The reduced or C^{3-} state of 4-Fe bacterial ferredoxins, such as those from *Bacillus stearothermophilus* and *polymyxa,* exhibit rhombic EPR signals with g values around 2.06, 1.92, and 1.88. A rather more complex EPR spectrum, centered around the same average g value, is observed for the reduced 8-Fe ferredoxins such as those from *Peptococcus aerogenes* and *Clostridium pasteurianum*; this is rationalized in terms of weak magnetic coupling between the two [4Fe-4S] clusters (53). The MCD results discussed in this section are for *Clostridium pasteurianum* ferredoxin, which is taken to have the same structure as that of *Peptococcus aerogenes* ferredoxin on the basis of analogous EPR and Mössbauer data. MCD spectra at three representative temperatures, namely 1.55, 4.22, and 18.7 K, are shown in Figure 10, along with room-temperature MCD and absorption spectra. The MCD shows pronounced temperature dependence, with the signal increasing by a factor of ~20 on cooling from 300 to 1.55 K. The detail of the low-temperature MCD spectrum, when compared to the room-temperature data, is potentially of great use in assigning the complex electronic transitions that make up the UV/visible absorption. However, for the present we confine its use to providing a fingerprint for [4Fe-4S] Fd-type clusters, since theoretical descriptions of the excited states of [4Fe-4S] centers are still controversial (79–81). The MCD magnetization curves, constructed for the prominent positive peak at 530 nm (see Fig. 11), show that at 4.22 K and below, the C^{3-} state magnetizes as a simple $S = \frac{1}{2}$ Kramers doublet. The data points fit a theoretical curve computed for $g_{\parallel} = 2.06$, $g_{\perp} = 1.90$, and $m_z/m_+ = -1.00$, and the intercept value gives a good estimate of the average g value. The intercluster coupling must be considerably less than 1 cm^{-1} to avoid disappearance of the EPR spectrum. This explains why such coupling is not manifested in the MCD magnetization curves, where the available Zeeman energies can attain 5 cm^{-1}. No low-temperature magnetic susceptibility data have been reported for proteins in the C^{3-} state. The room-temperature magnetic moment of reduced ferredoxin from *Bacillus polymyxa* has been determined to be 3.2–3.4 BM/4 Fe atoms (56). This is considerably higher than the value expected for an $S = \frac{1}{2}$ Kramers doublet and implies thermal population of excited electronic states at room temperature. However, it should be borne in mind that the susceptibility measurements may give erroneously high values

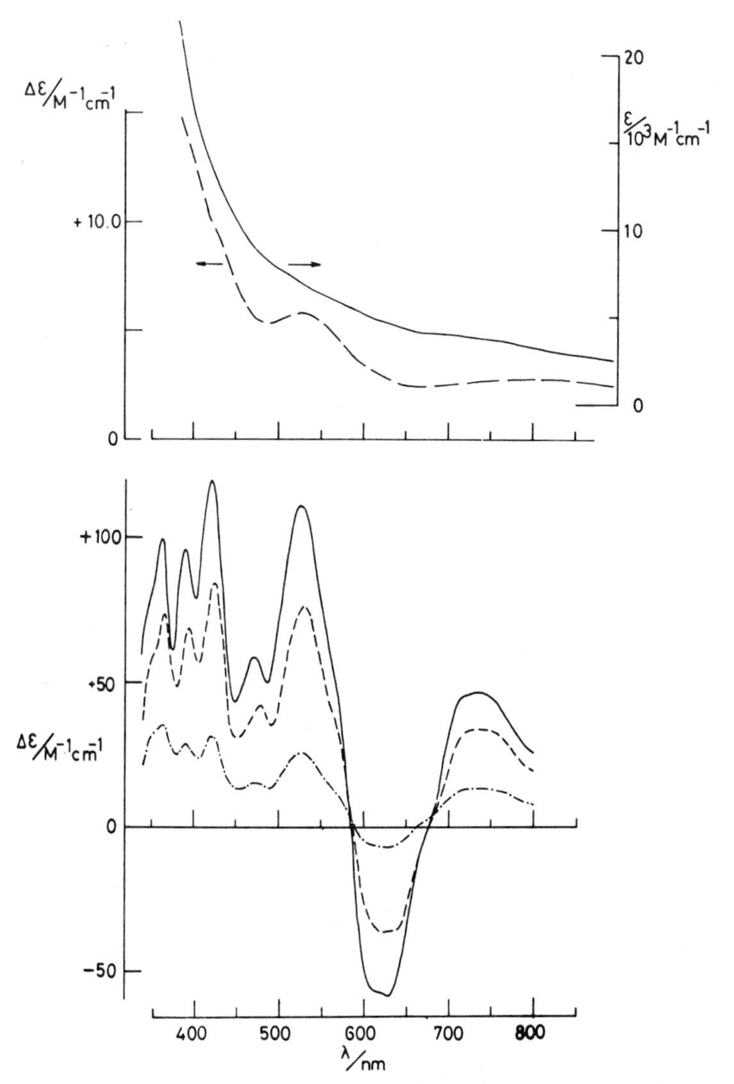

Figure 10. Reduced *Clostridium pasteurianum* Fd. Conditions: 50% *v/v* ethylene glycol, aqueous 0.8 *M* NaCl + 20 *mM* Tris/HCl, pH 8.5, buffer, excess sodium dithionite added anaerobically, protein concentration 360 μM. Upper panel: absorption spectrum at 295 K ——; MCD spectrum at 295 K, 5 Tesla ---. Lower panel: MCD spectra. Magnetic field = 5 Tesla, 1.55 K ——, 4.22 K ---, 18.7 K —·—, pathlength = 1.15 mm.

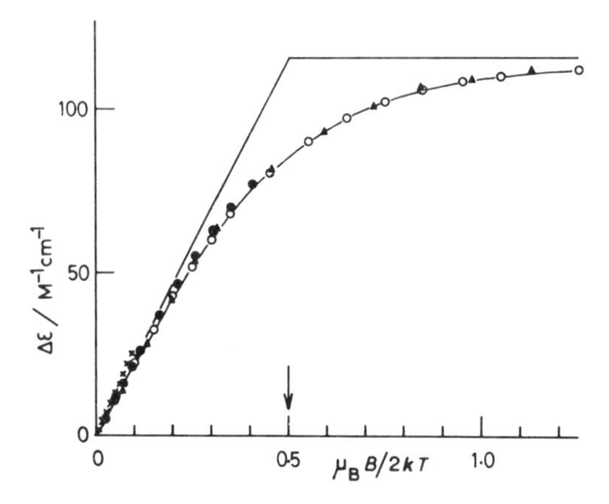

Figure 11. MCD magnetization curve for reduced *Clostridium pasteurianum* Fd. Wavelength = 530 nm. ▲) 1.52 K; ●) 4.22 K; ×) 18.7 K; at each temperature the magnetic fields used were 0.29, 0.59, 0.88, 1.16, 1.42, 2.06, 2.66, 3.24, 3.81, 4.39, and 5.10 Tesla. The open circles indicate the theoretical curve computed with parameters $g_{\parallel}$ = 2.06, $g_{\perp}$ = 1.90, m_z/m_+ = −1.0. The arrow indicates intercept value I = 0.50.

for the $[4Fe\text{-}4S]^{1+}$ cluster if the sample is contaminated by other clusters, especially the reduced [3Fe-3S] center with the relatively high spin quantum number $S = 2$. As discussed below, such contamination is highly likely. The slope for data points at 18.6 K is greater than the initial slope of the lower-temperature data in the MCD magnetization curve, (see Fig. 11). This suggests thermal population of low-lying electronic states at this temperature, although a more detailed study should be made over a wide range of temperatures. In contrast to the proteins, detailed magnetic data are available for model $[4Fe\text{-}4S]^{1+}$ clusters (57–59). Here there is substantial evidence, via EPR and magnetic susceptibility measurements, to indicate electronic states lying within 30 cm^{-1} of the ground state Kramers doublet. MCD spectra and magnetization curves now provide an excellent method of comparing and contrasting the magnetic properties of Fe-S clusters in proteins with those of synthetic models.

Although no detailed magnetic susceptibility studies are available, the Mössbauer spectra do show the oxidized or C^{2-} state of 4-Fe and 8-Fe ferredoxins to be predominantly diamagnetic (26). But the Mössbauer spectra have invariably been measured on material that has been reconstituted with ^{57}Fe. It is possible that this leads to a more homogeneous preparation than the extracted protein, which invariably exhibits a weak, almost isotropic signal in the EPR spectrum below 35 K at $g = 2.01$ (28). In our experience for *Clostridium pasteurianum* ferredoxin the magnitude of this signal varies for each preparation, but generally corresponds to <0.02 spins/[4Fe-4S] cluster. The temperature dependence of the MCD spectra for oxidized *Clostridium pasteurianum,* shown in the middle panel of Fig. 12, proves that a paramagnetic species is being detected. However, the magnitude of the signal compared to other paramagnetic Fe-S chromophores shows that the paramagnetic component is present in very low concentrations. The MCD spectrum at 49.0 K is seen to be very similar to the room-temperature spectrum except for greatly increased resolution. This is consistent with a large temperature-independent contribution, which dominates the spectrum at higher temperatures. The similarity between the 49.0 K MCD spectrum and the low-temperature MCD spectrum of reduced HIPIP (Fig. 9*b*) supports the validity of the three oxidation state hypothesis. By measuring the MCD difference spectrum between the 1.95 and 49.0 K spectra (bottom panel of Fig. 12), we obtain the MCD spectrum corresponding to the small paramagnetic component. This is the species that gives rise to the $g = 2.01$ EPR signal. Comparison with the low-temperature MCD spectra of other Fe-S proteins reveals that the MCD spectrum of this species corresponds most closely to a mixture of reduced plus oxidized Fd II from *D. gigas.* A [3Fe-3S] cluster is clearly implicated.

Sweeney and co-workers (31) demonstrated that the $g = 2.01$ signal of oxidized 8-Fe *Clostridial* ferredoxins increases ~100-fold on treatment with potassium ferricyanide, retaining the same g value and similar linewidth. This species, termed *superoxidized,* was assigned to a 4-Fe Fd-type cluster in the C^- oxidation level, that is, a cluster isoelectronic with that in oxidized HIPIP, namely $[4Fe\text{-}4S]^{3+}$. The 4.22 K MCD spectrum of ferricyanide-oxidized *Clostridium pasteurianum* is shown as the solid line in Figure 13. To facilitate comparison the corresponding spectrum of oxidized Fd II from *D. gigas* is shown as the broken line in the same figure. The overlap of bands is remarkable and unambiguously assigns superoxidized *Clostridium pasteurianum* as an oxidized [3Fe-3S] cluster. As pointed out in Chapter 4, Mössbauer spectroscopy cannot uniquely distinguish between oxidized [3Fe-3S] and oxidized HIPIP-type clusters; since both give EPR signals with $g_{av} > 2$, it is likely that low-temperature MCD provides the best spectroscopic means of discriminating between them. MCD magnetization curves verify that the EPR species at $g = 2.01$ dominates the low-temperature MCD spectrum (6). Further support for this conclusion is provided by resonance Raman experiments (see Chapter 11). Quantitative EPR experiments on *Clostridium pasteurianum* Fd indicate that, when the species is fully superoxidized, the $g = 2.01$ signal corresponds to 0.8 spins/molecule. This signal completely disappears on dithionite reduction and is replaced by a *very* weak rhombic signal (<0.04 spin/molecule) with g factors of 2.05, 1.92, and 1.88 (78). Such a signal is attributed to a small quantity of unreacted $[4Fe\text{-}4S]^{1+}$ clusters. On anaerobic addition of a large excess of Na_2S to the rereduced, ferricyanide-

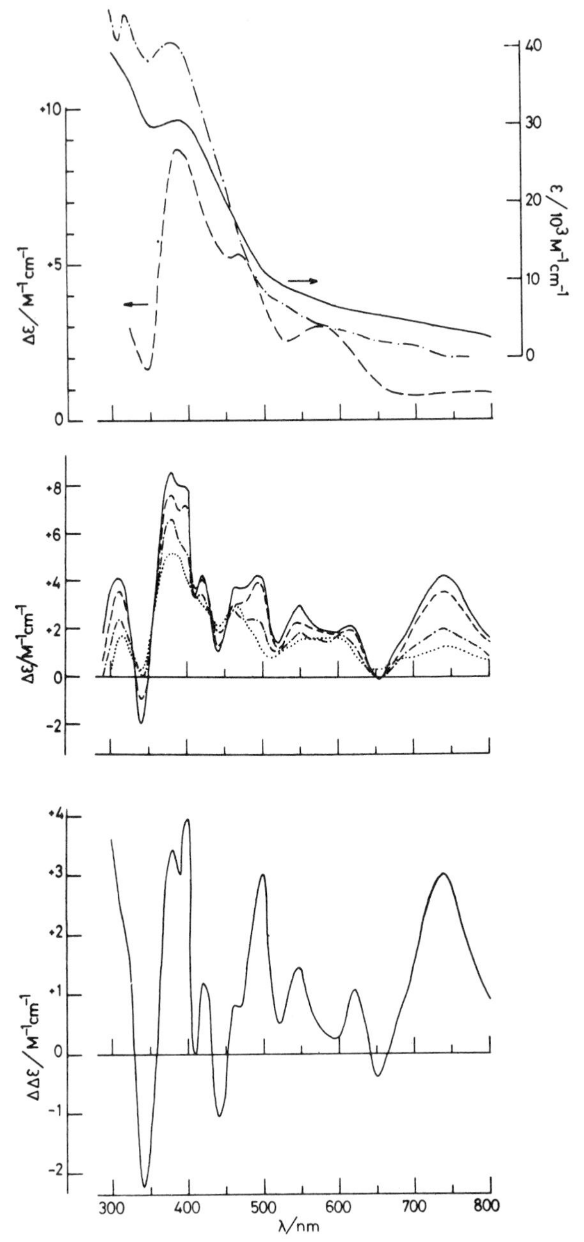

Figure 12. Oxidized *Clostridium pasteurianum* Fd. Conditions: 50% *v/v* ethylene glycol, aqueous 0.8 *M* NaCl + 20 *mM* Tris/HCl, pH 8.5, buffer. Protein concentration = 360 μM. Upper panel: absorption spectrum at 295 K —— and 1.5 K —·—·, MCD spectrum at 295 K, 5 Tesla, ---. Middle panel: MCD spectra, magnetic field = 5 Tesla, at 1.95 K ——, 4.22 K ---, 18.0 K —·—·, and 49.0 K ·······. Pathlength = 1.06 mm. Lower panel: difference between MCD spectra at 1.95 K and 49.0 K.

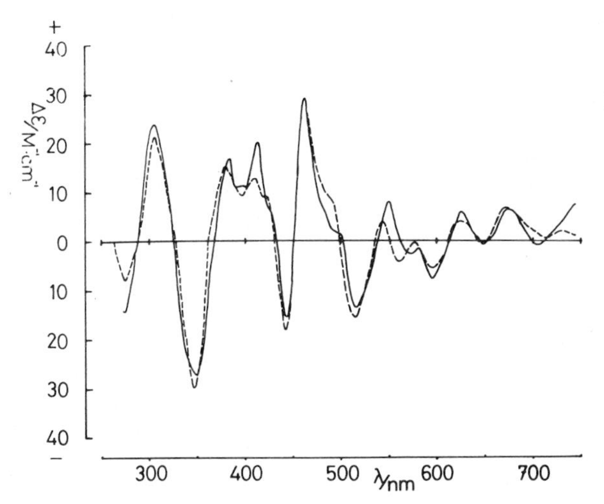

Figure 13. MCD spectra of ferricyanide-oxidized *Clostridium pasteurianum* Fd (——) and oxidized Fd II from *D. gigas* (---) at 4.22 K and 5.1 Tesla. The $\Delta\epsilon$ scale refers only to the spectrum of Fd II. The spectra have been normalized at 462 nm.

oxidized protein, the weak C^{3-}-type EPR signal is greatly enhanced, accounting for approximately 0.8 spins/molecule. This result is observed irrespective of whether the ferricyanide-oxidized sample was purified by chromatography and $(NH_4)_2SO_4$ precipitations. Therefore it is tentatively concluded that the fully ferricyanide-oxidized protein contains only 1 [3Fe-3S] cluster remaining from the 2 [4Fe-4S] clusters, which can subsequently undergo reconversion to a [4Fe-4S] cluster on anaerobic addition of an excess of sulfide ions (78). MCD and EPR studies on a rereduced sample of superoxidized *Clostridium pasterurianum* Fd indicated predominantly the presence of reduced [3Fe-3S] clusters with some residual $[4Fe\text{-}4S]^{1+}$ centers. The former are readily distinguished by the unique form of their magnetization curves (6). The ferricyanide-induced breakdown of [4Fe-4S] Fd-type clusters to [3Fe-3S] clusters does not appear to be unique to 8-Fe ferredoxins. Similar MCD results have been observed for the 4-Fe ferredoxin from *Bacillus stearothermophilus,* and more detailed studies are currently in progress (61). Air oxidation of this ferredoxin appears to produce [3Fe-3S] clusters rather readily (61).

These conclusions raise many interesting and important questions about the biological significance of [3Fe-3S] clusters and their structural chemistry. Are [3Fe-3S] clusters of physiological importance or are they produced by air oxidative damage during the extraction of Fe-S proteins? Our evidence is that the latter certainly takes place for the Fds from *Clostridium pasteurianum* and *Bacillus stearothermophilus*, and therefore the phenomenon should be widely suspected. Furthermore, Fd II from *D. gigas*, reconstituted from the apoprotein under reducing conditions, gives solely [4Fe-4S] centers (see Chapter 5). However, it has been shown that Fd II is more active than Fd I in the activation of sulfite reduction by sulfite reductase and hydrogenase (40). It could be that this type of assay is not very specific for cluster type but rather depends on the requirements of the correct reducing potential. It is interesting to note that another [3Fe-3S] protein, aconitase from beef heart, requires activation with a mixture of ascorbate, cysteine and ferrous Fe after the first few purification steps. The $g = 2.01$ signal remains in the presence of ascorbate and cysteine, but is lost on addition of the complete activation mixture or just Fe(II) ion alone (62). In light of the preceding discussion, it is tempting to speculate that the Fe(II) is required to reconstitute a [4Fe-4S] cluster, which is the physiological active species. We note that Mössbauer studies indicate conversion of the [3Fe-3S] center of aconitase to other structural forms when activated in the presence of Fe and dithiothreitol (see Chapter 4).

On the basis of the X-ray crystallographic studies of the ferredoxin from *Azotobacter vinelandii*, Stout and his co-workers have suggested that the important feature in a polypeptide sequence stabilizing a [3Fe-3S] core is cys-X-glu-X-cys-pro (47). This sequence is also found in the polypeptide chain of *D. gigas* ferredoxins (63) and in at least four other ferredoxins now suspected to contain [3Fe-3S] centers (47). However, such a sequence is not present in the polypeptide chain of the Fd from *Clostridium pasteurianum*. There is a cys-X-X-glu-cys-pro sequence, and the glu residue is invariant in the sequences of seven (64) anaerobic bacteria with the sole exception of that from *D. elsdenii* (65). This sequence may fulfill the same role as that in Fd II from *D. gigas*. However, the ligands binding the [3Fe-3S] cluster in Fd II have yet to positively be identified. Certainly, in contrast to the Fd from *Azotobacter vinelandii* there are sufficient cysteine residues available to accommodate two for each Fe atom. It is doubtful whether Mössbauer spectroscopy would distinguish between coordination involving 5 S atoms from cysteines and 1 O of undetermined origin and 6 S atoms from cysteines. The MCD spectrum may well be sensitive to such a change, and thus a low-temperature spectrum of *Azotobacter vinelandii* Fd is awaited with considerable interest.

4.5 Nitrogenase

Nitrogenase, the enzyme responsible for the biological reduction of N_2 to NH_3, is comprised of two dissociable protein components. Component I, the FeMo protein, contains between 28 and 38 Fe atoms/molecule and 2 Mo atoms/molecule together with acid-labile S and has a molecular weight of $\approx 220{,}000$.

Component II, the Fe protein, is smaller, having a molecular weight of ≈60,000, and contains 4 Fe atoms/molecule and 4 acid-labile S atoms/molecule (66, 67). Component II is considered to be a [4Fe-4S] cluster of the Fd type, having a rhombic EPR signal with $g = 2.05$, 1.94, and 1.86 as isolated and no signal on oxidation (66). The cluster composition of component I has presented a major challenge to the available spectroscopic techniques. Mössbauer spectroscopy has been particularly fruitful in this area, and is responsible for establishing the essential similarity of the Fe centers in the Fe-Mo proteins from a variety of different sources, namely *Klebsiella pneumoniae, Azotobacter vinelandii,* and *Clostridium pasteurianum* (68–72). For clarity, throughout this chapter we use the nomenclature for Fe cluster type devised by Münck and his co-workers in analyzing the Mössbauer data for Av1 and Cp1. This distinguishes three distinct Fe cluster types, termed M, P, and S centers. The superscripts N and OX refer to the native (or dithionite-reduced) and the thionine-oxidized proteins, respectively. These centers account for ≈40, ≈54, and ≈6% of the Fe, respectively. For the minor component, the S center, the magnetic properties are not well defined, but it appears as a quadrupole doublet and apparently remains unchanged as the redox state of the protein is changed. In the native or dithionite-reduced protein M^N is paramagnetic and P^N is diamagnetic. For the thionine-oxidized protein the situation is reversed, with M^{OX} diamagnetic and P^{OX} paramagnetic (71). Since low-temperature MCD is specifically a probe for paramagnets, by investigating the enzyme in these two oxidation states it is possible to probe separately the magnetic properties of both P and M centers. The results described in this chapter are for the Kp1 protein (5), and are discussed below in the light of the existing Mössbauer and EPR data.

4.5.1 Dithionite-Reduced Kp1. The paramagnetic component M^N is responsible for the characteristic EPR signal for native protein (g values of 4.32, 3.63, and 2.01). EPR and Mössbauer experiments demonstrate that it is analogous to the Mo-containing cofactor or FeMoco (74), which is extractable from the protein by treatment with *N*-methyl formamide (73). The magnetic state of this paramagnet is relatively well understood as a result of analysis of the EPR spectrum. The center has a spin $S = 3/2$, which is split by a predominantly axial crystal field to give two pairs of doublets, $M_s = \pm 1/2$ and $\pm 3/2$. The EPR signal originates from the lower $M_s = \pm 1/2$ doublet (68, 69, 75), and the observed g values can be generated theoretically by using a rhombicity parameter, $E/D = 0.055$, indicating slight ground state rhombic distortion (69). For such parameters the upper $M_s = \pm 3/2$ doublet is predicted to have g factors of 0.32, 0.34, and 6.0 (69), and there have been suggestions that a weak EPR signal at $g = 5.93$ can be detected in the extracted cofactor (74). In the case of Av1 the magnitude of the zero-field splitting has been estimated to be 11 cm^{-1} by utilizing the temperature dependence of the EPR signal at $g = 4.32$ (69).

The temperature dependence of the MCD signal for dithionite-reduced Kp1 demonstrates that the EPR-detected paramagnetic species is being observed (see Fig. 14). The magnetization curve constructed at 803 mm, shown in Figure 15, is nested, implying a complex ground state with zero-field components becoming thermally populated over the temperature range of the experiment. Similar curves have been obtained for the other prominent peaks in the spectrum. Using the curve for the lowest available temperature, 1.54 K, when only the ground state doublet is significantly populated, we obtain an intercept value $I = 0.28 \pm 0.02$. The uncertainty arises because the saturation limit is not clearly defined. Indeed, the fact that the MCD signal is not completely independent of magnetic field at the highest available fields implies some field-induced mixing of the zero-field components. Nevertheless the intercept value is appreciably different from that of ≈ 0.5, which is expected for an isolated $S = ½$ Kramers doublet ground state. If the system is approximated to be isotropic, this intercept value corresponds to a g value

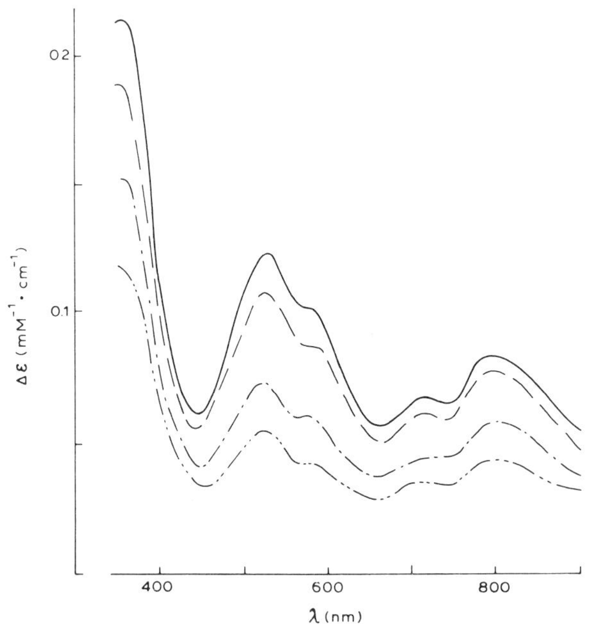

Figure 14. MCD spectra of dithionite-reduced Kp1, in aqueous pH 7.6 buffer, 50% v/v ethylene glycol, excess sodium dithionite. Protein concentration = 62 μM, magnetic field = 5.1 Tesla, 1.57 K (——), 4.22 K (---), 11.0 K (—·—·), and 20.0 K (—··—). Pathlength = 3.36 mm.

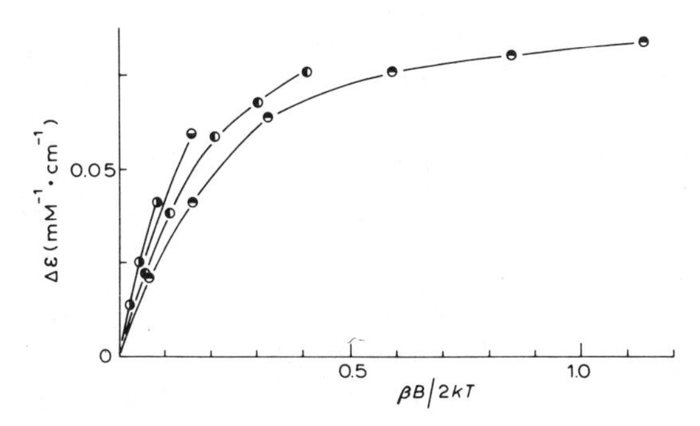

Figure 15. MCD magnetization curve of dithionite-reduced Kp1. Conditions of measurement as for Figure 14, wavelength = 803 nm. ◓) 1.52 K, 0.300, 0.735, 1.540, 2.665, 3.825, and 5.100 Tesla; ◐) 4.22 K, 0.735, 1.415, 2.665, 3.825, and 5.100 Tesla; ◒) 11.0 K, 5.100 Tesla; ◑) 20.0 K, 1.415, 2.665 and 5.100 Tesla.

of 3.6 ± 0.3, which encompasses the experimental g_{av} value. However, it should be emphasized that this simple procedure for estimating g values is not rigorously valid in this instance, since it ignores effects arising from the polarization of the electronic transition, rhombic splitting of the ground state, and field-induced mixing of the zero-field components. However, it is clear that the nature of the MCD magnetization curves is compatible with the EPR data, and detailed theoretical simulations using the published crystal field parameters are underway.

The form of the MCD spectrum in Figure 14 is unique in our experience and remarkable in that it possesses bands all of one sign. It provides an optical fingerprint of the center in the protein and thus new spectroscopic criteria to assess the validity of proposed Fe and Mo model complexes and structural analogs of the cofactor. The form of the low-temperature MCD spectrum of etracted cofactor, FeMoco, will be of considerable interest.

4.5.2 Thionine-Oxidized Kp1. In this redox state the Mössbauer spectrum shows the presence of a single type of paramagnetic cluster, P^{OX} (71). There is no EPR signal from this oxidation state of Kp1, although magnetic Mössbauer measurements clearly indicate that the paramagnet present has nonintegral spin. For oxidized Av1 and Cp1 detailed analysis of the magnetic Mössbauer spectra reveals the lowest component of the ground state to be a Kramers doublet with highly anisotropic g values such that $g_z/(g_x \approx g_y) \geq 15$. The first excited doublet is separated in energy by at least 5 cm^{-1} from the ground state. No definitive estimate of the total spin S of the ground state has been made by Mössbauer spectroscopy. However, it is suggested that $9/2 \geq S \geq 3/2$ and that this ground state has a peculiar combination of zero-field splitting parameters and spin relaxation times that renders it EPR undetectable (71). There is some preliminary evidence for oxidized Cp1 which suggests that the ground state has $S = 5/2$ (72).

The low-temperature MCD spectrum, shown in Figure 16, gives the first indication of the nature of the optical spectrum for this type of cluster. The three prominent positive peaks at 380, 520, and 813 nm indicate the energies of the three principal electronic transitions. The form of the spectrum, particularly the intense band at 813 nm, is significantly different from that of any other cluster type we have observed. Moreover, it is in the form of the magnetization curves that the uniqueness of the cluster really becomes apparent (see Fig. 17). The remarkable feature is the very steep initial slope; at 1.54 K the paramagnet is >90% magnetized at a field of only 1.5 Tesla. All the major bands show this property. This is an invaluable method of identifying this cluster type and clearly distinguishes the MCD signal of P^{OX} from that of M^N, the FeMoco cluster. That the lowest state is an electronic doublet is demonstrated by the fact that magnetic saturation is observed at low-temperatures as the field increases. Since magnetization curves at both 1.54 and 1.97 K overlap with good precision, it can be concluded that only this electronic doublet is populated below 2 K. At 4.22 K the magnetization curve departs from the lower-temperature curves above ≈3 Tesla, implying the thermal population of another zero-field component of the ground state. A plot of MCD intensity at a constant field of 5.1 Tesla against $1/T$ indicates that all Zeeman components of the ground state lie below 17 cm^{-1} and that there appear to be no other electronic states <85 cm^{-1} from the ground state (5).

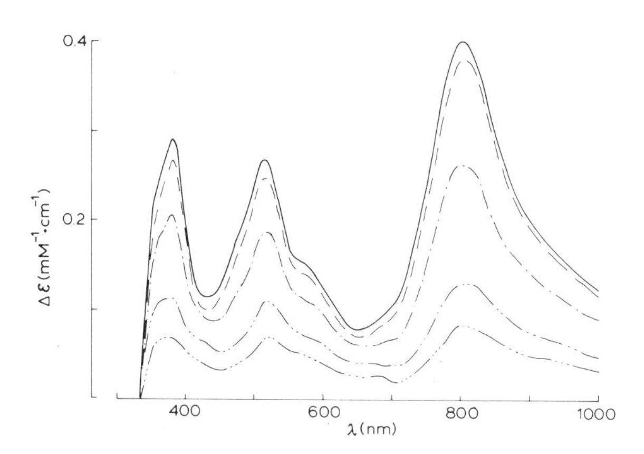

Figure 16. MCD spectra of thionine-oxidized Kp1 in aqueous pH 7.6 buffer, 50% *v/v* ethylene glycol, protein concentration = 77 μM, magnetic field = 5.1 Tesla. 1.54 K (——), 4.22 K (---), 20.0 K (—·—·), 40.0 K (—··—), 60.0 K (—···—). Pathlength = 2.13 mm.

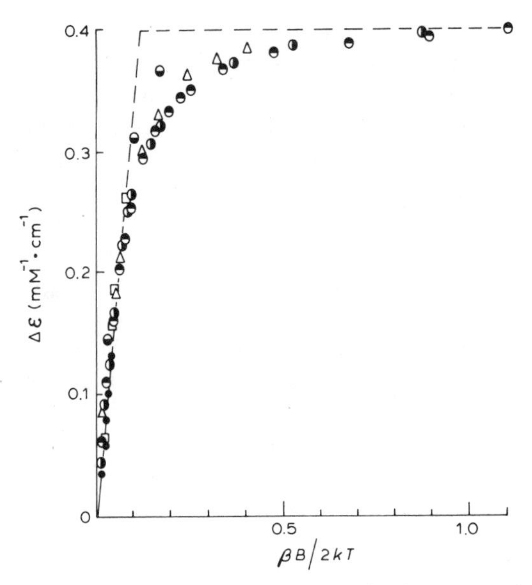

Figure 17. MCD magnetization curves of thionine-oxidized Kp1. Conditions of measurement as for Fig. 16, wavelength = 813 nm. ◓) 1.54 K, 0.075, 0.150, 0.225, 0.300, 0.370, 0.440, 0.585, 0.735, 0.880, 1.025, 1.155, 1.540, 2.157, 3.089, 4.095, and 5.100 Tesla; ◑) 1.972 K, magnetic fields as for 1.54 K; Δ) 4.22 K, 0.300, 0.586, 0.880, 1.155, 1.585, 2.157, 3.089, 4.095, and 5.100 Tesla; ◒) 10.0 K, 1.019, 3.089, and 5.100 Tesla; □) 20.0 K, 1.019, 3.089, and 5.100 Tesla; ●) 40 K, 50 K, 60 K, 80 K, 120 K at 5.100 Tesla.

The steepness of the initial slope of the magnetization curve is reflected in the intercept value for the lowest-temperature data, $I = 0.12$, by far the lowest value observed to date for any Fe-S cluster. If an axially distorted ground state is assumed, this intercept value gives $g_{\parallel} = 12.5$ and $g_{\perp} = 0$. For $S = 5/2$ and $S = 7/2$ systems, split under an axial field, leaving $M_s = \pm 5/2$ and $\pm 7/2$ as the lowest Kramers doublets, the theoretical ground state values of $g_{\parallel}$ are 10 and 14, respectively. That the experimental $g_{\parallel}$ value falls between these two possibilities is not surprising, since we are neglecting any rhombic distortions in the ground state or field-induced mixing of zero-field components. Therefore we conclude that the total spin of the ground state is $S = 5/2$ or $7/2$.

It has been argued that P clusters are comprised of 4 [4Fe-4S] centers (71, 72), although with ground state magnetic properties that are markedly different from those observed in HIPIP and bacterial ferredoxins. The argument is primarily based on three pieces of evidence. First, cluster extrusion experiments generate 4 classical [4Fe-4S] clusters on removal from the protein, none of which are derived from the cofactor center M. Second, the integration of the Mössbauer signals of the two types of Fe, called D and Fe^{2+}, which belong to this cluster, gives a ratio of 3:1 (71). Third, the quantitative redox titration of native Av1 monitored by EPR and Mössbauer spectroscopy suggested that $4e^-$/molecule could be removed before the M^N center starts to become oxidized (71). However, similar EPR experiments with the Kp1 protein gave a linear decrease in the M^N center EPR signal on oxidation with 1-6 oxidizing equivalents. The recent article by Smith and colleagues (67) provides a more detailed critique of known structural information for the FeMo protein, and points to an alternative interpretation of the data in terms of P clusters containing both [4Fe-4S] and [2Fe-2S] centers. However, both of these interpretations were made prior to the knowledge of the existence of [3Fe-3S] clusters. The low-temperature MCD data have little to add to this debate except to emphasize the uniqueness of the magnetic properties of P clusters when compared to other Fe-S centers.

Finally, we turn our attention to the S cluster. The analyses of the Mössbauer data agree in assigning 2 Fe atoms/molecule to the S cluster. This limits the possibilities to a [2Fe-2S] center or a pair of single Fe clusters. We have characterized the low-temperature MCD spectra of the paramagnetic states of oxidized and reduced rubredoxin and a variety of reduced [2Fe-2S] ferredoxins (see above). All of these signals are very intense at low-temperatures and should be easily recognized if present in an MCD spectrum of Kp1. The failure to recognize such centers is in accord with the absence of EPR signals in Kp1 that would be assigned to classical 1-Fe or [2Fe-2S] centers. The remaining possibility among the known Fe-S cluster types is that of an oxidized [2Fe-2S] center in both the thionine-oxidized and dithionite-reduced states of Kp1. However, it should be pointed out that [2Fe-2S] centers invariably give an intense characteristic CD in the visible region. This feature is not apparent in the CD of either oxidized or reduced Kp1 (77).

A room-temperature MCD study of the reduced and oxidized forms of Kp1 and Av1 has been made (77). Both diamagnetic and paramagnetic clusters are expected to give observable room-temperature MCD signals. Hence the problem of unraveling the contributions from individual clusters is very much more acute. Low-temperature MCD overcomes this problem and allows characterization of the metal centers in their paramagnetic oxidation levels.

ACKNOWLEDGMENTS

The work was greatly aided by the provision of the following samples from colleagues: *D. gigas* Fd I and II, rubredoxin, and desulforedoxin, A. V. Xavier, J. J. G. Moura, and I. Moura; *Cl. past.* Fd and Rd, *B. stearo.* Fd,

Chromatium HIPIP, spinach Fd, *S. maxima* Fd, and adrenodoxin, K. K. Rao, and D. O. Hall; Nitrogenase Kp1, B. E. Smith.

REFERENCES

1. A. J. Thomson and M. K. Johnson, *Biochem. J.*, **191,** 411-420 (1980).
2. J. C. Rivoal, B. Briat, R. Cammack, D. O. Hall, K. K. Rao, I. N. Douglas, and A. J. Thomson, *Biochim. Biophys. Acta,* **493,** 122-131 (1977).
3. M. K. Johnson, A. J. Thomson, A. E. Robinson, K. K. Rao, and D. O. Hall, *Biochim. Biophys. Acta,* **667,** 433-451 (1981).
4. A. J. Thomson, A. E. Robinson, M. K. Johnson, J. J. G. Moura, I. Moura, A. V. Xavier, and J. Le Gall, *Biochim. Biophys. Acta,* **670,** 93-100 (1981).
5. M. K. Johnson, A. E. Robinson, A. J. Thomson, and B. E. Smith, *Biochim. Biophys. Acta* **671,** 61-70 (1981).
6. A. J. Thomson, A. E. Robinson, M. K. Johnson, R. Cammack, K. K. Rao, and D. O. Hall, *Biochim. Biophys. Acta,* **637,** 423-432 (1981).
7. A. J. Thomson, M. K. Johnson, C. Greenwood, and P. E. Gooding, *Biochem. J.,* **193,** 687-697.
8. P. N. Schatz and A. J. McCaffery, *Q. Rev. Chem. Soc.*, **23,** 552-584 (1969).
9. P. J. Stephens, *Adv. Chem. Phys.*, **35,** 197-265 (1976)
10. P. N. Schatz, R. L. Mowery, and E. R. Krausz, *Mol. Phys.*, **35,** 1535-1557 (1978).
11. K. D. Watenpaugh, L. C. Sieker, J. R. Herriott, and L. H. Jenson, *Acta Crystallogr.*, **B29,** 943-956 (1973).
12. D. E. Sayers, A. E. Stern, and J. R. Herriott, *J. Chem. Phys.*, **64,** 427-428 (1979).
13. R. G. Shulman, P. Eisenberger, W. E. Blumberg, and N. A. Stombaugh, *Proc. Natl. Acad. Sci. USA,* **72,** 4003-4007.
14. J. Peisach, W. E. Blumberg, E. T. Lode, and M. J. Coon, *J. Biol. Chem.*, **246,** 5877-5881.
15. E. T. Lode and M. J. Coon, *J. Biol. Chem.*, **246,** 791-802 (1971).
16. W. D. Phillips, M. Poe, J. F. Weiher, C. C. McDonald, and W. Lovenberg, *Nature (London),* **227,** 574-577 (1970).
17. K. K. Rao, M. C. W. Evans, R. Cammack, D. O. Hall, C. L. Thompson, R. J. Jackson, and C. E. Johnson, *Biochem. J.*, **129,** 1063-1070 (1972).
18. W. A. Eaton and W. Lovenberg in *Iron-Sulfur Proteins,* Vol. II, W. Lovenberg, ed., Academic, New York, 1973, pp. 131-162.
19. I. Moura, M. Bruschi, J. LeGall, J. J. G. Moura, and A. V. Xavier, *Biochim. Biophys. Res. Commun.*, **75,** 1037-1044 (1977).
20. I. Moura, B. H. Huynh, R. P. Hausinger, J. LeGall, A. V. Xavier, and E. Münck, *J. Biol. Chem,* **255,** 2493-2498 (1980).
21. P. G. Debrunner, E. Münck, L. Que, and C. E. Schulz, in *Iron-Sulfur Proteins,* Vol. III, W. Lovenberg, ed., Academic, New York, 1977, pp. 381-417.
22. G. Palmer, in *Iron-Sulfur Proteins,* Vol. II, W. Lovenberg, ed., Academic, New York, 1973, pp. 285-325.

23. K. Fukuyama, T. Hase, S. Matsumoto, T. Tsukihara, Y. Katsube, N. Tanaka, M. Kakudo, K. Wada, and H. Matsubara, *Nature,* **286,** 522-524 (1980).
24. A. J. Thomson, R. Cammack, D. O. Hall, K. K. Rao, B. Briat, J. C. Rivoal, and J. Badoz, *Biochim. Biophys. Acta,* **493,** 132-141 (1977).
25. G. Palmer, W. R. Dunham, J. A. Fee, R. H. Sands, T. Iizuka, and T. Yonetani, *Biochim, Biophys. Acta,* **245,** 201-207 (1971).
26. R. Cammack, D. P. E. Dickson, and C. E. Johnson, in *Iron-Sulfur Proteins,* Vol. III, W. Lovenberg, ed., Academic, New York, 1977, pp. 283-330.
27. T. Kimura, A. Tasaki, and H. Watari, *J. Biol. Chem.,* **245,** 4450-4452 (1970).
28. W. H. Orme-Johnson and R. H. Sands, in *Iron-Sulfur Proteins,* Vol. II, W. Lovenberg, ed., Academic, New York, 1973, pp. 195-238.
29. M. H. Emptage, T. A. Kent, B. H. Huynh, J. Rawlings, W. H. Orme-Johnson, and E. Münck, *J. Biol. Chem.,* **255,** 1793-1796 (1980).
30. B. H. Huynh, J. J. G. Moura, I. Moura, T. A. Kent, J. LeGall, A. V. Xavier, and E. Münck, *J. Biol. Chem.,* **255,** 3242-3244 (1980).
31. W. V. Sweeney, A. J. Bearden, and J. C. Rabinowitz, *Biochem. Biophys. Res. Commun.,* **59,** 188-194 (1974).
32. T. Ohnishi, H. Blum, S. Sato, K. Nakazawa, K. Hon-nami, and T. Oshima, *J. Biol. Chem.,* **255,** 345-348 (1980).
33. M. G. Yates, M. J. O'Donnell, D. J. Lowe, and H. Bothe, *Eur. J. Biochem.,* **85,** 291-299 (1978).
34. D. C. Yoch, R. P. Carithers, and D. I. Arnon, *J. Biol. Chem.,* **252,** 7453-7460 (1977).
35. H. Berndt, D. J. Lowe, and M. G. Yates, *Eur. J. Biochem.,* **86,** 133-142 (1978).
36. T. Ohnishi, J. Lim, D. B. Winter, and T. E. King, *J. Biol. Chem.,* **251,** 2105-2109 (1976).
37. H. Beinert, B. A. C. Ackrell, E. B. Kearney, and T. P. Singer, *Eur. J. Biochem.,* **54,** 184-194 (1975).
38. W. J. Ingledew, and R. C. Prince, *Arch. Biochem. Biophys.,* **178,** 303-307 (1977).
39. P. Forget and D. V. DerVartanian, *Biochim. Biophys. Acta,* **256,** 600-606 (1972).
40. M. Bruschi, E. C. Hatchikian, J. LeGall, J. J. G. Moura, and A. V. Xavier, *Biochim. Biophys. Acta,* **449,** 275-284 (1976).
41. R. Cammack, K. K. Rao, D. O. Hall, J. J. G. Moura, A. V. Xavier, M. Bruschi, J. LeGall, A. Deville, and J-P. Gayda, *Biochim. Biophys. Acta,* **490,** 311-321 (1977).
42. M. K. Johnson, A. E. Robinson, and A. J. Thomson, unpublished observations.
43. D. Piszkiewicz, O. Gawron, and J. C. Sutherland, *Biochemistry,* **20,** 363-366 (1981).
44. C. W. Carter, J. Kraut, S. T. Freer, N. H. Xuong, R. A. Alden, and R. G. Bartsch, *J. Biol. Chem.,* **249,** 4212-4225 (1974).
45. C. W. Carter, J. Kraut, S. T. Freer, and R. A. Alden, *J. Biol. Chem.,* **249,** 6339-6346 (1974).
46. S. T. Freer, R. A. Alden, C. W. Carter, and J. Kraut, *J. Biol. Chem.,* **250,** 46-54 (1975).
47. D. Ghosh, W. Furey, S. O'Donnell, and C. D. Stout, *J. Biol. Chem.,* **256,** 4185-4192 (1981).
48. E. T. Adman, L. C. Sieker, and L. H. Jensen, *J. Biol. Chem.,* **248,** 3987-3996 (1973).
49. E. T. Adman, L. C. Sieker, and L. H. Jensen, *J. Biol. Chem.,* **251,** 3801-3806 (1976).
50. C. W. Carter, J. Kraut, S. T. Freer, R. A. Alden, L. C. Sieker, A. Adam, and L. H. Jensen, *Proc. Natl. Acad. Sci. USA,* **69,** 3526-3529 (1972).

51. R. Cammack, *Biochem. Biophys. Res. Commun.*, **54,** 548–554 (1973).
52. B. C. Antanaitis and T. H. Moss, *Biochim. Biophys. Acta,* **405,** 262–279 (1975).
53. R. Mathews, S. Charlton, R. H. Sands, and G. Palmer, *J. Biol. Chem.*, **249,** 4326–4328 (1974).
54. P. J. Stephens, A. J. Thomson, J. B. R. Dunn, T. A. Keiderling, J. Rawlings, K. K. Rao, and D. O. Hall, *Biochemistry,* **17,** 4770–4778 (1978).
55. H. Blum, J. C. Salerno, R. C. Prince, J. S. Leigh, and T. Ohnishi, *Biophys. J.*, **20,** 23–31 (1977).
56. W. D. Phillips, C. C. McDonald, N. A. Stombaugh, and W. H. Orme-Johnson, *Proc. Natl. Acad. Sci. USA,* **71,** 140–143 (1974).
57. E. J. Laskowski, R. B. Frankel, W. O. Gillum, G. C. Papaefthymiou, J. Renaud, J. A. Ibers, and R. H. Holm, *J. Am. Chem. Soc.*, **100,** 5322–5336 (1978).
58. E. J. Laskowski, J. G. Reynolds, R. B. Frankel, S. Foner, G. C. Papaefthymiou, and R. H. Holm, *J. Am. Chem. Soc.*, **101,** 6562–6570 (1979).
59. G. C. Papaefthymiou, R. B. Frankel, S. Foner, E. J. Laskowski, and R. H. Holm, *J. Phys. (Paris),* **41,** C1–493 (1980).
60. D. V. Pamphilis, B. A. Averill, T. Herskovitz, L. Que, Jr., and R. H. Holm, *J. Am. Chem. Soc.*, **96,** 4159–4167 (1974).
61. A. J. Thomson, M. K. Johnson, and A. E. Robinson, unpublished observations.
62. F. J. Ruzicka and H. Beinert, *J. Biol. Chem.*, **253,** 2514–2517 (1978).
63. M. Bruschi, G. Bovier-Lapierre, J. Bonicel, and P. Couchoud, *Biochem. Biophys. Res. Commun.*, **91,** 623–628 (1979).
64. M. Tanaka, T. Nakashima, A. M. Benson, H. Mower, and K. T. Yasunoba, *Biochem. Biophys. Res. Commun,* **16,** 422–427 (1964).
65. K. T. Yasunoba and M. Tanaka, in *Iron-Sulfur Proteins,* Vol. II, W. Lovenberg, ed., Academic, New York, 1973, pp. 27–130.
66. W. H. Orme-Johnson and L. C. Davis, in *Iron-Sulfur Proteins,* Vol. III, W. Lovenberg, ed., Academic, New York, 1977, pp. 15–60.
67. B. E. Smith, M. J. O'Donnell, G. Lang, and K. Spartalian, *Biochem. J.*, **191,** pp. 449–455 (1980).
68. B. E. Smith and G. Lang, *Biochem. J.*, **137,** 169–180 (1974).
69. E. Münck, H. Rhodes, W. H. Orme-Johnson, L. C. Davis, W. J. Brill, and V. K. Shah, *Biochim. Biophys. Acta,* **400,** 32–53 (1975).
70. B. H. Huynh, E. Münck, and W. H. Orme-Johnson, *Biochim. Biophys. Acta,* **527,** 192–203 (1979).
71. R. Zimmerman, E. Münck, W. J. Brill, V. K. Shah, M. T. Henzl, J. Rawlings, and W. H. Orme-Johnson, *Biochim. Biophys. Acta,* **537,** 185–207 (1978).
72. B. H. Huynh, M. T. Henzl, J. A. Christner, R. Zimmerman, W. H. Orme-Johnson, and E. Münck, *Biochim. Biophys. Acta,* **623,** 124–138 (1980).
73. V. K. Shah and W. J. Brill, *Proc. Natl. Acad. Sci. USA,* **74,** 3249–3253 (1977).
74. J. Rawlings, V. K. Shah, J. R. Chisnell, W. J. Brill, R. Zimmermann, E. Münck, and W. H. Orme-Johnson, *J. Biol. Chem.*, **253,** 1001–1004 (1978).
75. G. Palmer, J. S. Multani, W. C. Cretney, W. G. Zumft, and L. E. Mortenson, *Arch. Biochem. Biophys.*, **153,** 325–332 (1972).

76. D. M. Kurtz, R. S. McMillan, B. K. Burgess, L. E. Mortenson, and R. H. Holm, *Proc. Natl. Acad. Sci. USA*, **76**, 4986–4989 (1979).

77. P. J. Stephens, C. E. McKenna, B. E. Smith, H. T. N. Guyen, M.-C. McKenna, A. J. Thomson, F. Devlin, and J. B. Jones, *Proc. Natl. Acad. Sci. USA*, **76**, 2585–2589 (1979).

78. M. K. Johnson and T. G. Spiro, unpublished observations.

79. A. J. Thomson, *Biochem. Soc. Trans.*, **3**, 468–470 (1975).

80. C. Y. Yang, K. H. Johnson, R. H. Holm, and J. G. Norman, Jr., *J. Am. Chem. Soc.*, **97**, 6596–6598 (1975).

81. A. J. Thomson, *J. Chem. Soc., Dalton Trans.*, 1180–1189 (1981).

CHAPTER 11

Resonance Raman Spectra of Iron-Sulfur Proteins and Analogs

T. G. SPIRO
J. HARE
V. YACHANDRA
A. GEWIRTH
M. K. JOHNSON
E. REMSEN

Department of Chemistry, Princeton University
Princeton, New Jersey

CONTENTS

1 INTRODUCTION

Resonance Raman (RR) spectroscopy, a useful structure probe for chromophoric sites in complex molecules, is being increasingly applied to biological systems (1-3). Laser excitation within an electronic absorption band enhances those vibrational modes that contribute to the distortion in the electronic excited state (4, 5). The visible and UV absorption bands of Fe-S proteins arise from S → Fe charge-transfer transitions (6), which should lead to lengthened Fe-S bonds in the excited states. They are therefore expected to provide RR enhancement of Fe-S stretching modes, whose frequencies may be diagnostic of structural variations. Indeed the first RR spectrum of a biological molecule was that of rubredoxin, published by Long and co-workers (7).

Although this pioneering study appeared a decade ago, the promise of Fe-S RR spectroscopy has been slow in developing, and published results have been meager. Two preliminary reports have appeared from this laboratory, one on adrenodoxin (8) and the other on [4Fe-4S] molecules (9). An unpublished RR spectrum and normal coordinate analysis of rubredoxin from *P. elsdenii* by Yamamoto and co-workers has been cited (10). Finally, Adar and colleagues have reported RR features of adrenodoxin (11) and spinach ferredoxin (12) which were attributed to spin-ladder electronic transitions, but we have been unable to reproduce these features (see below).

The paucity of data arises from experimental difficulties. Fe-S complexes are not very stable, and are prone to fall apart under laser irradiation; this problem is even more severe for analog complexes than for proteins, whose ligands are less able to diffuse away from the Fe centers. Moreover, the RR enhancements are much lower for Fe-S complexes than for many other chromophores, such as hemes or polyenes, despite the quite large molar absorptivities for the visible bands. The reason for the disappointing scattering power is uncertain, but it may result from interferences (5) among the numerous closely spaced charge-transfer transitions that are no doubt responsible for the broad absorption bands. Finally, Fe-S protein samples frequently show unacceptably high fluorescence backgrounds, which are probably due to flavin contamination.

These formidable difficulties have slowly yielded to improved techniques, and we have recently been able to obtain spectra of sufficient quality to begin characterizing the vibrational modes of the various Fe-S species. These new results (13), which do give promise for monitoring structural variations, are the subject of this chapter.

2 4-Fe MOLECULES

Assignments are currently most secure for [4Fe-4S] molecules, despite their being the most complex of the currently known Fe-S structure classes (see Chapter 1).

In the limiting tetrahedral symmetry of the $(Fe_4S_4)(SR)_4$ clusters, the stretching modes of the four terminal Fe–SR bonds can be classified as having A_1 (Raman active) and T_2 (Raman and IR active) symmetry. They are expected to be nearly coincident, since the Fe–SR bonds are not connected by common atoms, and are therefore not coupled kinematically. The modes of the 12 bridging Fe–S bonds classify as A_1, $2T_2$, E (Raman active), and T_1 (inactive). The Fe-S modes are expected to be well separated, since coupling among the cluster bands should be appreciable. These expectations are borne out in the RR and IR spectra of $(Fe_4S_4)(SBz)_4^{2-}$ (Bz = benzyl), and of its ^{34}S-substituted forms, shown in Figure 1. The 358 cm^{-1} RR band and the 356 cm^{-1} IR band, which shift down on ^{34}SBz substitution, are identifiable as the Fe-SBz terminal modes. Two IR bands, at 382 and 247 cm^{-1}, shift on $^{34}S_4$ substitution, and have RR counterparts at 386 and 244 cm^{-1}; these are T_2 Fe-S modes. The strong, $^{34}S_4$-sensitive RR band at 335 cm^{-1} is clearly the A_1 Fe-S mode. A weaker RR mode at 272 cm^{-1}, whose $^{34}S_4$ shift is uncertain, is tentatively assigned to the E Fe-S mode. These assignments are summarized in Table 1. They supersede the suggestions made earlier (9) on the basis of partial RR spectra.

Three additional cluster modes (A_1, T_2, E) are expected at lower frequencies. They can equivalently be described in terms of Fe–Fe stretching, or of deformations of the Fe-S-Fe and S-Fe-S angles (14). The relatively short Fe–Fe distance, ≈ 2.75 Å (Chapter 1), and electronic structure calculations (15a) are suggestive of Fe–Fe bonding, which might be sufficiently perturbed in the charge-transfer excited states to provide RR enhancement of Fe modes. The RR spectrum (13) of $(Fe_4S_4)(SPh)_4^{2-}$ (Ph = phenyl) does show weak bands at 203 and 146 cm^{-1} which have appropriate frequencies to be Fe–Fe stretches [$\nu_{A_1}/\nu_{T_2}/\nu_E = 2/\sqrt{2}/1$ in the "simple cluster" approximation (15)]. Because of the redundancy between Fe–Fe stretching modes and cluster bending modes (14), however, nothing firm can be deduced about Fe–Fe bonding from these data.

Tetrahedral symmetry is only approximate for $(Fe_4S_4)(SR)_4$ species, as discussed in Chapter 1. For the dianions a slight elongation of the cube, along a $\bar{4}$ axis, is observed. The symmetry lowering (approximately D_{2d}) is expected to split the degenerate vibrational modes ($E \rightarrow A_1 + B_2$, $T_2 \rightarrow B_2 + E$, $T_1 \rightarrow A_2 + E$), and to activate one of the T_1 components (E). No pronounced effects are seen in the $(Fe_4S_4)(SBz)_4^{2-}$ spectra, however. A small but definite splitting can be seen in the 356 cm^{-1} IR band, assigned to Fe–SR stretching, but this might arise from solid state effects. Similar small splittings probably contribute to the width of the RR band, and might also account for slight mismatches between RR and IR band maxima of the nominally T_2 modes: 386 vs. 382 cm^{-1}, and 244 vs. 247 cm^{-1}.

The RR spectrum of oxidized ferredoxin from *C. pasteurianum* (C. Fd)

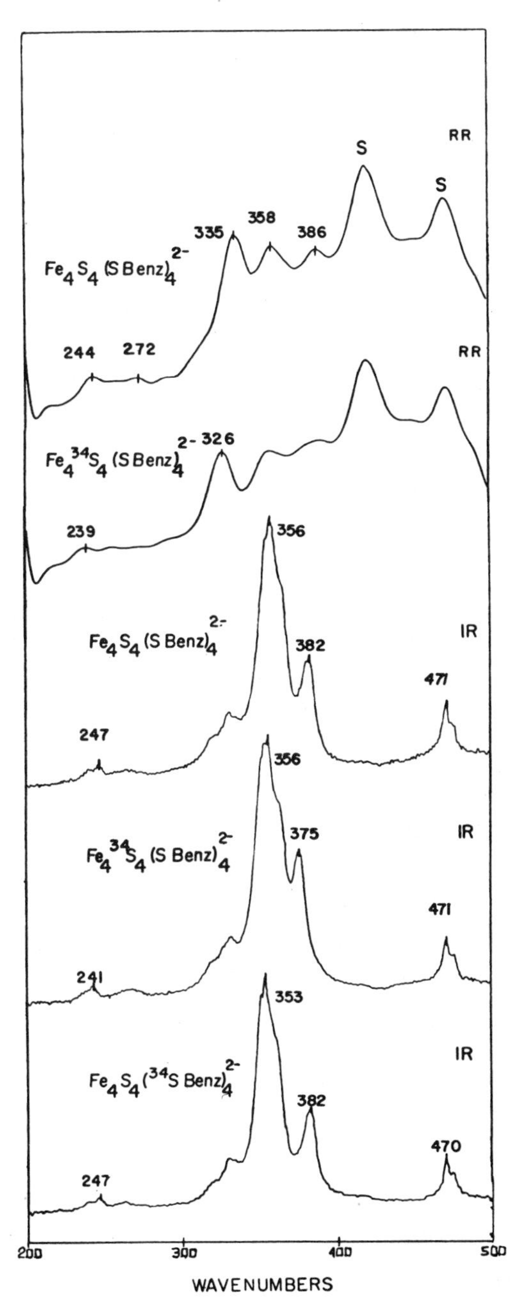

Figure 1. Resonance Raman (RR) and infrared (IR) spectra of $[Et_4N]_2[Fe_4S_4(CH_2Ph)_4]$, and ^{34}S derivatives. RR spectra obtained via backscattering on ≈ 4 *mM* solutions in dimethyl acetamide, contained in sealed spinning NMR tubes, cooled with cold N_2 gas; 4579 Å Ar^+ laser excitation, 100 mW, 10 cm^{-1} slit width. Spectra were recorded digitally (Spex 1401 double monochromator, with cooled RCA 31034 photomultiplier, with photon counting electronics) and smoothed with a Fourier transform filter. Solvent bands marked "S". IR spectra obtained with a Digilab FTIR, on Nujol mulls sealed between polyethylene plates.

Table 1 Vibrational Assignments for [4Fe-4S] Species

	Frequency (cm^{-1}) $(Fe_4S_4)(SBz)_4^{2-}$		C. Fd_{red}	C. Fd_{ox}	HP_{red}	HP_{ox}
Assignment (T_d)	RR	IR	RR	RR	RR	RR
$Fe\text{-}S_{br}(T_2)$	386	382(7)[a]	395	395		
$Fe\text{-}SR(A_1, T_2)$	358	356(3)[b]	356	357	362	370
$Fe\text{-}S_{br}(A_1)$	335(9)[a]		334	335	339	338
$Fe\text{-}S_{br}(E)$	272			277		
$Fe\text{-}S_{br}(T_2)$	244(5)[a]	247(6)[a]	248	248	247	

[a]Downshift on $^{34}S_{br}$ substitution.
[b]Downshift on ^{34}SR substitution.

(Fig. 2) is nearly indistinguishable from that of $(Fe_4S_4)(SBz)_4^{2-}$, consistent with the homology of these Fe-S structures. Since low-symmetry splittings are not resolved in the $(Fe_4S_4)(SBz)_4^{2-}$ spectrum, however, it is not clear whether small differences in structure between protein and analog would be detected. Reduction of C. Fd has surprisingly little effect on the RR spectrum. Frequencies are essentially unaltered for the terminal Fe–SR stretch, 357 cm^{-1}, or for three of the Fe–S bridging modes, 395, 335 and 248 cm^{-1}. However, the 277 cm^{-1} band, tentatively identified as the Fe–S(E) stretch, and an unassigned band at 298 cm^{-1} are abolished and replaced with shoulders on both sides of the 248 cm^{-1} band. These apparent shifts require further investigation.

The RR spectrum of the HIPIP from *Chromatium* (Fig. 3), although not presently as well defined as that of C. Fd, shows the two major bands, 362 cm^{-1} ($\nu_{Fe\text{-}SR}$) and 339 cm^{-1} [$\nu_{Fe\text{-}S}(A_1)$], and the moderate intensity 247 cm^{-1} [$\nu_{Fe\text{-}S}(T_2)$] band. Oxidation of HIPIP increases $\nu_{Fe\text{-}SR}$ by 8 cm^{-1}, suggesting a strengthening of the Fe–S (cys) bonds, but the cluster breathing mode is left unshifted, at 338 cm^{-1} (the 247 cm^{-1} mode is lost in the rising background).

3 3-Fe PROTEINS

As evident from several chapters in this volume, there is much excitement about the recent discovery of a new type of Fe-S site, containing 3 Fe and 3 labile S atoms. The structure of *Azotobacter* ferrodoxin (Chapter 3) shows these atoms to lie in a somewhat puckered trigonal plane, with a pair of pro-

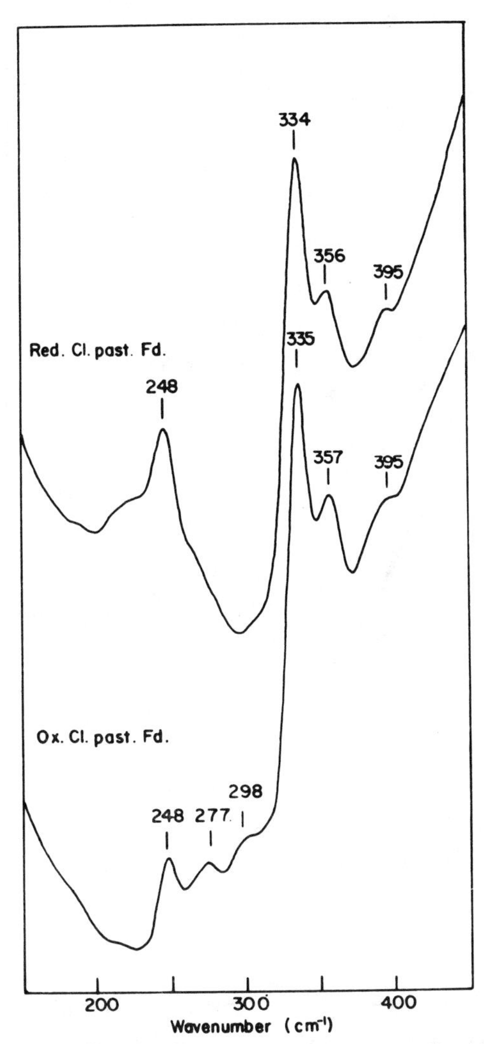

Figure 2. RR spectra of *C. pasteurianum* ferredoxin (0.4 *mM* in Tris, pH 7.6) in oxidized and reduced (via anaerobic addition of dithionite) forms. The spectra were obtained with 4579 Å Ar^+ laser excitation (100 mW) via backscattering from sealed NMR tubes cooled with cold N_2 gas, 6 (oxidized) and 10 (reduced) cm^{-1} slit width.

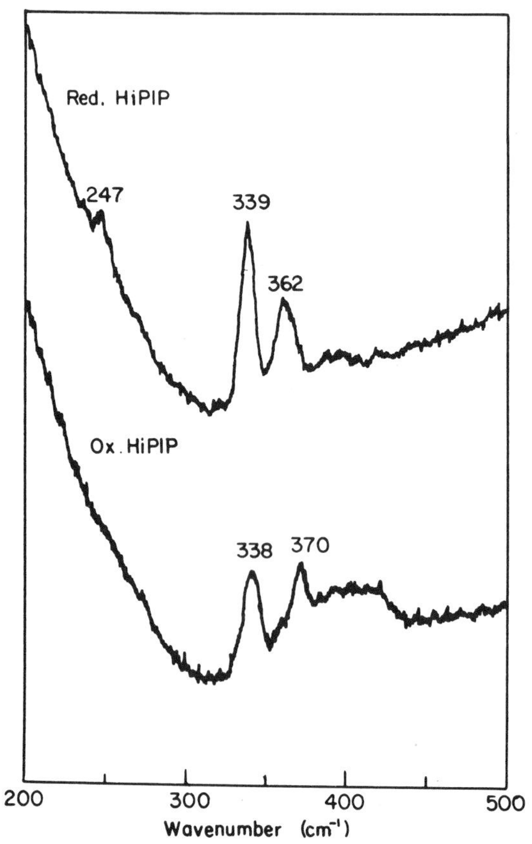

Figure 3. RR spectra of *Chromatium* HIPIP (≈ 1 mM in Tris, pH 8.0) in reduced and oxidized (via ferricyanide addition) forms. The spectra were obtained with 4545 Å Ar^+ laser excitation, in a spinning cell.

tein ligands bound to each Fe above and below the plane. Five of the six ligands are cysteine, while the sixth is probably glutamate.

Figure 4 shows the RR spectrum of *D. gigas* ferredoxin (Fd) II, (Chapter 5), whose EPR and Mössbauer spectra have unambiguously shown it to contain solely [3Fe-3S] clusters (Chapter 4). It is similar in appearance to the $(Fe_4S_4)(SR)_4$ spectra with somewhat upshifted frequencies. These shifts are sufficient to differentiate the two structure types. Thus the spectrum of *D. gigas* Fd I (Fig. 4), which is known (Chapter 5) to contain both 4-Fe and 3-Fe clusters, can be cleanly decomposed into the two types of spectra, as illustrated by the Fd I − Fd II difference spectrum (Fig. 4), which is very similar to that of C. Fd.

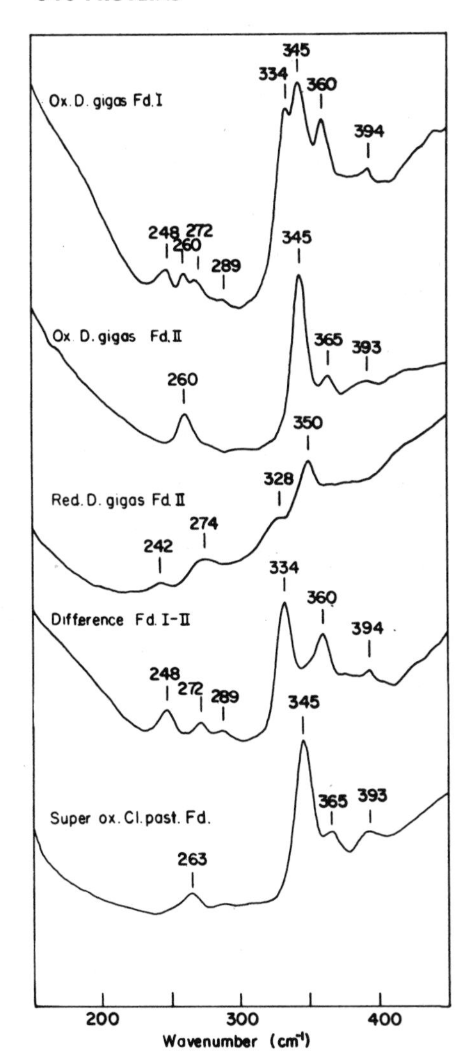

Figure 4. RR spectra of *D. gigas* ferredoxins I and II (≈1 mM in monomeric unit in Tris, pH 7.6). The third spectrum is of Fd II, reduced anaerobically with dithionite. The fourth spectrum shows the subtractions of the Fd II spectrum after equalization of the 345 cm^{-1} band intensities. The fifth spectrum is that of *C. pasteurianum* Fd, treated with a 20-fold excess of ferricyanide, for 10 hr; the ferricyanide was subsequently removed by passage through a G-25 sephadex column. All spectra (6 cm^{-1} slit widths) were obtained with 4579 Å Ar^+ laser excitation, via backscattering from sealed and cooled NMR tubes.

The idealized symmetry of the [3Fe-3S] cluster is D_{3h}, for which the six bridging Fe-S vibrations can be classified as A_1', $2E'$ (all Raman active), and A_2' [inactive, except via antisymmetric RR scattering (5)]. It is logical to assign the strong RR band at 345 cm^{-1} to the A_1' cluster mode, and the outlying weak bands at 393 and 260 cm^{-1} to the E' modes. The 365 cm^{-1} band is left as a candidate for Fe–SR stretching, 6 cm^{-1} higher than the Fe–SR Raman band for the 4-Fe cluster in C. Fd. For the pair of Fe–SR bonds at each Fe atom of the [3Fe-3S] cluster, the in and out of phase stretches should be separated by 10–20 cm^{-1} (see the following section); interactions across the cluster should be much smaller. A second Fe–SR mode might be hidden under the intense 345 cm^{-1} mode. If *D. gigas* Fd II has a glutamate ligand, the Fe–O stretching mode is expected to be in the 400–500 cm^{-1} region, but it is not expected to experience much RR enhancement via the S → Fe

It has recently been established via MCD spectroscopy (Chapter 10) that the reaction of C. Fd with ferricyanide, which had been thought on the basis of a $g = 2.01$ EPR signal to superoxidize the cluster to the redox level of oxidized HIPIP (16), in fact produces a 3-Fe cluster. This is nicely confirmed by the RR spectrum of superoxidized C. Fd (Fig. 4), which is identical to that of *D. gigas* Fd II.

Reduction of *D. gigas* Fd II produces a marked change in the RR spectrum (Fig. 4), in contrast to the slight redox effect seen for [4Fe-4S] clusters. The spectral pattern is sufficiently different that there is no obvious way to correlate the bands between the oxidized and reduced forms. The complex reduced spectrum may be related to the localization of the added electron to 2 of the 3 Fe atoms (Chapter 4).

4 2-Fe MOLECULES

The $(Fe_2S_2)(SR)_4$ species have four bridging Fe–S and four terminal Fe–SR modes. Tentative assignments are given in Table 2, based on a comparison of the spectra of two proteins and two analogs (Fig. 5) and on $^{34}S_2$ shifts for

Table 2 Vibrational Assignments for [2Fe-2S] Species

	Frequency (cm^{-1})				
Assignment (D_{2h})	Sp. Fd_{ox}	Ad_{ox}	$Fe_2S_2(S_2\text{-}o\text{-xyl})_2^{2-}$	$Fe_2S_2Cl_4^{2-}$	$Fe_2S_2Br_4^{2-}$
$Fe\text{-}S_{br}(B_{3u})$	425(5)[a]	418	417	415	
$Fe\text{-}S_{br}(A_{1g})$	395(6)[a]	392	392	401	395
$Fe\text{-}SR(B_{3g})$	336	347	347		
$Fe\text{-}SR(B_{2u})$			338	345	
$Fe\text{-}SR(B_{1u})$			326	334	
$Fe\text{-}SR(A_{1g})$	328	328	324	331	271
$Fe\text{-}S_{br}(B_{2g})$				318	316
$Fe\text{-}S_{br}(B_{1u})$	284(5)[a]	290	279	287	

[a]Downshift on $^{34}S_{br}$ substitution.

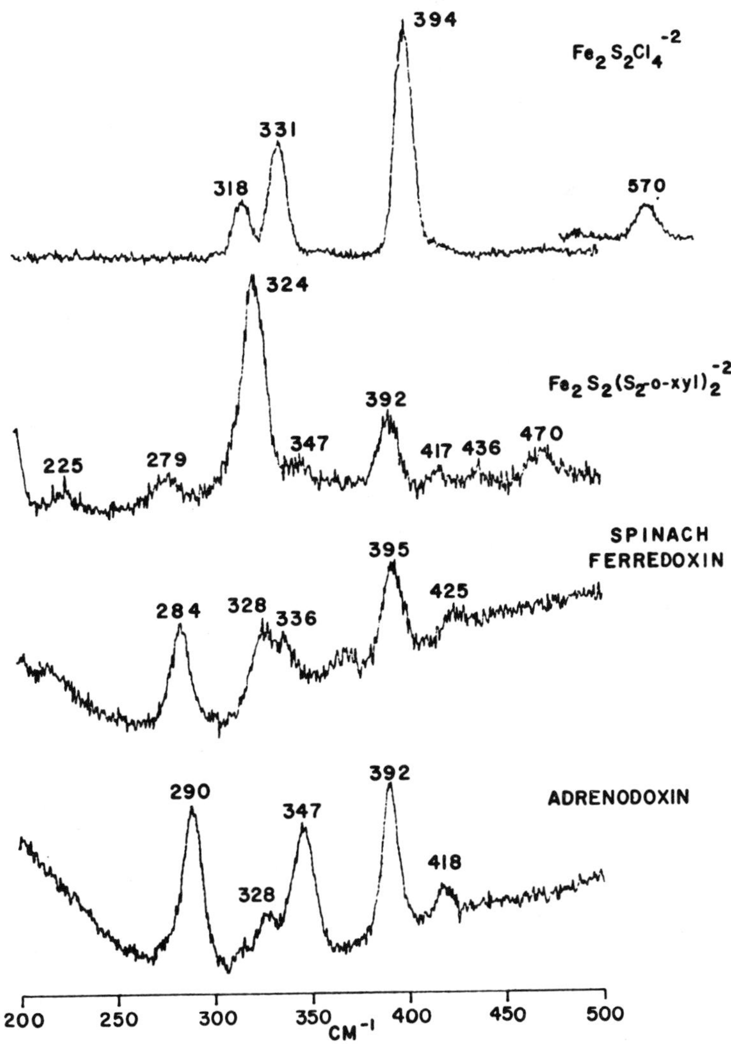

Figure 5. RR spectra of $Fe_2S_2X_4^{2-}$ species. $Fe_2S_2Cl_4^{2-}$ (λ_{ex} = 4545 Å) and $Fe_2S_2(S_2\text{-}o\text{-xyl})_2^{2-}$ (λ_{ex} = 4880 Å) were run as Et_4N^+ salts in KBr pellets in a liquid N_2 dewar, via backscattering. Slit width 10 cm^{-1}, ≈ 20 mW laser power. Spinach ferredoxin and adrenodoxin (both oxidized) were run as ≈ 2 *mM* solutions in pH 8 Tris buffer, in a flowing capillary cell. λ_{ex} = 4545 Å, 150 mW, slit width 6 cm^{-1}.

spinach ferredoxin (sp. Fd). To the extent that these structures conform to idealized D_{2h} symmetry, the two sets of modes should each divide into a Raman-active and an IR-active pair.

All species show an intense RR band near 395 cm^{-1}, which shifts 6 cm^{-1} on $^{34}S_2$ substitution in sp. Fd and is assigned to the cluster breathing mode, Fe–S (A_{1g}). Additional $^{34}S_2$ shifts are observed for sp. Fd for bands at 425 and 284 cm^{-1}. Bands at corresponding frequencies, seen also for adrenodoxin, are much weaker for the *o*-xylene dithiolate analog, and are absent for $(Fe_2S_2)Cl_4^{2-}$, whose structure (17) is very close to having D_{2h} symmetry. These modes are located in the (Fe_2S_2) Cl_4^{2-} IR spectrum at 415 and 287 cm^{-1}, and are assigned to the B_{2u} and B_{1u} bridging Fe–S vibrations. It is of interest that the overtone of the 287 cm^{-1} mode, which is Raman active ($2 \times B_{1u} = A_{1g}$), is observed in the RR spectrum (Fig. 5). The activation of these IR modes in the $(Fe_2S_2)(S_2\text{-}o\text{-xyl})_2^{2-}$ RR spectrum is attributable to symmetry lowering, which is observed in the crystal structure (18). In both proteins the ≈ 285 cm^{-1} band is as intense as the ≈ 395 cm^{-1} A_{1g} mode, implying a pronounced asymmetry of the cluster. No appreciable structural distortion appears likely in view of the similarity of the vibrational frequencies to those of the analogs. Rather, the strong activation of the ≈ 285 cm^{-1} IR mode is probably caused by an assymmetric disposition of hydrogen bonding or charged residues at the active site.

The location of the remaining cluster mode, B_{2g}, is less certain. All the RR spectra show a pair of bands of variable intensity near 340 cm^{-1}. At least one of these must be a terminal Fe–SR (or Fe–Cl) mode. When the RR spectrum of $(Fe_2S_2)Br_4^{2-}$ is examined, a terminal Fe–Br mode is seen at 271 cm^{-1}, and a weak band remains at 316 cm^{-1} as a plausible candidate for the B_{2g} Fe–S mode (the A_{1g} Fe–S mode appears at 395 cm^{-1}). This assignment is therefore suggested for the weak $Fe_2S_2Cl_4^{2-}$ RR band at 318 cm^{-1}. Presumably the B_{2g} mode occurs at a similar frequency for all the [2Fe-2S] species, but it may be weak and overlapped by other bands; no clear $^{34}S_2$ shift is seen in this region for sp. Fd [to the extent that the electronic symmetry conforms to D_{2h} symmetry, RR enhancement of the B_{2g} mode requires it to be active in mixing two nearby electronic transitions (5)].

Both the Raman and IR-active terminal modes should consist of in and out of phase combinations of Fe–X bond stretches. In the IR spectra of $(Fe_2S_2)(S_2\text{-}o\text{-xyl})_2^{2-}$ and $(Fe_2S_2)Cl_4^{2-}$ these pairs stand out clearly, at 326, 338, and 334, 345 cm^{-1}, respectively. (Since chloride and thiolate S are isoelectronic and have nearly the same mass, it is not surprising that the vibrational frequencies are nearly the same.) For both species the stronger of the RR bands in this region is close to the lower frequency IR band: 324 for $Fe_2S_2(S_2\text{-}o\text{-xyl})_2$ and 338 for $(Fe_2S_2)Cl_4^{2-}$. This must be the in phase, A_{1g}, mode. The out of phase mode, B_{3g}, is then expected to be ≈ 10 cm^{-1} higher, by analogy with the IR modes. $Fe_2S_2(S_2\text{-}o\text{-}xyl)_2^{2-}$ does show a weak band at 347 cm^{-1}, which may be the B_{3g} mode. In view of the absence of substantial $^{34}S_2$ shifts in this region, the sp. Fd RR doublet at 328, 336 cm^{-1} is probably assignable to the terminal Fe–SR pair. Adrenodoxin also shows a doublet, at 328, 347 cm^{-1}. It is of interest that strong enhancement is seen for the 347 cm^{-1} band, which the present assignments suggest is the out of phase B_{3g} mode. Its enhancement may be linked to the same symmetry lowering required to explain the enhancement of the 290 cm^{-1} B_{1u} Fe–S mode. These assignments are in accord with the earlier observation (8) that Se substitution for S* in adrenodoxin shifts the 392 and 290 cm^{-1} RR bands, but not the 347 cm^{-1} band.

As noted in the introduction, Adar and co-workers have reported RR peaks for sp. Fd (770, 1080, 1475, 1930 cm^{-1}) (12) and adrenodoxin (995 cm^{-1}) (11), which were interpreted as arising from electronic transitions among the levels in the spin manifold of the antiferromagnetically coupled pair of Fe^{3+} ions. A similar assignment has been suggested by Larrabee and Spiro (19) for a broad RR band at 1075 cm^{-1} in oxyhemocyanin, which contains a pair of coupled Cu^{2+} ions. We have searched the sp. Fd and adrenodoxin for the bands reported by Adar's group, but our spectra are blank in these regions. In view of their observation of C–H stretching modes ($\approx$2900 cm^{-1}) in the same spectra (11, 12), the possibility arises that Adar and his colleagues were seeing polypeptide modes at the reported frequencies, perhaps because of partially denatured protein. In our experience high laser powers can result in sample damage for static protein solutions in a capillary tube.

5 1-Fe MOLECULES

Rubredoxin (Rd) contains the simplest Fe–S structure, a single $Fe^{3+,2+}$ ion bound to four cysteinate ligands. It is well modeled by the analogs $Fe(S_2\text{-}o\text{-}xyl)_2^{1-,2-}$ and $Fe(SPh)_4^{2-}$ (Chapter 1). Both $Fe(S_2\text{-}o\text{-}xyl)_2^{1-}$ and Rd_{ox} have nearly tetrahedral FeS_4 coordination groups (bond angles 104–114°) with Fe–S distances (2.27 Å) that are equal to within experimental error. The Rd_{ox} RR spectrum, as originally reported by Long and co-workers (7), was satisfactorily interpreted as arising from a tetrahedral FeS_4 complex: a strong, polarized band at 314 cm^{-1}, due to FeS_4 breathing; a weaker, depolarized band at higher frequency, 368 cm^{-1}, assigned to the T_2 Fe–S stretch; and two lower frequency bands, at 150 and 126 cm^{-1}, corresponding to the two expected bending modes (T_2 and E). Indeed, the spectrum rather closely resembled that of the isoelectronic complex $FeCl_4^-$ (330, 385, 133, and 106 cm^{-1}) (20).

It therefore came as a considerable surprise to us to find that the RR spectrum of $Fe(S_2\text{-}o\text{-}xyl)_2^{1-}$ (Fig. 6) is far from being tetrahedral in appearance. It shows four bands in the Fe-S stretching region, two weak ones at 374 and 350 cm^{-1}, and two strong ones at 321 and 297 cm^{-1}.

Reexamination of the Rd spectrum (Fig. 6) shows that the apparently tetrahedral pattern is illusory. The band that had been assigned to the T_2 stretch is actually two bands, 371 and 359 cm^{-1}. This had been noted by Rimai and co-workers, who attributed the pair to a splitting of the T_2 mode (10). However, the intense 312 band is broad and asymmeteric, and quite clearly has an additional component on the high-frequency side. This can be resolved with the aid of polarization spectra (it is largely depolarized) and located at ≈ 325 cm^{-1}. The tetrahedral breathing mode, being nondegenerate, cannot, of course, be split. The high-frequency component at 325 cm^{-1} must consequently be correlated with the T_2 tetrahedral mode, which must then be split very widely. With this additional band resolution the Rd spectral pattern is similar to that of $Fe(S_2\text{-}o\text{-}xyl)_2^{1-}$ except that the pairs of bands are not as well separated, and the actual frequencies differ between the two spectra by as much as 15 cm^{-1}.

To explain these different and decidedly nontetrahedral spectra from structures with similar and nearly tetrahedral FeS_4 cores, we must look to vibrational coupling effects beyond the S atoms of the thiolate ligands. A precedent for such effects can be found in the variability of S-S stretching frequencies in dialkyl disulfides, and at least some of this has been shown to be explicable on the basis of coupling with the S-C-C bend (21). Rotation about the S-C bond alters this coupling, which is maximal when the S-S and C-C bonds are in line and minimal when they are staggered. Similar coupling can be expected between Fe-S stretching and S-C-C bending. The Rd_{ox} (22) and $Fe(S_2\text{-}o\text{-}xyl)_2^{1-}$ (23) crystal structures do indicate substantial conformational differences with respect to rotation about the S-C bonds. Normal mode calculations are underway to test the plausibilty of this coupling mechanism. Preliminary results indicate that coupling with S-C-C binding can induce appreciable frequency shifts, which differ among the four Fe-S modes, and that the frequency differences between Rd_{ox} and $Fe(S_2\text{-}o\text{-}xyl)_2^{1-}$ may be explicable on the basis of the known conformational differences.

The reality of ligand vibrational mode coupling is dramatically illustrated by the effects of deuterating the methylene positions on $Fe(S_2\text{-}o\text{-}xyl)_2^{1-}$ (Fig. 6). The bands at 321 and 350 both shift *up* by 10 cm^{-1}. Deuteration upshifts occur when there is coupling with a higher-lying hydrogenic mode, which is shifted on deuteration to a frequency lower than the band being monitored, so that coupling then occurs in the opposite sense. The IR spectrum of $Fe(S_2\text{-}o\text{-}xyl)_2^{1-}$ shows, in addition to the Fe-S modes, bands at 396, 361, and

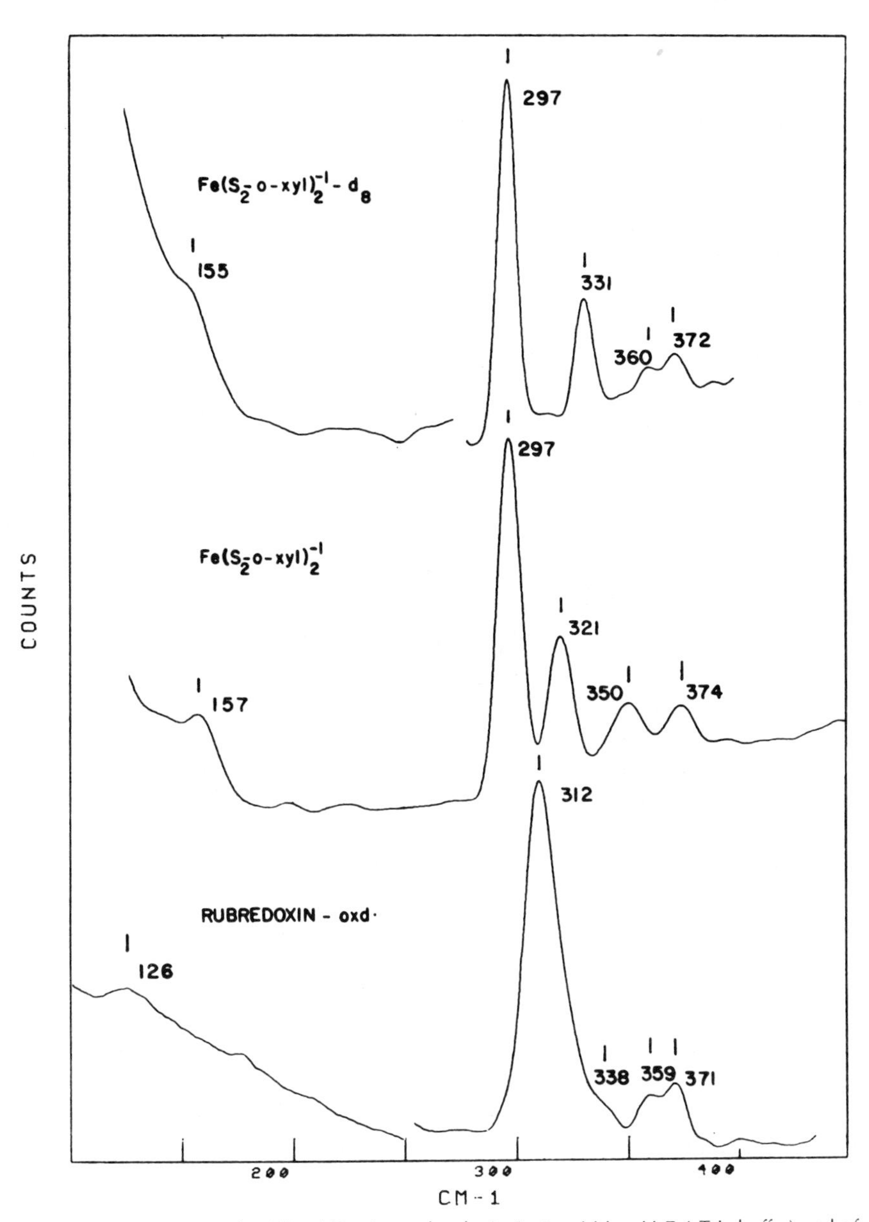

Figure 6. RR spectra of oxidized *D. gigas* rubredoxin ($\approx 1\ mM$ in pH 7.4 Tris buffer) and of (Et_4N) [Fe(o-xylyldithiolate)$_2$] and the analog, with deuterium substituted for the 8 methylene H atoms, in pyridine ($\approx 4\ mM$). The rubredoxin spectrum was obtained with a flowing capillary cell, using 4965 Å Ar^+ laser excitation (200 mW) and 4 cm^{-1} slit width. The analog spectra were obtained via backscattering from sealed NMR tubes, cooled with cold N_2 gas, using 4880 Å Ar^+ laser excitation (≈ 50 mW) and 6 cm^{-1} slit width.

344 cm^{-1}, which correlate with modes of the free ligand, *o*-xylyldithiol, at 395, 334, and 319 cm^{-1}. These modes shift down on deuteration, to 352, 337, and 325 cm^{-1}, crossing the upshifted Fe–S modes, and are likely candidates for the coupling effect.

6 CONCLUSIONS

The currently available data base provides a reasonable starting point for the characterization of Fe-S proteins by RR spectroscopy. For those molecules containing sulfide bridges the RR spectrum is dominated by the totally symmetric bridging mode, as might be expected from the likely dominance of bridging S* → Fe charge-transfer transitions in the visible region. These bridging A_l modes are found at frequencies characteristic of the cluster type: ≈335 cm^{-1} for [4Fe-4S], and ≈345 cm^{-1} for [3Fe-3S], and 395 cm^{-1} for [2Fe-2S]. Other, nominally nontotally symmetric bridging modes are seen, usually with much lower intensity, although in the [2Fe-2S] proteins examined so far an IR mode at ≈285 cm^{-1} is strongly enhanced, presumably via an electrostatic or H-bond asymmetry at the active site. For [4Fe-4S] and [3Fe-3S] clusters the nontotally symmetric modes are nominally degenerate, and their splittings may provide indicators for structural distortions. No large splittings of this sort have yet been observed, and because the intensities are low, better spectra will be needed to evaluate small splittings. For [4Fe-4S] molecules the cluster frequencies are nearly invariant to oxidation level. Reduction of [3Fe-3S] clusters produces large spectral changes, however; reduction of [2Fe-2S*] clusters is still to be explored.

Terminal Fe–SR frequencies are located near 360 cm^{-1} in [4Fe-4S], and probably in [3Fe-3S], species, and are moderately enhanced. For [2Fe-2S] species a pair of terminal Fe–SR bands is found at ≈325 and ≈340 cm^{-1}, with variable relative intensities. The surprising results of the analysis of $Fe(SR)_4^-$ spectra are that the Fe–SR stretching modes must be viewed as strongly coupled to modes of the thiolate ligand, and they are sensitive to orientation about the S–C bond. Although this coupling complicates the analysis, it also raises the possibility that the Fe–SR frequencies may prove useful in monitoring protein conformational differences.

ACKNOWLEDGMENTS

The work described in this chapter, carried out over several years, was greatly aided by provision of the following samples from colleagues: analogs, R. H. Holm; *D. gigas* ferredoxins I and II and rubredoxin, A. Xavier and

J. J. G. Moura; *C. pasteurianum* ferredoxin, L. Mortenson; *Chromatium* HIPIP, C. Carter; adrenodoxin, T. Kimura.

REFERENCES

1. P. R. Casey and V. R. Salares, in *Advances in Infrared and Raman Spectroscopy,* Vol. 7, R. J. H. Clark and R. E. Hester, eds., Heyden, London, 1980, Chapter 1.
2. T. G. Spiro and T. C. Loehr, in *Advances in Infrared and Raman Spectroscopy,* Vol. 1, R. J. H. Clark and R. E. Hester, eds., Heyden, London, 1975, Chapter 3.
3. T. G. Spiro and B. P. Gaber, *Ann. Rev. Biochem.,* **46,** 353 (1977).
4. A. Y. Hirakawa and M. Tsuboi, *Science,* **188,** 359 (1975).
5. T. G. Spiro and P. Stein, *Ann. Rev. Phys. Chem.,* **28,** 501 (1977).
6. W. A. Eaton and W. Lovenberg, in *Iron-Sulfur Proteins,* Vol. II, W. Lovenberg, ed., Academic, New York, 1973, Chapter 3.

7a. T. V. Long and T. M. Loehr, *J. Am. Chem. Soc.,* **92,** 6384 (1970).

b. T. V. Long, T. M. Loehr, J. R. Allkins, and W. Lovenberg, *J. Am. Chem. Soc.,* **93,** 1809 (1971).

8. S-P. W. Tang, T. G. Spiro, K. Mukai, and T. Kimura, *Biochem. Biophys. Res. Commun.,* **53,** 869 (1973).
9. S-P. W. Tang, T. G. Spiro, C. Antanaitis, T. H. Moss, R. H. Holm, T. Herskovitz, and L. E. Mortenson, *Biochem. Biophys. Res. Commun.,* **62,** 1 (1975).
10. T. Yamamoto, L. Rimai, M. E. Heyde, and G. Palmer, unpublished results, cited in reference 6 (1972).
11. F. Adar, H. Blum, J. S. Leigh, Jr., T. Ohnishi, J. C. Salerno, and J. Kimura, *FEBS Lett.,* **84,** 214 (1977).
12. H. Blum, F. Adar, J. C. Salerno, and J. S. Leigh, Jr., *Biochem. Biophys. Res. Commun.,* **77,** 650 (1977).
13. V. Yachandra, A. Gewirth, J. Hare, M. K. Johnson, and T. G. Spiro, manuscript in preparation.
14. P. A. Bulliner and T. G. Spiro, *Spectrochim. Acta,* **26A,** 1641 (1970).

15a. C. Y. Yang, K. H. Johnson, R. H. Holm, and J. G. Norman, Jr., *J. Am. Chem. Soc.,* **97,** 6596 (1975).

15b. T. G. Spiro, *Prog. Inorg. Chem.,* **11,** 1 (1970).

16. W. V. Sweeney, A. J. Bearden, and J. C. Rabinowitz, *Biochem. Biophys. Res. Commun.,* **59,** 188 (1974).
17. M. A. Bobrik, K. O. Hodgson, and R. H. Holm, *Inorg. Chem.,* **16,** 1851 (1977).
18. J. J. Mayer, S. E. Denmark, B. V. DePamphilis, J. A. Ibers, and R. H. Holm, *J. Am. Chem. Soc.,* **97,** 1032 (1975).
19. J. A. Larrabee and T. G. Spiro, *J. Am. Chem. Soc.,* **102,** 4217 (1980).
20. L. A. Woodward and M. J. Taylor, *J. Chem. Soc.,* 4473 (1960).
21. H. Sugeta, *Spectrochim. Acta,* **31A,** 1729 (1975).
22. K. D. Watenpaugh, L. C. Sieker, and L. H. Jensen, *J. Mol. Biol.,* **138,** 615 (1980).
23. R. W. Lane, J. A. Ibers, R. B. Frankel, G. C. Papaefthymiou, and R. H. Holm, *J. Am. Chem. Soc.,* **99,** 84 (1977).

Subject Index